SECOND EDITION

COMPOSITE MATERIALS

DESIGN AND APPLICATIONS

SECOND EDITION

COMPOSITE MATERIALS

DESIGN AND APPLICATIONS

Daniel Gay
Suong V. Hoa

CRC Press
Taylor & Francis Group
Boca Raton   London   New York

CRC Press is an imprint of the
Taylor & Francis Group, an **informa** business

CRC Press
Taylor & Francis Group
6000 Broken Sound Parkway NW, Suite 300
Boca Raton, FL 33487-2742

International Standard Book Number-10: 1-4200-4519-9 (Hardcover)
International Standard Book Number-13: 978-1-4200-4519-2 (Hardcover)

Library of Congress Cataloging-in-Publication Data

Gay, Daniel, 1942-
   [Matériaux composites. English]
   Composite materials : design and applications / Daniel Gay and Suong V. Hoa. -- 2nd ed.
     p. cm.
   Includes bibliographical references and index.
   ISBN 978-1-4200-4519-2 (alk. paper)
   1. Composite materials. I. Hoa, S. V. (Suong V.) II. Title.

TA418.9.C6G3913 2003
620.1'18--dc22
                                        2007061450

Visit the Taylor & Francis Web site at
http://www.taylorandfrancis.com

and the CRC Press Web site at
http://www.crcpress.com

CONTENTS

PART IV: APPLICATIONS

APPENDICES, BIBLIOGRAPHY, AND INDEX

PREFACE

"Composite Materials: Design and Applications" is a major work that contributes greatly to the domain of "composite structures" both for academic and industrial use. This new edition responds to the concerns of people that deal with the conceptualization and design of components made from composite materials, i.e., engineers or technicians, teachers, undergraduate or graduate students in universities.

There is a need by engineers working in composites for a practical source of reference for the design and application of composites. This book fulfills that need. In the educational sector, composite materials are now taught at many universities around the world. The topic usually covered is laminate theory. Composites design courses also exist in a few universities and institutes. The demand from students and also practitioners of composites for knowledge and training in the design of composites is increasing. However a good design book has not been available. The content of these design courses concentrates mostly on analysis, while applications still remain at the specimen level.

This book, initially written by Daniel Gay in French, has been distributed widely in France and in French-speaking countries. The authors are of the opinion that having the book in the English language would facilitate the training and dissemination of knowledge to the regions where composites are used the most. The book has been translated into English with modifications and updates. It is composed of three main parts. The technical level increases from one part to the next, and one can also use each part independently from the other parts. A fourth final part groups a large number of original case studies that are themselves totally formulated and classified according to different levels of difficulties.

- **The first part** presents an introduction to composite materials, the fabrication processes, the properties of a single ply, sandwich materials, conceptual design, assembly and applications of composites in aerospace and others areas. The principal ideas in the preliminary step, which consists in the sizing of a laminate makes up a novel method for design. This part can be used by itself to form a part of a course on advanced materials and associated designs.
- **The second part** presents the mechanics of composites materials. This consists of a discussion on elastic anisotropic properties, the directional dependence of different properties, and mechanical properties of thin laminates. This part can be used by itself to teach students and engineers about the mechanics of composite materials.
- **The third part** presents the orthotropic coefficients that may be conveniently used for design: The Hill–Tsai failure criterion, bending and torsion

of any cross-section composite beams, and bending of thick composite plates. The proposed method of analysis for the composite beams is original, as well as the proposed method for the analysis of thick laminated composite plates, which goes along the same principles as the composite beams. This part requires a knowledge of the strength of materials. The information presented here is intended to contribute to a better interpretation of the behavior of composite components.

■ **The "applications" part** provides more than 40 case studies with complete solutions. There are three levels of applications, each dealing with one of the three parts above. These cover the large majority of the practical cases encountered in industry. These problems have been posed in such a way as to allow the reader to get right into the essential part of the problem. In the international literature, there is no other work on the subject of composites that covers such a wide range of issues in a form that is easily exploitable in the applications domain.

This book can be used to teach students at the first year graduate level as well as the final year undergraduate level. It is also useful for practical engineers who want to learn, on the job, the guidelines for the use of composites in their applications. As in the previous edition, the authors chose to keep only a small number of reinforcements accompanied by their characteristic numerical values. This allows for the limiting of the number of tables that accompany the text. The adaptation of the technique to other reinforcements that are not included in the book does not create a problem. The reader can find all the necessary elements to construct either a computer program or a table to produce the performances for these new cases if necessary.

The authors hope that this volume will make a significant contribution to the training of future engineers who utilize composites.

Suong V. Hoa
Montreal, Quebec, Canada

Daniel Gay
Toulouse, France
September 2006

PART I

PRINCIPLES OF CONSTRUCTION

1

COMPOSITE MATERIALS, INTEREST, AND PROPERTIES

1.1 WHAT IS COMPOSITE MATERIAL?

As the term indicates, *composite material* reveals a material that is different from common heterogeneous materials. Currently *composite materials* refers to materials having strong fibers—continuous or noncontinuous—surrounded by a weaker matrix material. The matrix serves to distribute the fibers and also to transmit the load to the fibers.

Notes: Composite materials are not new. They have been used since antiquity. Wood, straw and mud have been everyday composites. Composites have also been used to optimize the performance of some conventional weapons. For example:

- In the Mongolian arcs, the compressed parts are made of corn, and the stretched parts are made of wood and cow tendons glued together.
- Japanese swords or sabers have their blades made of steel and soft iron: the steel part is stratified like a sheet of paste, with orientation of defects and impurities in the long direction[1] (see Figure 1.1), then formed into a U shape into which the soft iron is placed. The sword then has good resistance for flexure and impact.

One can see in this period the beginning of the distinction between the common composites used universally and the high performance composites.

The composite material as obtained is

- Very heterogeneous.
- Very "anisotropic." This notion of "anisotropy" will be illustrated later in Section 3.1 and also in Chapter 9. Simply put this means that the mechanical properties of the material depend on the direction.

---

[1] In folding a sheet of steel over itself 15 times, one obtains $2^{15} = 32,768$ layers.

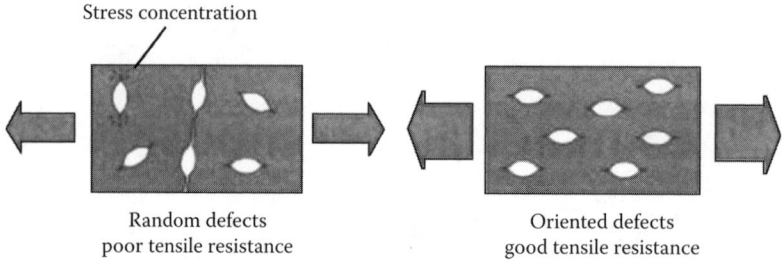

Stress concentration

Random defects
poor tensile resistance

Oriented defects
good tensile resistance

Figure 1.1 Effect of the Orientation of Impurities

1.2 FIBERS AND MATRIX

The bonding between fibers and matrix is created during the manufacturing phase of the composite material. This has fundamental influence on the mechanical properties of the composite material.

1.2.1 Fibers

Fibers consist of thousands of filaments, each filament having a diameter of between 5 and 15 micrometers, allowing them to be producible using textile machines;[2] for example, in the case of glass fiber, one can obtain two **semi-products** as shown in Figure 1.2. These fibers are sold in the following forms:

- Short fibers, with lengths of a few centimeters or fractions of millimeters are **felts, mats,** and short fibers used in injection molding.
- Long fibers, which are cut during time of fabrication of the composite material, are used as is or woven.

Principal fiber materials are

- Glass
- Aramid or **Kevlar**® (very light)
- Carbon (high modulus or high strength)
- Boron (high modulus or high strength)
- Silicon carbide (high temperature resistant)

In forming fiber reinforcement, the assembly of fibers to make fiber forms for the fabrication of composite material can take the following forms:

---

[2] One wants to have fibers as thin as possible because their rupture strength decreases as their diameter increases, and very small fiber diameters allow for effective radius of curvature in fiber bending to be on the order of half a millimeter. However, exception is made for boron fibers (diameter in the order of 100 microns), which are formed around a tungsten filament (diameter = 12 microns). Their minimum radius of curvature is 4 mm. Then, except for particular cases, weaving is not possible.

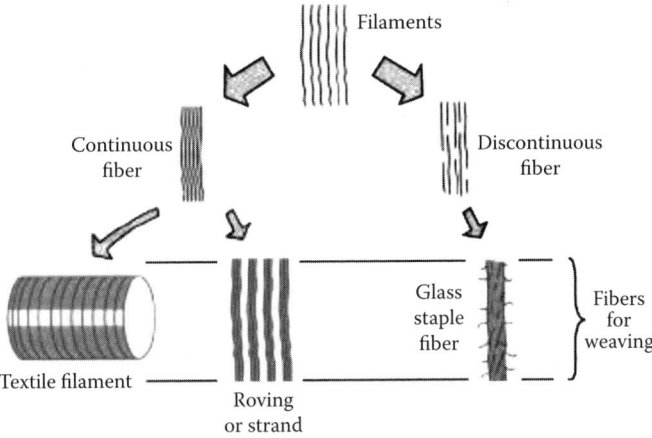

Figure 1.2 Different Fiber Forms

- **Unidimensional:** unidirectional tows, yarns, or tapes
- **Bidimensional:** woven or nonwoven fabrics (felts or mats)
- **Tridimensional:** fabrics (sometimes called *multidimensional fabrics*) with fibers oriented along many directions (>2)

Before the formation of the reinforcements, the fibers are subjected to a surface treatment to

- Decrease the abrasion action of fibers when passing through the forming machines.
- Improve the adhesion with the matrix material.

Other types of reinforcements, full or empty spheres (microspheres) or powders (see Section 3.5.3), are also used.

1.2.1.1 Relative Importance of Different Fibers in Applications

Figure 1.3 allows one to judge the relative importance in terms of the amount of fibers used in the fabrication of composites. One can immediately notice the industrial importance of fiber glass (produced in large quantities). Carbon and Kevlar fibers are reserved for high performance components.

Following are a few notes on the fibers:

- Glass fiber: The filaments are obtained by pulling the glass (silicon + sodium carbonate and calcium carbonate; $T > 1000°C$) through the small orifices of a plate made of platinum alloy.
- Kevlar fiber: This is an aramid fiber, yellowish color, made by DuPont de Nemours (USA). These are aromatic polyamides obtained by synthesis at $-10°C$, then fibrillated and drawn to obtain high modulus of elasticity.
- Carbon fiber: Filaments of polyacrylonitrile or pitch (obtained from residues of the petroleum products) are oxidized at high temperatures (300°C), then heated further to 1500°C in a nitrogen atmosphere. Then only the hexagonal

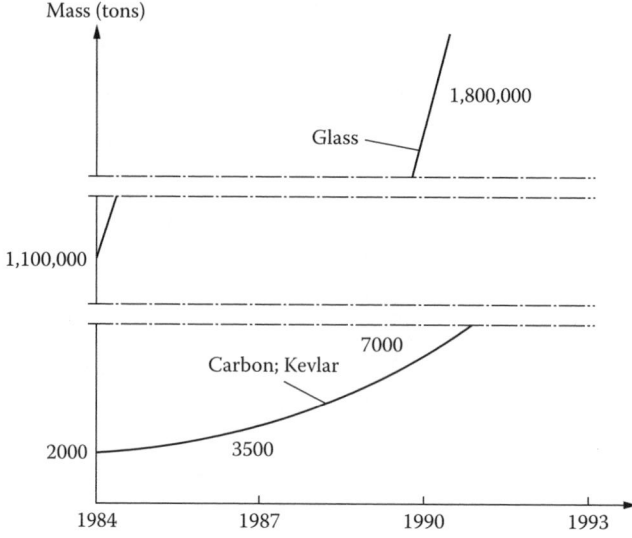

Figure 1.3 Relative Sale Volume of Different Fibers

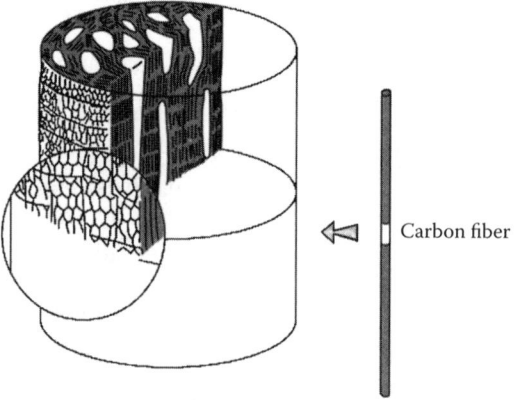

Carbon fiber

Figure 1.4 Structure of Carbon Fiber

carbon chains, as shown in Figure 1.4, remain. Black and bright filaments are obtained. High modulus of elasticity is obtained by drawing at high temperature.

- Boron fiber: Tungsten filament (diameter 12 µm) serves to catalyze the reaction between boron chloride and hydrogen at 1200°C. The boron fibers obtained have a diameter of about 100 µm (the growth speed is about 1 micron per second).
- Silicon carbide: The principle of fabrication is analogous to that of boron fiber: chemical vapor deposition (1200°C) of methyl trichlorosilane mixed with hydrogen.

The principal physical–mechanical properties of the fibers are indicated in Table 1.3. Note the very significant disparity of the prices per unit weight.

1.2.2 Matrix Materials

The matrix materials include the following:

- **Polymeric matrix:** thermoplastic resins (polypropylene, polyphenylene sulfone, polyamide, polyetheretherketone, etc.) and thermoset resins (polyesters, phenolics, melamines, silicones, polyurethanes, epoxies). Their principal physical properties are indicated in the Table 1.4.
- **Mineral matrix:** silicon carbide, carbon. They can be used at high temperatures (see Sections 2.2.4, 3.6, 7.1.10, 7.5).
- **Metallic matrix:** aluminum alloys, titanium alloys, oriented eutectics.

1.3 WHAT CAN BE MADE USING COMPOSITE MATERIALS?

The range of applications is very large. A few examples are shown below.

- Electrical, Electronics
  - Insulation for electrical construction
  - Supports for circuit breakers
  - Supports for printed circuits
  - Armors, boxes, covers
  - Antennas, radomes
  - Tops of television towers
  - Cable tracks
  - Windmills
- Buildings and Public Works
  - Housing cells
  - Chimneys
  - Concrete molds
  - Various covers (domes, windows, etc.)
  - Swimming pools
  - Facade panels
  - Profiles
  - Partitions, doors, furniture, bathrooms
- Road Transports
  - Body components
  - Complete body
  - Wheels, shields, radiator grills,
  - Transmission shafts
  - Suspension springs
  - Bottles for compressed petroleum gas
  - Chassis
  - Suspension arms
  - Casings
  - Cabins, seats
  - Highway tankers, isothermal trucks
  - Trailers

- Rail transports:
  - Fronts of power units
  - Wagons
  - Doors, seats, interior panels
  - Ventilation housings
- Marine Transports:
  - Hovercrafts
  - Rescue crafts
  - Patrol boats
  - Trawlers
  - Landing gears
  - Anti-mine ships
  - Racing boats
  - Pleasure boats
  - Canoes
- Cable transports:
  - Telepherique cabins
  - Telecabins
- Air transports
  - All composite passenger aircrafts
  - All composite gliders
  - Many aircraft components: radomes, leading edges, ailerons, vertical stabilizers, wings, …
  - Helicopter blades, propellers
  - Transmission shafts
  - Aircraft brake discs
- Space Transports
  - Rocket boosters
  - Reservoirs
  - Nozzles
  - Shields for atmosphere reentrance
- General mechanical applications
  - Gears
  - Bearings
  - Housings, casings
  - Jack body
  - Robot arms
  - Fly wheels
  - Weaving machine rods
  - Pipes
  - Components of drawing table
  - Compressed gas bottles
  - Tubes for offshore platforms
  - Pneumatics for radial frames
- Sports and Recreation
  - Tennis and squash rackets
  - Fishing poles
  - Skis

- Poles used in jumping
- Sails
- Surf boards
- Roller skates
- Bows and arrows
- Javelins
- Protection helmets
- Bicycle frames
- Golf clubs
- Oars

1.4 TYPICAL EXAMPLES OF INTEREST ON THE USE OF COMPOSITE MATERIALS

In the domain of commercial aircraft, one can compare the concerns of manufacturers with the principal characteristic properties of composite materials. The concerns of the manufacturers are performance **and** saving. The characteristics of composite components include the following:

- Weight saving leads to fuel saving, increase in payload, or increase in range which improves performances.
- Good fatigue resistance leads to enhanced life which involves saving in the long-term cost of the product.
- Good corrosion resistance means fewer requirements for inspection which results in saving on maintenance cost.

Moreover, taking into account the cost of the composite solution as compared with the conventional solution, one can state that composites fit the demand of aircraft manufacturers.

1.5 EXAMPLES ON REPLACING CONVENTIONAL SOLUTIONS WITH COMPOSITES

Table 1.1 shows a few significant cases illustrating the improvement on price and performance that can be obtained after replacement of a conventional solution with a composite solution.

1.6 PRINCIPAL PHYSICAL PROPERTIES

Tables 1.2 through 1.5 take into account the properties of only individual components, reinforcements, or matrices. The characteristics of composite materials resulting from the combination of reinforcement and matrix depend on

- The proportions of reinforcements and matrix (see Section 3.2)
- The form of the reinforcement (see Section 3.2)
- The fabrication process

Table 1.1 Some Significant Cases

Application	Price of Previous Construction	Price of Composite Construction
65 m³ reservoir for chemicals	Stainless steel + installation: 1.	0.53
Smoke stack for chemical plant	Steel: 1.	0.51
Nitric acid vapor washer	Stainless steel: 1.	0.33
Helicopter stabilizer	Light alloys + steel (16 kg): 1.	Carbon/epoxy (9 kg): 0.45
Helicopter winch support	Welded steel (16 kg): 1.	Carbon/epoxy (11 kg): 1.2
Helicopter motor hub	(Mass: 1): 1.	Carbon/Kevlar/epoxy (mass: 0.8): 0.4
X–Y table for fabrication of integrated circuits	Cast aluminum: Rate of fabrication: 30 plates/hr	Carbon/epoxy honeycomb sandwich: 55 plates/hr
Drum for drawing table	Speed of drawing: 15 to 30 cm/sec	Kevlar/epoxy: 40 to 80 cm/sec
Head of welding robot	Aluminum: Mass = 6 kg	Carbon/epoxy: Mass = 3 kg
Weaving machine rod	Aluminum: Rate = 250 shots/minute	Carbon/epoxy: Rate = 350 shots/minute
Aircraft floor	(Mass = 1): 1.	Carbon/Kevlar/epoxy (mass: 0.8): 1.7

These characteristics may be observed in Figure 1.5, which shows the tensile strength for different fiber fractions and different forms of reinforcement for the case of glass/resin composite, and Figure 1.6, which gives an interesting view on the specific resistance of the principal composites as a function of temperature. (The specific strength is defined as the strength divided by the density σ_{rupt}/ρ.)

Other remarkable properties of these materials include the following:

■ Composite materials **do not yield** (their elastic limits correspond to the rupture limit; see Section 5.4.5).

■ Composite materials are very **fatigue resistant** (see Section 5.1).

■ Composite materials **age** subject to humidity (epoxy resin can absorb water by diffusion up to 6% of its mass; the composite of reinforcement/resin can absorb up to 2%) and heat.

■ Composite materials **do not corrode**, except in the case of contact "aluminum with carbon fibers" in which galvanic phenomenon creates rapid corrosion.

■ Composite materials are not sensitive to the common chemicals used in engines: grease, oils, hydraulic liquids, paints and solvents, petroleum. However, cleaners for paint attack the epoxy resins.

■ Composite materials have medium to low level impact resistance (inferior to that of metallic materials).

■ Composite materials have excellent fire resistance as compared with the light alloys with identical thicknesses. However, the smokes emitted from the combustion of certain matrices can be toxic.

Table 1.2 Properties of Commonly Used Metals and Alloys and Silicon

Metals and Alloys	Density ρ (kg/m³)	Elastic Modulus E (MPa)	Shear Modulus G (MPa)	Poisson Ratio ν	Tensile Strength σ_{ult} (Mpa)	Elongation (%)	Coefficient of Thermal Expansion at 20°C α (°C⁻¹)	Coefficient of Thermal Conductivity at 20°C λ(W/m°C)	Heat Capacity c(J/kg°C)	Useful Temperature Limit T_{max} (°C)
Steels	7800	205,000	79,000	0.3	400 to 1600	1.8 to 10	1.3×10^{-5}	20 to 100	400 to 800	800
Aluminum Alloy 2024	2800	75,000	29,000	0.3	450	10	2.2×10^{-5}	140	1000	350
Titanium Alloy TA 6V	4400	105,000	40,300	0.3	1200	14	0.8×10^{-5}	17	540	700
Copper	8800	125,000	48,000	0.3	200 to 500		1.7×10^{-5}	380	390	650
Nickel	8900	220,000			500 to 850			70	500	900
Beryllium	1840	294,000		0.05	200		1.2×10^{-5}	150 (20°C) 90 (800°C)	1750 (20°C) 3000 (800°C)	900
Silicon	2200	95,000				5		1.4 (20°C) 3 (1200°C)	750 (20°C) 1200(500°C)	1300

Table 1.3 Properties of Commonly Used Reinforcements

Reinforcements	Fiber Diameter d(μm)	Density ρ(kg/m³)	Modulus of Elasticity E(Mpa)	Shear Modulus G(Mpa)	Poisson Ratio ν	Tensile Strength σUlt (Mpa)	Elongation E(%)	Coefficient of Thermal Expansion α(°C⁻¹)	Coefficient of Thermal Conductivity λ(W/M°C)	Heat Capacity c(J/kg°C)	Useful Temperature Limit Tmax (°C)	Price ($/kg)
"R" glass, high performance	10	2500	86,000		0.2	3200	4	0.3×10^{-5}	1	800	700	14
"E" glass, common applications	16	2600	74,000	30,000	0.25	2500	3.5	0.5×10^{-5}	1	800	700	2
Kevlar 49	12	1450	130,000	12,000	0.4	2900	2.3	-0.2×10^{-5}	0.03	1400		70
"HT" graphite, high strength	7	1750	230,000	50,000	0.3	3200	1.3	0.02×10^{-5}	200 (20°C) 60 (800°C)	800	>1500	70
"HM" graphite, high modulus	6.5	1800	390,000	20,000	0.35	2500	0.6	0.08×10^{-5}	200 (20°C) 60 (800°C)	800	>1500	140
Boron	100	2600	400,000			3400	0.8	0.4×10^{-5}			500	500
Aluminum	20	3700	380,000			1400	0.4		50 (20°C) 7 (800°C)	900	>1000	
Aluminum silicate	10	2600	200,000			3000	1.5					
Silicon carbide	14	2550	200,000			2800	1.3	0.5×10^{-5}			1300	600
Polyethylene		960	100,000			3000					150	

Table 1.4 Properties of Commonly Used Resins

Resins	Density ρ (kg/m³)	Elastic Modulus E(Mpa)	Shear Modulus G(Mpa)	Poisson Ratio ν	Tensile Strength σ_{Ult} (Mpa)	Elongation E%	Coefficient of Thermal Expansion $\alpha(°C^{-1})$	Coefficient of Thermal Conductivity λ(W/m°C)	Heat Capacity C(J/kg°C)	Useful Temperature Limit T_{max} (°C)	Price ($/kg)
Thermosets											
Epoxy	1200	4500	1600	0.4	130	2 (100°C) 6 (200°C)	11×10^{-5}	0.2	1000	90 to 200	6 to 20
Phenolic	1300	3000	1100	0.4	70	2.5	1×10^{-5}	0.3	1000	120 to 200	
Polyester	1200	4000	1400	0.4	80	2.5	8×10^{-5}	0.2	1400	60 to 200	2.4
Polycarbonate	1200	2400		0.35	60		6×10^{-5}		1200	120	
Vinylester	1150	3300			75	4	5×10^{-5}			>100	4
Silicone	1100	2200		0.5	35					100 to 350	
Urethane	1100	700 to 7000			30	100				100	4
Polyimide	1400	4000 to 19,000	1100	0.35	70	1	8×10^{-5}	0.2	1000	250 to 300	
Thermoplastics											
Polypropylene (pp)	900	1200		0.4	30	20 to 400	9×10^{-5}		330	70 to 140	
Polyphenylene sulfone (pps)	1300	4000			65	100	5×10^{-5}			130 to 250	
Polyamide (pa)	1100	2000		0.35	70	200	8×10^{-5}		1200	170	6
Polyether sulfone (pes)	1350	3000			85	60	6×10^{-5}			180	25
Polyetherimide (pei)	1250	3500			105	60	6×10^{-5}	0.2		200	20
Polyether-ether-ketone (peek)	1300	4000			90	50	5×10^{-5}	0.3		140 to 250	96

Table 1.5 Properties of Commonly Used Core Materials

Cores	Density $\rho(Kg/M^3)$	Modulus of Elasticity $E(Mpa)$	Shear Modulus $G(Mpa)$	Poisson Ratio ν	Compressive Strength $\sigma_{Ult}(Mpa)$	Elongation $E\%$	Coefficient of Thermal Expansion $\alpha(°C^{-1})$	Coefficient of Thermal Conductivity $\lambda(W/M°C)$	Heat Capacity $C(J/Kg°C)$	Useful Temperature Limit T_{max} (°C)	Price ($/Kg)
Balsa	100 to 190	2000 to 6000	100 to 250		8 to 18			0.05			11
Polyurethane foam	30 to 70	25 to 60		0.4						75	
Polystyrene foam	30 to 45	20 to 30		0.4	0.25 to 1.25					75	
Honeycombs											
Impregnated carton			50 to 350								
Impregnated glass fabric			100 to 600								
Aluminum	15 to 130		130 to 910		0.2 to 8						
Steel			550 to 1250								
Nomex®	25 to 50		10 to 40		0.2 to 2.5						

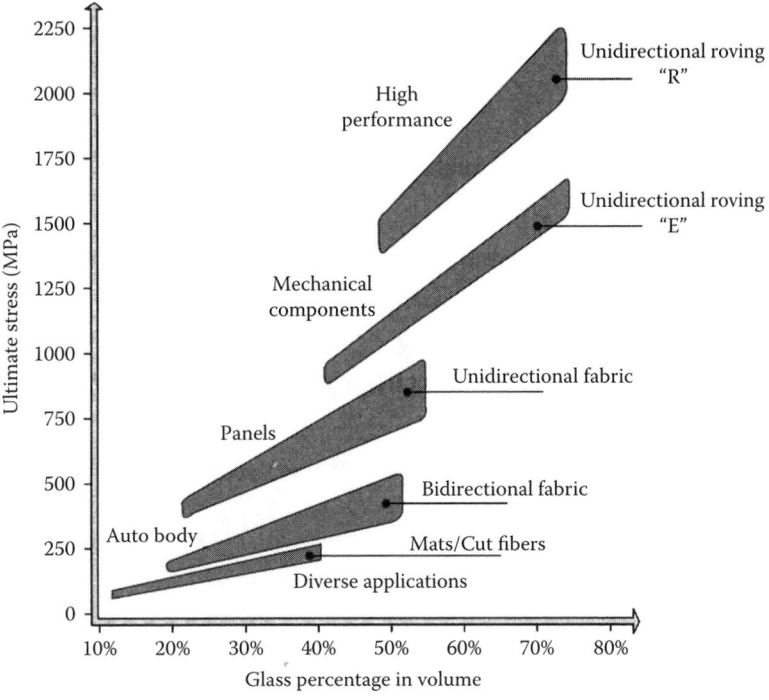

Figure 1.5   Tensile Strength of Glass/Resin Composites

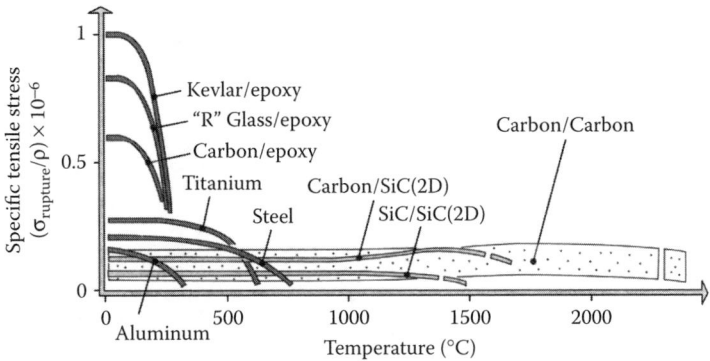

Figure 1.6   Specific Strength of Different Composites

2

FABRICATION PROCESSES

The mixture of reinforcement/resin does not really become a composite material until the last phase of the fabrication, that is, when the matrix is hardened. After this phase, it would be impossible to modify the material, as in the way one would like to modify the structure of a metal alloy using heat treatment, for example.

In the case of polymer matrix composites, this has to be polymerized, for example, polyester resin. During the solidification process, it passes from the liquid state to the solid state by copolymerization with a monomer that is mixed with the resin. The phenomenon leads to hardening. This can be done using either a chemical (accelerator) or heat. The following pages will describe the principal processes for the formation of composite parts.

2.1 MOLDING PROCESSES

The flow chart in Figure 2.1 shows the steps found in all molding processes. Forming by molding processes varies depending on the nature of the part, the number of parts, and the cost. The mold material can be made of metal, polymer, wood, or plaster.

2.1.1 Contact Molding

Contact molding (see Figure 2.2) is open molding (there is only one mold, either male or female). The layers of fibers impregnated with resin (and accelerator) are placed on the mold. Compaction is done using a roller to squeeze out the air pockets. The duration for resin setting varies, depending on the amount of accelerator, from a few minutes to a few hours. One can also obtain parts of large dimensions at the rate of about 2 to 4 parts per day per mold.

2.1.2 Compression Molding

With compression molding (see Figure 2.3), the countermold will close the mold after the impregnated reinforcements have been placed on the mold. The whole assembly is placed in a press that can apply a pressure of 1 to 2 bars. The polymerization takes place either at ambient temperature or higher.

The process is good for average volume production: one can obtain several dozen parts a day (up to 200 with heating). This has application for automotive and aerospace parts.

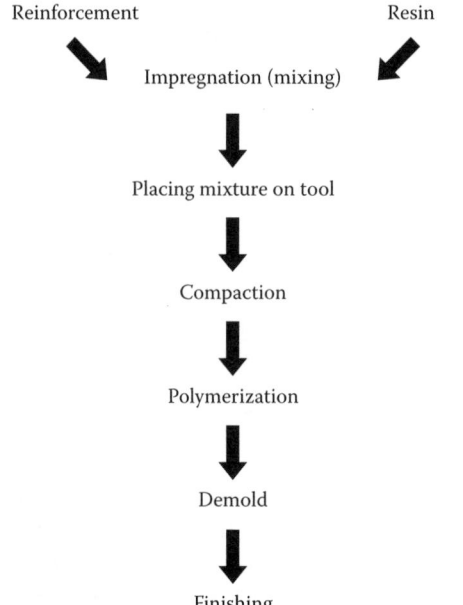

Figure 2.1   Steps in Molding Process

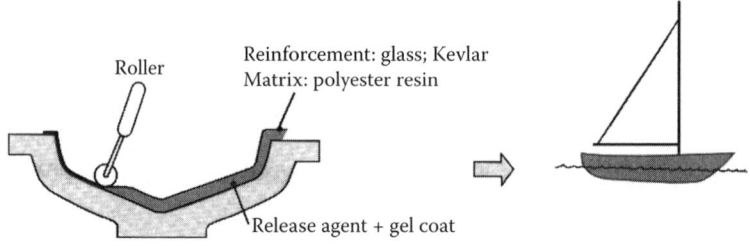

Figure 2.2   Contact Molding

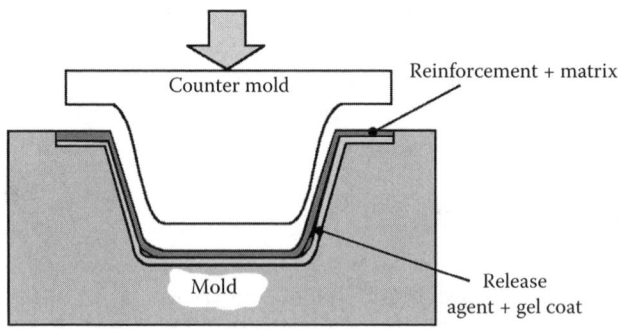

Figure 2.3   Compression Molding

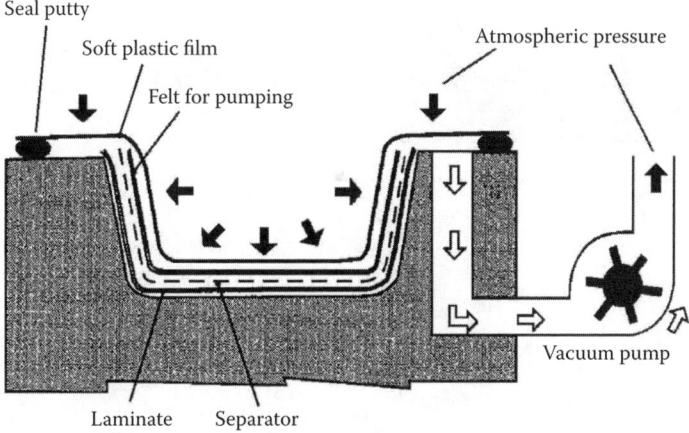

Figure 2.4 Vacuum Molding

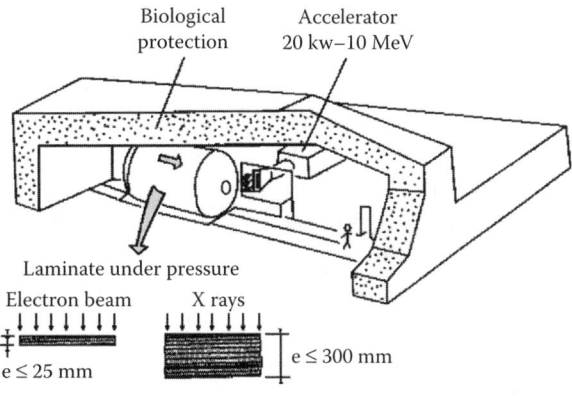

Figure 2.5 Electron Beam or X-ray Molding

2.1.3 Molding with Vacuum

This process of molding with vacuum is still called **depression molding** or **bag molding**. As in the case of contact molding described previously, one uses an open mold on top of which the impregnated reinforcements are placed. In the case of sandwich materials, the cores are also used (see Chapter 4). One sheet of soft plastic is used for sealing (this is adhesively bonded to the perimeter of the mold). Vacuum is applied under the piece of plastic (see Figure 2.4). The piece is then compacted due to the action of atmospheric pressure, and the air bubbles are eliminated. Porous fabrics absorb excess resin. The whole material is polymerized by an oven or by an autoclave under pressure (7 bars in the case of carbon/epoxy to obtain better mechanical properties), or with heat, or with electron beam, or x-rays; see Figure 2.5). This process has applications for aircraft structures, with the rate of a few parts per day (2 to 4).

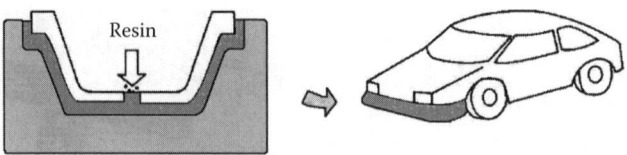

Figure 2.6 Resin Injection Molding

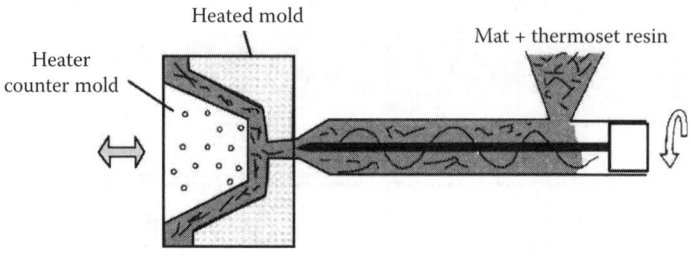

Figure 2.7 Injection of Premixed

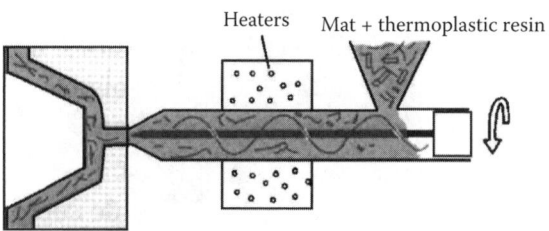

Figure 2.8 Injection of Thermoplastic Premixed

2.1.4 Resin Injection Molding

With resin injection molding (see Figure 2.6), the reinforcements (mats, fabrics) are put in place between the mold and countermold. The resin (polyester or phenolic) is injected. The mold pressure is low. This process can produce up to 30 pieces per day. The investment is less costly and has application in automobile bodies.

2.1.5 Molding by Injection of Premixed

The process of molding by injection of premixed allows automation of the fabrication cycle (rate of production up to 300 pieces per day).

- **Thermoset resins:** Can be used to make components of auto body. The schematic of the process is shown in Figure 2.7.
- **Thermoplastic resins:** Can be used to make mechanical components with high temperature resistance, as shown in Figure 2.8.

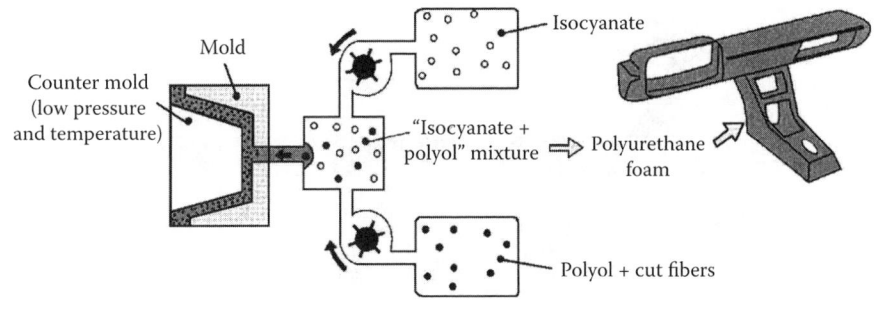

Figure 2.9   Foam Injection

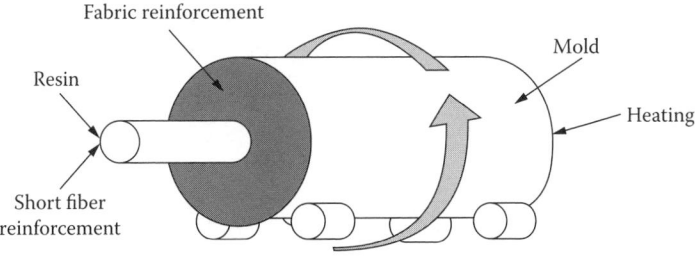

Figure 2.10   Centrifugal Molding

2.1.6 Molding by Foam Injection

Molding by foam injection (see Figure 2.9) allows the processing of pieces of fairly large dimensions made of polyurethane foam reinforced with glass fibers. These pieces remain stable over time, with good surface conditions, and have satisfactory mechanical and thermal properties.

2.1.7 Molding of Components of Revolution

The process of **centrifugal molding** (see Figure 2.10) is used for the fabrication of tubes. It allows homogeneous distribution of resin with good surface conditions, including the internal surface of the tube. The length of the tube depends on the length of the mold. Rate of production varies with the diameter and length of the tubes (up to 500 kg of composite per day).

The process of **filament winding** (see Figure 2.11) can be integrated into a continuous chain of production and can fabricate tubes of long length. The rate of production can be up to 500 kg of composite per day. These can be used to make missile tubes, torpillas, containers, or tubes for transporting petroleum.

For pieces which must revolve around their midpoint, winding can be done on a mandrel. This can then be removed and cured in an autoclave (see Figure 2.12). The fiber volume fraction is high (up to 85%). This process is used to fabricate components of high internal pressure, such as reservoirs and propulsion nozzles.

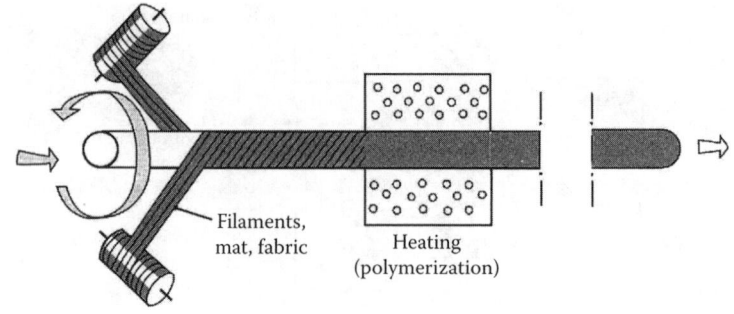

Figure 2.11   Filament Winding

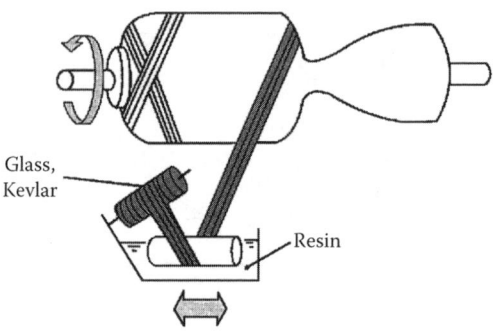

Figure 2.12   Filament Winding on Complex Mandrel

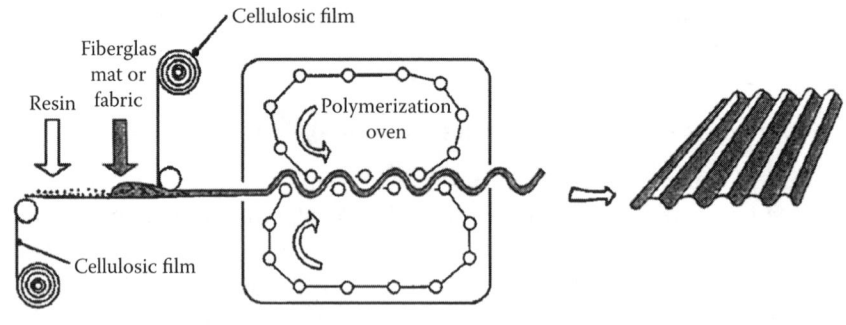

Figure 2.13   Sheet Forming

2.2 OTHER FORMING PROCESSES

2.2.1 Sheet Forming

This procedure of sheet forming (see Figure 2.13) allows the production of plane or corrugated sheets by corrugation or ribs.

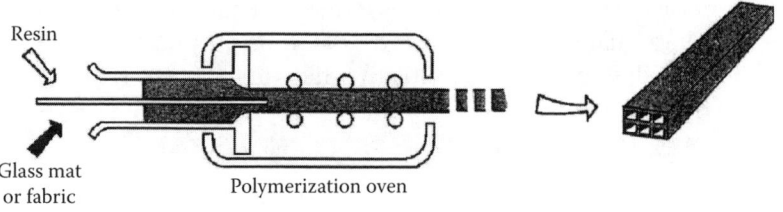

Figure 2.14 Profile Forming

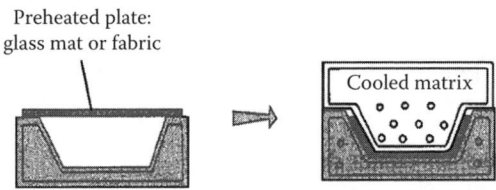

Figure 2.15 Stamp Forming

2.2.2 Profile Forming

The piece shown in Figure 2.14 is made by pultrusion. This process makes possible the fabrication of continuous open or closed profiles. The fiber content is important for high mechanical properties. The rate of production varies between 0.5 and 3 m/ minute, depending on the nature of the profile.

2.2.3 Stamp Forming

Stamp forming (see Figure 2.15) is only applicable to thermoplastic composites. One uses preformed plates, which are heated, stamped, and then cooled down.

2.2.4 Preforming by Three-Dimensional Assembly

Example: Carbon/carbon. The carbon reinforcement is assembled by depositing the woven tows along several directions in space. Subsequently the empty space between the tows is filled by "impregnation." The following two techniques are used:

- **Impregnation using liquid:** Pitch is used under a pressure of 1000 bars, followed by carbonization.
- **Impregnation using gas:** This involves chemical vapor deposition using a hot gaseous hydrocarbon atmosphere.

Example: Silicon/silicon. The reinforcement is composed of filaments of silicon ceramics. The silicon matrix is deposited in the form of liquid solution of colloidal silicon, followed by drying under high pressure and high temperature (2000 bars, 2000°C). The preforms are then machined. The phases of development of these composites, such as the densification (formation of the matrix) are long and delicate. These make the products very onerous. Applications include missile and launcher nozzles, brake disks, ablative tiles for reentry body of spacecraft into the atmosphere.

2.2.5 Cutting of Fabric and Trimming of Laminates

Some components need a large number of fabric layers (many dozens, can be hundreds). For the small and medium series, it can be quite expensive to operate manually for

- following the form of a cut.
- respect the orientation specified by the design (see Chapter 5).
- minimizing waste.

There is a tendency to produce a cut or a drape automatically with the following characteristics:

- a programmed movement of the cutting machine
- a rapid cutting machine, such as an orientable vibrating cutting knife or a laser beam with the diameter of about 0.2 mm and a cutting speed varying from 15 to 40 meters/minute, depending on the power of the laser and the thickness of the part.

Example: With a draping machine MAD Forest-Line (FRA), the draping is done in two steps by means of two distinct installations:

- A cutting machine that produces a roller to which the cut pieces are attached (cassettes)
- A depositing machine which uses the cassette of cut pieces to perform the draping.

The two operations are shown schematically in Figure 2.16.

2.3 PRACTICAL HINTS ON MANUFACTURING PROCESSES

2.3.1 Acronyms

To describe the modes of fabricating the composite pieces, the professionals use many abbreviations. Each is detailed below, with reference to the paragraph number where the process is explained.

> **B.M.C.:** "Bulk Molding Compound." Matrix: resins. Reinforcement: cut fibers; additional fillers (powder). Pressure: 5–10 *MPa*. Temperature: 120–150°C.
>
> **Centrifugation:** Matrix: resins. Reinforcement: cut fibers, mat, fabrics; see Section 2.1.7.

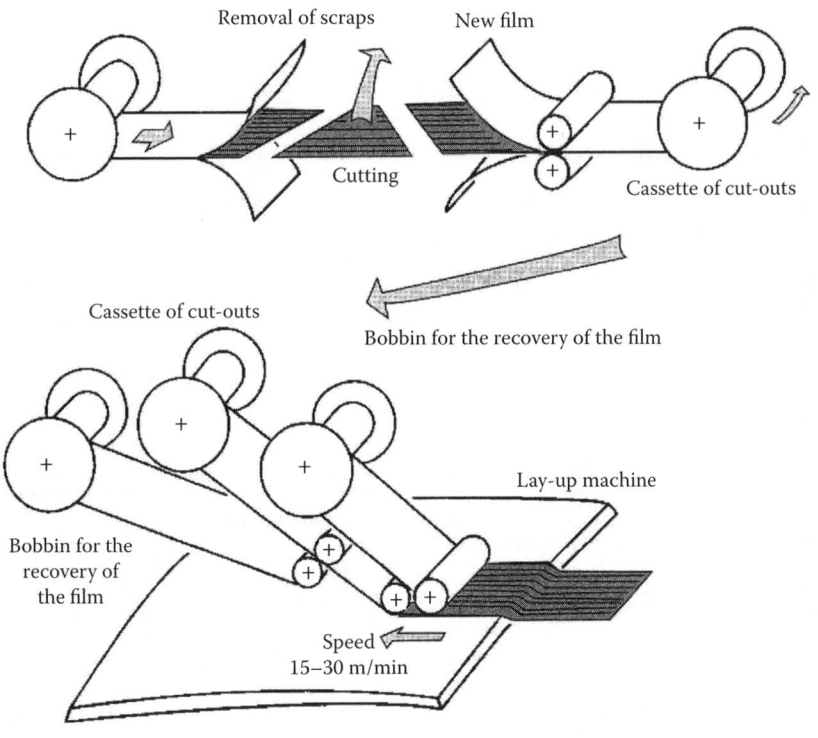

Figure 2.16 Draping Process

Contact molding: Matrix: resins. Reinforcement: mat, fabrics; see Section 2.1.1.

Filament winding: Matrix: resins. Reinforcement: continuous fibers; see Section 2.1.7.

Compression molding: Matrix: resins. Reinforcement: fabrics or unidirectionals; see Section 2.1.2.

Autoclave molding: Matrix: resins. Reinforcement: fabrics or unidirectionals; under pressure in an autoclave; see Section 2.1.3.

Pultrusion: Matrix: resins. Reinforcement: mat, fabrics, continuous fibers; see Section 2.2.2.

R-RIM: "Reinforced–Reaction Injection Molding" (there is expansion in the mold). Pressure: 0.5 MPa. Temperature: 50–60° C; see Section 2.1.6.

S-RIM: "Structural Reaction Injection Molding" (components for structure, particularly in automobiles). Similar to R-RIM, injection of liquid thermoset resins consists of two highly reactive constituents.

RTM: "Resin Transfer Molding." Matrix: resins. Reinforcement: Preforms of cut fibers or fabrics. Pressure: low (in vacuum or 0.1–0.3 *MPa*). Temperature: 80°C.

SMC: "Sheet Molding Compound." Matrix: liquid resin with addition of magnesia. Reinforcement: mat, unidirectionals. Pressure: 5–10 *MPa*. Temperature: 120–150°C.

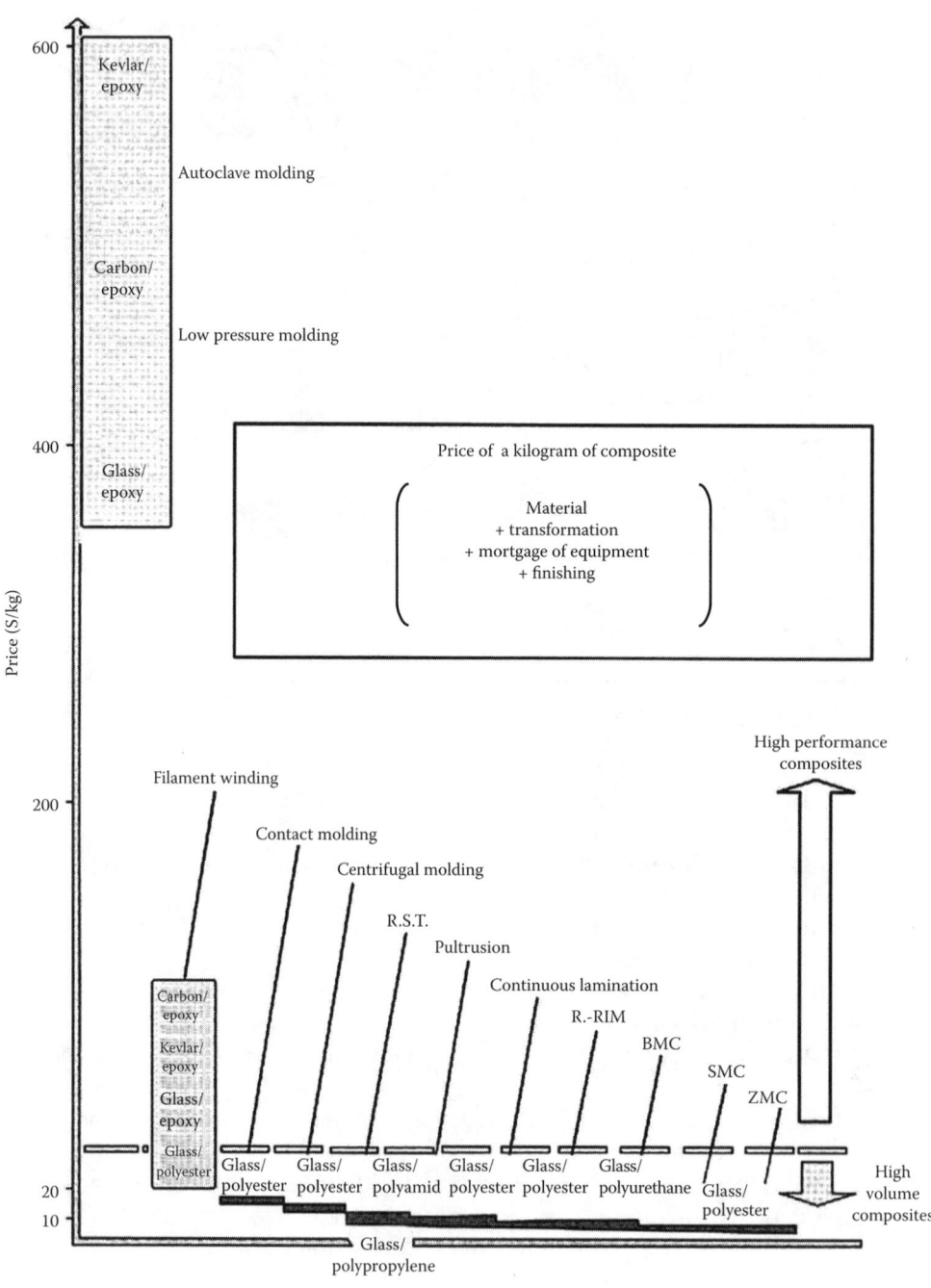

Figure 2.17 Cost Comparison for Different Processes

RTP: "Reinforced Thermoplastics." Matrix: thermoplastic resins. Reinforcement: cut fibers. Pressure: 50 to 150 *MPa*. Temperature: 120–150°C.

RST: "Reinforced Stamped Thermoplastics." Pressure: 15–20 *MPa*. Initial temperature: ≈200°C; see Section 2.2.3.

ZMC: Matrix: resin. Reinforcement: cut fibers. Pressure: 30–50 *MPa*. Temperature: 120–150°C.

TMC: Similar to "SMC" but with higher amount of glass fibers (a few millimeters in thickness).

XMC: Similar to "SMC" but with specific orientation of the fibers.

2.3.2 Cost Comparison

The diagram in Figure 2.17 allows the comparison of the cost to fabricate composite products. One needs to note the important difference between the cost of composites produced in large volume and the cost of high performance composites.

3

PLY PROPERTIES

It is of fundamental importance for the designer to understand and to know precisely the geometric and mechanical characteristics of the "fiber + matrix" mixture which is the basic structure of the composite parts. The description of these characteristics is the object of this chapter.

3.1 ISOTROPY AND ANISOTROPY

When one studies the mechanical behavior of elastic bodies under load (elasticity theory), one has to consider the following:

- An **elastic** body subjected to stresses deforms in a **reversible** manner.
- At each point within the body, one can identify the **principal planes** on which there are only **normal stresses**.
- The **normal directions** on these planes are called the **principal stress directions**.
- A small **sphere** of material surrounding a point of the body becomes an **ellipsoid** after loading.

The spatial position of the ellipsoid relative to the principal stress directions enables us to characterize whether the material under study is isotropic or anisotropic. Figure 3.1 illustrates this phenomenon.

Figure 3.2 illustrates the deformation of an isotropic sample and an anisotropic sample. In the latter case, the oblique lines represent the preferred directions along which one would place the fibers of reinforcement. One can consider that a longitudinal loading applied to an isotropic plate would create an extension in the longitudinal direction and a contraction in the transverse direction. The same loading applied to an anisotropic plate creates an angular distortion, in addition to the longitudinal extension and transversal contraction.

In the simple case of plane stress, one can obtain the elastic constants using stress–strain relations.

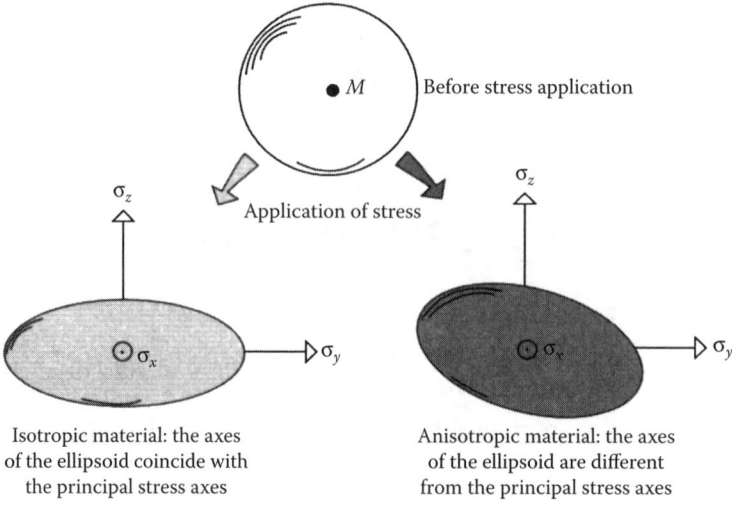

Figure 3.1 Schematic of Deformation

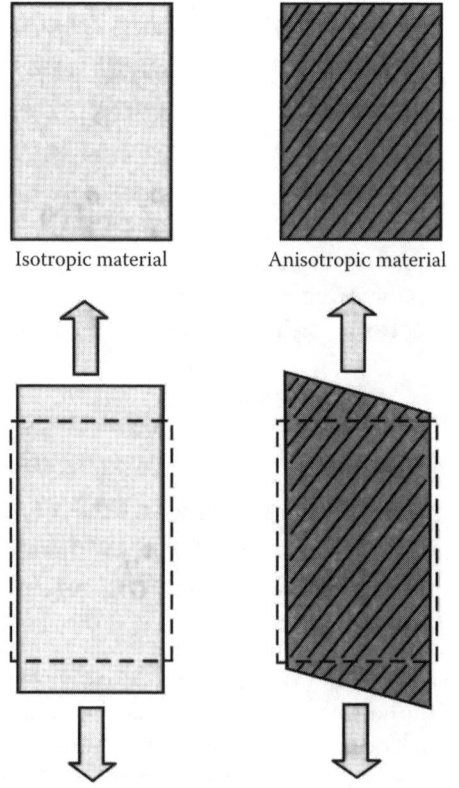

Figure 3.2 Comparison between Deformation of an Isotropic and Anisotropic Plate

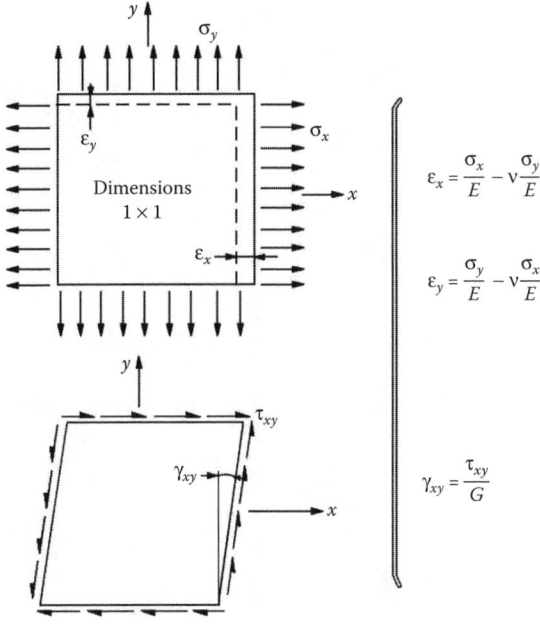

Figure 3.3   Stress–Strain Behavior in an Isotropic Material

3.1.1 Isotropic Materials

The following relations are valid for a material that is elastic and isotropic.

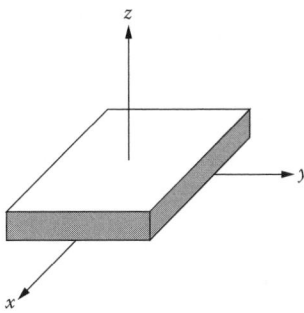

One can write the stress–strain relation (see Figure 3.3) in matrix form as[1]

$$
\left\{ \begin{array}{c} \varepsilon_x \\ \varepsilon_y \\ \gamma_{xy} \end{array} \right\} = \begin{bmatrix} \dfrac{1}{E} & -\dfrac{\nu}{E} & 0 \\ -\dfrac{\nu}{E} & \dfrac{1}{E} & 0 \\ 0 & 0 & \dfrac{1}{G} \end{bmatrix} \left\{ \begin{array}{c} \sigma_x \\ \sigma_y \\ \tau_{xy} \end{array} \right\}
$$

---

[1]   In these equations, $\varepsilon_x, \varepsilon_y, \gamma_{xy}$ are also the small strains that are obtained in a classical manner from the displacements u_x and u_y as: $\varepsilon_x = \partial u_x / \partial x$; $\varepsilon_y = \partial u_y / \partial y$; $\gamma_{xy} = \partial u_x / \partial y + \partial u_y / \partial x$.

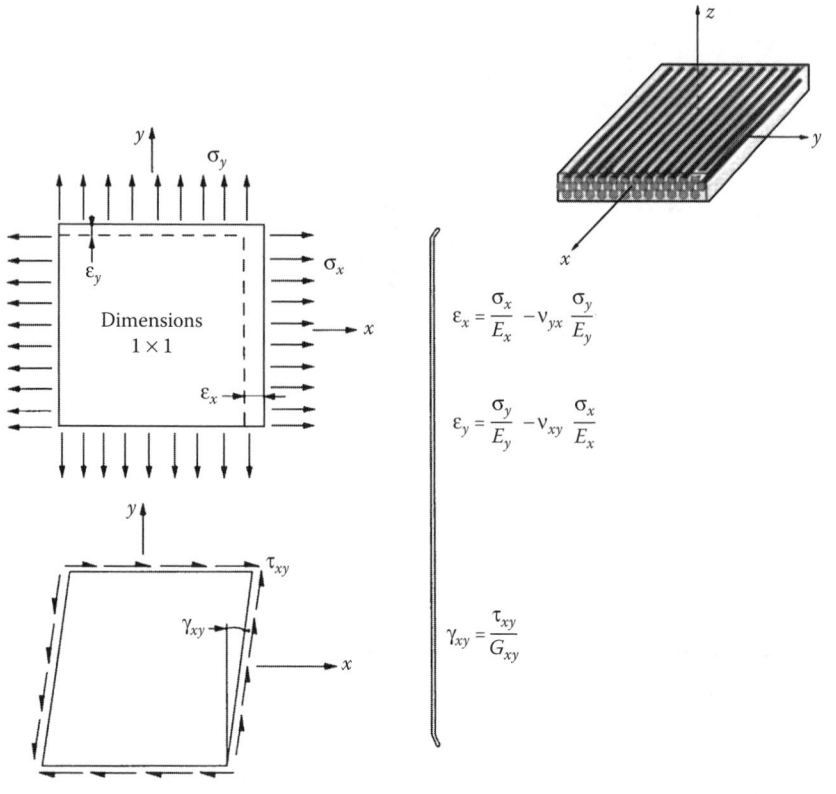

Figure 3.4   Deformation in an Anisotropic Material

There are three elastic constants: E, v, G. There exists a relation among them as:

$$G = \frac{E}{2(1 + v)}$$

The above relation shows that a material that is isotropic and elastic can be characterized by two independent elastic constants: E and v.

3.1.2 Anisotropic Material

The matrix equation for anisotropic material (see Figure 3.4) is

$$
\begin{Bmatrix} \varepsilon_x \\ \varepsilon_y \\ \gamma_{xy} \end{Bmatrix}
=
\begin{bmatrix}
\dfrac{1}{E_x} & -\dfrac{v_{yx}}{E_y} & 0 \\[2ex]
-\dfrac{v_{xy}}{E_x} & \dfrac{1}{E_y} & 0 \\[2ex]
0 & 0 & \dfrac{1}{G_{xy}}
\end{bmatrix}
\begin{Bmatrix} \sigma_x \\ \sigma_y \\ \tau_{xy} \end{Bmatrix}
$$

In the figure the following relations appear:

$$\varepsilon_x = \frac{\sigma_x}{E_x} - v_{yx}\frac{\sigma_y}{E_y}$$

$$\varepsilon_y = \frac{\sigma_y}{E_y} - v_{xy}\frac{\sigma_x}{E_x}$$

$$\gamma_{xy} = \frac{\tau_{xy}}{G_{xy}}$$

Note that the stress–strain matrix above is symmetric.[2] The number of distinct elastic constants is five:

- Two moduli of elasticity: E_x and E_y,
- Two Poisson coefficients: v_{yx} and v_{xy}, and
- One shear modulus: G_{xy}.

In fact there are only four independent elastic constants:[3] E_x, E_y, G_{xy}, and v_{yx} (or v_{xy}). The fifth elastic constant can be obtained from the others using the symmetry relation:

$$v_{xy} = v_{yx} \frac{E_x}{E_y}$$

3.2 CHARACTERISTICS OF THE REINFORCEMENT–MATRIX MIXTURE

We denote as **ply** the semi-product "reinforcement + resin" in a quasi-bidimensional form.[4] This can be

- A tape of unidirectional fiber + matrix,
- A fabric + matrix, or
- A mat + matrix.

These are examined more below.

3.2.1 Fiber Mass Fraction

Fiber mass fraction is defined as

$$M_f = \frac{\text{Mass of fibers}}{\text{Total mass}}$$

In consequence, the mass fraction of matrix is

$$M_m = \frac{\text{Mass of matrix}}{\text{Total mass}}$$

with

$$M_m = 1 - M_f$$

---

[2] To know more about the development on this point, refer to Section 9.2 and Exercise 18.1.2.
[3] Refer to Section 13.2.
[4] Such condition exists in the commercial products. These are called *preimpregnated* or *SMC* (sheet molding compound). One can also find non-preformed mixtures of short fibers and resin. These are called *premix* or *BMC* (bulk molding compound).

Table 3.1   Common Fiber Volume Fractions in Different Processes

Molding Process	Fiber Volume Fraction
Contact Molding	30%
Compression Molding	40%
Filament Winding	60%–85%
Vacuum Molding	50%–80%

3.2.2 Fiber Volume Fraction

Fiber volume fraction is defined as

$$V_f = \frac{\text{Volume of fiber}}{\text{Total volume}}$$

As a result, the volume fraction of matrix is given as

$$V_m = \frac{\text{Volume of matrix}}{\text{Total volume}}$$

with[5]

$$V_m = 1 - V_f$$

Note that one can convert from mass fraction to volume fraction and vice versa. If ρ_f and ρ_m are the specific mass of the fiber and matrix, respectively, we have

$$V_f = \frac{\dfrac{M_f}{\rho_f}}{\dfrac{M_f}{\rho_f} + \dfrac{M_m}{\rho_m}} \qquad M_f = \frac{V_f \rho_f}{V_f \rho_f + V_m \rho_m}$$

Depending on the method of fabrication, the common fiber volume fractions are as shown in Table 3.1.

3.2.3 Mass Density of a Ply

The mass density of a ply can be calculated as

$$\rho = \frac{\text{total mass}}{\text{total volume}}$$

---

[5] In reality, the mixture of fiber/matrix also includes a small volume of voids, characterized by the *porosity* of the composite. One has then $V_m + V_f + V_p = 1$, in which V_p denotes the percentage of "volume of void/total volume." V_p is usually much less than 1 (See Exercise 18.1.11).

Table 3.2 Ply Thicknesses of Some Common Composites

	M_f	h
E glass	34%	0.125 mm
R glass	68%	0.175 mm
Kevlar	65%	0.13 mm
H.R. Carbon	68%	0.13 mm

The above equation can also be expanded as

$$\rho = \frac{\text{mass of fiber}}{\text{total volume}} + \frac{\text{mass of matrix}}{\text{total volume}}$$

$$= \frac{\text{volume of fiber}}{\text{total volume}}\rho_f + \frac{\text{volume of matrix}}{\text{total volume}}\rho_m$$

or

$$\rho = \rho_f V_f + \rho_m V_m$$

3.2.4 Ply Thickness

The ply thickness is defined starting from the number of **grams** of mass of fiber m_{of} per m^2 of area. The ply thickness, denoted as h, is such that:

$$h \times 1(m^2) = \text{total volume} = \text{total volume} \times \frac{m_{of}}{\text{fiber volume} \times \rho_f}$$

or

$$h = \frac{m_{of}}{V_f \rho_f}$$

One can also express the thickness in terms of mass fraction of fibers rather than in terms of volume fraction.

$$h = m_{of}\left[\frac{1}{\rho_f} + \frac{1}{\rho_m}\left(\frac{1-M_f}{M_f}\right)\right]$$

Table 3.2 shows a few examples of ply thicknesses.

3.3 UNIDIRECTIONAL PLY

3.3.1 Elastic Modulus

The mechanical characteristics of the fiber/matrix mixture can be obtained based on the characteristics of each of the constituents. In the literature, there are theoretical as well as semi-empirical relations. As such, the results from these relations may not always agree with experimental values. One of the reasons is

Table 3.3 Fiber Elastic Modulus

		Glass E	Kevlar	Carbon H.R.	Carbon H.M.
	fiber longitudinal modulus in ℓ direction Ef_ℓ (MPa)	74,000	130,000	230,000	390,000
	fiber transverse modulus in t direction Ef_t (MPa)	74,000	5400	15,000	6000
	fiber shear modulus $Gf_{\ell t}$ (Mpa)	30,000	12,000	50,000	20,000
	fiber Poisson ratio $vf_{\ell t}$	0.25	0.4	0.3	0.35
		Isotropic		Anisotropic	

because the fibers themselves exhibit some degree of anisotropy. In Table 3.3, one can see small values of the elastic modulus in the transverse direction for Kevlar and carbon fibers, whereas glass fiber is isotropic.[6]

With the definitions in the previous paragraph, one can use the following relations to characterize the unidirectional ply:

■ **Modulus of elasticity along the direction of the fiber E_l is given by**[7]

$$E_\ell = E_f V_f + E_m V_m$$

or

$$E_\ell = E_f V_f + E_m(1 - V_f)$$

In practice, this modulus depends essentially on the longitudinal modulus of the fiber, E_f, because $E_m \ll E_f$ (as $E_{m\ \text{resin}}/E_{f\ \text{glass}} \# 6\%$).

■ **Modulus of elasticity in the transverse direction to the fiber axis, E_t:**
In the following equation, E_{ft} represents the modulus of elasticity of the

---

[6] This is due to the drawing of the carbon and Kevlar fibers during fabrication. This orients the chain of the molecules.

[7] Chapter 10 gives details for the approximate calculation of the moduli E_ℓ, E_t, $G_{\ell t}$ and $v_{\ell t}$ which lead to these expressions.

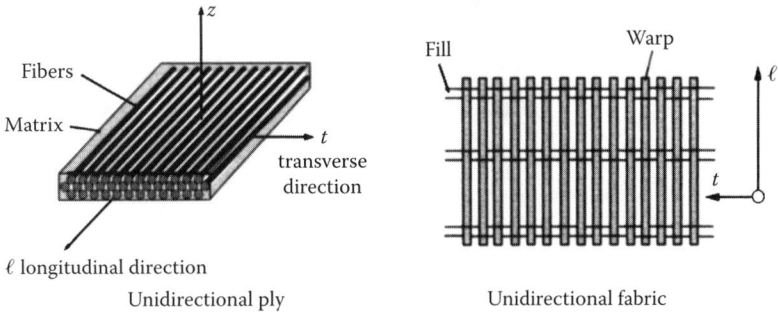

Figure 3.5   Orientations in Composite Layers

fiber in the direction that is transverse to the fiber axis as indicated in Table 3.3.

$$E_t = E_m \left[\frac{1}{(1 - V_f) + \frac{E_m}{E_{ft}} V_f} \right]$$

■ **Shear modulus $G_{\ell t}$:** An order of magnitude of this modulus is given in the following expression, in which $G_{f\ell t}$ represents the shear modulus of the fiber (as shown in Table 3.3).

$$G_{\ell t} = G_m \left[\frac{1}{(1 - V_f) + \frac{G_m}{G_{f\ell t}} V_f} \right]$$

■ **Poisson coefficient $v_{\ell t}$:** The Poisson coefficient represents the contraction in the transverse direction t when a ply is subjected to tensile loading in the longitudinal direction ℓ (see Figure 3.5).

$$v_{\ell t} = v_f V_f + v_m V_m$$

■ **Modulus along any direction:** The modulus along a certain direction in plane ℓt, other than along the fiber and transverse to the fiber,[8] is given in the expression below, where $c = \cos\theta$ and $s = \sin\theta$. Note that this modulus decreases rapidly as one is moving away from the fiber direction (see Figure 3.6).

$$E_x = \frac{1}{\frac{c^4}{E_\ell} + \frac{s^4}{E_t} + 2c^2 s^2 \left(\frac{1}{2G_{\ell t}} - \frac{v_{\ell t}}{E_\ell} \right)}$$

---

[8] The calculation of these moduli is shown in details in Chapter 11.

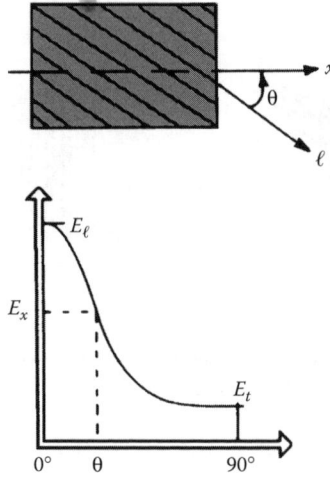

Figure 3.6 Off-axis Modulus

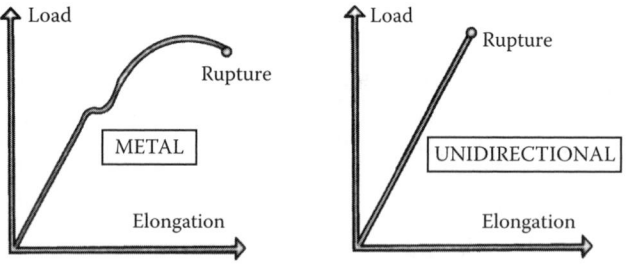

Figure 3.7 Loading Curves of Metal and Unidirectional Composite

3.3.2 Ultimate Strength of a Ply

The curves in Figure 3.7 show the important difference in the behavior between classical metallic materials and the unidirectional plies. These differences can be summarized in a few points as

- There is lack of plastic deformation in the unidirectional ply. (This is a disadvantage.)
- Ultimate strength of the unidirectional ply is higher. (This is an advantage.)
- There is important elastic deformation for the unidirectional ply. (This can be an advantage or a disadvantage, depending on the application; for example, this is an advantage for springs, arcs, or poles.)

When the fibers break before the matrix during loading along the fiber direction, one can obtain the following for the composite:

$$\sigma_{\ell \atop rupture} = \sigma_{f \atop rupture}\left[V_f + (1 - V_f)\frac{E_m}{E_f}\right]$$

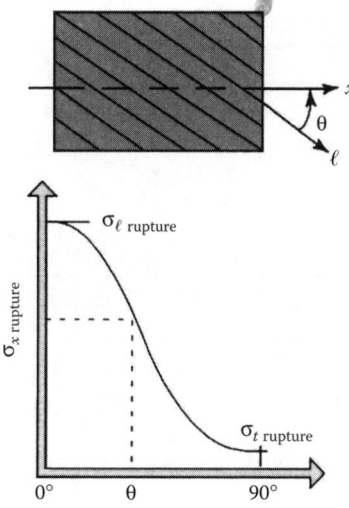

Figure 3.8   Off-axis Rupture Strength

or, approximately

$$\sigma_{\ell \atop rupture} \approx \sigma_{f \atop rupture} \times V_f$$

The **ultimate strength along any direction**[9] is given by the following relation (see Figure 3.8), where

$\sigma_{l,rupture}$ = Fracture strength in the direction of the fibers,
$\sigma_{t,rupture}$ = Fracture strength transverse to the direction of the fibers,
$\tau_{\ell t,rupture}$ = Shear strength in the plane (ℓ,t) of the ply

$$c = \cos\theta$$

$$s = \sin\theta$$

$$\sigma_{x \atop rupture} = \cfrac{1}{\sqrt{\dfrac{c^4}{\sigma_{\ell \atop rupture}^2} + \dfrac{s^4}{\sigma_{t \atop rupture}^2} + \left(\dfrac{1}{\tau_{\ell t \atop rupture}^2} - \dfrac{1}{\sigma_{\ell \atop rupture}^2}\right)c^2 s^2}}$$

3.3.3   Examples

Table 3.4 gives the properties of the fibers/epoxy unidirectional ply at 60% fiber volume fraction.[10]

---

[9]   Detailed calculation is shown in Section 14.3.

Table 3.4 Properties of Fiber/Epoxy Plies

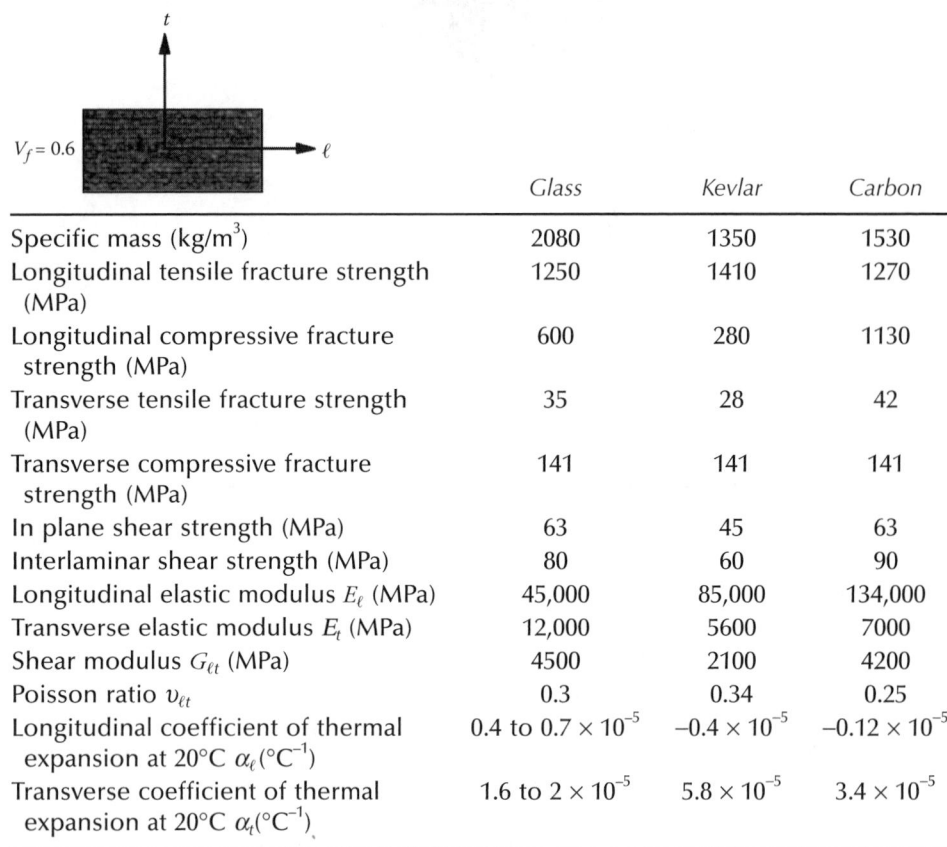

	Glass	Kevlar	Carbon
Specific mass (kg/m³)	2080	1350	1530
Longitudinal tensile fracture strength (MPa)	1250	1410	1270
Longitudinal compressive fracture strength (MPa)	600	280	1130
Transverse tensile fracture strength (MPa)	35	28	42
Transverse compressive fracture strength (MPa)	141	141	141
In plane shear strength (MPa)	63	45	63
Interlaminar shear strength (MPa)	80	60	90
Longitudinal elastic modulus E_ℓ (MPa)	45,000	85,000	134,000
Transverse elastic modulus E_t (MPa)	12,000	5600	7000
Shear modulus $G_{\ell t}$ (MPa)	4500	2100	4200
Poisson ratio $\nu_{\ell t}$	0.3	0.34	0.25
Longitudinal coefficient of thermal expansion at 20°C $\alpha_\ell(°C^{-1})$	0.4 to 0.7×10^{-5}	-0.4×10^{-5}	-0.12×10^{-5}
Transverse coefficient of thermal expansion at 20°C $\alpha_t(°C^{-1})$	1.6 to 2×10^{-5}	5.8×10^{-5}	3.4×10^{-5}

The compression strength along the longitudinal direction is smaller than the tensile strength along the same direction due to the **micro buckling** phenomenon of the fibers in the matrix.

3.3.4 Examples of "High Performance" Unidirectional Plies

The unidirectionals in Table 3.5 have $V_f = 50\%$ boron fibers. The boron/aluminum composite mentioned above belongs to the group of metal matrix composites (see Section 3.7), among these one can find the following:

- For fibers, these can be
  - Glass
  - Silicon carbide
  - Aluminum
  - Other ceramics
- For matrices, these can be
  - Magnesium and its alloys

---

[10] The values assigned in this table can vary depending on the fabrication procedure.

Table 3.5 Properties of Unidirectional Plies Made of Boron Fibers

	Boron/Epoxy	Boron/ Aluminum
Specific mass (kg/m^3)	1950	2650
Longitudinal tensile strength (MPa)	1400	1400
Longitudinal compressive strength (MPa)	2600	3000
Transverse tensile strength (MPa)	80	120
Longitudinal elastic modulus E_ℓ (MPa)	210,000	220,000
Transverse elastic modulus E_t (MPa)	12,000	140,000
Shear modulus $G_{\ell t}$ (MPa)		7500
Longitudinal coefficient of thermal expansion at 20°C, α_ℓ (°C^{-1})	0.5×10^{-5}	0.65×10^{-5}

$V_f = 0.5$

■ Aluminum
■ Ceramics

3.4 WOVEN FABRICS

3.4.1 Forms of Woven Fabric

The fabrics are made of fibers oriented along two perpendicular directions: one is called the **warp** and the other is called the **fill** (or weft) direction. The fibers are woven together, which means the fill yarns pass over and under the warp yarns, following a fixed pattern. Figure 3.9a shows a **plain weave** where each fill goes over a warp yarn then under a warp yarn and so on. In Figure 3.9b, each fill yarn goes over 4 warp yarns before going under the fifth one. For this reason, it is called a "5-harness satin." Figure 3.9c shows a twill weave.

For an approximation (about 15%) of the elastic properties of the fabrics, one can consider these to consist of two plies of unidirectionals crossing at 90° angles with each other. One can use the following notation:

e = total layer thickness
n_1 = number of warp yarns per meter
n_2 = number of fill yarns per meter
$k = \dfrac{n_1}{n_1 + n_2}$
V_f = volume fraction of fibers

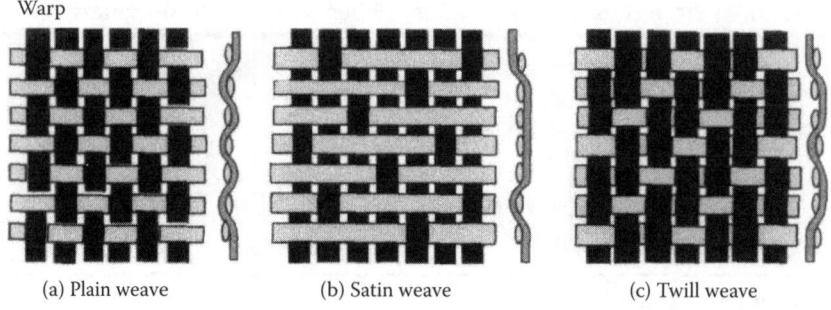

(a) Plain weave　　　　　(b) Satin weave　　　　　(c) Twill weave

Figure 3.9　Forms of Woven Fabrics

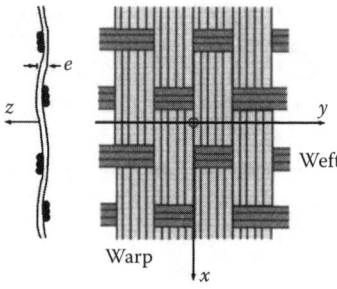

Figure 3.10　Notation for a Fabric Layer

One can deduce the thickness of the equivalent unidirectional plies (see Figure 3.10) as

$$e_{warp} = e \times \frac{n_1}{n_1 + n_2} = k \times e$$

$$e_{fill} = e \times \frac{n_2}{n_1 + n_2} = (1 - k) \times e$$

3.4.2　Elastic Modulus of Fabric Layer

For approximate values, the two plies of reinforcement can be considered separately or together.

■ **Separately**, the fabric layer is replaced by two unidirectional plies crossed at 90° with each other, with the following thicknesses:

$$e_{warp} = k \times e \qquad e_{fill} = (1 - k) \times e$$

The average fiber volume fraction V_f is already known, and the mechanical properties E_ℓ, E_t, $G_{\ell t}$, $v_{\ell t}$ of these plies can be determined according Section 3.3.1.

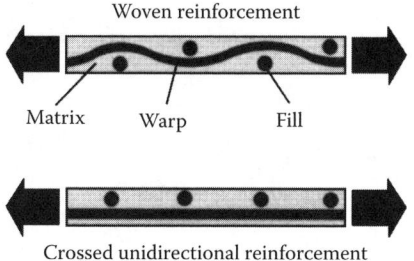

Figure 3.11   Cross Section of a Layer with Fibers Crossed at 90°

■ **Together**, the fabric layer is replaced by one single anisotropic layer with thickness e, x being along the warp direction and y along the fill direction (see Figure 3.9). One can therefore obtain[11]

$$E_x \approx k \times E_\ell + (1 - k) \times E_t$$

$$E_y \approx (1 - k) \times E_\ell + k \times E_t$$

$$G_{xy} = G_{\ell t}$$

$$v_{xy} \approx v_{\ell t} \bigg/ \left(k + (1 - k)\frac{E_\ell}{E_t} \right)$$

Notes: The stiffness obtained with a woven fabric is less than what is observed if one were to superpose two cross plies of unidirectionals. This is due to the curvature of the fibers during the weaving operation (see Figure 3.11). This curvature makes the woven fabric more deformable than the two cross plies when subjected to the same loading. (There exist fabrics that are of "high modulus" where the unidirectional layers are not connected with each other by weaving. The unidirectional plies are held together by stitching fine threads of glass or polymer.)

■ There is also a stronger tensile strength of a woven fabric and a lower compressive strength, as compared with the strengths obtained by superposing two cross plies.[12]

3.4.3 Examples of Balanced Fabrics/Epoxy

A fabric is said to be **balanced** when the number of yarns along the warp and fill directions are the same. The material is therefore identical along the two directions.

---

[11] For the corresponding calculations, cf. Section 12.1.2 and also Chapter 18, Application 18.2.12.

[12] For example, compare the strengths in tension and in compression in Table 3.6, and in Tables 5.1, 5.6, and 5.11 of Section 5.4 for proportions of 50% for 0° and 50% for 90° (taking into account the difference in fiber volume fraction).

Table 3.6 Properties of Balanced Fabric/Epoxy Composites

	E Glass	Kevlar	Carbon
Fiber volume fraction V_f (%)	50	50	45
Specific mass (kg/m³)	1900	1330	1450
Tensile fracture strength along x or y (MPa)	400	500	420
Compressive fracture strength along x or y (MPa)	390	170	360
In plane shear strength (MPa)		150	55
Elastic modulus E_x (= E_y) (MPa)	20,000	22,000	54,000
Shear modulus G_{xy} (MPa)	2850		4000
Poisson coefficient υ_{xy}	0.13		0.045
Coefficient of thermal expansion $\alpha_x = \alpha_y(°C^{-1})$		-0.2×10^{-5}	0.05×10^{-5}
Maximum elongation (%)		2.1	1.0
Price (relative)	1	4.2	7.3

The warp and fill directions play equal roles affecting the thermomechanical properties. The material in Table 3.6 is of epoxy matrix.

3.5 MATS AND REINFORCED MATRICES

3.5.1 Mats

Mats are made of cut fibers (fiber lengths between 5 and 10 cm) or of continuous fibers making a bidimensional layer. Mats are isotropic within their plane (x, y). Therefore they can be characterized by only two elastic constants, as identified in Section 3.1.

If E_ℓ and E_t are the elastic moduli (along the longitudinal and transverse directions) of an unidirectional ply with the same volume fraction of V_f, one has

$$E_{mat} \approx \frac{3}{8}E_\ell + \frac{5}{8}E_t$$

$$G_{mat} \approx \frac{E_{mat}}{2(1 + \nu_{mat})}$$

$$\nu_{mat} \approx 0.3$$

For example, mats with cut fibers made of glass/epoxy have the following characteristics:

Fiber volume fraction	28%
Specific mass (kg/m^3)	1800
Elastic modulus (MPa)	14,000
Tensile fracture strength (MPa)	140
Heat capacity (J/g × °C)	1.15
Coefficient of heat conduction (W/m × °C)	0.25
Linear coefficient of thermal expansion (°C^{-1})	2.2×10^{-5}

3.5.2 Summary Example of Glass/Epoxy Layers

Figures 3.12 and 3.13 give a summary of the principal characteristics of different types of layers (unidirectionals, fabrics, and mats) with the variation of the fiber volume fraction V_f.

3.5.3 Spherical Fillers

Spherical fillers are reinforcements associated with polymer matrices (see Figure 3.14). They are in the form of **microballs**, either solid or hollow, with diameters between 10 and 150 μm. They are made of glass, carbon, or polystyrene.

- The filler volume fraction V_f can reach up to 50%.
- The filler properties are such that $E_f \gg E_m$.

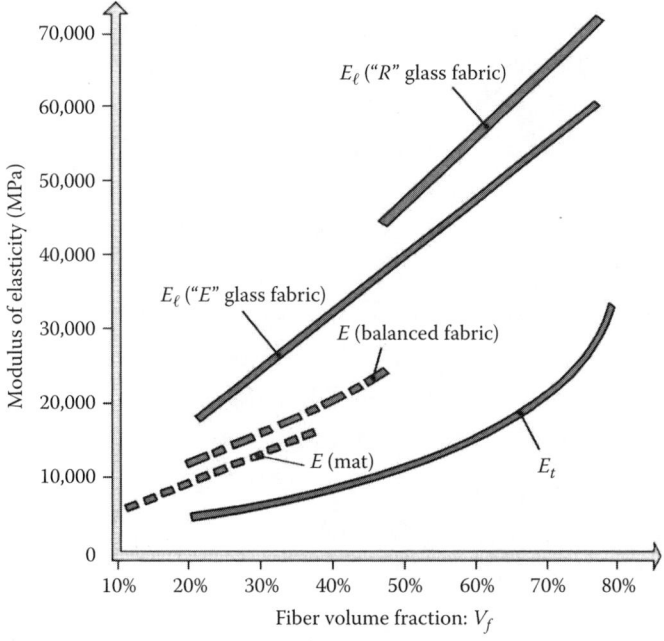

Figure 3.12   Elastic Modulus of Glass/Epoxy Layers

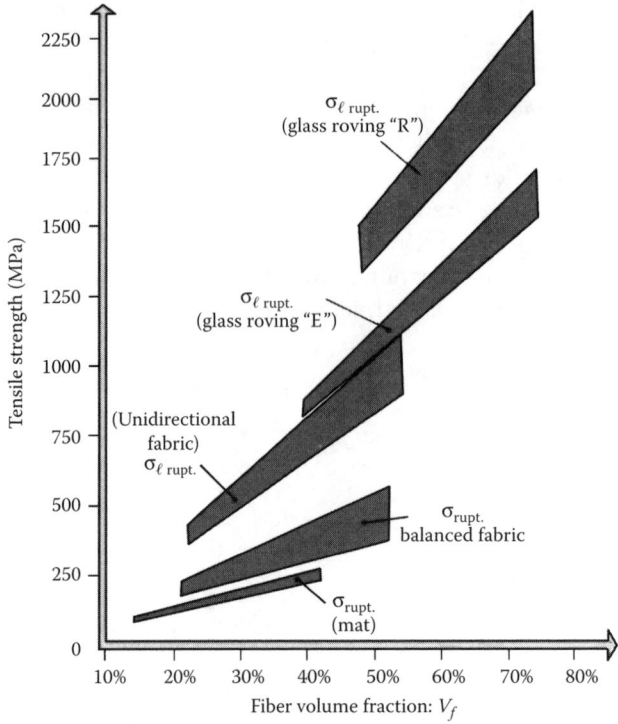

Figure 3.13 Tensile Strength of Glass/epoxy Layers

The composite (matrix + filler) is isotropic, with elastic properties E, G, ν given by the following relations:

Defining: $K = \dfrac{E_m}{3(1-2\nu_m)}\left[1 + 3\left(\dfrac{1-\nu_m}{1+\nu_m}\right)\dfrac{V_f}{(1-V_f)}\right]$

$E \approx \dfrac{9KG}{3K+G}$

$G \approx \dfrac{E_m}{2(1+\nu_m)}\left[1 + \dfrac{15}{2}\left(\dfrac{1-\nu_m}{4-5\nu_m}\right)\dfrac{V_f}{(1-V_f)}\right]$

$\nu \approx \dfrac{1}{2}\left(\dfrac{3K-2G}{3K+G}\right)$

3.5.4 Other Reinforcements

One may also use reinforcements in the form of crushed fibers, flakes (see Figure 3.15), or powders made of the following:

■ Glass
■ Graphite
■ Metals

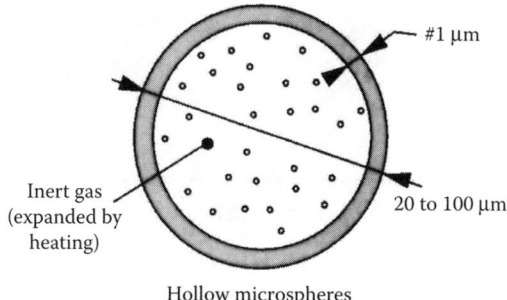

#1 μm

Inert gas
(expanded by
heating)

20 to 100 μm

Hollow microspheres

Figure 3.14   Spherical Fillers

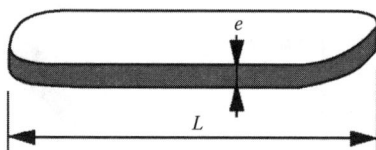

e

L

Figure 3.15   Form of Flakes

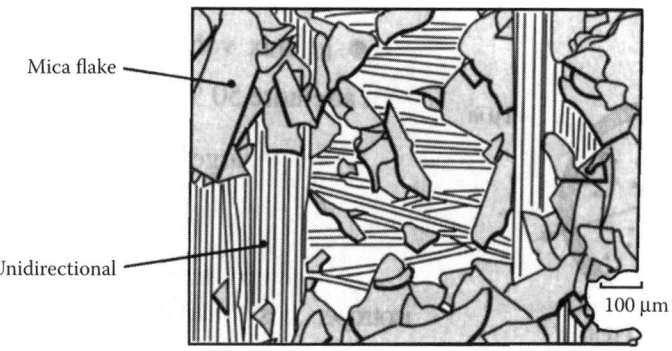

Mica flake

Unidirectional

100 μm

Figure 3.16   Mica Flakes Arrangement

- Aluminum
- Mica ($L \approx 100$ μm)
- Talc ($L \approx 10$ μm)

Immersed in a resin with stratified fibers, mica flakes have the arrangement as shown in the Figure 3.16. One can observe that the modulus of the resin is increased as[13]

$$E = \left[1 - \frac{Ln(1+u)}{u} \right] \times E_{mica} V_{mica} + E_m V_m; \qquad u = \frac{L}{e} \sqrt{\frac{G_m}{E_{mica}} \times \frac{V_{mica}}{V_m}}$$

---

[13] For more details, see "Interaction Effects in Fiber Composites," which is listed in the Bibliography at the end of the book.

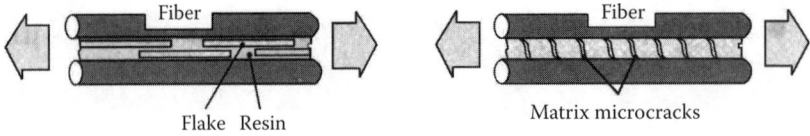

Figure 3.17   Cross Section with and without Mica

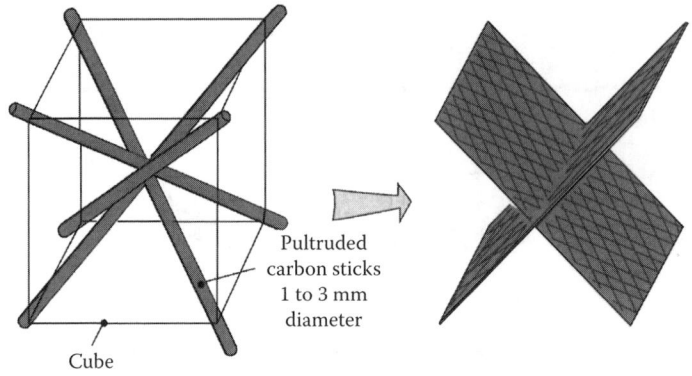

Figure 3.18   4D Architecture

The average properties of mica are

$$E_{mica} = 170,000 \text{ MPa} \quad \text{and} \quad \rho_{mica} = 2,800 \text{ kg/m}^3.$$

Figure 3.17 shows the increase resistance against the microfracture of the resin.

3.6 MULTIDIMENSIONAL FABRICS

An example of "4D" architecture of carbon reinforcement from the European society of propulsion (FRA) has the reinforcement assembled according to pre-established directions (see Figure 3.18). The fiber volume fraction is on the order of 30%. The matrix comes to fill the voids between the fibers.[14] The principal advantages of these types of composites are

- The supplementary connection (as compared with the bidimensional plies) makes the composites safe from delamination.
- The mechanical resistance is conserved—and even improved—at high temperatures (up to 3,000°C for carbon/carbon).
- The coefficient of expansion remains small.
- They are resistant against thermal shock.

---

[14] See Section 2.2.4.

Table 3.7   Properties of Three-Dimensional Carbon/Carbon

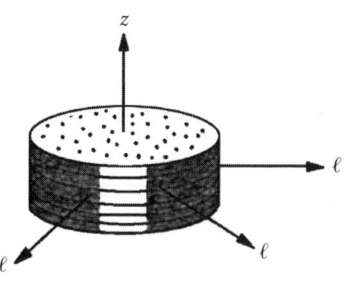

	Aerolor 41[16]	Septcarb 4[17]
Specific mass (kg/m^3)	1700 to 2000	1500 to 2000
Longitudinal tensile fracture strength (MPa)	40 to 100	95 and increasing, up to 2000°C
Longitudinal compressive fracture strength (MPa)	80 to 200	65
Tensile strength in z direction (MPa)	>10	3
Compressive strength in z direction (MPa)	80 to 200	120
Shear strength in ℓz plane (MPa)	20 to 40	10
Longitudinal elastic modulus E_ℓ (MPa)	30,000	16,000
Elastic modulus E_z (MPa)		5000
Shear modulus $G_{\ell z}$ (MPa)		2200
Shear modulus $G_{\ell\ell}$ (MPa)		5700
Poisson ratio $v_{z\ell}$		0.17
Poisson ratio $v_{\ell\ell}$		0.035
Thermal expansion coefficient α_ℓ(°C^{-1}) at 1000°C at 2,500°C	0.7×10^{-6}; 3×10^{-6}	3×10^{-6}; 4×10^{-6}
Thermal expansion coefficient α_z(°C^{-1}) at 1000°C at 2,500°C	6×10^{-6}; 6×10^{-6}	7×10^{-6}; 9×10^{-6}
Coefficient of thermal conductivity (W/m × °C)	300	

- There is a high coefficient of thermal conductivity.
- They have small density.
- The silicon/silicon is transparent to radio-electric waves.

Table 3.7 gives the characteristics of two composites made of tridimensional carbon/carbon. The mechanical properties are the same along all directions noted by ℓ in the following figure. The composite is **transversely isotropic**.[15]

---

[15] This notion is shown in detail in Section 13.2.
[16] Product of the Aerolor company (FRA).
[17] Product of European company in propulsion (FRA).

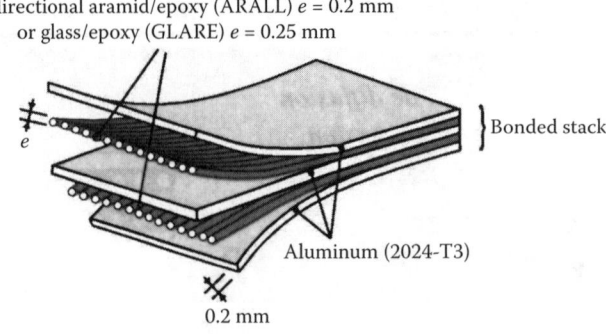

Figure 3.19   Layers of ARALL and GLARE

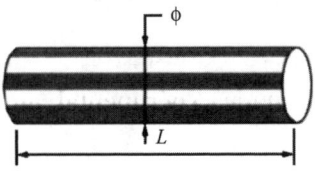

Figure 3.20   SiC Whisker

3.7 METAL MATRIX COMPOSITES

There are a number of products including:

- Matrices: aluminum, magnesium, titanium (see also Section 7.4) and
- Reinforcements (fibers): aramid, carbon, boron, silicon carbide (SiC)

Example: ARALL (Aluminum Reinforced Aramid) and **GLARE** (Aluminum Reinforced Glass).[18] The essential advantage is better impact damage tolerance because of better resistance to failure due to thin metallic layers and resistance against the crack propagation from one layer to the other (see Figure 3.19).

Example: Short silicon carbide fibers (whiskers)/aluminum. This is a so-called "incompatible" composite because the big difference between the thermo-mechanical properties of the constituents leads to high stress concentrations and debonding between the fibers and the matrix (see Figure 3.20). This has good applications at high temperatures. The diameter of the whisker is about 20 μm. Slenderness ratio $L/\varnothing \approx 5$ and fiber volume fraction $V_f \approx 0.3$.

Example: Boron/aluminum. This is used in aerospace applications (see Section 7.5.4). The technique to obtain these materials is shown schematically in Figure 3.21. This composite can be used continuously at temperatures on the order of 300°C, while maintaining notable mechanical properties (see Section 1.6 for the properties of boron).

---

[18] AKZO Fibers/DELFT University (Holland). ®Structural Laminates Company/New Kensington (USA).

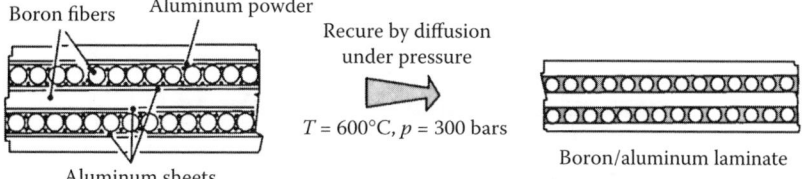

Figure 3.21 Boron/Aluminum Composite

Some characteristics of unidirectional plies made of aluminum matrix (6061) include:

	HR Carbon	Aluminum	Silicon Carbide
Fiber volume fraction V_f (%)	50	50	50
Specific mass (kg/m³)	2300	3100	2700
Longitudinal tensile strength (MPa)	800	550	1400
Longitudinal compressive strength (MPa)	600	3100	3000
Longitudinal elastic modulus (MPa)	200,000	190,000	140,000

3.8 TESTS

The relations cited on the previous pages for the calculation of the modulus and Poisson coefficients of the composites only allow obtaining an order of magnitude for the mechanical properties. Some of these relations are not quite accurate, particularly for the shear modulus. Also, these properties are very sensitive to the fabrication conditions. It is necessary for the design engineer to have the results provided by the suppliers of the reinforcement and the matrix materials or to better the results obtained by tests in the labs. Some of these tests have been standardized, for example, tensile tests, flexure tests, and impact tests.

A tensile test (NF T 51–034, ASTM D 3039) on the specimen in Figure 3.22, instrumented with electrical strain gage, allows the measurement of the strength and the elongation to fracture. A delamination test (NF T 57–104) on a specimen which has a small span for bending (see Figure 3.23). It fails by delamination under the effect of interlaminar shear stress. One can obtain the interlaminar shear strength.[19]

There are other tests, not yet standardized, that are very useful for the fabrication of high performance composites, for example, for control of the fiber volume fraction. In fact, during the phase of polymerization under pressure of a polymer matrix composite, the resin gets absorbed into an absorbing fabric, in variable quantity depending on the cycle of pressure and temperature. In consequence,

---

[19] This is a simplified way that little reflects the complexity of the state of real stress due to the presence of very close concentrated loads.

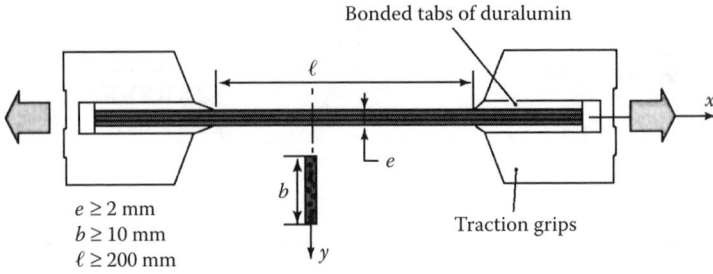

Figure 3.22   Tensile Test

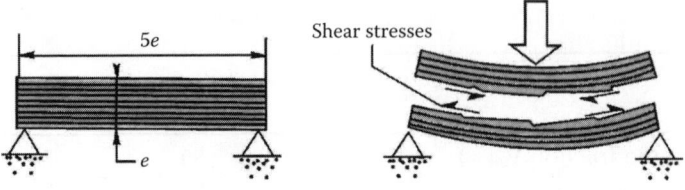

Figure 3.23   Short Beam Shear Test

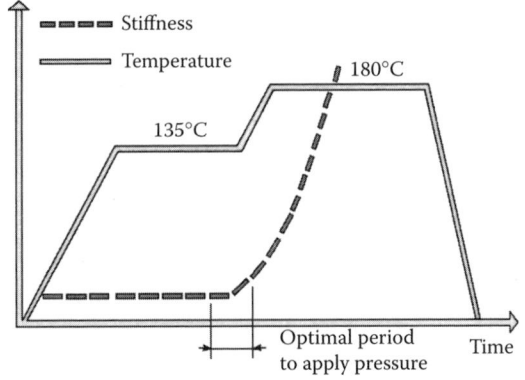

Figure 3.24   Variation In Stiffness During Curing

the fiber volume fraction V_f varies, and as a result, the dimensional characteristics of the piece (thickness) also vary. To deal with this problem, one may want to evaluate by test the optimal moment for the application of pressure, by the measurement of the flexural rigidity of a specimen as a function of time of fabrication (see Figure 3.24).

4

SANDWICH STRUCTURES

Sandwich structures occupy a large proportion of composite materials design. They appear in almost all applications. Historically they were the first light and high-performance structures.[1] In the majority of cases, one has to design them for a specific purpose. Sandwich structures usually appear in industry as semi-finished products. In this chapter we will discuss the principal properties of sandwich structures.

4.1 WHAT IS A SANDWICH STRUCTURE?

A sandwich structure results from the assembly by bonding—or welding—of two thin facings or skins on a lighter core that is used to keep the two skins separated (see Figure 4.1).

Their properties are astonishing. They have

- **Very light weight.** As a comparison, the mass per unit area of the dome of the Saint Peter's Basilica in Rome (45 meter diameter) is 2,600 kg/m^2, whereas the mass per surface area of the same dome made of steel/polyurethane foam sandwich (Hanover) is only 33 kg/m^2.
- **Very high flexural rigidity.** Separation of the surface skins increases flexural rigidity.
- **Excellent thermal insulation characteristics.**

However, be careful:

- Sandwich materials are not dampening (no acoustic insulation).
- Fire resistance is not good for certain core types.
- The risk of **buckling** is greater than for classical structures.

The facing materials are diverse, and the core materials are as light as possible. One can denote couples of compatible materials to form the sandwich (see Figure 4.2).

Be careful: Polyester resins attack polystyrene foams.

---

[1] See Section 7.1.

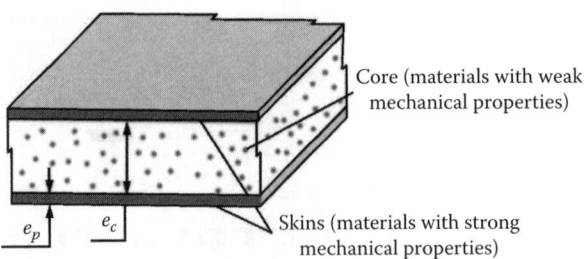

Core (materials with weak mechanical properties)

e_p e_c

Skins (materials with strong mechanical properties)

Figure 4.1 Sandwich Structure ($10 \leq e_c/e_p \leq 100$)

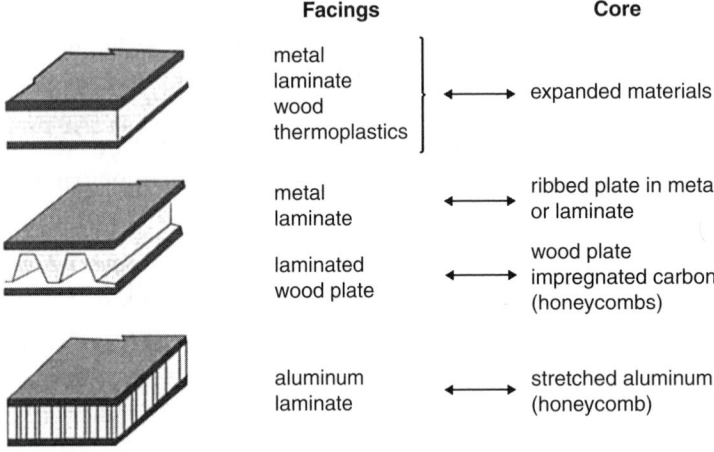

	Facings		Core
	metal laminate wood thermoplastics	⟷	expanded materials
	metal laminate	⟷	ribbed plate in metal or laminate
	laminated wood plate	⟷	wood plate impregnated carbon (honeycombs)
	aluminum laminate	⟷	stretched aluminum (honeycomb)

Figure 4.2 Constituents of Sandwich Materials

The **assembly** of the facings to the core is carried out using bonding adhesives. In some exceptional cases, the facings are welded to the core. The quality of the bond is fundamental for the performance and life duration of the piece. In practice we have

$$0.025 \text{ mm} \leq \text{adhesive thickness} \leq 0.2 \text{ mm}$$

4.2 SIMPLIFIED FLEXURE

4.2.1 Stresses

Figure 4.3 shows in a simple manner the main stresses that arise due to the application of bending on a sandwich beam.[2] The beam is clamped at its left end, and a force T is applied at its right end. Isolating and magnifying one elementary segment of the beam, on a cross section, one can observe the **shear stress resultant** T and the **moment resultant M**. The shear stress resultant T causes shear stresses τ and the moment resultant causes normal stresses σ.

---

[2] For more details on these stresses, see Chapters 15 and 17, and also Applications 18.3.5 and 18.3.8.

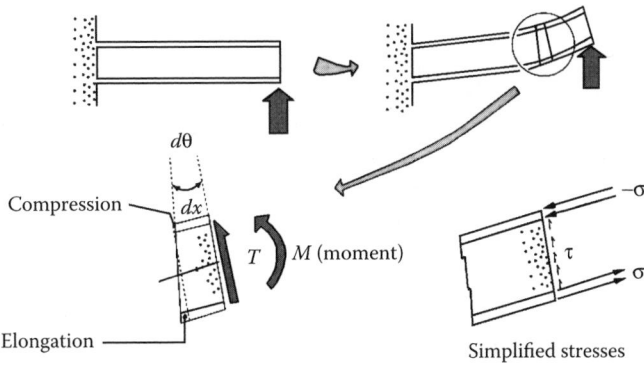

Figure 4.3 Bending Representation

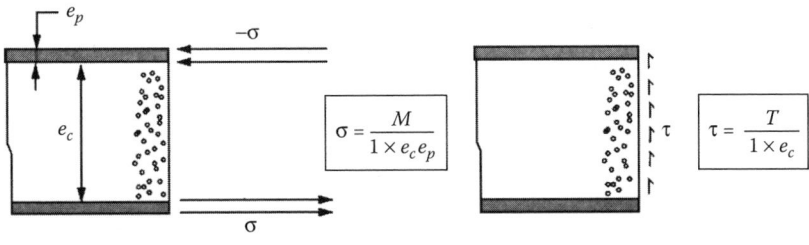

Figure 4.4 Stresses in Sandwich Structure

To evaluate τ and σ, one makes the following simplifications:

- The normal stresses are assumed to occur in the facings only, and they are uniform across the thickness of the facings.
- The shear stresses are assumed to occur in the core only, and they are uniform in the core.[3]

One then obtains immediately the expressions for τ and σ for a beam of **unit width** and thin facings shown in Figure 4.4.

4.2.2 Displacements

In the following example, the displacement Δ is determined for a sandwich beam subjected to bending as a consequence of

- Deformation due to normal stresses σ and
- Deformation created by shear stresses τ (see Figure 4.5).

---

[3] See Section 17.7.2 and the Applications 18.2.1 and 18.3.5 for a better approach.

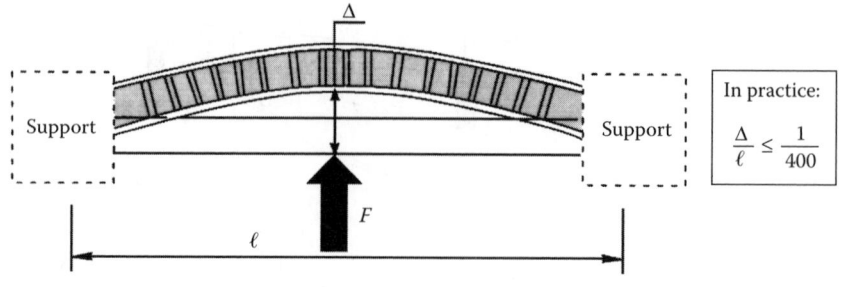

Figure 4.5   Bending Deflection

To evaluate Δ, one can, among other methods,[4] use the Castigliano theorem

$$W = \frac{1}{2}\int \frac{M^2}{\langle EI \rangle}\,dx + \frac{1}{2}\int \frac{k}{\langle GS \rangle}\,T^2\,dx$$

elastic
energy   contribution   contribution
from bending   from shear

$$\underset{\text{deflection}}{\Delta} = \underset{\text{load}}{\frac{\partial W}{\partial F}}\text{ energy}$$

where the following notations[5] are used for a beam of unit width:

M = Moment resultant

T = Shear stress resultant

E_p = Modulus of elasticity of the material of the facings

G_c = Shear modulus of the core material

$$\langle EI \rangle \# E_p e_p \times 1 \times \frac{(e_c + e_p)^2}{2}; \quad k/\langle GS \rangle = 1/G_c(e_c + 2e_p) \times 1.$$

Example: A cantilever sandwich structure treated as a sandwich beam (see Figure 4.6). Elastic energy is shown by

$$W = \frac{1}{2}\int_0^\ell \frac{F^2(\ell - x)^2}{\langle EI \rangle}\,dx + \frac{1}{2}\int_0^\ell \frac{k}{\langle GS \rangle}F^2\,dx$$

$$W = \frac{F^2}{2}\left(\frac{\ell^3}{3\langle EI \rangle} + \frac{k}{\langle GS \rangle}\ell\right)$$

---

[4]   See Equation 15.16 that allows one to treat this sandwich beam like a homogeneous beam. One can also use the classical strength of materials approach.

[5]   See Application 18.2.1 or Chapter 15.

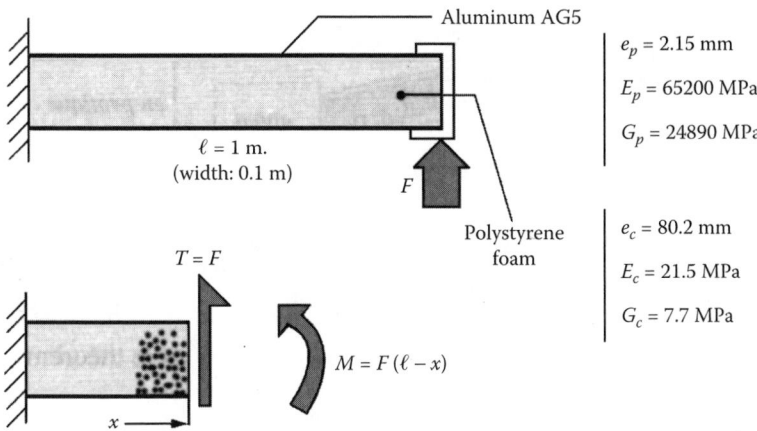

Figure 4.6   Cantilever Beam

where

$$\langle EI \rangle = 475 \times 10^2 (N \cdot m^2); \quad \frac{\langle GS \rangle}{k} = 650 \times 10^2 (N)$$

The end displacement Δ can be written as

$$\Delta = \frac{\partial W}{\partial F}$$

Then for an applied load of 1 Newton

$$\Delta = \underset{\text{Flexure}}{0.7 \times 10^{-2} \text{ mm/N}} + \underset{\text{Shear}}{1.54 \times 10^{-2} \text{ mm/N}}$$

Remark: Part of the displacement Δ due to shear appears to be higher than that due to bending, whereas in the case of classical homogeneous beams, the shear displacement is very small and usually neglected. Thus, this is a specific property of sandwich structures that strongly influences the estimation of the bending displacements.

4.3 A FEW SPECIAL ASPECTS

4.3.1 Comparison of Mass Based on Equivalent Flexural Rigidity (EI)

Figure 4.7 allows the comparison of different sandwich structures having the same flexural rigidity $\langle EI \rangle$. Following the discussion in the previous section, this accounts for only a part of the total flexural deformation.

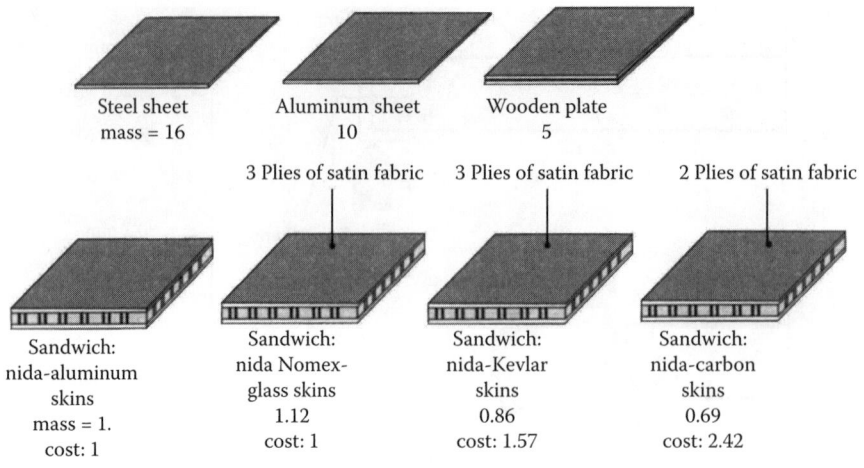

Figure 4.7   Comparison of Plates Having Similar Flexural Rigidity *EI*

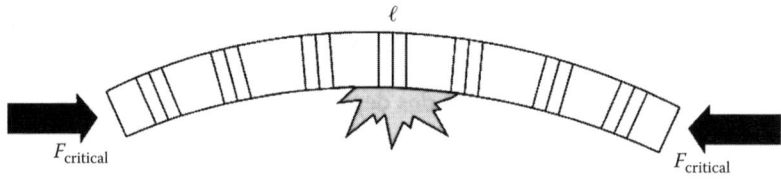

Figure 4.8   Buckling of Sandwich Structure

4.3.2 Buckling of Sandwich Structures

The compression resistance of all or part of a sandwich structure is limited by the so-called **critical values** of the applied load, above which the deformations become large and uncontrollable. This phenomenon is called **buckling** of the structure (see Figure 4.8). Depending on the type of loading, one can distinguish different types of buckling which can be global or local.

4.3.2.1 Global Buckling

Depending on the supports, the critical buckling load F_c is given[6] by

$$F_{cr} = K \frac{\pi^2 \langle EI \rangle}{\ell^2 + \pi^2 \dfrac{\langle EI \rangle}{\langle GS \rangle} kK}$$

$K = 1$ $K = 4$ $K = 2.04$ $K = 0.25$

---

[6] See Application 18.3.4.

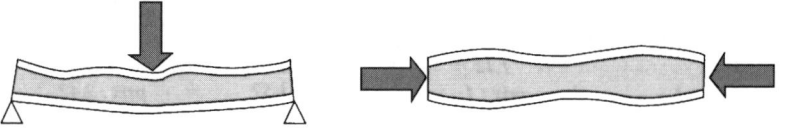

Figure 4.9   Local Buckling of Facings

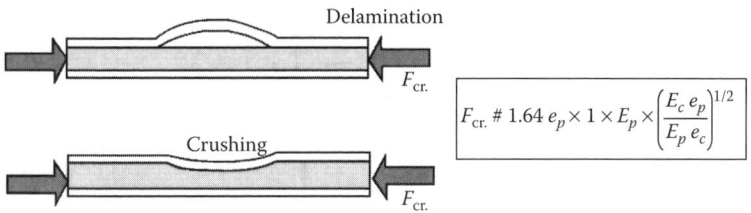

Delamination

$F_{cr.}$

Crushing

$F_{cr.}$

$$F_{cr.} \# 1.64 \, e_p \times 1 \times E_p \times \left(\frac{E_c \, e_p}{E_p \, e_c} \right)^{1/2}$$

Figure 4.10   Damage by Local Buckling

4.3.2.2   Local Buckling of the Facings

The facings are subject to buckling due to the low stiffness of the core. Depending on the type of loading, one can find the modes of deformation as shown in Figure 4.9.

The critical compression stress is given in the equation below where v_c is the Poisson coefficient of the core.

$$\sigma_{cr} = a \times (E_p \times E_c^2)^{1/3}$$

with

$$a = 3\{12(3 - v_c)^2(1 + v_c)^2\}^{-1/3}$$

The critical load to cause local damage by local buckling of a facing and the types of damage are shown in Figure 4.10.

4.3.3   Other Types of Damage

Local crushing: This is the crushing of the core material at the location of the load application (see figure below).

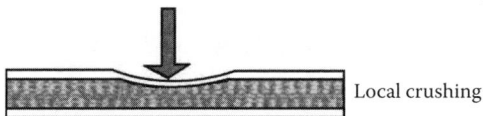

Local crushing

Compression rupture: In this case (see figure below), note that the weak compression resistance of Kevlar fibers[7] leads to a compression strength about two times less than for sandwich panels made using glass fibers.

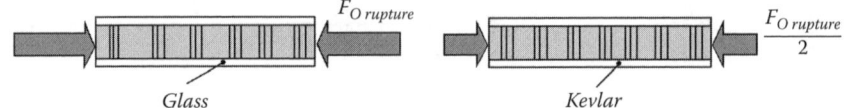

4.4 FABRICATION AND DESIGN PROBLEMS

4.4.1 Honeycomb: An Example of Core Material

These well-known materials are made of hexagonal cells that are regularly spaced. Such geometry can be obtained using a technique that is relatively simple. Many thin sheets are partially bonded. Starting from stacked bonded sheets, they are expanded as shown in Figure 4.11.

The honeycomb material can be metal (light alloy, steel) or nonmetal (carton impregnated with phenolic resin, polyamide sheets, or impregnated glass fabrics).

Metallic honeycombs are less expensive and more resistant. Nonmetallic honeycombs are not sensitive to corrosion and are good thermal insulators. Table 4.1 shows the mechanical and geometric characteristics of a few current honeycombs, using the notations of Figure 4.11.

Table 4.1   Properties of Some Honeycomb

	Bonded Sheets of Polyamide: Nomex[a]	Light Alloy AG3	Light Alloy 2024
Dia. (D): inscribed circle (mm)	6; 8; 12	4	6
Thickness e (mm)		0.05	0.04
Specific mass (kg/m^3)	64	80	46
Shear strength $\tau_{xz\,rup}$ (MPa)	1.7	3.2	1.5
Shear modulus: G_{xz} (MPa) # 1.5 G_{mat}(e/D)	58	520	280
Shear strength $\tau_{yz\,rup}$ (MPa)	0.85	2	0.9
Shear modulus: G_{yz} (MPa)	24	250	140
Compression strength: $\sigma_{z\,rup}$ (MPa)	2.8	4.4	2

[a] Nomex® is a product of Du Pont de Nemours.

---

[7] See Section 3.3.3.

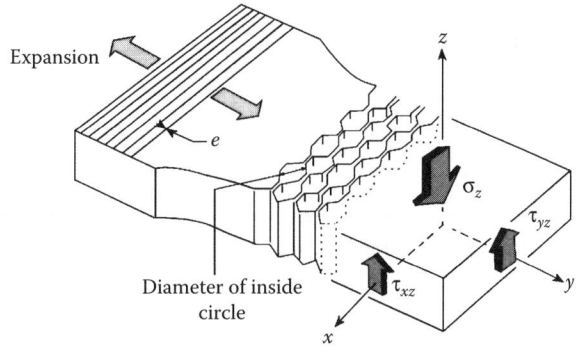

Figure 4.11 Honeycomb

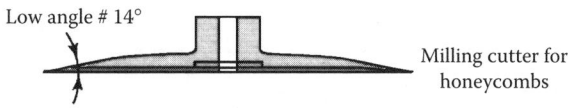

Figure 4.12 Processing of Honeycomb

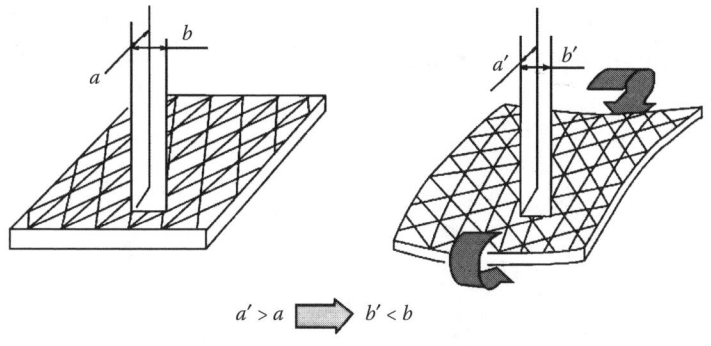

Figure 4.13 Deformation of Honeycomb

4.4.2 Processing Aspects

The processing of the honeycomb is done with a diamond disk (peripheral speed in the order of 30 m/s). The honeycomb is kept on the table of the machine by an aluminum sheet to which it is bonded. Below the aluminum sheet, a depression anchors it to the table (see Figure 4.12).

One can also **deform** the honeycomb. It is important to constrain it carefully, because the deformation behavior is complex. For example, a piece of honeycomb under cylindrical bending shows two curvatures as illustrated in Figure 4.13.[8]

---

[8] This phenomenon is due to the Poisson effect, particularly sensitive here (see Section 12.1.4).

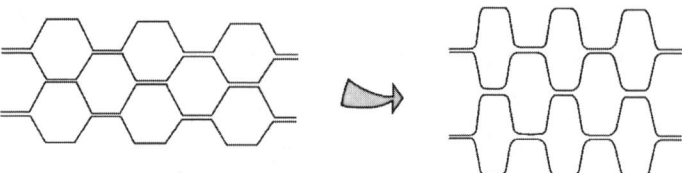

Figure 4.14   Over-Expansion of Honeycomb

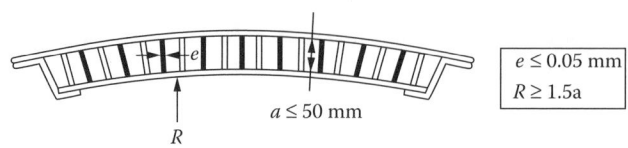

Figure 4.15   Curvature of Honeycomb

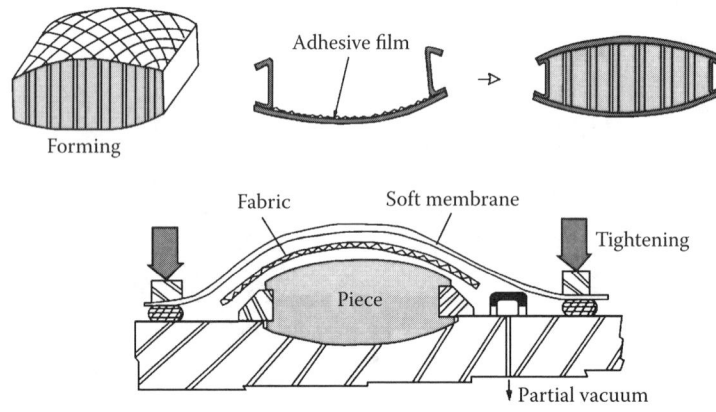

Figure 4.16   Processing of a Sandwich Piece of a Structural Part

The processing can be facilitated using the method of **overexpansion** which modifies the configuration of the cells as shown in Figure 4.14.

At limit of **curvature**, R is the radius of the contour, and e is the thickness of the sheets which consitute the honeycombs (see Figure 4.15). **Nomex** honeycombs (sheets of bonded polyamide) must be processed at high temperature. The **schematic** for the processing of a structural part of sandwich honeycomb is as in Figure 4.16. For **moderate loadings** (for example, bulkheads), it is possible to fold a sandwich panel following the schematic in Figure 4.17.

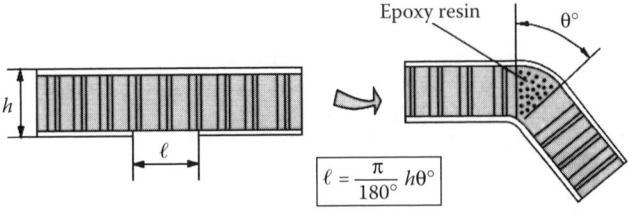

Figure 4.17 Folding of Honeycomb

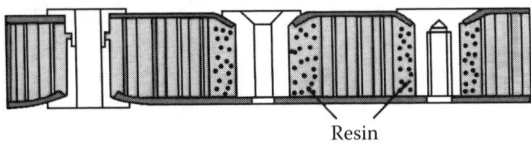

Figure 4.18 Incorporation of Inserts

4.4.3 Insertion of Attachment Pieces

When it is necessary to transmit local loadings, depending on the intensity of these loads, it is convenient to distribute these over one or many **inserts,** as indicated in Figure 4.18.[9] The filling resin of epoxy type, shown in Figure 4.18, can be lightened by incorporation of phenolic microspheres with resulting density for the lightened resin of 700 to 900 kg/m^3 and crush strength # 35 MPa (see Figure 4.19).

4.4.4 Repair of Laminated Facings

For sandwich materials of the type "honeycombs/laminates," the repair of local damage is relatively easy. It consists of **patching** the plies of the laminate. The configuration of the repair zone appears as in Figure 4.20.

4.5 NONDESTRUCTIVE QUALITY CONTROL

Apart from using the classical methods for controlling the surface defects, which allows the repair of external delaminations of laminated facings, using the following techniques allows the identification and repair of internal defects due to fabrication or due to damages in service. These defects can entail imperfect bonding, delaminations, and inclusions. Principal nondestructive detection methods are illustrated in Figure 4.21.

---

[9] See Sections 6.2.4 and 6.3.

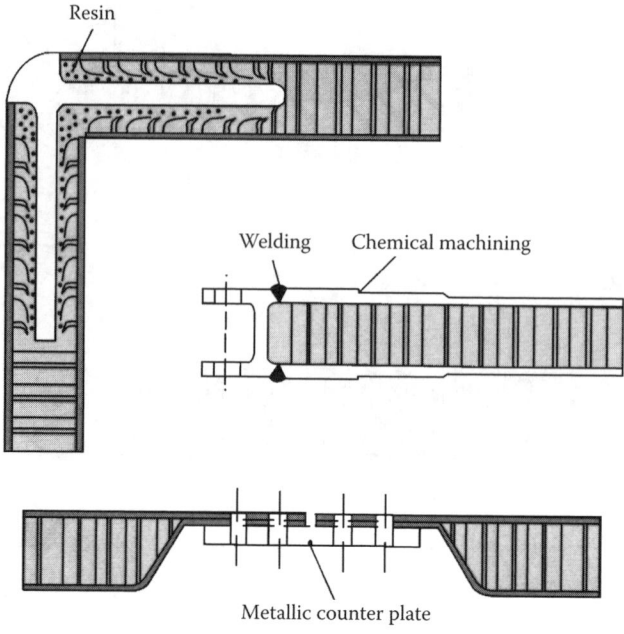

Figure 4.19   Some Links for Sandwich Structures

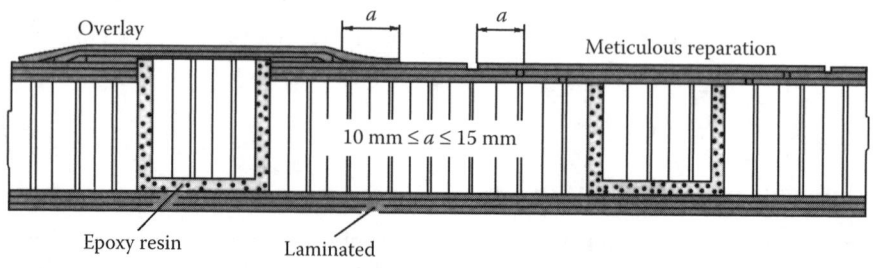

Figure 4.20   Honeycomb Repair

When a composite structure (for example, a reservoir under pressure) is subjected to loading, many microcracks can occur within the piece. Microcracking in the resin, fiber fracture, and disbond between fiber and matrix can exist even within the admissible loading range. These ruptures create acoustic waves that propagate to the surface of the piece. They can be detected and analyzed using **acoustic emission** sensors (see Figure 4.22).

The number of peaks as well as the duration and the amplitude of the signal can be used to indicate the integrity of the piece. In addition, the accumulated number of peaks may be used to predict the fracture of the piece (i.e., the change of slope of the curve in Figure 4.23).

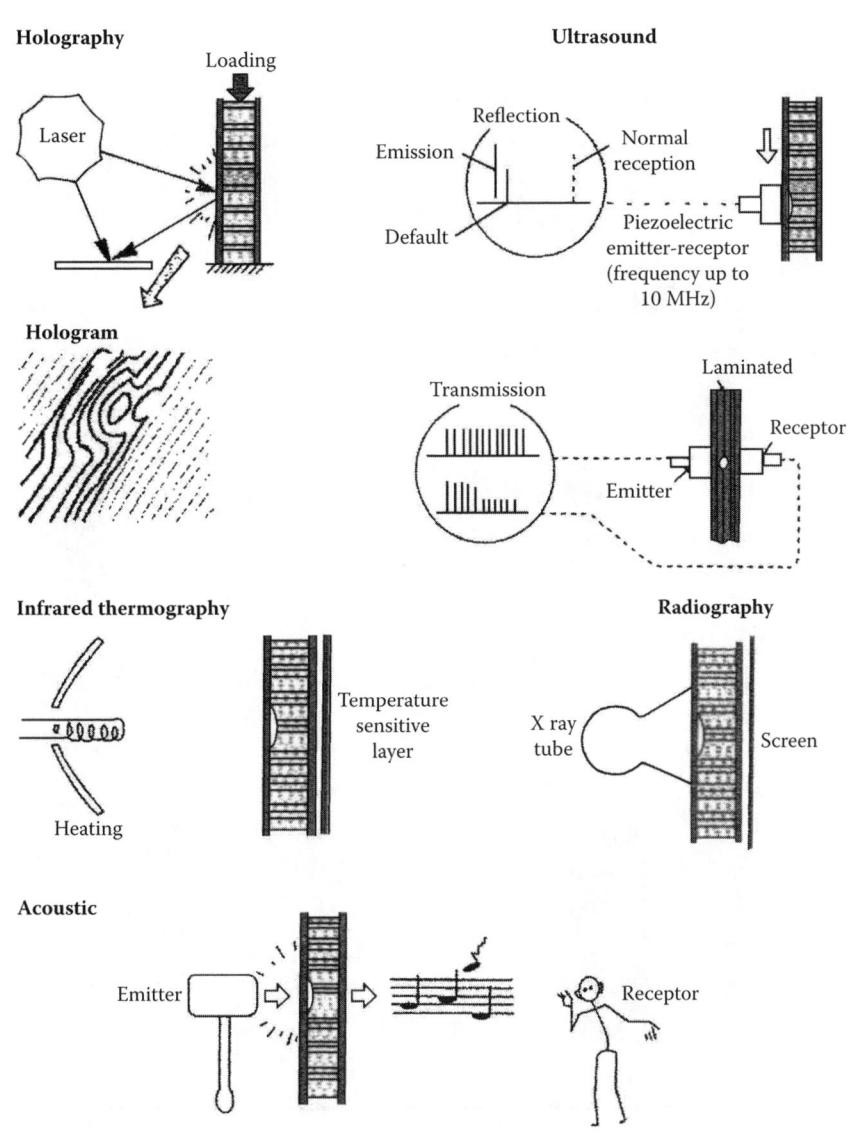

Figure 4.21   Principal Nondestructive Testing Methods

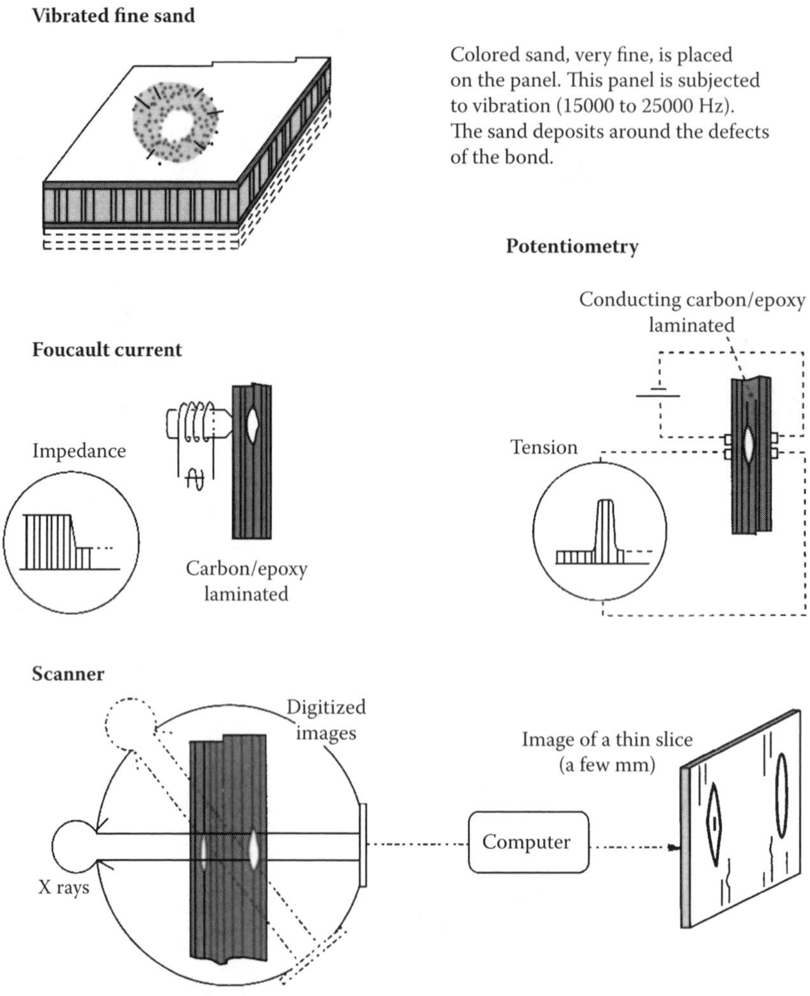

Vibrated fine sand

Colored sand, very fine, is placed on the panel. This panel is subjected to vibration (15000 to 25000 Hz). The sand deposits around the defects of the bond.

Potentiometry

Conducting carbon/epoxy laminated

Foucault current

Impedance

Carbon/epoxy laminated

Tension

Scanner

Digitized images

Image of a thin slice (a few mm)

Computer

X rays

Figure 4.21 (Continued).

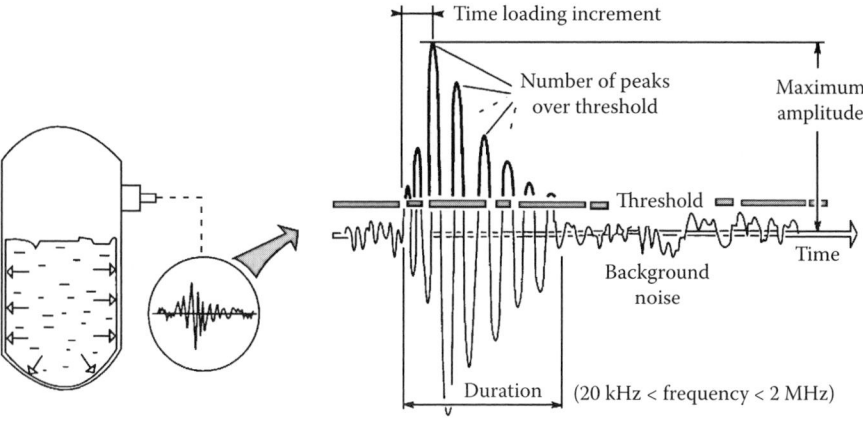

Figure 4.22 Acoustic Emission Technique (AE)

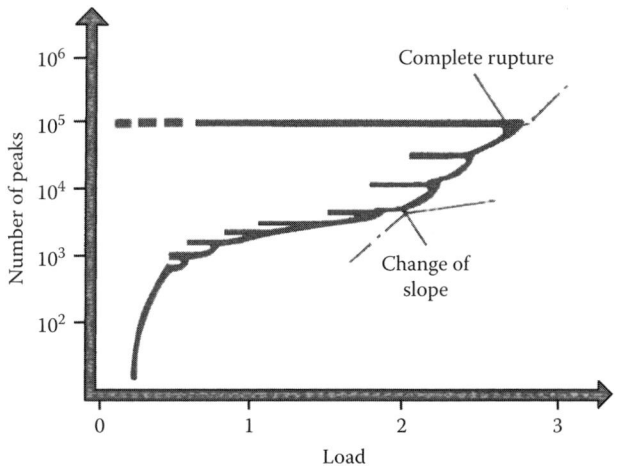

Figure 4.23 Plotting of AE Events

5

CONCEPTION AND DESIGN

A different paradigm: As every mechanical part, a composite part has to withstand loadings. In addition, the conception process has to extend over a range much larger than for a component made of "pre-established" material. In fact,

- For isotropic materials, the classical process of conception consists of selection of an existing material and then design of the piece.
- For a component made of composites, the designer "creates" the material based on the functional requirements. The designer chooses the reinforcement, the matrix, and the process for curing.

Following that the designer must define the component architecture, i.e., the arrangement and dimensions of plies, the representation of these on the designs, etc. These subjects are covered in this chapter.

5.1 DESIGN OF A COMPOSITE PIECE

The following characteristic properties always have to be kept in mind by the designer:

- Fiber orientation enables the optimization of the mechanical behavior along a specific direction.
- The material is elastic up to rupture. It **cannot yield** by local plastic deformation as can classical metallic materials.
- Fatigue resistance is excellent.

A Very Good Fatigue Resistance

The specific fatigue resistance is expressed by the ratio (σ/ρ), with ρ being the specific mass. For composite materials, this specific resistance is three times higher than for aluminum alloys and two times higher than that of high strength steel and titanium alloys because the fatigue resistance is equal to 90% of the static fracture strength for a composite, instead of 35% for aluminum alloys and 50% for steels and titanium alloys (see Figure 5.1).[1]

---

[1] See Section 5.4.4.

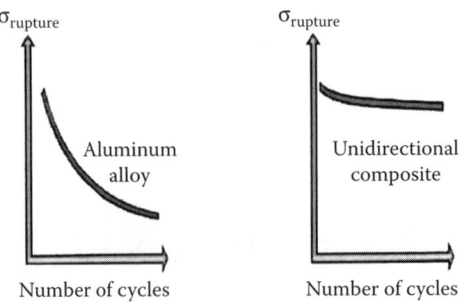

Figure 5.1   Comparison of Fatigue Behavior Between Composite and Aluminum

- The percent elongation is not the same as that for metals (attention should be paid to the metal/composite joints).
- Complex forms can be easily molded.
- It is possible to reduce the number of parts and to limit the amount of processing work.
- One must adapt the classical techniques of attachments and take into account their induced problems: fragility, bearing, fatigue, thermal stresses.

5.1.1   Guidelines for Values for Predesign

Figure 5.2 shows a comparison between different materials, which can help in the choice of composite in the predesign phase.

Figure 5.3 allows the comparison of principal specific properties of fibers which make up the plies. The specific modulus and specific strength are presented in the spirit of lightweight structural materials.

The safety factors are defined to take care of uncertainties on

- The magnitude of mechanical characteristics of reinforcement and matrix
- The stress concentrations
- The imperfection of the hypotheses for calculation
- The fabrication process
- The aging of materials

The orders of magnitude of safety factors are as follows:

High volume composites:		
Static loading	short duration:	2
	long duration:	4
Intermittent loading over long term:		4
Cyclic loading:		5
Impact loading:		10
High performance composites:		1.3 to 1.8

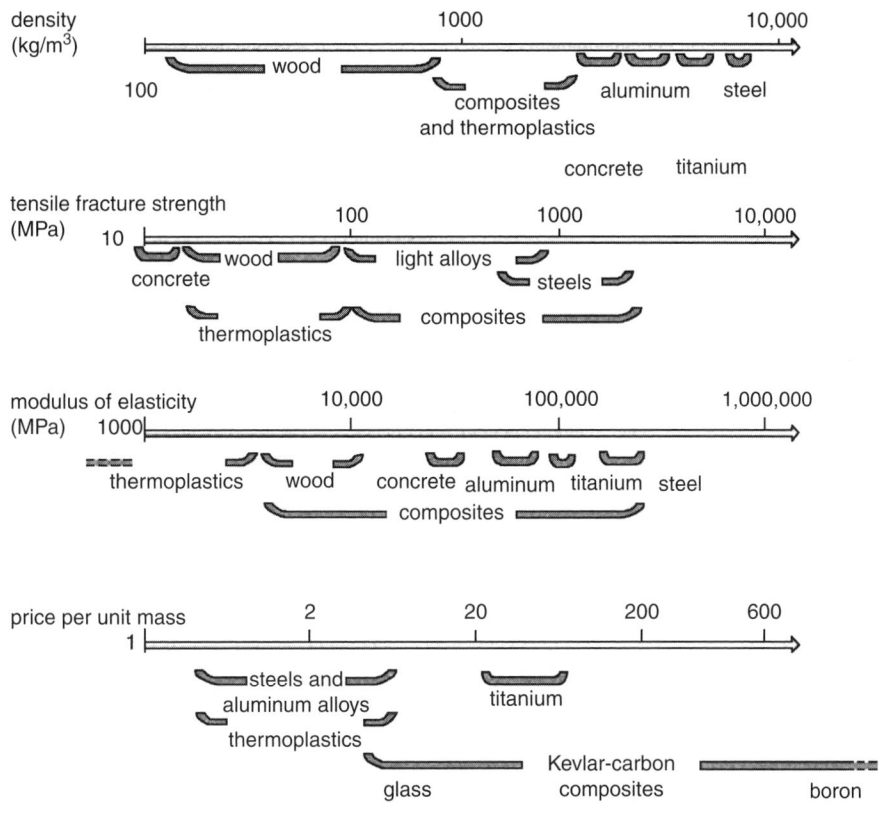

Figure 5.2   Comparison of Characteristics of Different Materials

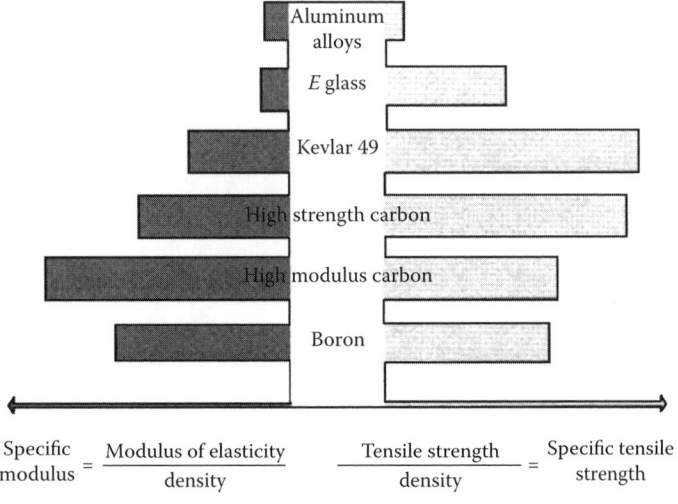

$$\underset{\text{modulus}}{\text{Specific}} = \frac{\text{Modulus of elasticity}}{\text{density}} \qquad \frac{\text{Tensile strength}}{\text{density}} = \underset{\text{strength}}{\text{Specific tensile}}$$

Figure 5.3   Specific Characteristics of Different Fibers

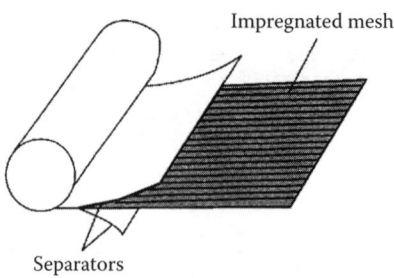

Figure 5.4   Unidirectional Layer

5.2 THE LAMINATE

Recall that laminates result in the superposition of many layers, or plies, or sheets, made of unidirectional layers, fabrics or mats, with proper orientations in each ply. This is the operation of **hand-lay-up**.

5.2.1   Unidirectional Layers and Fabrics

Unidirectional layers are as shown in Figure 5.4. The advantages of unidirectional layers are:

- They have high rigidity (maximum number of fibers in one direction).
- The ply can be used to wrap over long distance. Then the load transmission of the fibers is continuous over large distance.
- They have less waste.

The disadvantages of unidirectional layers are

- The time for wrapping is long.
- One cannot cover complex shapes using wrapping.

   Example: Carbon/epoxy unidirectionals: Width 300 or 1000 mm, preimpregnated with resin; usable over a few years when stored at cold temperature (−18°C).

   Fabrics can be found in rolls in dry form or impregnated with resin (Figure 5.5). The advantages of fabrics are

- Reduced wrapping time
- Possibility to shape complex form using the deformation of the fabric
- Possibility to combine different types of fibers in the same fabric

The disadvantages of fabrics are

- Lower modulus and strength than the case of unidirectionals
- Larger amount of waste material after cutting
- Requirement of joints when wrapping large parts

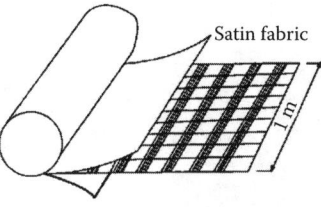

Figure 5.5   A Fabric Layer

5.2.2 Importance of Ply Orientation

One of the fundamental advantages of laminates is their ability to adapt and control the orientation of fibers so that the material can best resist loadings. It is therefore important to know how the plies contribute to the laminate resistance, taking into account their relative orientation with respect to the loading direction. Figures 5.6 through 5.9 show the favorable situations and those that should be avoided.

Recall the Mohr circle:

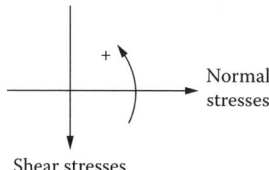

cf. for example the stress state below and the associated Mohr circle

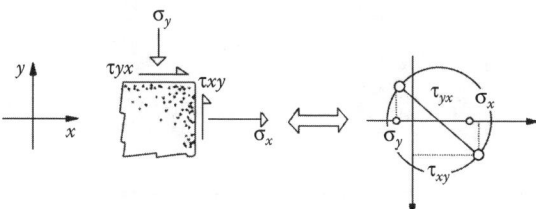

In Figure 5.7, the Mohr circle for stresses shows that the 45° fibers support the compression, $\sigma_1 = -\tau$ (τ is the arithmetic value of shear stress), while the resin supports the tension, $\sigma_2 = \tau$, with low fracture limit. The fibers in Figure 5.8 support the tension, $\sigma_1 = \tau$, whereas the resin supports the compression, $\sigma_2 = -\tau$. In Figure 5.9, one has deposited the fibers at 45° and −45°. Taking into account the previous remarks, one observes that the 45° fibers can support the tension $\sigma_1 = \tau$, whereas the −45° fibers can support the compression, $\sigma_2 = -\tau$. The resin is less loaded than previously.

5.2.3 Code to Represent a Laminate

5.2.3.1 Normalized Orientation

Considering the working mode of the plies as discussed in the previous section, the most frequently used orientations are represented as in Figure 5.10. The direction called "0°" corresponds to either the main loading direction, a preferred direction of the piece under consideration, or the axis of the chosen coordinates.

Note: One also finds in real applications plies with orientations ±30° and ±60°.

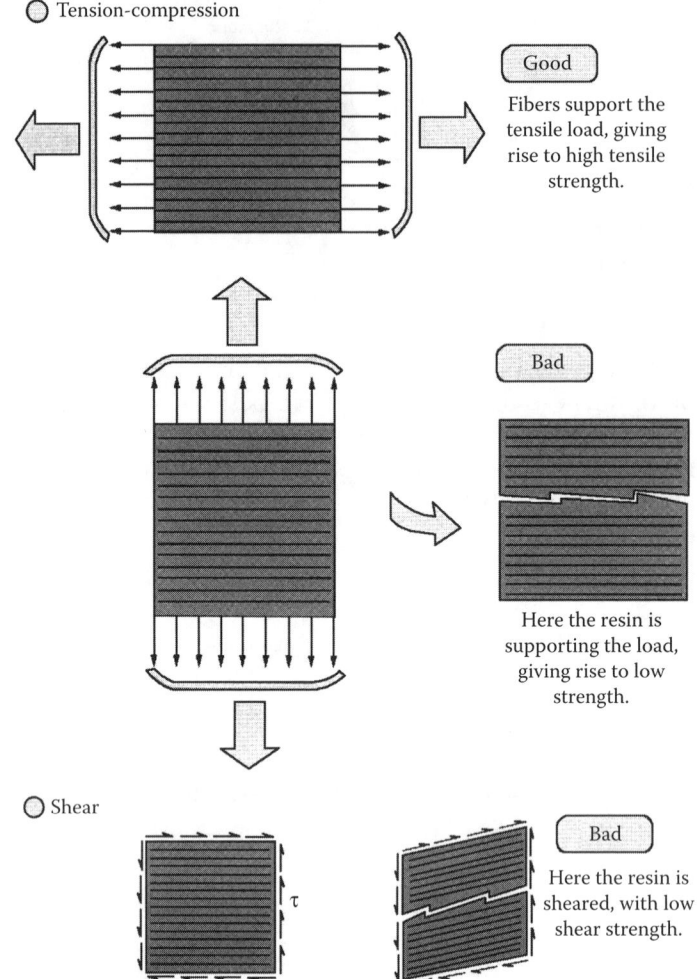

Figure 5.6   Effect of Ply Orientation

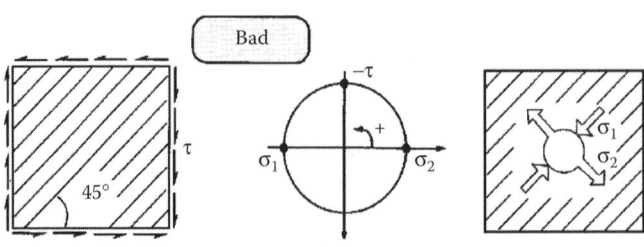

Figure 5.7   Bad Design

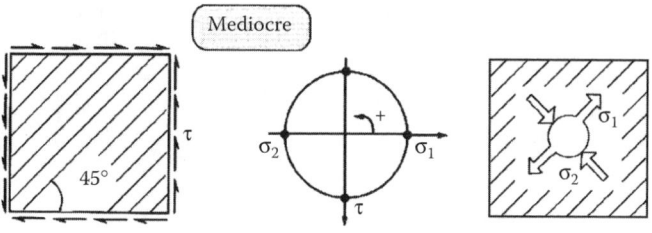

Figure 5.8 Mediocre Design

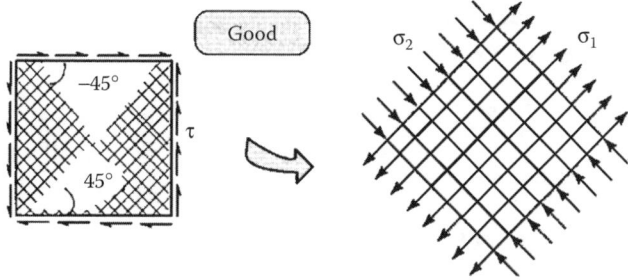

Figure 5.9 Good Design

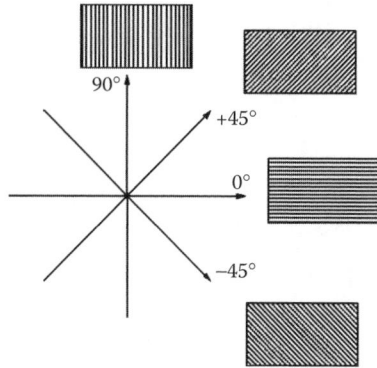

Figure 5.10 Common Orientations

5.2.3.2 Middle Plane

By definition the middle plane is the one that separates two half-thicknesses of the laminate. In Figure 5.11, the middle plane is the plane x–y. On this plane, $z = 0$.

5.2.3.3 Description of Plies

The description of plies is done by beginning with the lowest ply on the side $z < 0$ and proceeding to the uppermost ply of the side $z > 0$. In so doing,

- Each ply is noted by its orientation.
- The successive plies are separated by a slash "/".

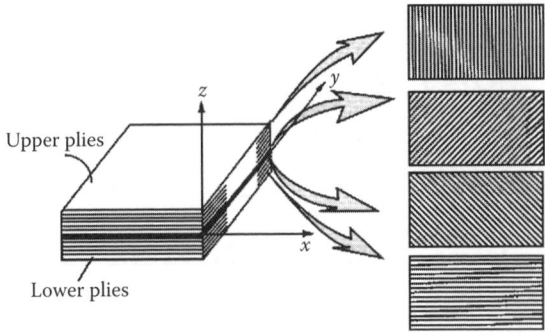

Figure 5.11   Laminate and Its Middle Plane x–y

■ One must avoid the grouping of too many plies of the same orientation.[2] However, when this occurs, an index number is used to indicate the number of these identical plies.

5.2.3.4 Midplane Symmetry

One notes that a laminate has midplane symmetry or is symmetric when the stacking of the plies on both sides starting from the middle plane is identical.

Example:

Ply number		Orientation	Conventional notation	Symbol
10		90°		
9		0°		
8		0°		
7		−45°		
6	Mid	+45°	$(90/0_2/-45/45)_s$	
5	plane	+45°		
4		−45°		
3		0°		
2		0°		
1		90°		

Example:

Ply number		Orientation	Conventional notation	Symbol
7		0°		
6		+45°		
5		−45°		
4	Mid	−90°	$(0/45/-45/\overline{90})_s$	
3	plane	−45°		
2		+45°		
1		0°		

---

[2] This is to limit the interlaminar stresses (see Section 5.4.4 and Chapter 17). This precaution applies also to the fabrics (for example, no more than four consecutive fabric layers of carbon/epoxy along one direction).

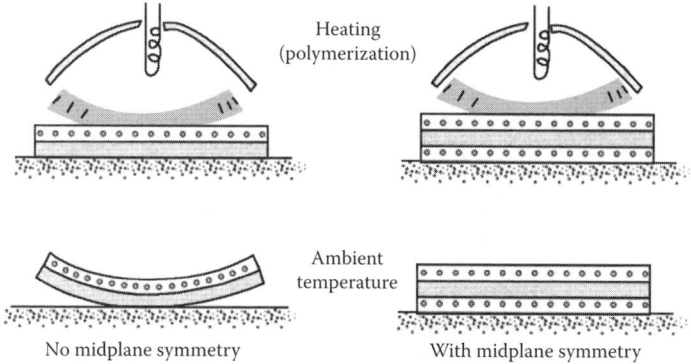

Figure 5.12   Effect of Laminate Lay-up on Deformation

5.2.3.4.1 What Is the Need of Midplane Symmetry

For the construction of laminated pieces, the successive impregnated plies are stacked at ambient temperature, then they are placed within an autoclave for curing. At high temperature, the extension of the whole laminate takes place without warping. However, during cooling, the plies have a tendency to contract differently depending on their orientations. From this, thermal residual stresses occur.

When midplane symmetry is utilized, it imposes the symmetry on these stresses and prevents the deformations of the whole part, for example, warping as shown in Figure 5.12.

5.2.3.5   Particular Cases of Balanced Fabrics

Some laminates are made partially or totally of layers of balanced fabric. One then needs to describe on the drawing the composition of the laminate.

Example:

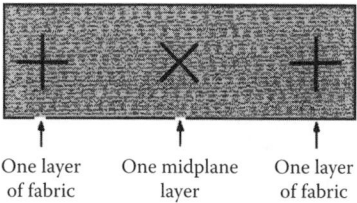

One layer   One midplane   One layer
of fabric   layer   of fabric

The previous laminate, made up of three layers of balanced fabric, has midplane symmetry. In effect, if one considers one woven fabric layer as equivalent to two series of unidirectional layers crossed at 90°, it also has midplane symmetry.[3]

---

[3] If this hypothesis is to be verified for a plain weave or a taffeta (see Section 3.4.1), and even for a ribbed twill, it becomes worse as long as the pitch of the weaving machine increases (pitch of the plain weave: 2; ribbed twill: 3; 4-harness satin: 4; 5-harness satin: 5; etc.). If one supposes that this pitch is increasing towards infinity, then the woven fabric becomes the superposition of two unidirectional layers crossed at 90°. It then does not possess midplane symmetry any more. This property can be observed on a unique ply of 5-harness satin of carbon/epoxy as it is cured in an autoclave, which deforms (curved surface) on demolding (see Application 18.2.17).

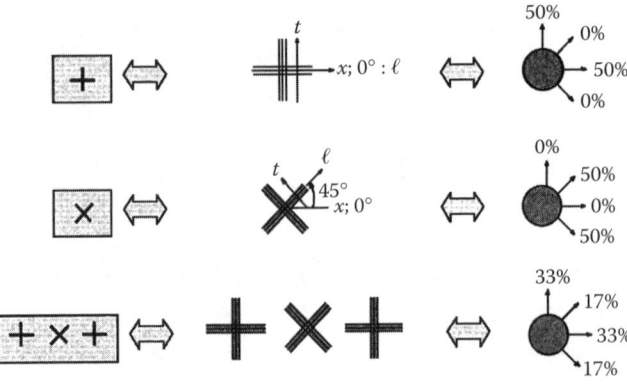

Figure 5.13 Laminate with Balanced Fabrics; Representation 1

As indicated in Section 3.4.2, one can consider the resulting laminate in two different ways[4]:

> (a) Each layer of fabric is replaced by two identical plies crossed at 90°, each with thickness equal to half the thickness e of the fabric layer and each with known elastic properties. This representation is convenient for the determination of the elastic properties of the laminate. One then has the equivalencies shown in Figure 5.13.
>
> (b) Each layer of fabric is replaced by one anisotropic ply with thickness e for which one knows the elastic properties and failure strengths. This representation is useful for the determination of the rupture stress of the laminate. One then has the equivalencies shown in Figure 5.14.

5.2.3.6 Technological Minimum

Generally one uses a minimum amount of plies (from 5 to 10%) for each direction: 0°, 90°, 45°, −45°. The minimum thickness of a laminate[5] should be of the order of one millimeter, for example, eight unidirectional layers, or three to four layers of balanced fabric of carbon/epoxy.

5.2.4 Arrangement of Plies

The proportion and the number of plies to place along each of the directions — 0°, 90°, 45°, −45° — take into account the mechanical loading that is applied to the laminate at the location under consideration. A current case consists of loading

---

[4] See Exercises 18.2.9 and 18.2.10.
[5] Apart from space applications, where thicknesses are very small: then, the skins of sandwich plates are laminates which do not have separately midplane symmetry.

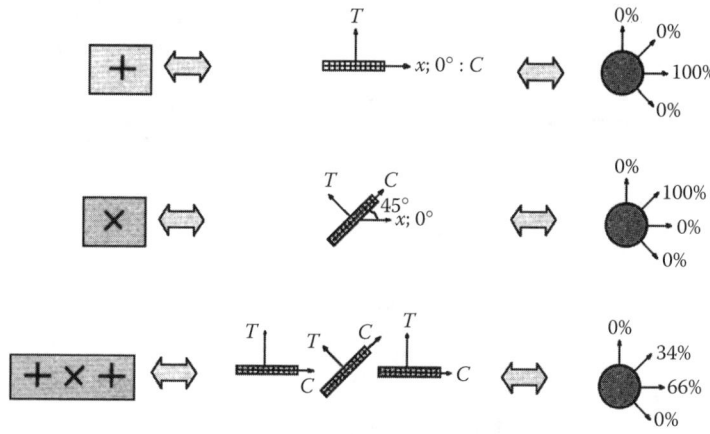

Figure 5.14 Laminate with Balanced Fabrics; Representation 2

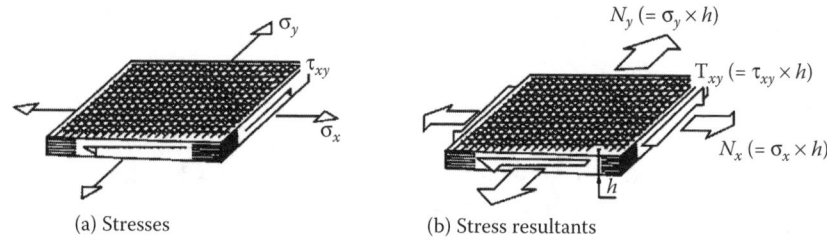

(a) Stresses   (b) Stress resultants

Figure 5.15 Stresses and Stress Resultants

of the laminate in its plane. This is called **membrane loading**.[6] The mechanical loadings can take the form of stresses (σ_x, σ_y, τ_{xy} in Figure 5.15a) or **stress resultants** (N_x, N_y, T_{xy} in Figure 5.15b). The stress resultants are the products of the stresses with the thickness h of the laminate.

Generally, three criteria should be considered by the designer for the ply configuration:

1. Support the loading without deterioration of the laminate
2. Limit the deformation of the loaded piece
3. Minimize the weight of the material used

These criteria do not always work together. For example, searching for minimum thickness might not be compatible with high rigidity. Searching for high rigidity might not be compatible with minimum weight. One will see in Section 5.4

---

[6] The laminate can also work in bending. This is studied in Chapters 12 and 17.

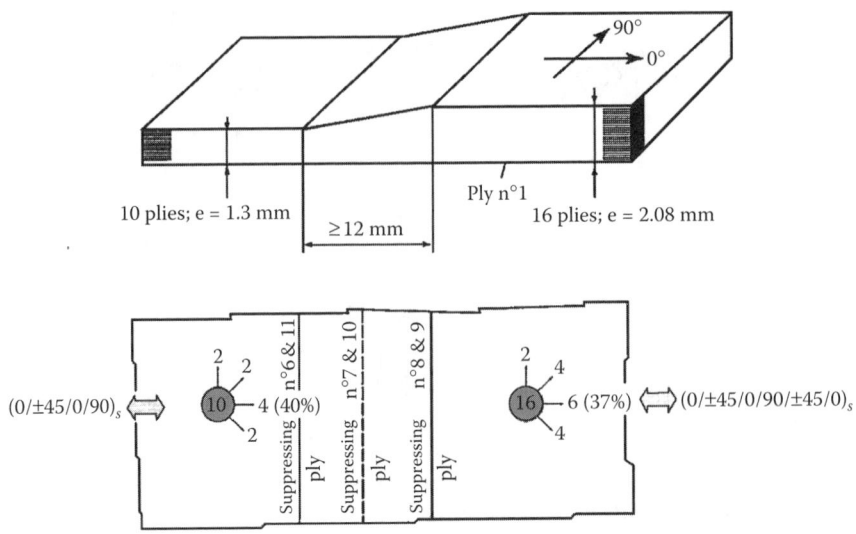

Figure 5.16 Example of Representation

guidelines for proportions values that allow a laminate with minimum laminate thickness to support specified mechanical loading without damage.

Once a laminate is defined (number of layers and orientations), one must respect the following conditions (without forgetting the technological minimum indicated at the end of the previous paragraph) as much as possible:

■ 90° plies placed on the surface, then 45° and −45° plies, when the predominant stress resultant is oriented along the 0° direction
■ No more than 4 consecutive plies along the same direction

5.2.4.1 Example of Representation

The plies are progressively terminated to obtain a gradual change in thickness (maximum 2 plies for each 6 mm interval). The symbols for the composition of the laminate are shown on **plan** view (see Figure 5.16).

5.2.4.2 The Case of Sandwich Structure

The description of the sandwich material is done as in Figure 5.17.

5.3 FAILURE OF LAMINATES

5.3.1 Damages

Figure 5.18 shows schematically different types of failure leading to damage of a laminate.

The main modes of damage, when the loads exceed the critical limits, are illustrated in Figure 5.19.

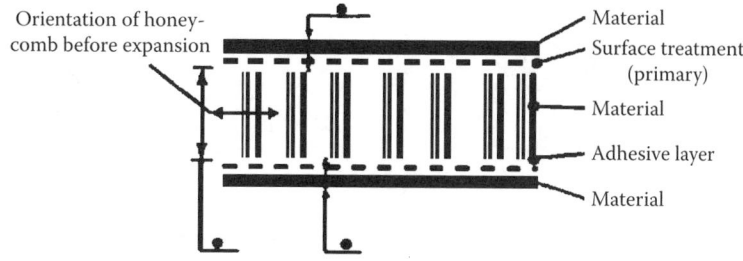

Figure 5.17 Description of a Sandwich Material

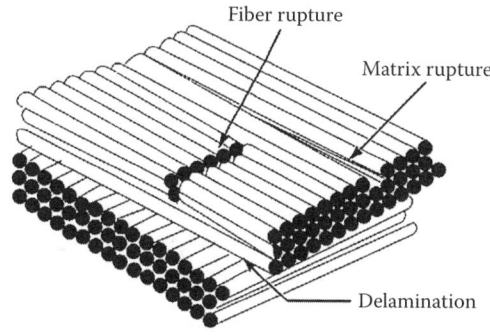

Figure 5.18 Different Modes of Failure

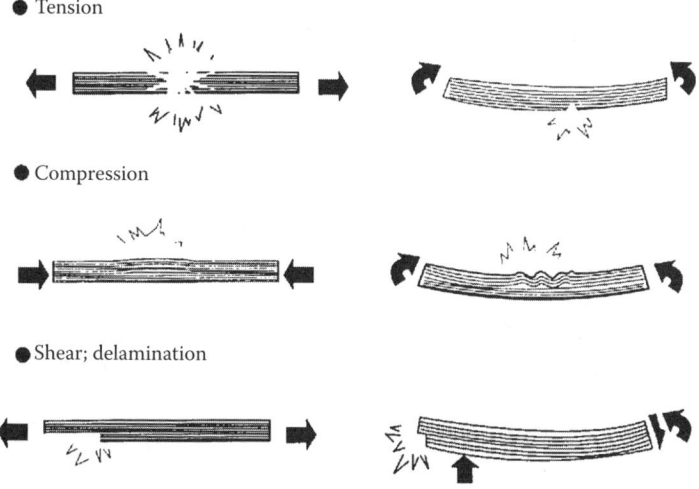

Figure 5.19 Modes of Damage

One cannot be satisfied with the classical maximum stress criterion

Figure 5.20 shows a unidirectional laminate loaded successively in two different manners. In the two cases, the maximum normal stress has the same value denoted

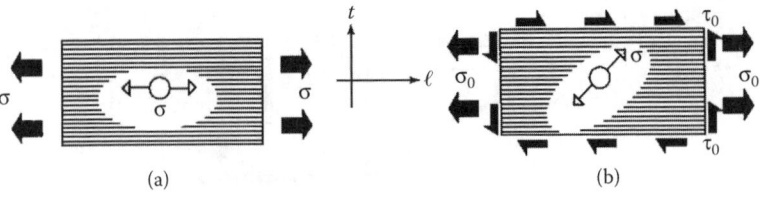

Figure 5.20   Stresses and Fiber Orientation

as σ. In the loading case (a), the unidirectional specimen will rupture when

$$\sigma > \sigma_{\text{rupture along } \ell}$$

This is the maximum stress criterion.

In the loading case (b), the maximum normal stress occurs in a direction that is different from that of the fibers (one can obtain this by tracing the Mohr's circle as discussed previously). We have seen (Section 3.3.2) that the rupture resistance decreases. It is weaker than the situation of case (a). The unidirectional laminate therefore ruptures when

$$\sigma < \sigma_{\text{rupture along } \ell}$$

This phenomenon is more evident if the unidirectional laminate is loaded in a direction transverse to the fibers t. In this case, the laminate rupture resistance is that of the matrix, which is much less than that of the fibers.

Taking into consideration the evolution of the rupture resistance with the loading direction, one can not use a simple maximum stress criterion as for the classical metallic materials.

5.3.2   Most Frequently Used Criterion: Hill–Tsai Failure Criterion[7]

One can apply this criterion successively to **each ply** of the laminate, that is for each one of the orientations 0°, 90°, ±45° that have been considered. As has been discussed in Chapter 3, the axes of a unidirectional ply are denoted as ℓ for the direction along the fibers, and t for the transverse direction. The stresses are denoted as σ_ℓ in the fiber direction, σ_t in the direction transverse to the fibers, and $\tau_{\ell t}$ for the shear stress (see Figure 5.21 below).

One denotes the **Hill–Tsai number,** the number α such that

- If $\alpha < 1$: no ply rupture occurs.
- If $\alpha \geq 1$: rupture occurs in the ply considered. Generally, this deterioration is due to the rupture of the resin. The mechanical properties of a broken ply become almost negligible, except for those along the fiber direction (modulus of elasticity and rupture resistance)

---

[7] For more detailed study of this criterion, see Chapter 14.

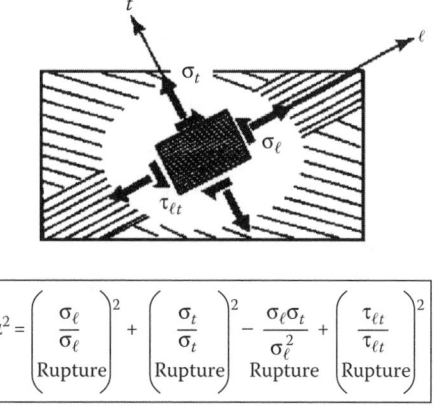

$$\alpha^2 = \left(\frac{\sigma_\ell}{\sigma_\ell^{\text{Rupture}}}\right)^2 + \left(\frac{\sigma_t}{\sigma_t^{\text{Rupture}}}\right)^2 - \frac{\sigma_\ell \sigma_t}{\sigma_\ell^2 {}^{\text{Rupture}}} + \left(\frac{\tau_{\ell t}}{\tau_{\ell t}^{\text{Rupture}}}\right)^2$$

Figure 5.21   Hill–Tsai Number

5.3.2.1   Notes

Attention: The rupture resistance σ_{rupture} does not have the same value in tension and in compression (see, for example, Section 3.3.3). It is therefore useful to place in the denominators of the previous Hill–Tsai expression the rupture resistance values corresponding to the mode of loading (tension or compression) that appear in the numerator.

- Using this criterion, when one detects the rupture of one of the plies (more precisely the rupture of the plies along one of the four orientations), this does not necessarily lead to the rupture of the whole laminate. In most cases, the degraded laminate continues to resist the applied stress resultants. In increasing these stress resultants, one can detect which orientation can produce new rupture. This may—or may not—lead to complete rupture of the laminate. If complete rupture does not occur, one can still increase the admissible stress resultants.[8] In this way one can use a multiplication factor on the initial critical loading to indicate the ratio between the first ply rupture and the ultimate rupture.
- As a consequence of the previous remark it appears possible to work with a laminate that is partially degraded. It is up to the designer to consider the finality of the application, to decide whether the partially degraded laminate can be used.

One can make a parallel — in a gross way — with the situation of classical metallic alloys as represented in Figure 5.22.

5.3.2.2   How to Determine σ_ℓ, σ_t, $\tau_{\ell t}$ in Each Ply

Consider for example the laminate shown in Figure 5.23, consisting of identical plies. The following characteristics are known:

---

[8]   See Exercise 18.2.7.

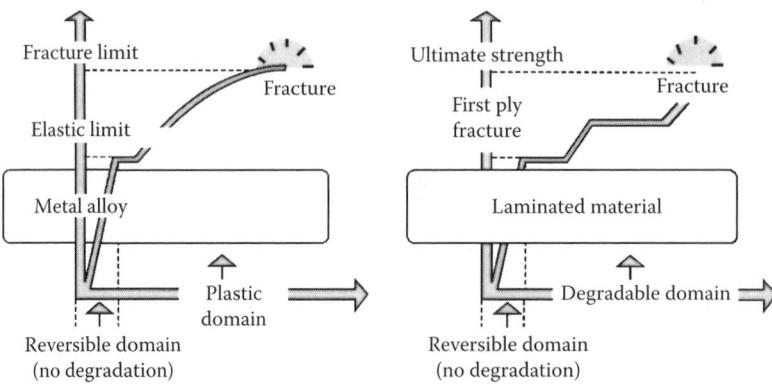

Figure 5.22   Comparison of Behavior until Failure Between Metal and Laminated Material

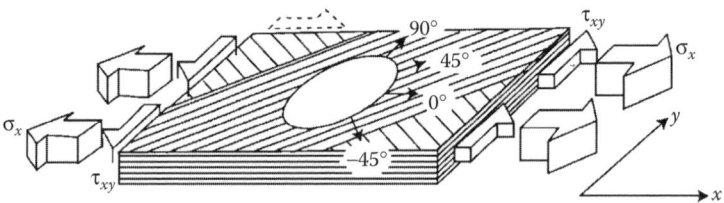

Figure 5.23   Average Stresses

- the mechanical properties of the basic ply
- the proportions (percentages) of plies in each of the directions (0°, 90°, 45°, −45°)
- the global values of the applied stresses, here, for example, σ_x and τ_{xy}

One considers this case of loading as consisting of the superposition of two simple loading cases: σ_x only, and then τ_{xy} only. For each of these cases of elementary loadings, one looks for the stresses σ_ℓ, σ_t, $\tau_{\ell t}$ in each ply. Manual calculation is usually too long.[9] It should be replaced by using a computer. Appendix 1 has tables for stress values for carbon/epoxy plies with 60% fiber volume fraction.

Subsequently, one finds, always **for each ply**, the sum of the stresses σ_ℓ, σ_t, and $\tau_{\ell t}$, respectively, due to each of the simple loadings σ_x and τ_{xy}. It is then possible to calculate the Hill–Tsai number to verify the integrity of each of the plies. In the Application 18.1.6, there is an example to determine the thickness of a laminate subject to this type of combined loading.

---

[9] The procedure for this calculation is described in Section 12.1.3.

Conception and Design ■ 85

5.4 SIZING OF THE LAMINATE

5.4.1 Modulus of Elasticity. Deformation of a Laminate

For the varied proportions of plies in the 0°, 90°, ±45°, the tables that follow allow the determination of the deformation of a laminated plate subject to the applied stresses. For this one uses a stress–strain relation similar to that described in Section 3.1 for an anisotropic plate, which is repeated below:

$$\left\{ \begin{array}{c} \varepsilon_x \\ \varepsilon_y \\ \gamma_{xy} \end{array} \right\} = \left[\begin{array}{ccc} \dfrac{1}{E_x} & -\dfrac{v_{yx}}{E_y} & 0 \\ -\dfrac{v_{xy}}{E_x} & \dfrac{1}{E_y} & 0 \\ 0 & 0 & \dfrac{1}{G_{xy}} \end{array} \right] \left\{ \begin{array}{c} \sigma_x \\ \sigma_y \\ \tau_{xy} \end{array} \right\}$$

E_x, E_y, G_{xy}, v_{xy}, v_{yx} are the modulus of elasticity and Poisson ratios of the laminate,[10] and ε_x, ε_y, γ_{xy} are normal and shear strains in the plane xy.

Example: What are the elastic moduli and thermal expansion coefficients for a glass/epoxy laminate ($V_f = 60\%$) with the following ply configuration?

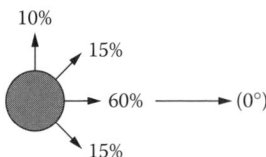

Answer: Table 5.14 indicates the following values for this laminate:

E_x = 33,100 MPa
E_y = 17,190 MPa (this value is obtained by permuting the proportions of
0° and 90°)
v_{xy} = 0.34
v_{yx} = 0.17

Table 5.15 shows $G_{xy} = 6,980$ MPa.
One then obtains the strains ε_x, ε_y, γ_{xy}, when the stresses are known, using the matrix relation mentioned above.

For the coefficient of thermal expansion, Table 5.14 shows: $\alpha_x = 0.64 \times 10^{-5}$ and $\alpha_y = 1.21 \times 10^{-5}$ by permuting the proportions of 0° and 90°.

---

[10] Recall (Sections 3.1 and 3.2) that $v_{xy}/E_x = v_{yx}/E_y$.

5.4.2 Case of Simple Loading

The laminate is subjected to only one single stress: σ_x or σ_y or τ_{xy}. Depending on the percentages of the plies in the four directions, one would like to know the order of magnitude of the stresses that can cause first ply failure in the laminate.

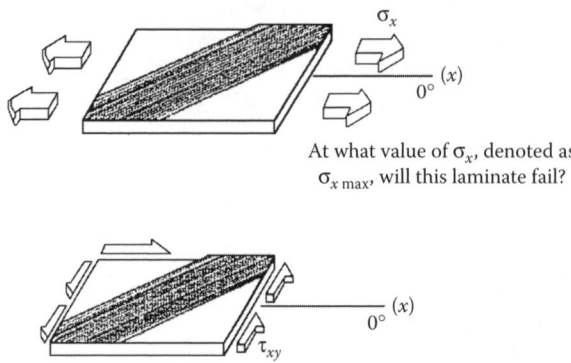

At what value of σ_x, denoted as $\sigma_{x\,max}$, will this laminate fail?

At what value of τ_{xy}, denoted as $\tau_{xy\,max}$, will this laminate fail?

Tables 5.1 through 5.15 indicate the maximum stresses as well as the elastic characteristics and the coefficients of thermal expansion for the laminates having the following characteristics:

■ Materials include **carbon, Kevlar, glass/epoxy** with $V_f = 60\%$ fiber volume fraction.
■ All have identical plies (same unidirectionals, same thickness).
■ The laminate is balanced (same number of 45° and −45° plies). The midplane symmetry is realized.
■ The percentages of plies along the 4 directions 0°, 90°, ±45° vary in increments of 10%.

Calculation of the maximum stresses $\sigma_{x\,max}$, $\sigma_{y\,max}$, $\tau_{xy\,max}$ is done based on the Hill–Tsai failure criterion.[11]

Example of how to use the tables:

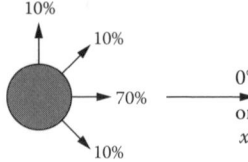

Which maximum tensile stress along the 0° direction can be applied to a Kevlar/epoxy laminate containing 60% fiber volume with the orientation distribution as shown in the above figure?

Answer: Table 5.6 indicates the maximum stress in the 0° direction (or x). For the percentages given, one has:

---

[11] See Application 18.2.2.

Table 5.1 Carbon/Epoxy Laminate: $V_f = 60\%$, Ply Thickness = 0.13 mm

Percentage
of 90° plies

100%	t 42										
	c 141										
90%	t 61	t 118									
	c 108	c 166									
80%	t 80	t 138	t 195								
	c 139	c 192	c 273								
70%	t 97	t 157	t 215	t 271							
	c 168	c 217	c 300	c 380							
60%	t 113	t 174	t 234	t 292	t 348						
	c 193	c 240	c 324	c 407	c 487						
50%	t 127	t 190	t 252	t 312	t 370	t 425					
	c 213	c 260	c 346	c 431	c 514	c 595					
40%	t 137	t 203	t 268	t 330	t 391	t 448	t 502				
	c 226	c 275	c 364	c 451	c 537	c 621	c 702				
30%	t 143	t 212	t 279	t 345	t 409	t 470	t 527	t 580			
	c 229	c 283	c 375	c 466	c 555	c 643	c 728	c 809			
20%	t 141	t 213	t 284	t 354	t 422	t 488	t 550	t 608	t 659		
	c 217	c 276	c 372	c 468	c 564	c 657	c 747	c 834	c 917		
10%	t 126	t 201	t 275	t 349	t 422	t 495	t 565	t 632	t 692	t 741	
	c 185	c 246	c 341	c 438	c 538	c 640	c 744	c 846	c 939	c 1024	
0%	t 123	t 223	t 324	t 426	t 529	t 633	t 740	t 848	t 957	t 1059	t 1270
	c 125	c 182	c 265	c 351	c 438	c 530	c 627	c 733	c 851	c 990	c 1130
	0%	10%	20%	30%	40%	50%	60%	70%	80%	90%	100%
	100%	90%	80%	70%	60%	50%	40%	30%	20%	10%	0%

$\sigma_{x\,max}$
t = tension
c = compression

→ 0°

Percentage of 0° plies

Percentage of ±45° plies

Maximum stress $\sigma_{x\,max}$ (MPa) as a function of the ply percentages in the directions 0°, 90°, +45°, −45°.
(More information on modulus and strength of a basic ply: see Section 3.3.3)

$$\sigma_{x\,max(tension)} = 308 \text{ MPa}$$

Example:

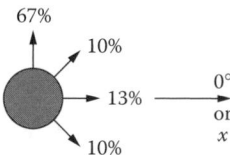

67%
10%
13% → 0° or x
10%

Which maximum compression stress along the 90° direction (or y) can be applied to a carbon/epoxy laminate containing 60% fiber volume fraction with the orientation distribution as shown in the above figure?

Table 5.2   Carbon/Epoxy Laminate: V_f = 60%, Ply Thickness = 0.13 mm

Percentage of 90° plies

% 90°										
100%	t 1270 / c 1130									
90%	t 1059 / c 990	t 741 / c 1024								
80%	t 957 / c 733	t 692 / c 939	t 659 / c 917							
70%	t 848 / c 733	t 632 / c 846	t 608 / c 834	t 580 / c 809						
60%	t 740 / c 627	t 565 / c 744	t 550 / c 747	t 527 / c 728	t 502 / c 702					
50%	t 633 / c 530	t 495 / c 640	t 488 / c 657	t 470 / c 643	t 448 / c 621	t 425 / c 595				
40%	t 529 / c 438	t 422 / c 538	t 422 / c 564	t 409 / c 555	t 391 / c 537	t 370 / c 514	t 348 / c 487			
30%	t 426 / c 351	t 349 / c 438	t 354 / c 468	t 345 / c 466	t 330 / c 451	t 312 / c 431	t 292 / c 407	t 271 / c 380		
20%	t 324 / c 265	t 275 / c 341	t 284 / c 372	t 279 / c 375	t 268 / c 364	t 252 / c 346	t 234 / c 324	t 215 / c 300	t 195 / c 273	
10%	t 223 / c 182	t 201 / c 246	t 213 / c 276	t 212 / c 283	t 203 / c 275	t 190 / c 260	t 174 / c 240	t 157 / c 217	t 138 / c 192	t 118 / c 166
0%	t 123 / c 125	t 126 / c 185	t 141 / c 217	t 143 / c 289	t 137 / c 226	t 127 / c 213	t 113 / c 193	t 97 / c 168	t 80 / c 139	t 61 / c 108

(bottom right) t 42 / c 141

Percentage of 0° plies — 0% 10% 20% 30% 40% 50% 60% 70% 80% 90% 100%

Percentage of ±45° plies — 100% 90% 80% 70% 60% 50% 40% 30% 20% 10% 0%

$\sigma_{y\,max}$
t = tension
c = compression

0°

Maximum stress $\sigma_{y\,max}$ (MPa) as a function of the ply percentages in the directions 0°, 90°, +45°, −45°.
(More information on modulus and strength of a basic ply: see Section 3.3.3)

Answer: Table 5.2 shows the maximum stresses in the 90° direction. For this configuration, one has

$$\sigma_{y\,max} = \sigma_{13/67/10/10} = \sigma_{10/60/15/15} + \Delta\sigma = 744 + \Delta\sigma$$

Denoting $p^{0°}$ and $p^{90°}$ as the proportions of the plies along the 0° and 90° directions, one has

$$\Delta\sigma = \frac{\partial\sigma}{\partial p^{0°}} \times \Delta p^{0°} + \frac{\partial\sigma}{\partial p^{90°}} \times \Delta p^{90°}$$

Table 5.3   Carbon/Epoxy Laminate: V_f = 60%, Ply Thickness = 0.13 mm

Percentage
of 90° plies

	100%	90%	80%	70%	60%	50%	40%	30%	20%	10%	0%
100%	63										
90%	83	63									
80%	118	83	63								
70%	153	118	83	63							
60%	188	153	118	83	63						
50%	223	188	153	118	83	63					
40%	258	223	188	153	118	83	63				
30%	293	258	223	188	153	118	83	63			
20%	327	293	258	223	188	153	118	83	63		
10%	362	327	293	258	223	188	153	118	83	63	
0%	397	362	327	293	258	223	188	153	118	83	63

Percentage of 0° plies

Horizontal axis (Percentage of ±45° plies, top row): 0% 10% 20% 30% 40% 50% 60% 70% 80% 90% 100%

(bottom row): 100% 90% 80% 70% 60% 50% 40% 30% 20% 10% 0%

$\tau_{xy\,max}$ — $0°$

Percentage of ±45° plies

Maximum stress $\tau_{xy\,max}$ (MPa) as a function of the ply percentages in the directions 0°, 90°, +45°, −45°.
(More information on modulus and strength of a basic ply: see Section 3.3.3)

One obtains by linear interpolation:

$$\Delta\sigma = (747 - 744) \times \frac{3}{10} + (846 - 744) \times \frac{7}{10} = 72 \text{ MPa}$$

Therefore,

$$\sigma_{y\,max} = 744 + 72 = 816 \text{ MPa}$$

Remark: The plates that show the maximum stresses are not usable for the balanced fabrics. In effect, the compression strength values of a layer of balanced

Table 5.4   Carbon/Epoxy Laminate: V_f = 60%, Ply Thickness = 0.13 mm

Percentage
of 90° plies

90° plies											
100%	3.4 / 0.013 / 7000										
90%	2.19 / 0.04 / 10181	1.03 / 0.014 / 19739									
80%	1.57 / 0.07 / 13179	0.82 / 0.04 / 22902	0.52 / 0.016 / 32477								
70%	1.2 / 0.1 / 15942	0.66 / 0.076 / 25855	0.43 / 0.05 / 35618	0.25 / 0.02 / 45215							
60%	0.94 / 0.14 / 18404	0.54 / 0.11 / 28533	0.35 / 0.08 / 38513	0.24 / 0.05 / 48326	0.17 / 0.02 / 57952						
50%	0.76 / 0.19 / 20466	0.44 / 0.16 / 30844	0.28 / 0.13 / 41076	0.19 / 0.10 / 51143	0.13 / 0.06 / 61022	0.093 / 0.025 / 70687					
40%	0.6 / 0.25 / 21986	0.35 / 0.22 / 32651	0.22 / 0.19 / 43178	0.14 / 0.16 / 53545	0.093 / 0.12 / 63729	0.006 / 0.08 / 73699	0.036 / 0.03 / 83419				
30%	0.5 / 0.33 / 22739	0.27 / 0.3 / 33735	0.16 / 0.27 / 44606	0.093 / 0.23 / 55333	0.052 / 0.19 / 65888	0.024 / 0.15 / 76239	0.006 / 0.1 / 86343	−0.006 / 0.04 / 96146			
20%	0.39 / 0.44 / 22360	0.19 / 0.41 / 33730	0.093 / 0.38 / 45002	0.038 / 0.34 / 56155	0.004 / 0.3 / 67163	−0.017 / 0.25 / 77993	−0.03 / 0.2 / 88598	−0.038 / 0.13 / 100590	−0.04 / 0.054 / 108860		
10%	0.27 / 0.58 / 20211	0.093 / 0.56 / 31979	0.013 / 0.53 / 43689	−0.03 / 0.5 / 55325	−0.057 / 0.46 / 66869	−0.072 / 0.42 / 78292	−0.08 / 0.36 / 89552	−0.083 / 0.29 / 100590	−0.08 / 0.21 / 111307	−0.07 / 0.09 / 121541	
0%	0.09 / 0.79 / 15055	−0.057 / 0.78 / 27152	−0.11 / 0.77 / 39240	−0.14 / 0.76 / 51315	−0.16 / 0.74 / 63373	−0.17 / 0.71 / 75407	−0.17 / 0.68 / 87405	−0.17 / 0.64 / 99345	−0.16 / 0.57 / 111186	−0.15 / 0.46 / 122830	0.25 / 134000 ; −0.12
	0%	**10%**	**20%**	**30%**	**40%**	**50%**	**60%**	**70%**	**80%**	**90%**	**100%**
	100%	**90%**	**80%**	**70%**	**60%**	**50%**	**40%**	**30%**	**20%**	**10%**	**0%**

(Each cell lists: $\alpha_x \times 10^5$ / ν_{xy} / E_x)

Percentage of 0° plies (top axis) / Percentage of ±45° plies (bottom axis)

Modulus E_x (MPa), Poisson ratio ν_{xy} and coefficient of thermal expansion α_x as a function of the ply percentages in the directions 0°, 90°, +45°, −45°.

(More information on modulus and strength of a basic ply: see Section 3.3.3)

fabric are smaller than what is obtained when one superimposes the unidirectional plies crossed at 0° and 90° in equal quantities in these two directions.[12]

5.4.3 Case of Complex Loading—Approximate Orientation Distribution of a Laminate

When the normal and tangential loadings are applied **simultaneously** onto the laminate, the previous tables are not valid because they were established for the

---

[12] Also see remarks in Section 3.4.2.

Table 5.5   Carbon/Epoxy Laminate: V_f = 60%, Ply Thickness = 0.13 mm

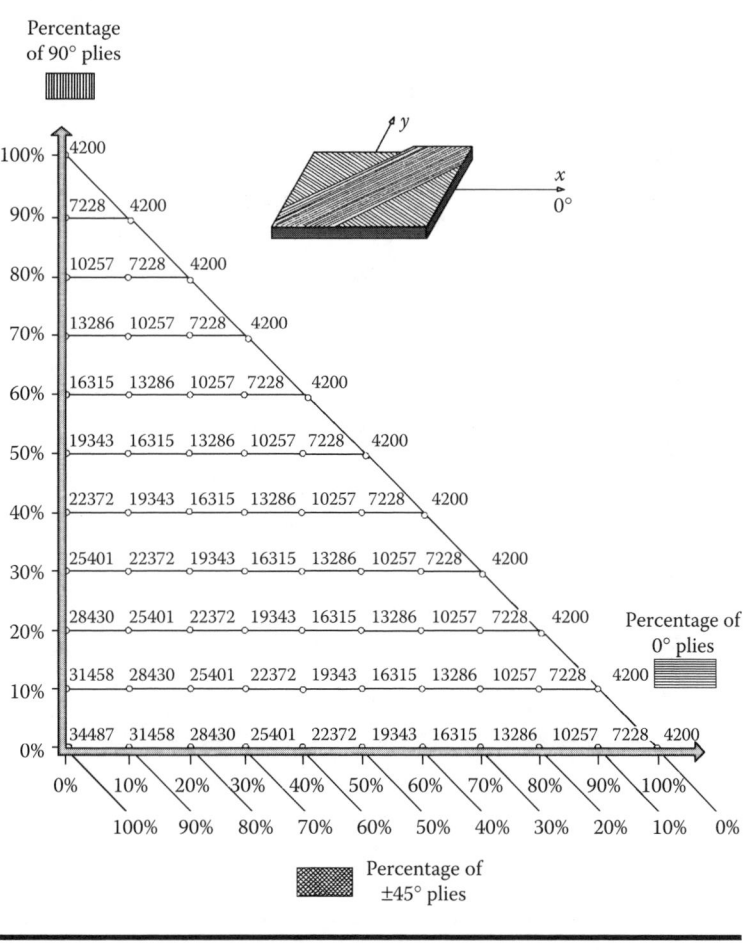

Shear modulus G_{xy} (MPa) as a function of the ply percentages in the
directions 0°, 90°, +45°, −45°.
(More information on modulus and strength of a basic ply: see Section
3.3.3)

cases of simple stress states. However, one can still use them to effectively obtain a first estimate of the proportions of plies along the four orientations.[13]

The principle is as follows: Consider the case of complex loading and replacing the stresses with the stress resultants N_x, N_y, T_{xy} which were defined in Section 5.2.4. In general these stress resultants constitute the initial numerical data that are given by some previous studies. One then can assume that each one of the three stress resultants is associated with an appropriate orientation of the plies following the remarks made in Section 5.2.2.

---

[13] Attention: What follows is for the determination of *proportions*, and not *thicknesses*.

Table 5.6   Kevlar/Epoxy Laminate: $V_f = 60\%$, Ply Thickness = 0.13 mm

Percentage of 90° plies

```
        t 28
100% ─  c 141
        t 37    t 68
 90% ─  c 49    c 44
        t 46    t 77    t 108
 80% ─  c 62    c 51    c 71
        t 54    t 86    t 118   t 148
 70% ─  c 75    c 56    c 77    c 97
        t 61    t 94    t 127   t 158   t 188
 60% ─  c 87    c 61    c 82    c 103   c 124
        t 66    t 101   t 134   t 167   t 199   t 229
 50% ─  c 100   c 65    c 87    c 108   c 129   c 150
        t 69    t 104   t 139   t 173   t 207   t 239   t 269
 40% ─  c 111   c 68    c 91    c 113   c 134   c 156   c 176
        t 67    t 103   t 139   t 175   t 211   t 246   t 280   t 310
 30% ─  c 121   c 70    c 93    c 116   c 138   c 160   c 182   c 203
        t 60    t 95    t 131   t 168   t 206   t 244   t 282   t 319   t 351
 20% ─  c 127   c 68    c 92    c 116   c 140   c 163   c 182   c 207   c 229
        t 47    t 79    t 112   t 147   t 184   t 223   t 265   t 308   t 353   t 392
 10% ─  c 122   c 62    c 87    c 112   c 137   c 161   c 186   c 210   c 233   c 255
        t 87    t 176   t 264   t 353   t 442   t 529   t 613   t 691   t 757   t 793   t 1410
  0% ─  c 86    c 49    c 73    c 73    c 123   c 148   c 174   c 200   c 227   c 254   c 280

        0%   10%   20%   30%   40%   50%   60%   70%   80%   90%  100%
           100%  90%  80%  70%  60%  50%  40%  30%  20%  10%  0%
```

$\sigma_{x\,max}$
t = tension
c = compression

→ 0°

Percentage of 0° plies

Percentage of ±45° plies

Maximum stress $\sigma_{x\,max}$ (MPa) as a function of the ply percentages in the directions 0°, 90°, +45°,−45°.
(More information on modulus and strength of a basic ply: see Section 3.3.3)

Using this hypothesis, N_x, assumed to be supported by the 0° plies (or along x), requires a thickness e_x for these plies such that:

$$e_x = \frac{N_x}{\sigma_{\ell\ rupture}}$$

where $\sigma_{\ell\ rupture}$ is the rupture stress of a unidirectional ply in the long direction. In the same manner, N_y is supposed to be supported by the 90° plies (or along y), and requires a thickness for these plies of

$$e_y = \frac{N_y}{\sigma_{\ell\ rupture}}$$

Table 5.7   Kevlar/Epoxy Laminate: V_f = 60%, Ply Thickness = 0.13 mm

Percentage
of 90° piles

100%	t 1410 c 280									
90%	t 793 c 254	t 392 c 255								
80%	t 757 c 227	t 353 c 233	t 351 c 229							
70%	t 691 c 200	t 308 c 210	t 319 c 207	t 310 c 203						
60%	t 613 c 174	t 265 c 186	t 282 c 185	t 280 c 182	t 269 c 150					
50%	t 529 c 148	t 223 c 161	t 244 c 163	t 246 c 160	t 239 c 156	t 229 c 150				
40%	t 442 c 123	t 184 c 137	t 206 c 140	t 211 c 138	t 207 c 134	t 199 c 129	t 148 c 124			
30%	t 353 c 98	t 147 c 112	t 168 c 116	t 175 c 116	t 173 c 113	t 167 c 108	t 158 c 103	t 148 c 97		
20%	t 264 c 73	t 112 c 87	t 131 c 92	t 139 c 93	t 139 c 91	t 134 c 87	t 127 c 82	t 118 c 77	t 108 c 71	
10%	t 176 c 49	t 79 c 62	t 95 c 68	t 103 c 70	t 104 c 70	t 101 c 65	t 94 c 61	t 86 c 56	t 77 c 51	t 68 c 44
0%	t 87 c 86	t 47 c 122	t 60 c 127	t 67 c 121	t 69 c 111	t 66 c 100	t 61 c 87	t 54 c 75	t 46 c 62	t 37 c 49
	0%	10%	20%	30%	40%	50%	60%	70%	80%	90%

t 28
c 141 (at 100%)

| | 100% | 90% | 80% | 70% | 60% | 50% | 40% | 30% | 20% | 10% | 0% |

$\sigma_{y\,max}$
t = tension
c = compression

Percentage
of 0° piles

Percentage of
±45° piles

Maximum stress σ_{ymax} (MPa) as a function of the ply percentages in the directions 0°, 90°, +45°, −45°.
(More information on modulus and strength of a basic ply: see Section 3.3.3)

Finally, T_{xy} is assumed to be supported by the ±45° plies and requires a thickness for these plies of

$$e_{xy} = \frac{T_{xy}}{\tau_{\text{rupture}}}$$

where τ_{rupture} is the maximum stress that a ±45° laminate can support.

Table 5.8   Kevlar/Epoxy Laminate: V_f = 60%, Ply Thickness = 0.13 mm

Maximum stress $\tau_{xy\,max}$ (MPa) as a function of the ply percentages in the directions 0°, 90°, +45°, −45°.

(More information on modulus and strength of a basic ply: see Section 3.3.3)

One then can retain for the complete laminate the proportions indicated below.

Table 5.9   Kevlar/Epoxy Laminate: V_f = 60%, Ply Thickness = 0.13 mm

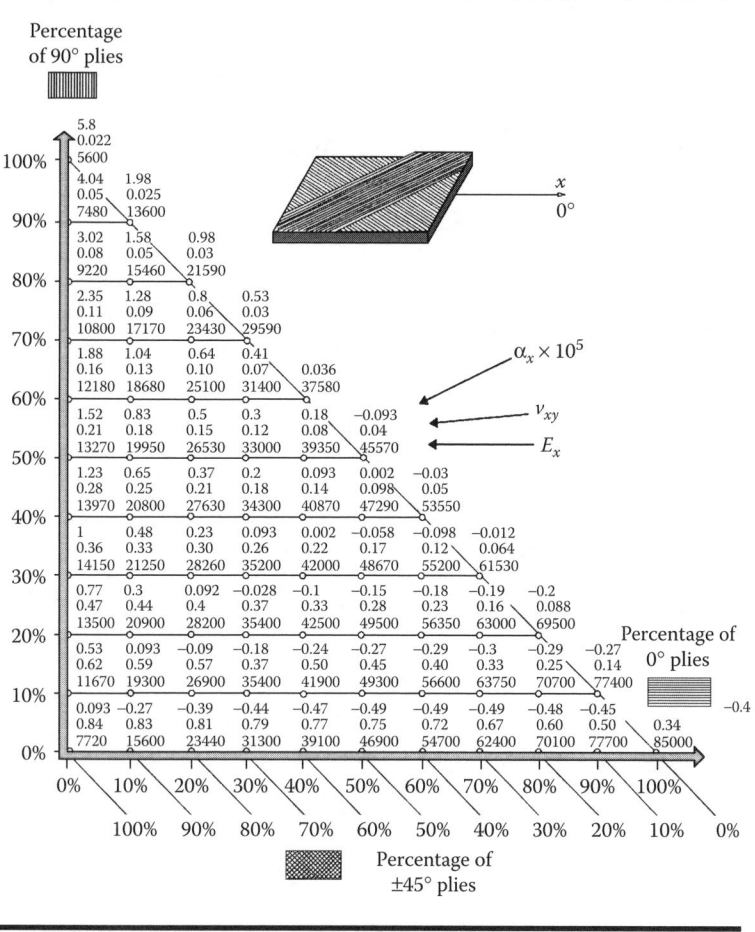

Longitudinal modulus E_x (MPa), Poisson ratio v_{xy} and coefficient of thermal expansion α_x as a function of the ply percentages in the directions 0°, 90°, +45°, −45°.

(More information on modulus and strength of a basic ply: see Section 3.3.3)

Example: Determine the composition of a laminate made up of unidirectional plies of carbon/epoxy (V_f = 60%) to support the stress resultants N_x = −800 N/mm, N_y = −900 N/mm, T_{xy} = −340 N/mm. The compression strength $\sigma_{\ell \text{ rupture}}$ is 1,130 MPa (see Section 3.3.3, or Table 5.1 for 100% of 0° plies). Then:

$$e_x = \frac{800}{1130} = 0.71 \text{ mm}; \quad e_y = \frac{900}{1130} = 0.8 \text{ mm}$$

Table 5.10 Kevlar/Epoxy Laminate: $V_f = 60\%$, Ply Thickness = 0.13 mm

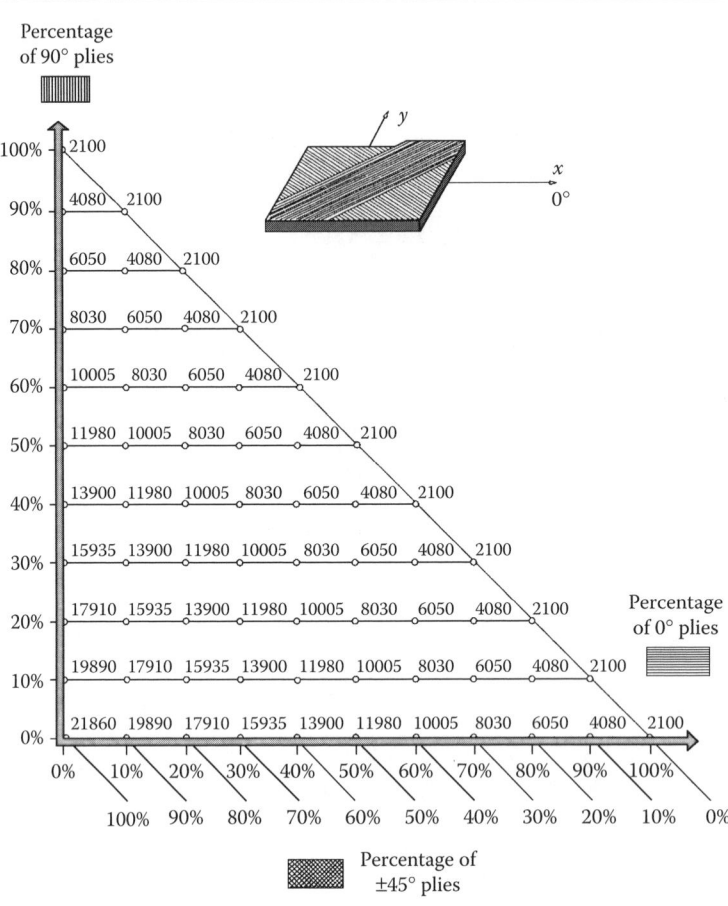

Shear modulus G_{xy} (MPa) as a function of the ply percentages in the directions 0°, 90°, +45°, −45°.
(More information on modulus and strength of a basic ply: see Section 3.3.3)

The optimum shear strength τ_{rupture} is given in Table 5.3 for 100% ±45°, then:

$$\tau_{\text{rupture}} = 397 \text{ MPa}$$

from which:

$$e_{xy} = \frac{340}{397} = 0.86 \text{ mm}$$

Table 5.11   Glass/Epoxy Laminate: V_f = 60%, Ply Thickness = 0.13 mm

Percentage
of 90° plies

100%	t 35 / c 141									
90%	t 37 / c 128	t 45 / c 156								
80%	t 39 / c 134	t 47 / c 162	t 54 / c 190							
70%	t 41 / c 139	t 49 / c 167	t 57 / c 195	t 64 / c 223						
60%	t 43 / c 143	t 51 / c 171	t 59 / c 200	t 67 / c 228	t 75 / c 257					
50%	t 45 / c 145	t 53 / c 174	t 61 / c 203	t 69 / c 232	t 77 / c 261	t 85 / c 290				
40%	t 46 / c 146	t 55 / c 175	t 64 / c 204	t 72 / c 234	t 80 / c 263	t 88 / c 293	t 95 / c 322			
30%	t 48 / c 144	t 57 / c 174	t 66 / c 203	t 74 / c 233	t 82 / c 263	t 90 / c 293	t 98 / c 323	t 106 / c 354		
20%	t 48 / c 140	t 58 / c 170	t 67 / c 200	t 76 / c 230	t 85 / c 260	t 93 / c 290	t 101 / c 321	t 109 / c 352	t 116 / c 384	
10%	t 48 / c 133	t 58 / c 162	t 68 / c 192	t 77 / c 222	t 87 / c 252	t 96 / c 283	t 104 / c 314	t 112 / c 346	t 120 / c 378	t 128 / c 411
0%	t 94 / c 122	t 114 / c 150	t 134 / c 179	t 152 / c 209	t 170 / c 239	t 186 / c 269	t 201 / c 300	t 214 / c 332	t 226 / c 365	t 235 / c 399	t 1250 / c 600

$\sigma_{x\,max}$
t = tension
c = compression

Percentage
of 0° plies

Top axis: 0% 10% 20% 30% 40% 50% 60% 70% 80% 90% 100%

Middle axis: 100% 90% 80% 70% 60% 50% 40% 30% 20% 10% 0%

Percentage of
±45° plies

Maximum stress $\sigma_{x\,max}$ (MPa) as a function of the ply percentages in the directions 0°, 90°, +45°, –45°.
(More information on modulus and strength of a basic ply: see Section 3.3.3)

One obtains for the proportions at

$$0°: \quad \frac{e_x}{e_x + e_y + e_{xy}} = 0.3$$

$$90°: \quad \frac{e_y}{e_x + e_y + e_{xy}} = 0.34$$

$$\pm45°: \quad \frac{e_{xy}}{e_x + e_y + e_{xy}} = 0.36$$

Table 5.12 Glass/Epoxy Laminate: V_f = 60%, Ply Thickness = 0.13 mm

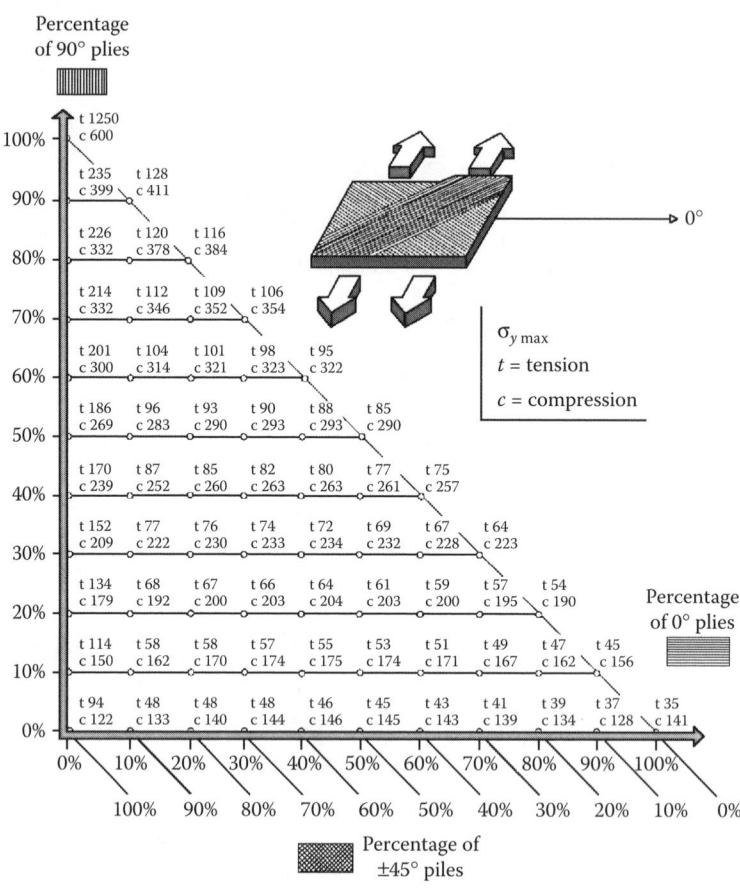

Maximum stress $\sigma_{y\,max}$ (MPa) as a function of the ply percentages in the directions 0°, 90°, +45°, –45°.

(More information on modulus and strength of a basic ply: see Section 3.3.3)

One can then retain for the composition of the laminate the following approximate values:

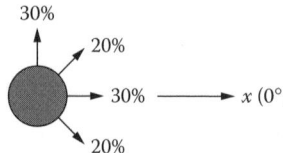

Remark: The thicknesses e_x, e_y, e_{xy} evaluated above only serve to determine the proportions. After that, they **should not be kept**. In effect each orientation really

Table 5.13   Glass/Epoxy Laminate: $V_f = 60\%$, Ply Thickness = 0.13 mm

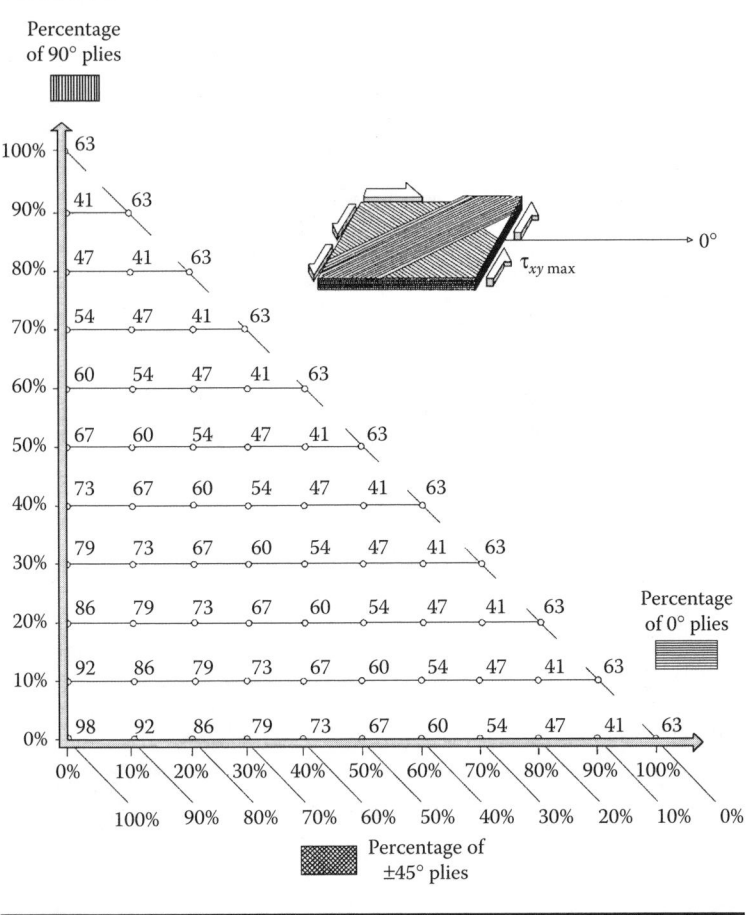

Maximum stress $\tau_{xy\,max}$ (MPa) as a function of the ply percentages in the directions 0°, 90°, +45°, −45°.
(More information on modulus and strength of a basic ply: see Section 3.3.3)

supports a part of each stress resultant. For example, the 0° plies cover the major part of stress resultant N_x, but they also support a part of stress resultant N_y and a part of stress resultant T_{xy}. This then results in a more unfavorable situation for each orientation as compared with what has been assumed previously. The minimum necessary thickness of the laminate will in fact be larger than the previous result $(e_x + e_y + e_{xy})$, which therefore appears to be **dangerously optimistic**. The practical determination of the minimum thickness of the laminate is determined based on the Hill–Tsai failure criterion, as indicated at the end of Section 5.3.2, and explained in details in Application 18.1.6. Also, with the same stress resultants and proportions as in the previous example, one finds a minimum thickness of 2.64 mm (see Application 18.1.6, in Chapter 18), whereas the previous sum $(e_x + e_y + e_{xy})$ gives a thickness of 2.37 mm, 10% lower than the required minimum thickness (2.64 mm).

Table 5.14 Glass/Epoxy Laminate: V_f = 60%, Ply Thickness = 0.13 mm

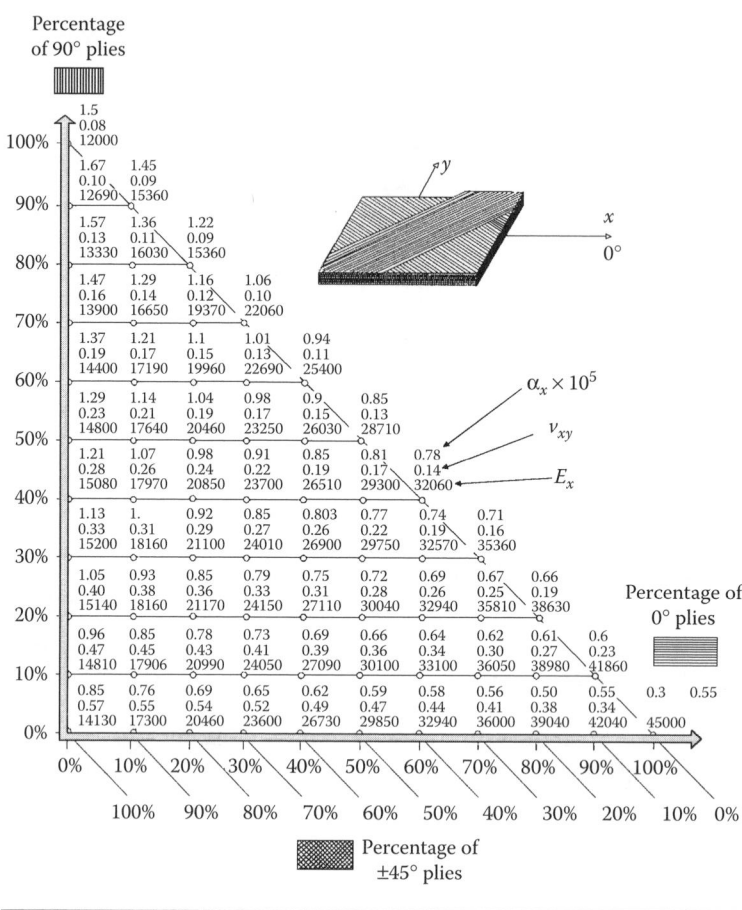

Longitudinal modulus E_x (MPa), Poisson ratio v_{xy}, and coefficient of thermal expansion α_x as a function of the ply percentages in the directions 0°, 90°, +45°, −45°.

(More information on modulus and strength of a basic ply: see Section 3.3.3)

5.4.4 Case of Complex Loading: Optimum Composition of a Laminate

Estimation of the proportions in the previous paragraph does not lead to an optimum laminate in general. An optimum laminate is the one with the smallest thickness among all laminates of different compositions that can support the given combined stress resultants N_x, N_y, T_{xy}.

Tables 5.16 through 5.19, based on Hill-Tsai criterion,[14] give the optimum compositions of laminates based on unidirectionals of carbon/epoxy for the various stress resultants N_x, N_y, T_{xy}. The indicated compositions correspond to

---

[14] See Section 5.3.2.

Table 5.15   Glass/Epoxy Laminates. V_f = 60%. Ply Thickness = 0.13 mm

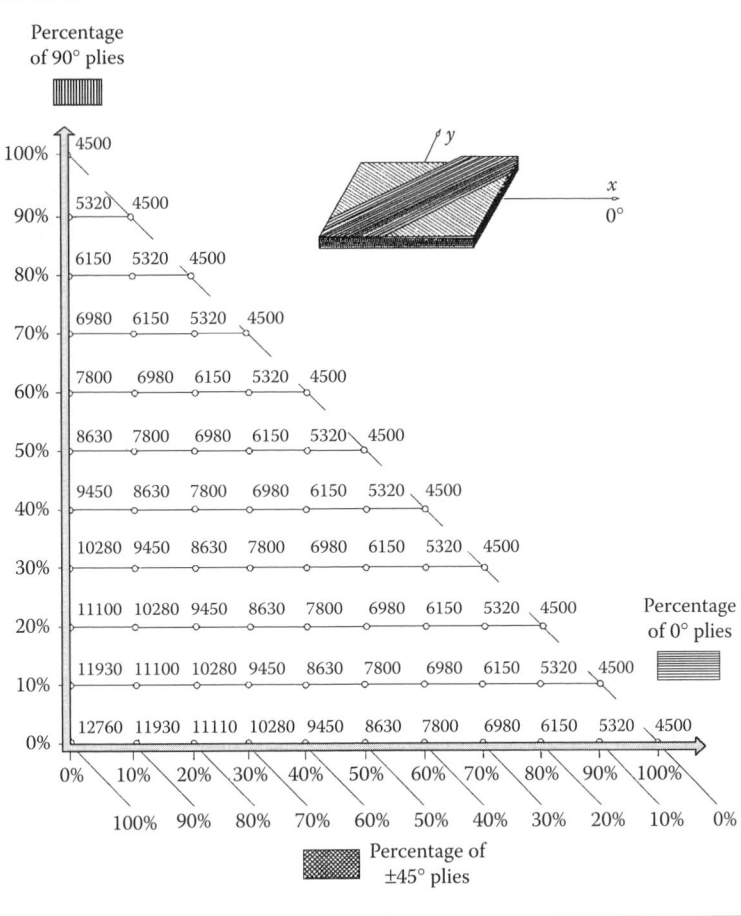

Shear modulus G_{xy} as a function of percentages of plies in directions 0°, 90°, +45°, −45°.
(More information on modulus and strength of a basic ply, see Section 3.3.3)

laminates that are capable of supporting the specified stress resultants and at the same time keeping a minimum thickness. One can see this number, in millimeters, within the circles, when the arithmetic sum of the stress resultants is equal to 100 N/mm.

Also represented in the tables are

▪ The direction along which the first ply failure will occur.
▪ The multiplication factor for the stress resultants in order to go from first ply failure to ultimate fracture of the laminate.
▪ The two compositions that are closest to the optimum composition, obtained by varying the indicated composition along the direction of the arrows.

First, the arrows in increasing solid line or decreasing solid line denote the increase or decrease of 5% in terms of the proportions marked. Next, the arrows in increasing broken line or decreasing broken line denote the increase or decrease of 5% in terms of the proportions marked.

Example: Given the stress resultants:

$$N_x = 720 \text{ N/mm}; \quad N_y = 0; \quad T_{xy} = 80 \text{ N/mm}$$

one can then deduce values of the reduced stress resultants:

$$\bar{N}_x = 720/(720 + 80) = 0.9; \quad \bar{N}_y = 0; \quad \bar{T}_{xy} = 80/(720 + 80) = 0.1$$

One then uses Table 5.16 (all stress resultants are positive), where one can obtain for these values of reduced stress resultants the following figure:

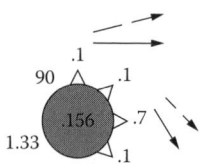

This can be interpreted as follows:

- Optimal composition of the laminate
  - 70% of 0° plies (along x direction)
  - 10% of 90° plies
  - 10% of plies in 45°, 10% of plies in −45°
- Critical thickness of the laminate: 0.156 mm when the arithmetic sum of the 3 stress resultants is equal to 100 N/mm. For this thickness, the first ply failure occurs in the 90° plies. However, one can continue to load this laminate until it reaches 1.33 times the critical load, as:

$$N_x = 1.33 \times 720 = 957 \text{ N/mm}; \quad N_y = 0$$

$$T_{xy} = 1.33 \times 80 = 106 \text{ N/mm}$$

At this point, there is complete rupture of the laminate.

Returning to our example, the arithmetic sum of the stress resultants is equal to 720 + 80 = 800 N/mm. Then, the thickness of the laminate has to be more than:

$$8 \times 0.156 = 1.25 \text{ mm}$$

Neighboring compositions: The second smallest thickness in the vicinity is obtained by modifying the indicated composition in the direction specified by the arrows in **solid** line, as

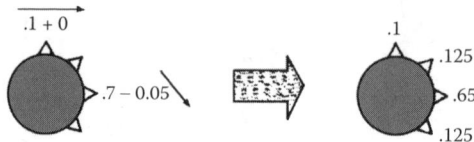

Table 5.16 Optimum Composition of a Carbon/Epoxy Laminate

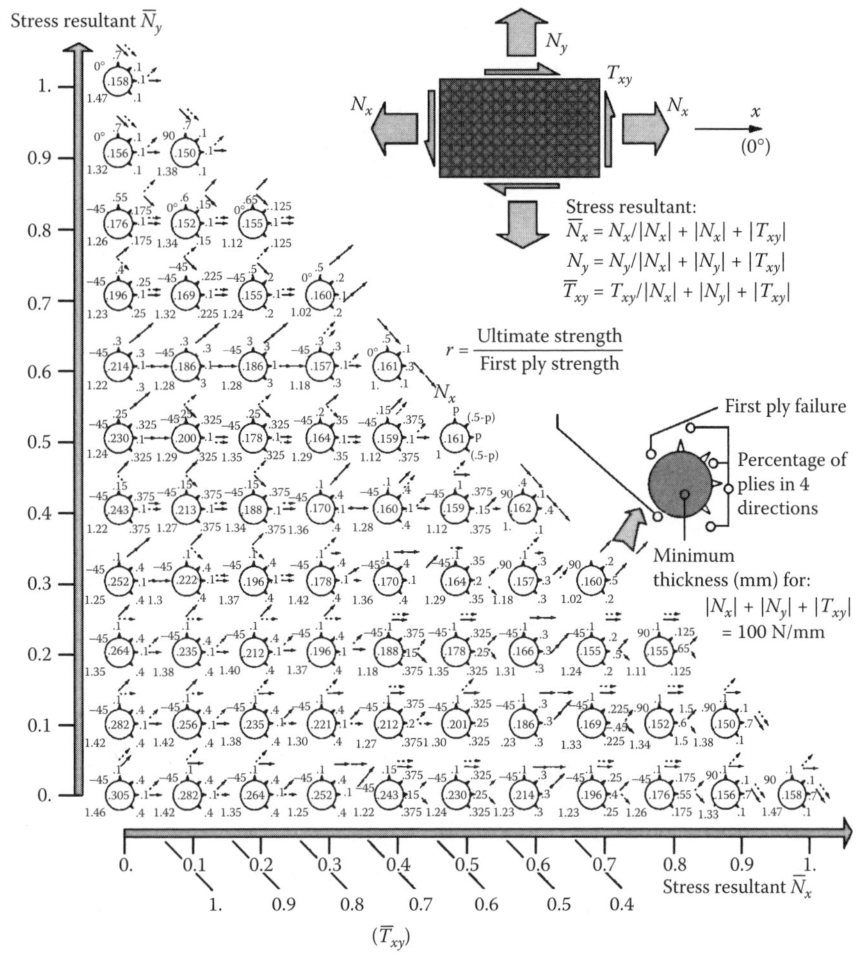

Stress resultant $\overline{N}_y$
N_y
T_{xy}
N_x
N_x
x
(0°)
Stress resultant:
$\overline{N}_x = N_x/|N_x| + |N_x| + |T_{xy}|$
$N_y = N_y/|N_x| + |N_y| + |T_{xy}|$
$\overline{T}_{xy} = T_{xy}/|N_x| + |N_y| + |T_{xy}|$
$r = \frac{\text{Ultimate strength}}{\text{First ply strength}}$
First ply failure
Percentage of plies in 4 directions
Minimum thickness (mm) for:
$|N_x| + |N_y| + |T_{xy}|$ = 100 N/mm
Stress resultant $\overline{N}_x$
($\overline{T}_{xy}$)

$V_f = 0.6$, 10% minimum in each direction of 0°, 90°, +45°, −45°. (Ply characteristics: see Appendix 1 or Section 3.3.3)

One then obtains (not shown on the plate) a thickness of 0.160 mm (increase of 2.5% relative to the previous value) and a multiplication factor for the loading equal to 1.35.

Continuing in the direction of increasing thickness, the third smallest thickness in the immediate vicinity is obtained by modifying the indicated composition in the direction specified by the arrows in **broken** line, as:

.1 + 0.05
.7 − 0.05
.15
.1
.65
.1

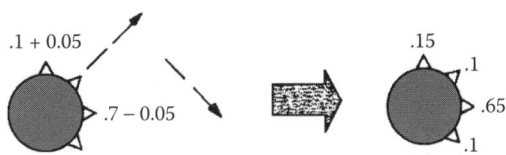

Table 5.17   Optimum Composition of a Carbon/Epoxy Laminate

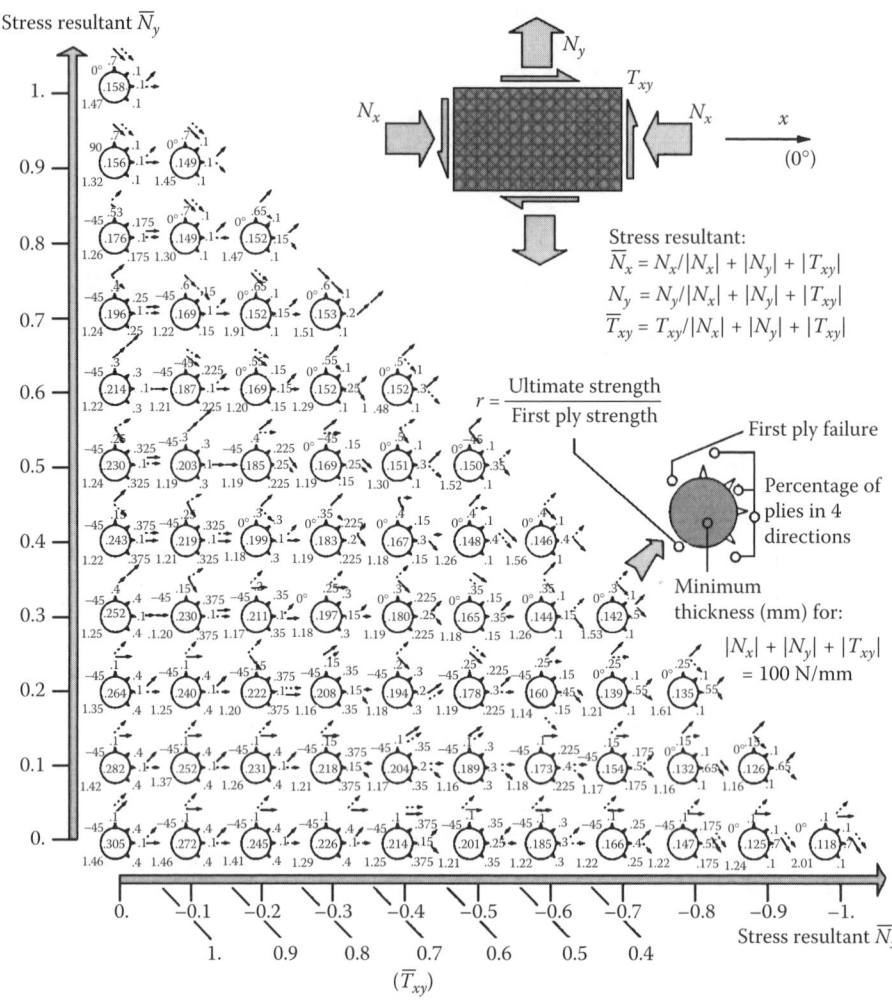

$V_f = 0.6$, 10% minimum in each direction of 0°, 90°, +45°, −45°. (Ply characteristics: see Appendix 1 or Section 3.3.3)

One then obtains a thickness (not shown on the plate) of 0.165 mm (increase of 6%) and a multiplication factor of 1.3 for the load.

Example: Given the stress resultants

$$N_x = 600 \text{ N/mm}; \ N_y = -300 \text{ N/mm}; \ T_{xy} = 100 \text{ N/mm}$$

The corresponding reduced stress resultants are

$$\overline{N}_x = .6; \ \overline{N}_y = -.3; \ \overline{T}_{xy} = 0.1 \text{ N/mm}$$

One obtains from Table 5.18:

Table 5.18 Optimum Composition of a Carbon/Epoxy Laminate

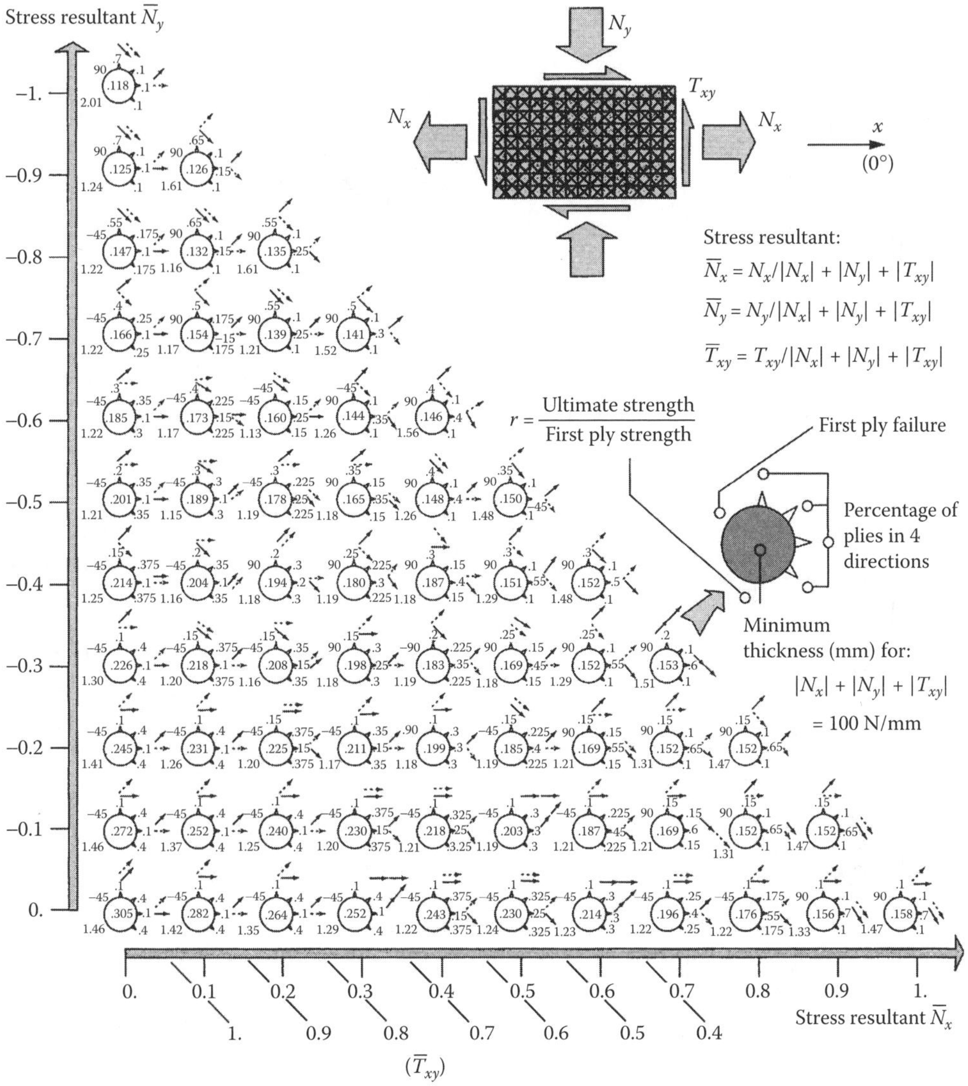

Stress resultant $\overline{N}_y$
N_y
T_{xy}
N_x
N_x
x
(0°)
Stress resultant:
$\overline{N}_x = N_x/|N_x| + |N_y| + |T_{xy}|$
$\overline{N}_y = N_y/|N_x| + |N_y| + |T_{xy}|$
$\overline{T}_{xy} = T_{xy}/|N_x| + |N_y| + |T_{xy}|$
$r = \frac{\text{Ultimate strength}}{\text{First ply strength}}$
First ply failure
Percentage of plies in 4 directions
Minimum thickness (mm) for:
$|N_x| + |N_y| + |T_{xy}|$
= 100 N/mm
Stress resultant $\overline{N}_x$
$(\overline{T}_{xy})$

V_f = 0.6, 10% minimum in each direction of 0°, 90°, +45°, −45°. (Ply characteristics: see Appendix 1 or Section 3.3.3).

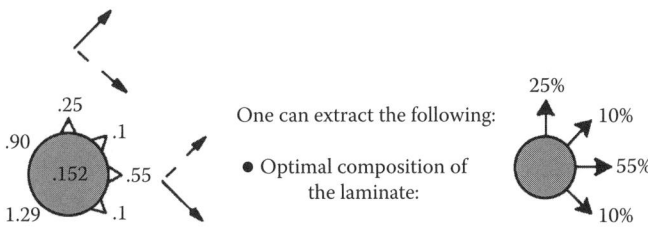

.25
.90
.152
.1
.55
1.29
.1
One can extract the following:
• Optimal composition of the laminate:
25%
10%
55%
10%

Table 5.19 Optimum Composition of a Carbon/Epoxy Laminate

Stress resultant:

$$\overline{N}_x = N_x/|N_x| + |N_y| + |T_{xy}|$$

$$\overline{N}_y = N_y/|N_x| + |N_y| + |T_{xy}|$$

$$\overline{T}_{xy} = T_{xy}/|N_y| + |N_y| + |T_{xy}|$$

$$r = \frac{\text{Ultimate strength}}{\text{First ply strength}}$$

$V_f = 0.6$, 10% minimum in each direction of 0°, 90°, +45°, −45°. (Ply characteristics: see Appendix 1 or Section 3.3.3)

where

- Critical thickness is 10 × 0.152 = 1.52 mm
- These are the 90° plies that fail first.
- Complete rupture of the laminate occurs when:

$$N_x = 1.29 \times 600 = 774 \text{ N/mm}$$

$$N_y = 1.29 \times -300 = -387 \text{ N/mm}$$

$$T_{xy} = 1.29 \times 100 = 129 \text{ N/mm}$$

- The closest critical thicknesses (in increasing order) are obtained with the following successive compositions:

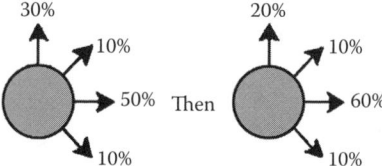

Remarks: A few loading cases can lead to several distinct optimum compositions, but with identical thicknesses. For example the reduced stress resultants:

$$\bar{N}_x = \bar{N}_y = 0.5; \ \bar{T}_{xy} = 0$$

This is a case of isotropic loading, the Mohr circle is reduced to one point (see figure below).

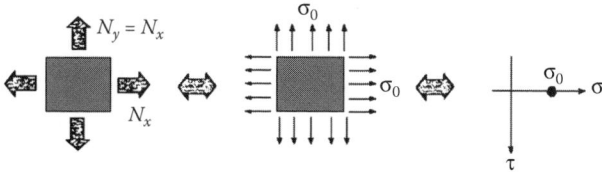

Table 5.16 indicates

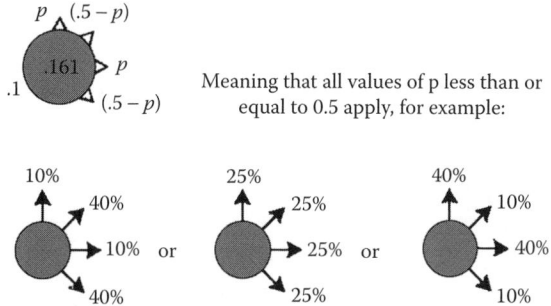

Meaning that all values of p less than or equal to 0.5 apply, for example:

One obtains in this case a unique critical thickness of 0.161 mm (corresponding to a sum $N_x + N_y = 100$ N/mm) **independent** of the proportion p.[15] The isotropic composition (25%, 25%, 25%, 25%) in the directions 0°, 90°, +45°, −45°, which appears intuitive, can in fact be replaced by compositions that present the values of modulus of elasticity that are varied and adaptable to the designer in the directions 0°, 90° 45°, or −45°,[16] or even α, $\alpha + \pi/2$ with a certain α.

---

[15] See Section 18.2.8.

[16] See Section 5.4.2, Table 5.4.

In some loading cases, one finds from the table only arrows in a solid line. For example, for the following reduced stress resultants:

$$\bar{N}_x = 0.3; \qquad \bar{N}_y = 0; \qquad \bar{T}_{xy} = 0.7$$

one finds from Table 5.16 the following figure:

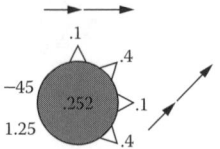

The three neighboring optimum compositions in increasing order are

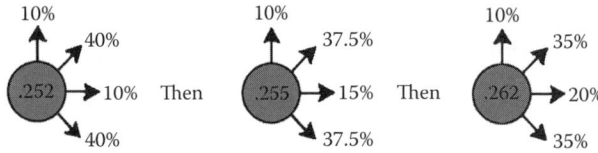

(The thicknesses of 0.255 mm and 0.262 mm are not indicated on the plate.) The third composition, characterized by an increase in thickness of 0.252 to 0.262 mm, or 6%, leads to an increase in modulus of elasticity in the x (0°) by 36% (see Section 5.4.2, Table 5.4).

One can note finally that for the majority of cases, the optimum compositions indicated in Tables 5.16 to 5.19 are not easy to postulate using intuition.[17]

5.4.5 Practical Remarks: Particularities of the Behavior of Laminates

- The **fabrics** are able to cover the double-curved surfaces[18] due to the possibility of pushing action in warp and fill directions.
- The radii of the mold must not be too small. This concerns particularly the inner radius R_i as shown in Figure 5.24. The graph gives an idea for the minimum value required for the inner and outer radii.
- The **thickness** of a polymerized ply is not more than **0.8** to **0.85** times that of a ply before polymerization. This value of the final thickness must also take into account a margin of uncertainty on the order of 15%.
- When one unidirectional sheet **does not cover** the whole surface required to constitute a ply, it is necessary to take precautions when cutting the different pieces of the sheet. A few examples of wrapping are given in Figure 5.25.
- The unidirectional sheets do not fit well into sharp corners in the fiber direction. The schematic in Figure 5.26 shows the dispositions to accommodate sudden changes in draping directions.

---

[17] See Exercise 18.1.6.

[18] It is more difficult for the plain weave than for the satins, due to the mode of weaving (see Section 3.4.1).

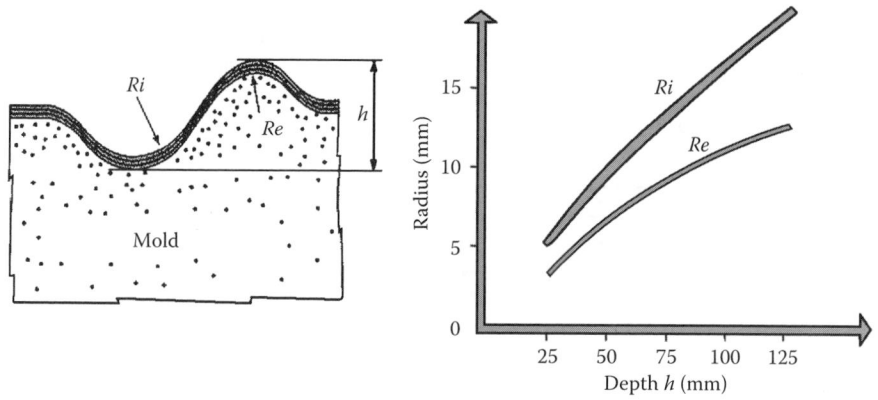

Figure 5.24   Minimum Required for Inner and Outer Radii of Mold

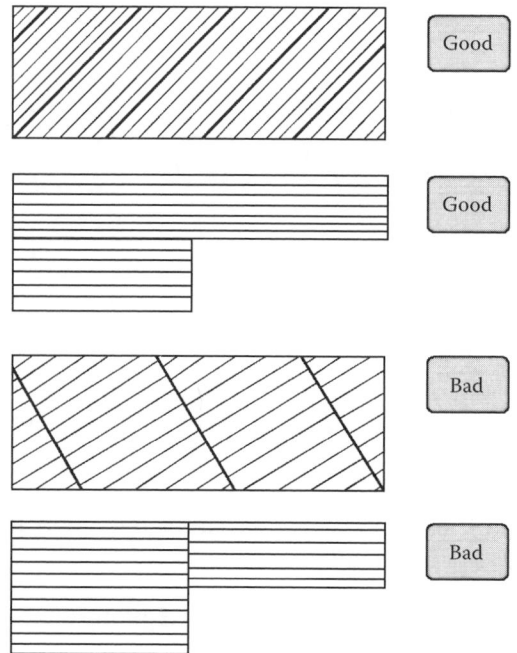

Figure 5.25   Recommended Arrangement for Cutting

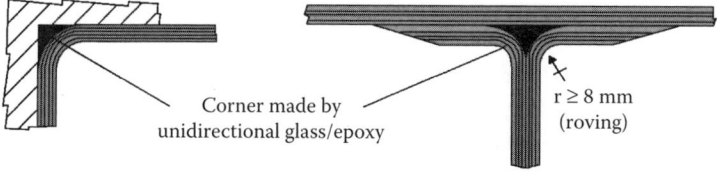

Corner made by
unidirectional glass/epoxy

r ≥ 8 mm
(roving)

Figure 5.26   Method to Lay Up at Corner

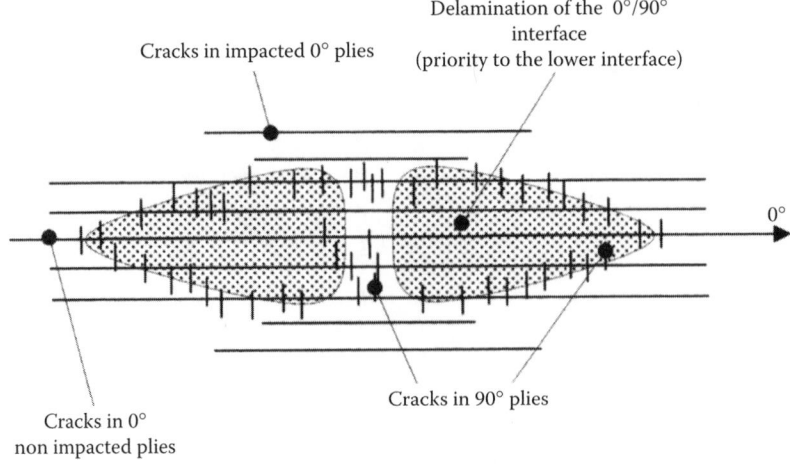

Figure 5.27a   Impact on a $0°_n/90°_n/0°_n$ laminate

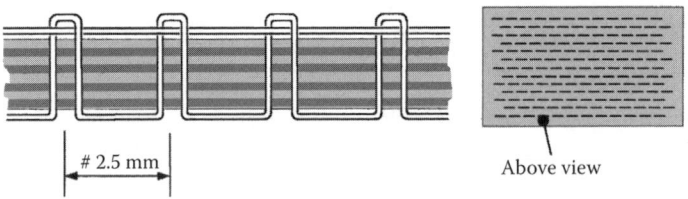

Figure 5.27b   Stich of a laminate

Delaminations: When the plies making up the laminate separate from each other, it is called **delamination**. Many causes are susceptible to provoke this deterioration:

- An **impact** that does not leave apparent traces on the surface and can lead to internal delaminations.

Example: Impact of a projectile on a layered plate $0°_n/90°_n/0°_n$
Order of the shock magnitude: mass: several kilograms; speed: several m/sec

The damage in the impacted area is shown in Figure 5.27a. An improvement: the stitch of the laminate. In view of reducing the imapact damage, one prevents the delamination by making a stitch (Figure 5.27b):

> On prepregs
> On dry preforms before injection molding
> - A mode of loading that leads to the disbond of the plies (tension over the interface) as shown in Figure 5.27(a).
> - **Shear** stresses at the interfaces between different plies, very near the edges of the laminates, which one can make evident as follows (with a three-ply laminate):

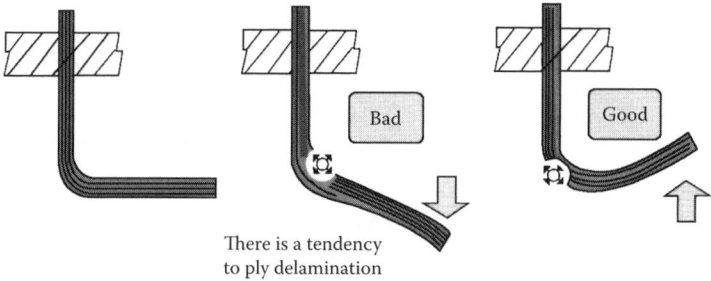

There is a tendency
to ply delamination

Figure 5.27c   Corner Situations

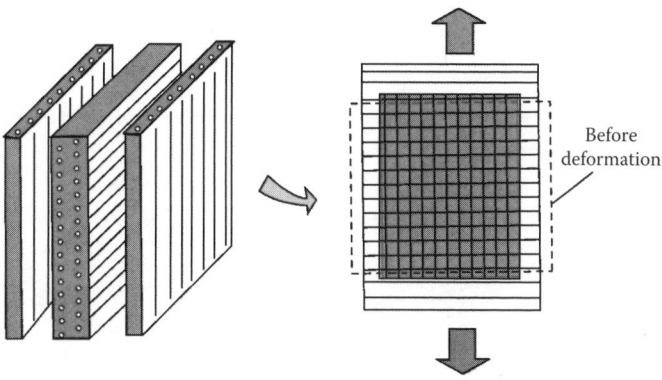

Before
deformation

Figure 5.28a   Three Plies in Separate Positions

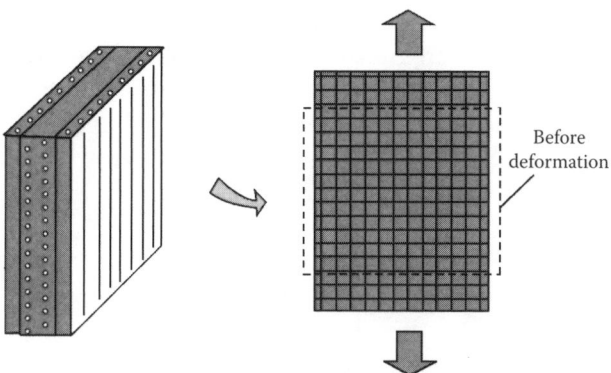

Before
deformation

Figure 5.28b   Three Plies, Together

1. Consider the three plies in Figure 5.28(a), separated. Under the effect of loading (right-hand-side figure), they are deformed independently and do not fit with each other when they are put together.
2. Now the plies constitute a balanced laminate. Under the same loading, they deform together, without distorsion, as shown in Figure 5.28(b).

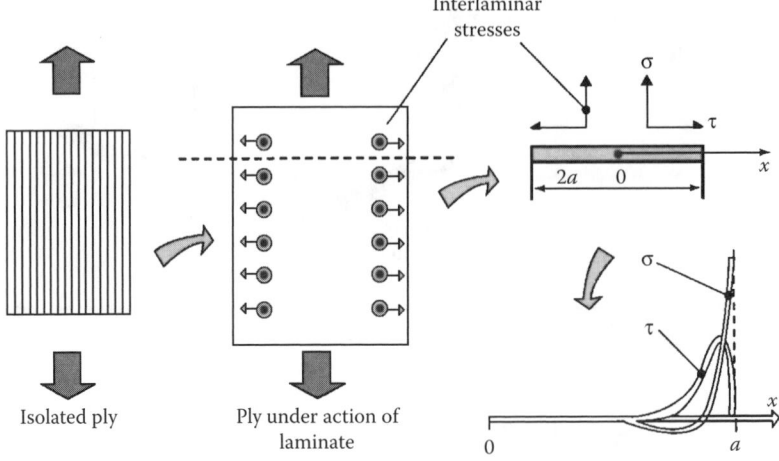

Figure 5.28c   Stresses at Free Edge

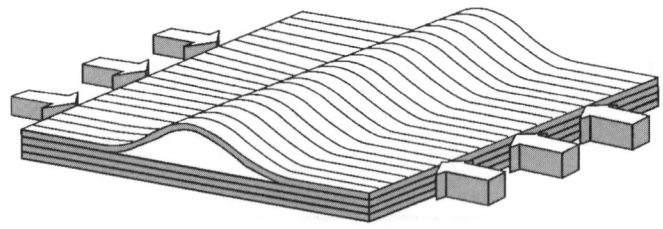

Figure 5.29   Delamination Due to Buckling at Interface

    3. This is because interlaminar stresses occur on the bonded faces. One can show that these stresses are located very close to the edges of the laminate, as illustrated in Figure 5.28(c).

■ A complex state of stresses at the interface, due to local buckling, for example (see Figure 5.29).

Practical as well as theoretical studies of these interlaminar stresses are difficult, and the phenomenon is still not well understood.

Why Is Fatigue Resistance so Good?

Paradox: Glass is a very brittle material (no plastic deformation). The resin is also often a brittle material (for example, epoxy). However, the association of reinforcement/matrix constituted of these two materials opposes to the propagation of cracks and makes the resultant composite to endure fatigue remarkably.

    Explanation: When the cracks propagate, for example, in the unidirectional layer shown schematically in Figure 5.30 in the form of alternating of fibers and resin, the initial stress concentration at the end of the crack leads to failure in the resin. The fibers, then debonded, benefit from a relaxation of the stresses. There is no more stress concentration as in a homogeneous material.

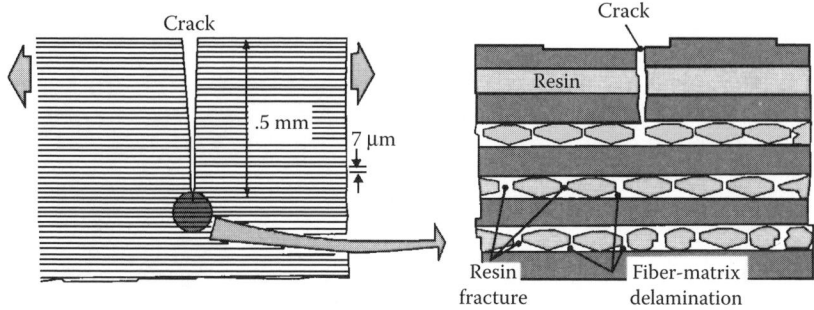

Figure 5.30   Crack Propagation in Composites

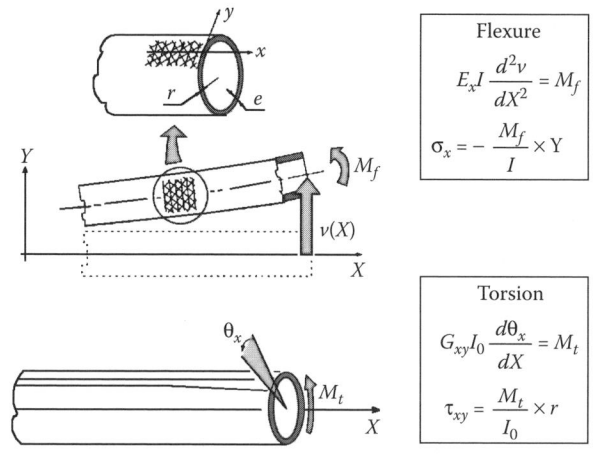

Flexure

$$E_x I \frac{d^2v}{dX^2} = M_f$$

$$\sigma_x = -\frac{M_f}{I} \times Y$$

Torsion

$$G_{xy} I_0 \frac{d\theta_x}{dX} = M_t$$

$$\tau_{xy} = \frac{M_t}{I_0} \times r$$

Figure 5.31   Composite Tube Relations

Laminated tubes can be obtained by winding of filaments, rubans of unidirectionals or fabrics. In a first approximation, one can[19] estimate the strains and stresses in flexure and in torsion from the relations in Figure 5.31 in which

> E_x and G_{xy} are the moduli of elasticity in the tangent plane x,y.
> I and I_0 are respectively the quadratic moment of inertia and polar moment of inertia ($I_0 = 2I$).
> Y is the coordinate of a point in the section (in the undeformed position) in the X, Y, Z coordinates.
> r is the average radius of the tube.

---

[19] For a complete study of the flexure and torsion of the composite beams, See Chapters 15 and 16.

6

JOINING AND ASSEMBLY

We have seen previously how to design a laminate to support loads. A second fundamental aspect of the design of a composite piece consists of the design for the attachment of the composite to the rest of the structure. Here we will examine the assembly problems involving riveting, bolting, and bonding:

- of a composite part to another composite part and
- of a composite part to a metallic part.

6.1 RIVETING AND BOLTING

- In all mechanical components, the introduction of holes gives stress concentration factors. Specifically in composite pieces, the introduction of holes (for molded-in holes or holes made by drilling) induces weakening of the fracture resistance in comparison with the region without holes by a factor of
    40 to 60% in tension
    15% in compression

Example: Figure 6.1 presents the process of degradation before rupture of a glass/epoxy laminate containing a free hole, under uniaxial stress.

Causes of hole degradation:

- **Stress concentration factors:** The equilibrium diagrams shown in Figure 6.2 demonstrate the increase in stress concentration in the case of a laminate. For the case of slight (and usually neglected) press-fit of the rivet, the stresses shown in these figures are:

$$\sigma'_M > \sigma$$

in a region where:

$$\sigma_{\text{local rupture}} < \sigma_{\text{laminate rupture}}$$

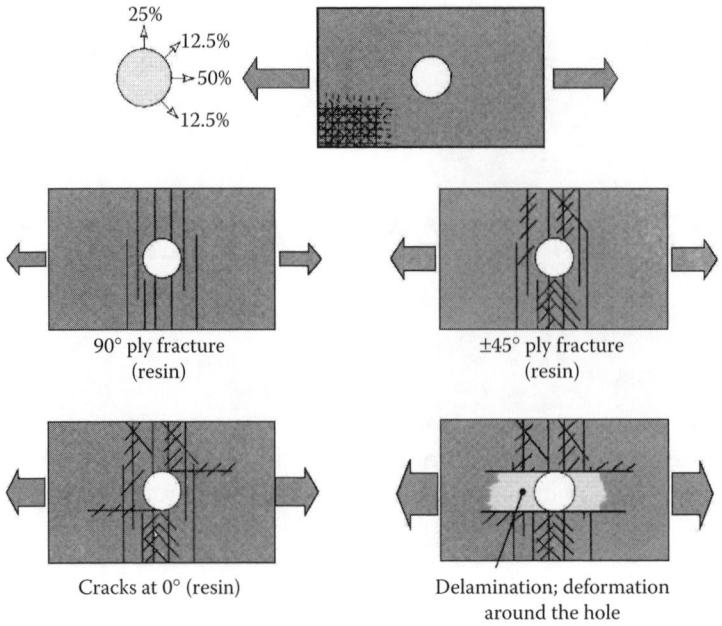

Figure 6.1 Cracks in a Laminate with Hole when Load Increases

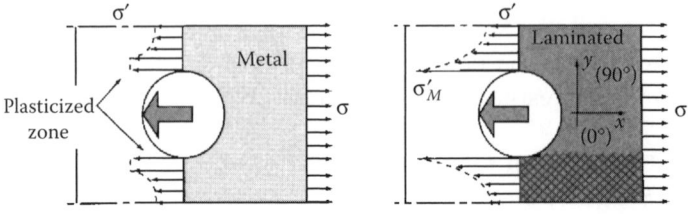

Figure 6.2 Stress Concentration Factors

with the maximum stress σ'_M in the laminate given as:

$$\sigma'_M = \sigma' \times \left\{ 1 + \sqrt{2\left(\sqrt{\frac{E_x}{E_y}} - v_{xy}\right) + \frac{E_x}{G_{xy}}} \right\}$$

where

E_x and E_y are the moduli of elasticity in the 0° and 90° directions
G_{xy} is the shear modulus
v_{xy} is the Poisson ratio

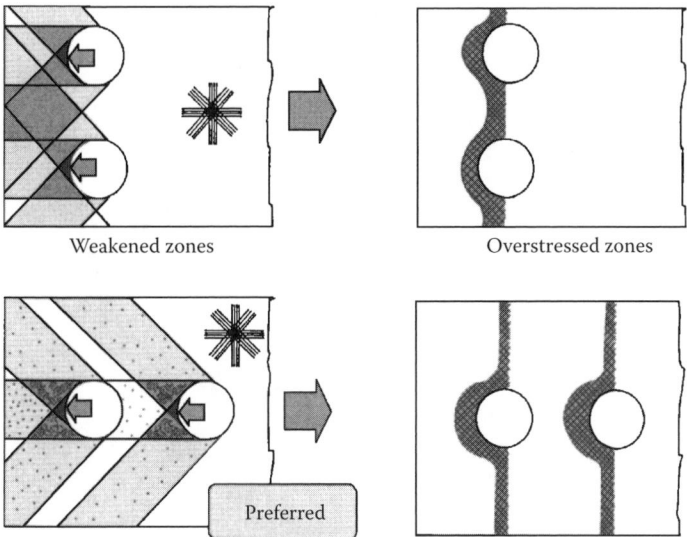

Figure 6.3 Weakened Zones Due to Presence of Holes

■ **Bearing due to lateral pressure:** This is the contact pressure between the shaft of the assembly device (rivet or bolt) and the wall of the hole. When this pressure is excessive, it leads to mushrooming and **delamination** of the laminate. In consequence:

The resistance of a hole occupied by the rivet or bolt is weaker than that of an empty hole: decrease on order of 40%.

■ **Fracture of fibers** during the hole cutting process, or the misalignment of fibers if the hole is made before polymerization: Figure 6.3 illustrates the correlation between the weakened zones consecutive to rupture of fibers and the "overstressed" zones.

6.1.1 Principal Modes of Failure in Bolted Joints for Composite Materials

These are represented in Figure 6.4.

6.1.2 Recommended Values

■ **Pitch, edge distance, thickness** (see Figure 6.5)
■ **Orientation of plies:** Recommendation for percentages of plies near the holes (see Figure 6.6).

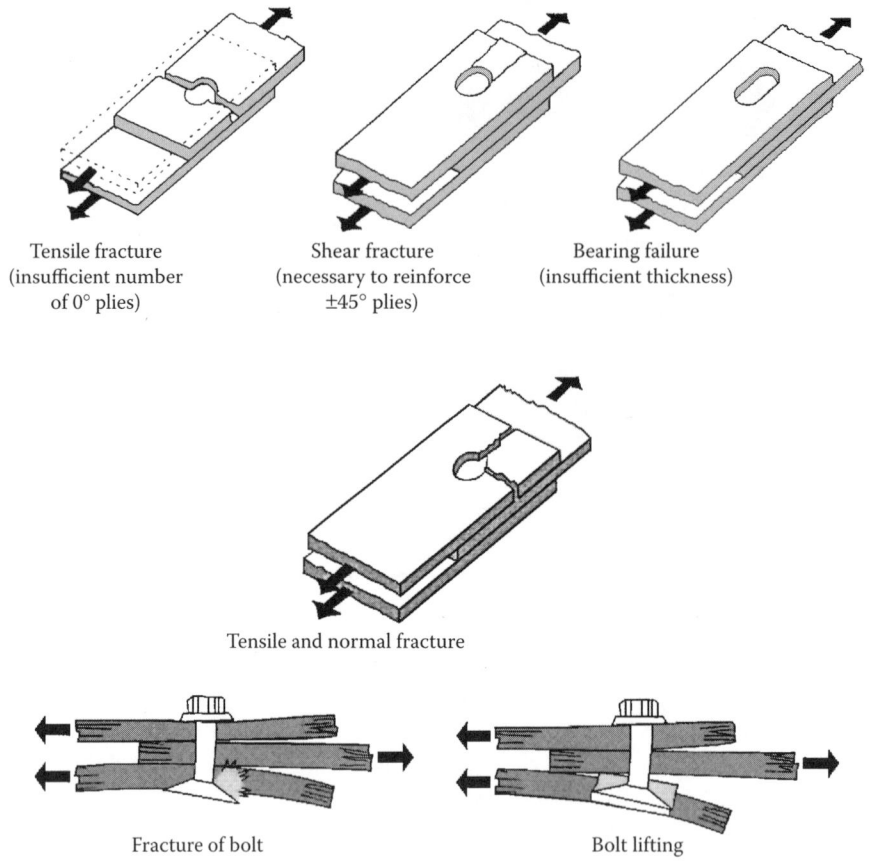

Tensile fracture
(insufficient number
of 0° plies)

Shear fracture
(necessary to reinforce
±45° plies)

Bearing failure
(insufficient thickness)

Tensile and normal fracture

Fracture of bolt

Bolt lifting

Figure 6.4   Different Types of Bolt Joint Failures

■ **Condition of nonbearing pressure:** In Figure 6.7, F and T designate the normal and shear loads respectively, that are applied on the assembly over a width equal to one pitch distance.

The equivalent bearing pressure which leads to the crushing of the wall of the hole of diameter Ø, is $F/(\text{Ø} \times e)$. It must remain smaller than an admissible maximum, as:

$$\frac{F}{\phi e} \leq \sigma_{admissible\ bearing\ pressure} \quad \text{carbon:} \quad \sigma_{admissible\ bearing} = 500 \text{ MPa}$$

$$\text{glass:} \quad \sigma_{admissible\ bearing} = 300 \text{ MPa}$$

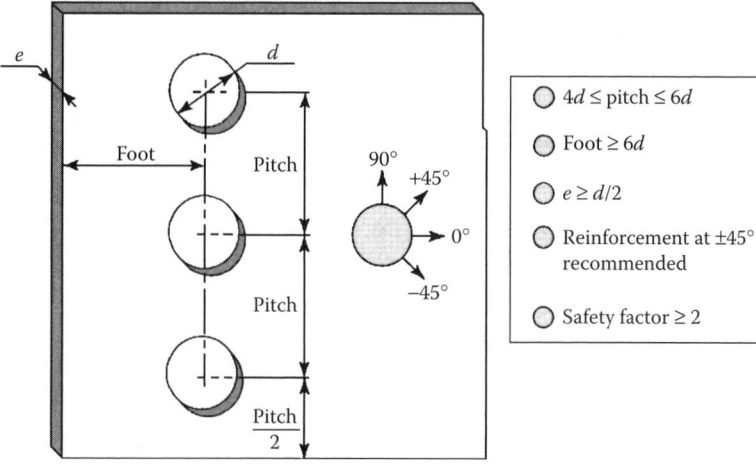

Figure 6.5   Recommended Pitch, Edge Distance, and Thickness

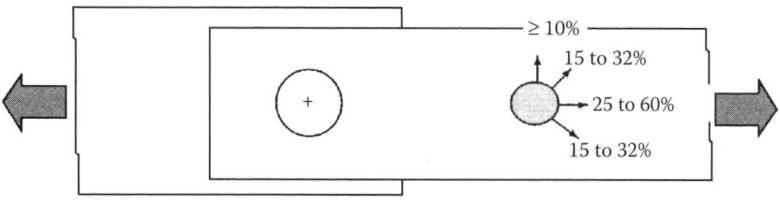

Figure 6.6   Recommended Orientation

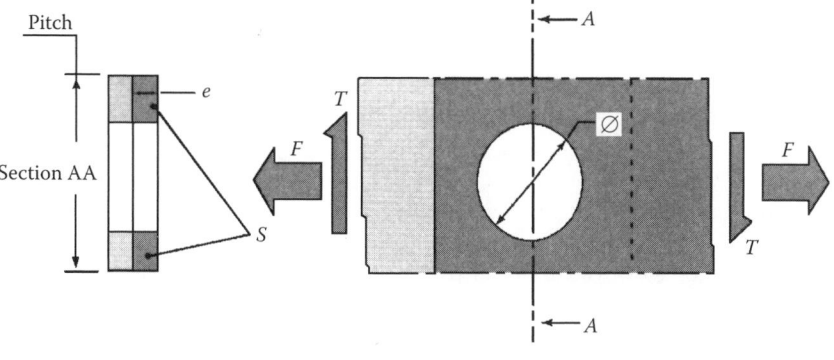

Figure 6.7   Normal and Shear Loads on Assembly

■ **Evaluation of the admissible stresses:** The principle of calculation consists of magnifying the stresses that are given by elementary considerations, by means of the empirical coefficients of magnification[1]:

---

[1] When one takes into account the aging of the piece, an additional 10% is applied to the maximum stresses.

- Due to the presence of the hole and
- Due to pressure of contact or bearing on the wall of the hole (rivet, bolt).

With the notations of Figure 6.7, one has:

$$\sigma_{\text{magnified}} = \frac{1}{\alpha}\left(\frac{F}{S} + 0.2\frac{F}{\phi e}\right)$$

tension: $\alpha = 0.6$

compression: $\alpha = 0.8$

$$\tau_{\text{magnified}} = \frac{1}{0.7}\frac{T}{S}$$

One must also verify that these stresses are admissible (that is, they do not lead to the fracture of the ply) by using the method of verification of fracture described in Paragraph 5.3.2.

6.1.3 Riveting

The relative specifics and recommendations for riveting the composite parts can be presented as follows:

- **Do not hit the rivets** as this can lead to poor resistance to impact of the laminates.
- **Pay attention to the risk of "bolt lifting" of the bolt heads** due to small thickness of the laminates.
- **Note the necessity to assure the galvanic compatibility** between the rivet and the laminates to be assembled.
- **Riveting accompanied by bonding** of the surfaces to be assembled provides a gain in the mechanical resistance on the order of 20 to 30%. On the other hand, the disassembly of the joint becomes impossible, and the weight is increased.

Characteristics of rivets for composites are shown in Figure 6.8.

6.1.4 Bolting

Examine a current example that requires a bolted joint.

> **Example:** Junction of a panel by bolted joint (simple case)[2]: Consider a sandwich panel fixed to a support component that is subjected to simple loadings that can be represented by a shear load and a bending moment (see Figure 6.9).

One expects an attachment using bolt. As shown in the schematics of Figure 6.10, even if the bolt is not tightened, it is able to act to equilibrate the bending moment. However, action of the shear load will separate the facings.

---

[2] A more complete case on the fixation of the panel is examined in the application in Paragraph 18.1.6.

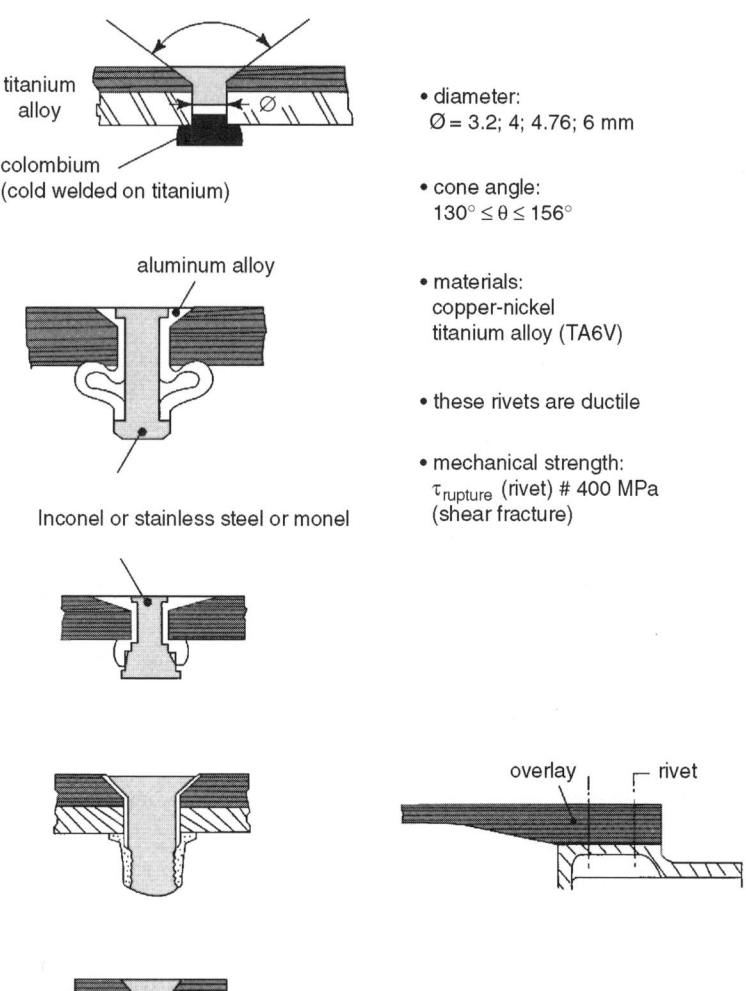

- diameter:
  Ø = 3.2; 4; 4.76; 6 mm

- cone angle:
  $130° \leq \theta \leq 156°$

- materials:
  copper-nickel
  titanium alloy (TA6V)

- these rivets are ductile

- mechanical strength:
  $\tau_{rupture}$ (rivet) # 400 MPa
  (shear fracture)

Figure 6.8   Different Types of Riveting

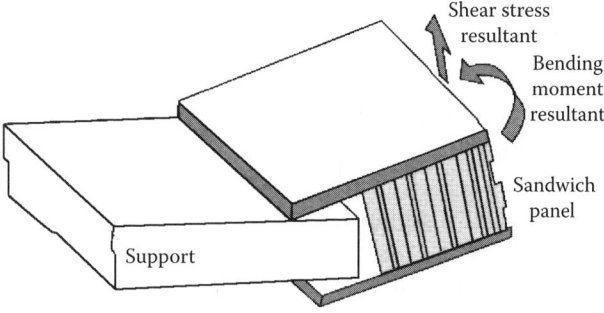

Figure 6.9   Junction of a Panel Using Bolted Joint

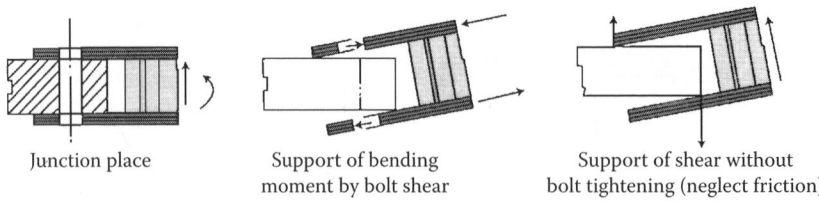

Junction place　　　　Support of bending　　　　Support of shear without
　　　　　　　　　moment by bolt shear　　　bolt tightening (neglect friction)

Figure 6.10　Local Behavior without Bolt Tightening

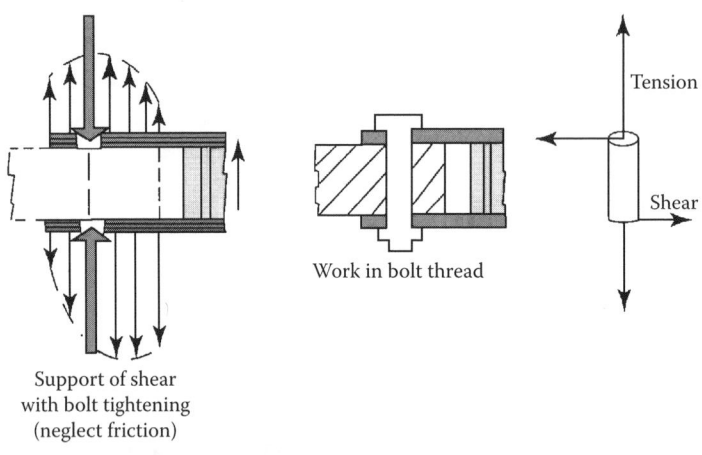

Tension

Shear

Work in bolt thread

Support of shear
with bolt tightening
(neglect friction)

Figure 6.11　Bolt Tightening Reduces the Possibility of Damage

It is the tightening of the bolt that will lead to a distribution of contact pressure between the support component and the facings. The sum of the forces due to this contact pressure will balance out the shear load, while suppressing the risk of separating the facings (see Figure 6.11).

The tightening of the bolt is therefore indispensable. However, the laminated facings being fragile **cannot admit** high contact pressures that are localized under the bolt head and under the nut. This leads to the insertion of metallic washers as shown in Figure 6.12.

The bolting accompanied by bonding of the surfaces provides a gain in mechanical resistance on the order of 20 to 30%. On the other hand, the joint cannot be disassembled, and there is an increase in weight.

6.2　BONDING

Remember briefly that this assembly technique consists of the adhesion by molecular attraction between two parties to be bonded and an adhesive that must be able to transfer the loads. One can cite the **principal advantages** of this mode of joining:

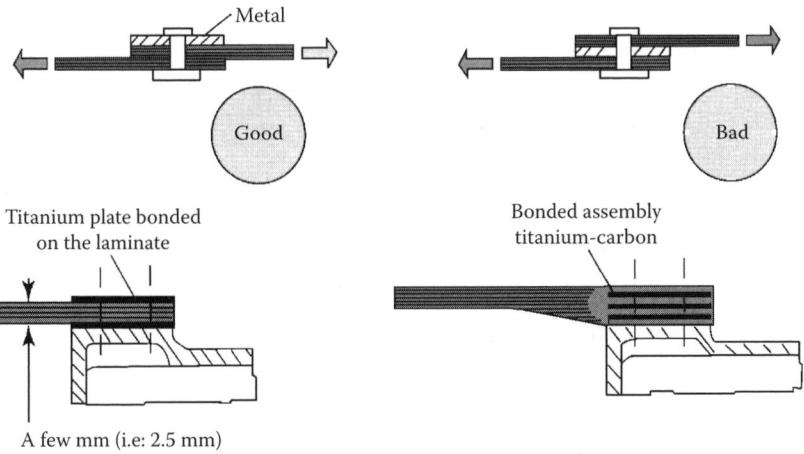

Figure 6.12   Configuration for Bolted Joints

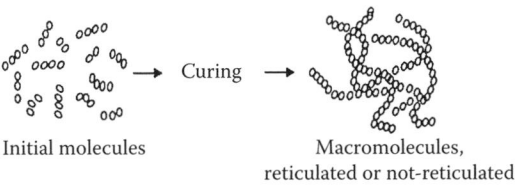

Figure 6.13   Curing of Adhesive

- distribution of stresses over an important surface
- possibility to optimize the geometry and dimensions of bonding
- light weight of the assembly
- insulation and sealing properties of adhesive

6.2.1 Adhesives Used

The adhesives used include:

- epoxies
- polyesters
- polyurethanes
- methacrylates

In all cases, the mechanism of curing is shown schematically in Figure 6.13.

- The adhesives are resistant simultaneously to
  - high temperatures (>180°C)
  - humidity
  - a number of chemical agents

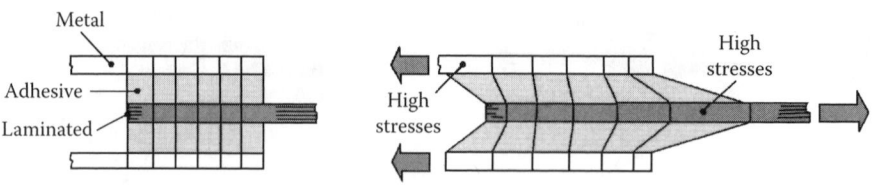

Figure 6.14   Stresses in Bolted Joint

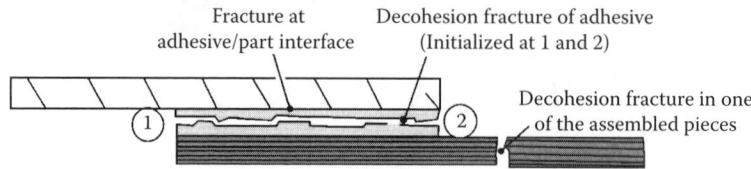

Figure 6.15   Fracture Modes in a Bonded Joint

- The pieces to be assembled have to be surface treated. This consists of three steps:
    - degreasing
    - surface cleaning
    - protection of cleaned surface
- The case of metal–laminate bond:

The differences in physical properties of the constituents requires that the adhesive must compensate for the differences in

- thermal dilatations
- elongation under stress

The schematic in Figure 6.14 indicates in an exaggerated manner the deformed configuration of a double bonded joint. This shows the role of the adhesive and the gradual transmission of the load from the central piece to the external support.[3]

Fracture of a bonded assembly can take different forms, as indicated in Figure 6.15.

6.2.2 Geometry of the Bonded Joints

One must, as much as possible, envisage the joint geometries that allow the following specifications:

- the adhesive joint must work in shear in its plane
- tensile stresses in the joint must be avoided

Consequently, the transmission of the loads will be dependent on the geometries, as shown in Figure 6.16. A double sided joint with increasing thickness is shown in Figure 6.17.

---

[3] See Application 18.3.2.

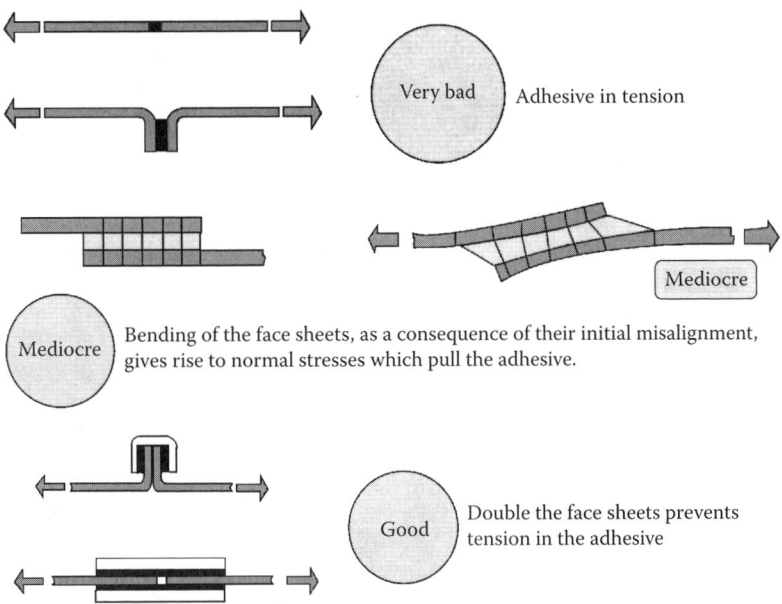

Figure 6.16   Different Designs for Bonded Joints

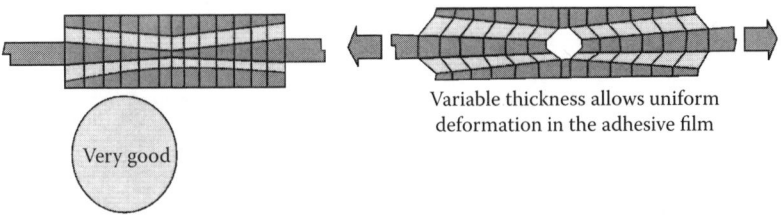

Figure 6.17   Double Sided Lap Joint

■ Transmission of couples is shown in Figure 6.18.

6.2.3   Sizing of Bonded Surfaces

The resistance of the adhesive is characterized by its shear strength τ_{rupture}. This resistance varies with the process of bonding (cold bonding or hot bonding). For epoxy adhesive, one can cite the following values:

■ **Cold bonding:** (Araldite):

Adhesive thickness = 0.2 mm

$$\tau_{\text{rupture}} = 10 \text{ MPa at } 20°C$$

$$= 3 \text{ MPa at } 80°C$$

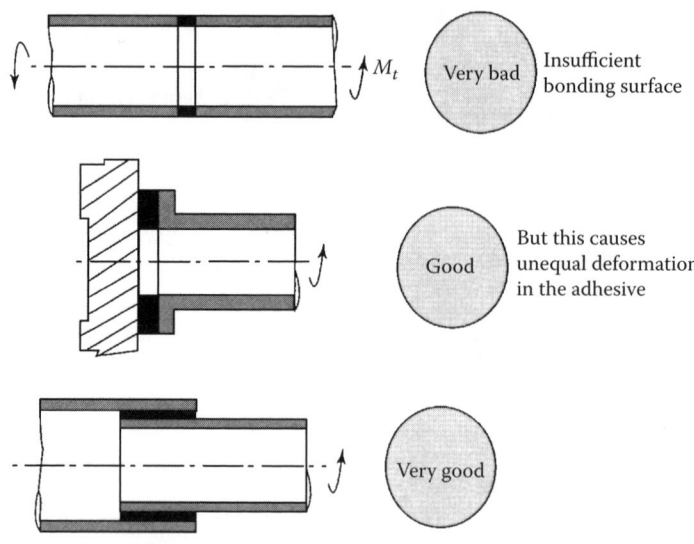

Figure 6.18   Design for the Transmission of Couples (see Application 18.3.1)

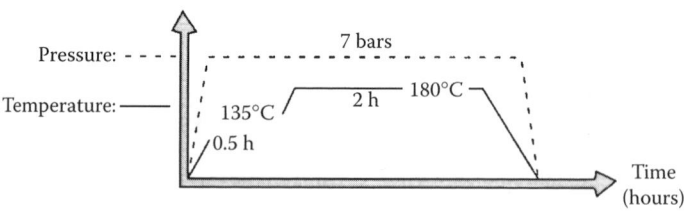

Figure 6.19   Curing Cycle of Epoxy Adhesive

■ **Hot bonding:** Polymerization temperature is between 120°C and 180°C:

$$\tau_{\text{rupture}} = 15 \text{ to } 30 \text{ MPa from } 20°C \text{ to } 100°C$$

The diagram in Figure 6.19 shows, for example, the cycle of polymerization of an epoxy adhesive "REDUX 914."

■ Denoting e_c for the **thickness of the adhesive joint**, one has for an order of magnitude:

$$0.1 \text{ mm} \leq e_c \leq 0.3 \text{ mm}$$

when the joints are very thick, one adds adhesive glass powder or cut fibers.

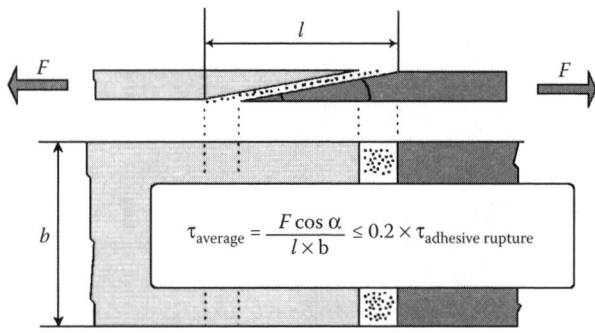

Figure 6.20   Scarf Joint

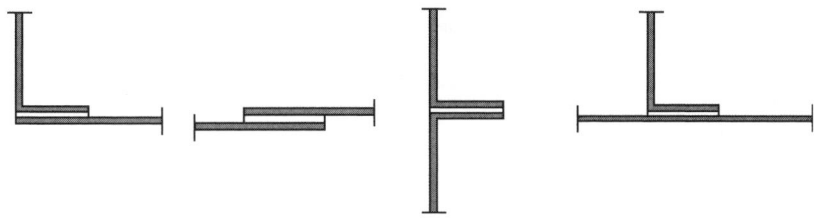

Figure 6.21   Configurations of Parallel Joint

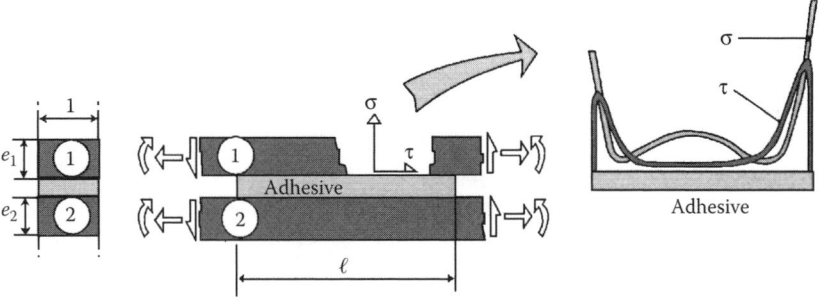

Figure 6.22   Stresses in Adhesive

- **Scarf joint:** This joint (see Figure 6.20) allows one to obtain a sufficient bonding surface, with weak tensile stress.
- **Parallel joint:** As illustrated in Section 6.2.2, there is bending in the bonded parts. The geometric configurations are varied (see Figure 6.21).

When one isolates the bonded zone, the stress variation is shown in the figure on the right-hand side of Figure 6.22 (the bond width is assumed to be equal to unity)

The stresses in the adhesive (Figure 6.22) consists essentially of

- a shear stress τ and
- a normal stress called "**peel stress**" σ.

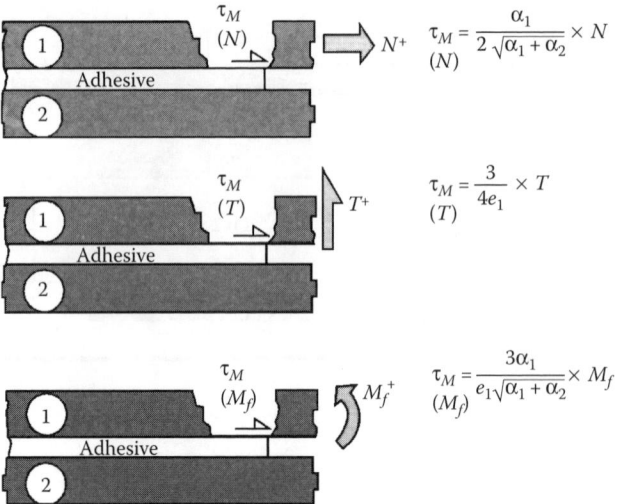

Figure 6.23 Maximum Shear Stress

These stresses present maximum values σ_M and τ_M very close to the edges of the adhesive. These maxima can be approached by superposition of the partial maxima created by each of the resultants N, T, M_f, by means of the following expressions in which E_c is the modulus of the adhesive, and E_1 and E_2 are the moduli along the horizontal direction of the bonded parts 1 and 2. One can also write:

$$\alpha_1 = \frac{G_c}{E_1 e_1 e_c}; \quad \alpha_2 = \frac{G_c}{E_2 e_2 e_c}; \quad \beta_1 = \frac{12E_c}{E_1 e_1^3 e_c}; \quad \beta_2 = \frac{12E_c}{E_2 e_2^3 e_c}$$

- Maximum shear stresses are illustrated in Figure 6.23
- Maximum peel stress is shown in Figure 6.24.

Remarks:

- The resultants N, T, M_f are evaluated per unit width of the bond.
- When several resultants coexist, one obtains the total maximum shear stress by superposition of the partial maxima of shear stresses and the maximum peel stress by superposition of the partial maxima of peel stresses.
- When the lower piece is also subjected to the resultants, the previously obtained relations are usable, by means of permuting the indices 1 and 2, and by changing the sign of the second member
- Limits for the relations[4]

---

[4] For more details, see Bibliography: "Elastic Analysis and Engineering Design Formulae for Bonded Joints."

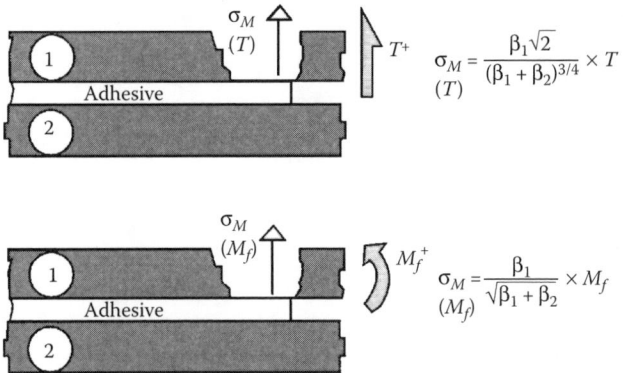

Figure 6.24   Maximum Peel Stress

$$0.6 \leq \frac{\alpha_1}{\alpha_2} \quad \text{and} \quad \frac{\beta_1}{\beta_2} \leq 2$$

$$(\alpha_1 + \alpha_2) \times \ell^2 \geq 9$$

$$(\beta_1 + \beta_2) \times \ell^4 \geq 4 \times 6^4$$

Example: For the simple lap joint below, one has (with the notations used previously)

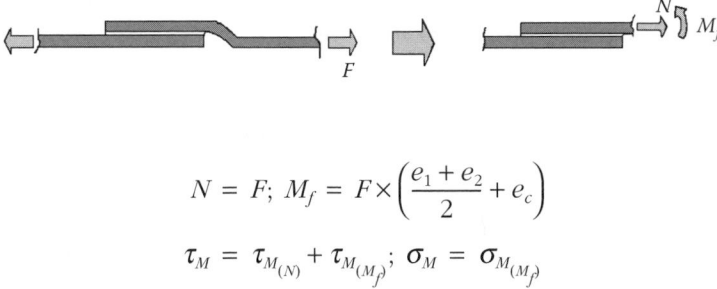

$$N = F; \; M_f = F \times \left(\frac{e_1 + e_2}{2} + e_c \right)$$

$$\tau_M = \tau_{M_{(N)}} + \tau_{M_{(M_f)}}; \; \sigma_M = \sigma_{M_{(M_f)}}$$

This is valuable if α_1, α_2, β_1, β_2 respect the limits of utilization written above.

- **Collar** (see Figure 6.25)
- **Cylindrical sleeve**[5] (see Figure 6.26)
- **In a laminate, orientation of the plies** that are in contact with the joint influences strongly the failure by fiber–resin decohesion. This can be easily understood through Figure 6.27. A tensile load in plies that are in contact with the adhesive requires that fiber orientation in these plies must be along the direction of the load.

---

[5] For different thicknesses and different materials to be assembled, see Exercise 18.3.1.

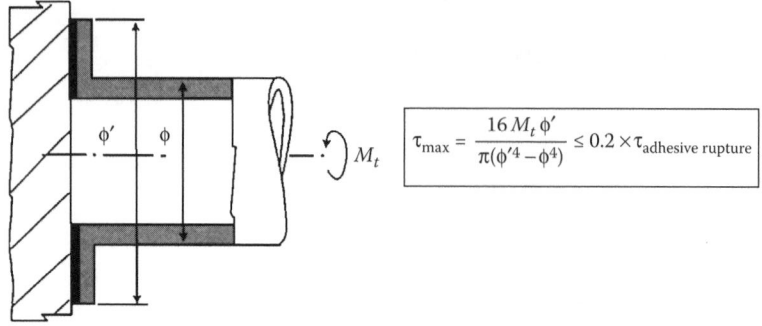

Figure 6.25 Shear Stresses in Simple Collar

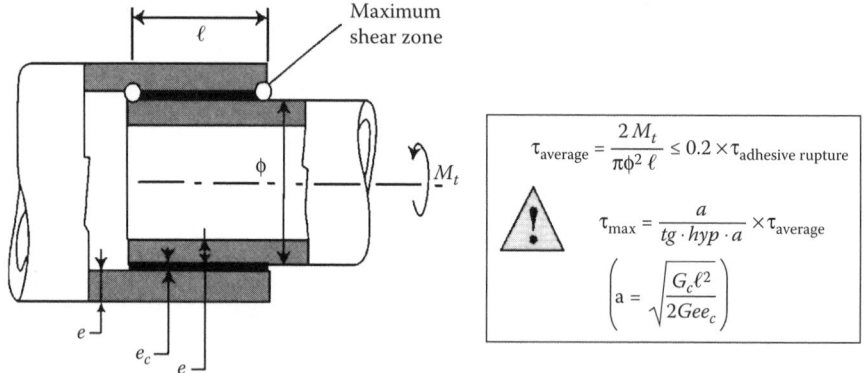

Figure 6.26 Shear Stresses in Cylindrical Sleeve

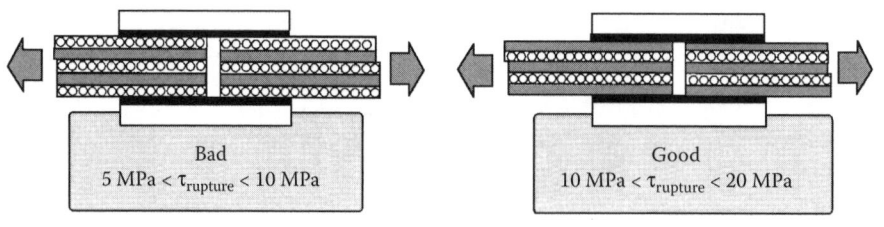

Figure 6.27 Ply Orientation in Bonded Laminates

6.2.4 Examples of Bonding

■ Laminates

One notes in Figure 6.28 the use of steps that gradually decrease the thickness of titanium piece. Note also that the design allows one to separate the stress concentration effects localized at the beginning of each step.

■ Sandwiches (see Figure 6.29)

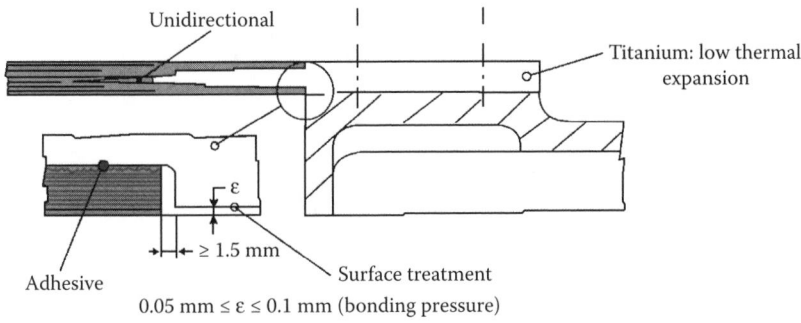

Figure 6.28 An Example of Laminate Bonding

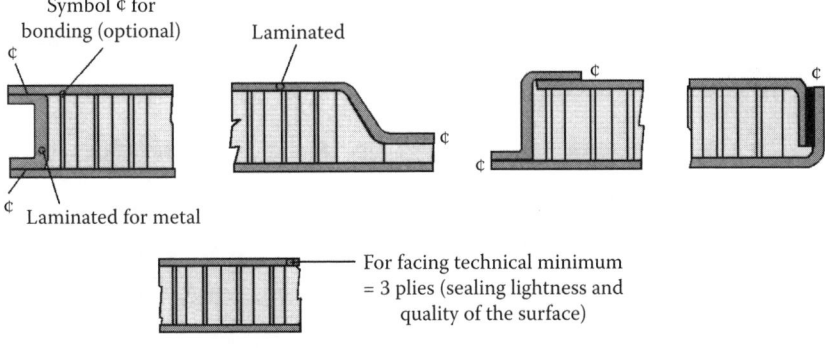

Figure 6.29 Bonding of Sandwich Facings

The bonding at the borders of sandwich panels must be done in a simple manner (especially for the preparation of the core) and with the best possible contact for the bonded parts, similar to the cases shown in Figure 6.30.

6.3 INSERTS

It seems necessary to include in composite parts reinforcement pieces, or "inserts," which may be used to attach to the surrounding structure. The inserts decrease the transmitted stresses to admissible values for the composite part.

- ■ The case of **sandwich pieces:** One frequently finds the metallic inserts following the schematics in Figure 6.31.
- ■ The case of pieces under uniaxial loads:
  - ■ **Tensile load** (see Figure 6.32)
  - ■ **Compression load** (see Figure 6.33)
  - ■ **Tension–compression load** (see Figure 6.34)

Arrangement that allows the increase of the bonded surfaces is shown in Figure 6.35.

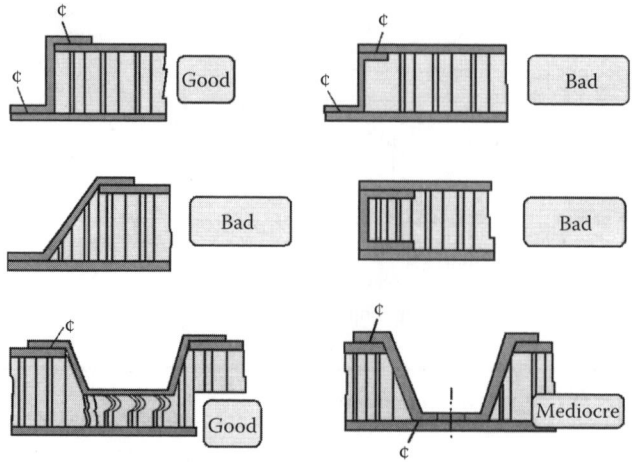

Figure 6.30 Different Sandwich Facing Designs

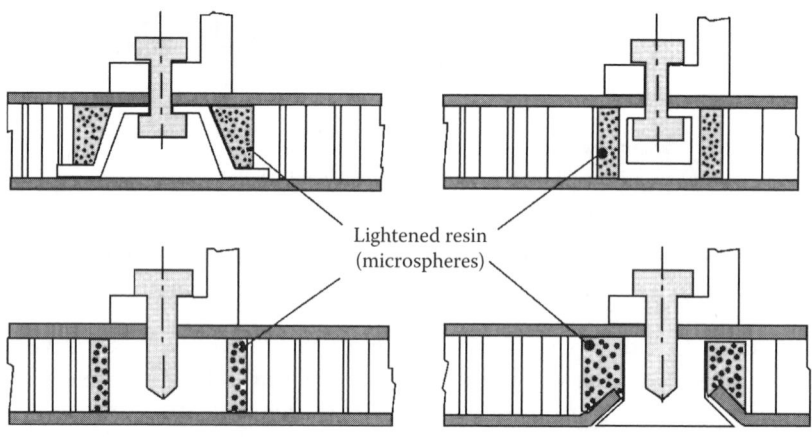

Figure 6.31 Inserts in Sandwich Construction

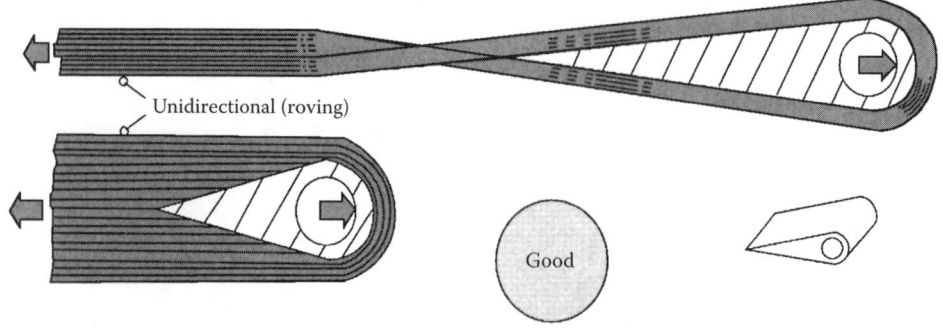

Figure 6.32 Composite Piece under Tensile Load

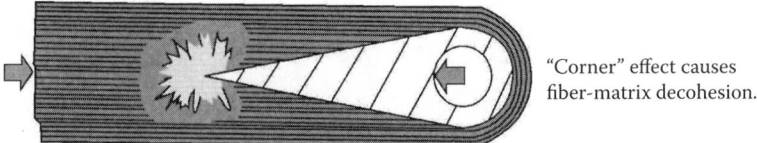

"Corner" effect causes
fiber-matrix decohesion.

Figure 6.33 Composite Piece under Compression Load

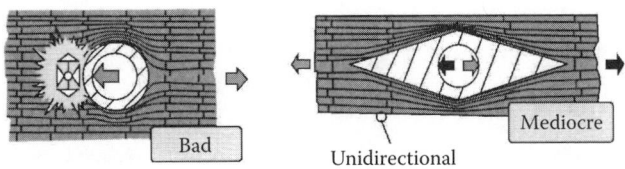

Bad

Unidirectional

Mediocre

Figure 6.34 Composite Piece under Tension-compression Load

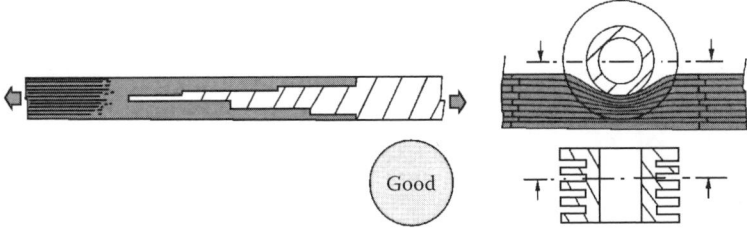

Good

Figure 6.35 Arrangement to Increase Bond Surface

7

COMPOSITE MATERIALS AND AEROSPACE CONSTRUCTION

Aeronautical constructors have been looking for light weight and robustness from composites since the earlier times. As a brief history:

- In 1938, the Morane 406 plane (FRA) utilized sandwich panels with wood core covered with light alloy skins.
- In 1943, composites made of hemp fiber and phenolic resin were used on the Spitfire (U.K.) airplane.
- Glass/resin has been used since 1950, with honeycombs. This allows the construction of the fairings with complex forms.
- Boron/epoxy was introduced around 1960, with moderate development since that time.
- Carbon/epoxy has been used since 1970.
- Kevlar/epoxy has been used since 1972.

Experiences have proved that the use of composites allows one to obtain weight reduction varying from 10% to 50%, with equal performance, together with a cost reduction of 10% to 20%, compared with making the same piece with conventional metallic materials.

7.1 AIRCRAFT

7.1.1 Composite Components in Aircraft

Currently a large variety of composite components are used in aircrafts. Following the more or less important role that composites play to assure the integrity of the aircraft, one can cite the following:

- **Primary structure components** (integrity of which is vital for the aircraft):

135

Wing box
Empennage box
Fuselage
■ **The control components:**
Ailerons
Control components for direction and elevation
High lift devices
Spoilers
■ **Exterior components:**
Fairings
"Karmans"
Storage room doors
Landing gear trap doors
Radomes, front cauls
■ **Interior components:**
Floors
Partitions, bulkheads
Doors, etc.

Example: The vertical stabilizer of the Tristar transporter (Lockheed Company, USA)

■ With classical construction, it consists of 175 elements assembled by 40,000 rivets.
■ With composite construction, it consists only with 18 elements assembled by 5,000 rivets.

7.1.2 Characteristics of Composites

One can indicate the qualities and weak points of the principal composites used. These serve to justify their use in the corresponding components.

7.1.2.1 Glass/Epoxy, Kevlar/Epoxy

These are used in fairings, storage room doors, landing gear trap doors, karmans, radomes, front cauls, leading edges, floors, and passenger compartments.

■ **Pluses:**
High rupture strength[1]
Very good fatigue resistance
■ **Minuses:**
High elastic elongation
Maximum operating temperature around 80°C
Nonconducting material

---

[1] See Section 3.3.3.

7.1.2.2 Carbon/Epoxy

This is used in wing box, horizontal stabilizers, fuselage, ailerons, wings, spoilers (air brakes) vertical stabilizers, traps, and struts.

- ■ **Pluses:**
  High rupture resistance
  Very good fatigue strength
  Very good heat and electricity conductor
  High operating temperature (limited by the resin)
  No dilatation until 600°C
  Smaller specific mass than that of glass/epoxy
- ■ **Minuses:**
  More delicate fabrication
  Impact resistance two or three times less than that of glass/epoxy
  Material susceptible to lightning

7.1.2.3 Boron/Epoxy

This is used for vertical stabilizer boxes and horizontal stabilizer boxes.

- ■ **Pluses:**
  High rupture resistance
  High rigidity
  Very good compatibility with epoxy resins
  Good fatigue resistance
- ■ **Minuses:**
  Higher density than previous composites[2]
  Delicate fabrication and forming
  High cost

7.1.2.4 Honeycombs

Honeycombs are used for forming the core of components made of sandwich structures.

- ■ **Pluses:**
  Low specific mass
  Very high specific modulus and specific strength
  Very good fatigue resistance
- ■ **Minuses:**
  Susceptible to corrosion
  Difficult to detect defects

---

[2] See Section 3.3.3.

7.1.3 A Few Remarks

The construction using only glass fibers is less and less favored in comparison with a combination of Kevlar fibers and carbon fibers for weight saving reasons:

- If one would like to have maximum strength, use Kevlar.
- If one would like to have maximum rigidity, use carbon.
- Kevlar fibers possess excellent vibration damping resistance.
- Due to bird impacts, freezing rain, impact from other particles (sand, dirt), one usually avoids the use of composites in the leading edges without metallic protection.[3]

Carbon/epoxy composite is a good electrical conductor and susceptible to lightning, with the following consequences:

- Damages at the point of impact: delamination, burning of resin
- Risk of lightning in attachments (bolts)
- The necessity to conduct to the mass for the electrical circuits situated under the composite element

Remedies consist of the following:

- Glass fabric in conjunction with a very thin sheet of aluminum (20 μm)
- The use of a protective aluminum film (aluminum flam spray)

Temperature is an important parameter that limits the usage of epoxy resins. A few experimental components have been made of bismaleimide resins (thermosets that soften[4] at temperatures higher than 350°C rather than 210°C for epoxies). One other remedy would be to use a thermoplastic resin with high temperature resistance such as poly-ether-ether-ketone "peek"[5] that softens at 380°C. Laminates made of carbon/peek are more expensive than products made of carbon/epoxy. However, they present good performance at higher operating temperatures (continuously at 130°C and periodically at 160°C) and have the following additional advantages:

- Superior impact resistance
- Negligible moisture absorption
- Very low smoke generation in case of fire

---

[3] The impacts can create internal damages that are invisible from the outside. This can also happen on the wing panels (for example, drop of tools on the panels during fabrication or during maintenance work).

[4] The mechanical properties of the thermoset resins diminish when the temperature reaches the "glass transition temperature."

[5] See Section 1.6 for the physical properties.

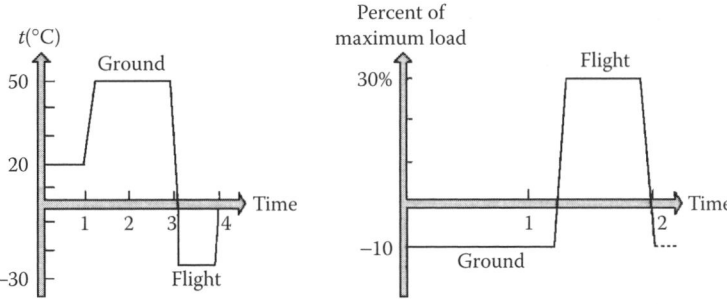

Figure 7.1 Temperature Cycle and Load Cycle for Components of an Aircraft

7.1.4 Specific Aspects of Structural Resistance

- One must apply to composite components the technique called *fail safe* in aerodynamics, which consists of foreseeing the mode of rupture (delamination, for example) and acting in such a manner that this does not lead to the destruction of the component during the period between inspections.
- Composite components are repairable. Methods of reparation are analogous to those for laminates made of unidirectionals or fabrics.[6]
- Considering the very important reduction of the number of rivets used as compared with the conventional construction, one obtains a smother surface, which can lead to better aerodynamic performance.
- One also considers that the attack of the environment and the cycles of fatigue over the years do not lead to significant deterioration of the composite pieces (shown in Figure 7.1 are two types of fatigue cycles for the components of aircraft structure).
- The failure aspect subject to a moderate impact is more problematic with the structures made of composite materials, because the energy absorbed by plastic deformation does not exist.
- For the cabins, one uses phenolic resins. These have good fire resistance, with low smoke emission. For the same reason, one prefers replacing Kevlar fibers with a combination of glass/carbon (lighter than glass alone and less expensive than carbon alone).
- It is possible to benefit from anisotropy of the laminates for the control of dynamic and aeroelastic behavior of the wing structures.[7]

7.1.5 Large Carriers

The following examples give an idea on the evolution of use of composites in aircrafts over two decades:

Example: Aerospatiale (FRA); Airbus Industry (EU) (Figure 7.2)
Example: Boeing (USA) (Figure 7.3)

---

[6] See Section 4.4.4.
[7] See Section 7.1.8.

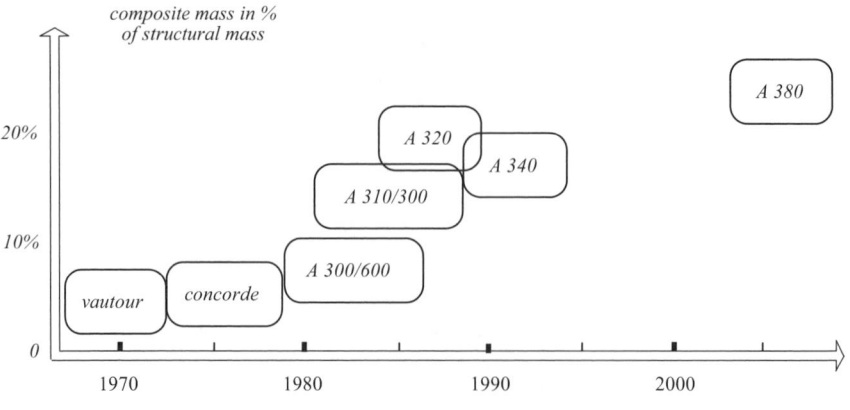

Figure 7.2   Evolution of Mass of Composites in Aircraft

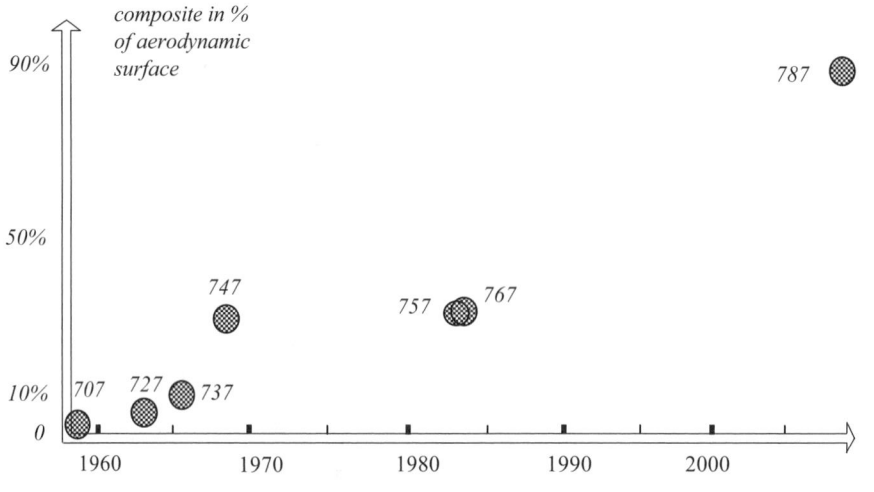

Figure 7.3   Use of Composite in Boeing Aircraft

How to evaluate the gains:

In theory: For example, a study was made by Lockheed Company (USA) for the design of a large carrier having the following principal characteristics: payload 68 tons, range 8300 km. This study gives the following significant results:

- for an aircraft made using conventional metallic construction:
  total mass at take-off: 363 tons
  mass of the structure: 175 tons
- for an aircraft made using "maximum" composite construction:
  total mass at take-off: 245 tons
  mass of the structure: 96 tons.

Such a difference can be explained by the cascading consequences that can be illustrated as in Figure 7.4.

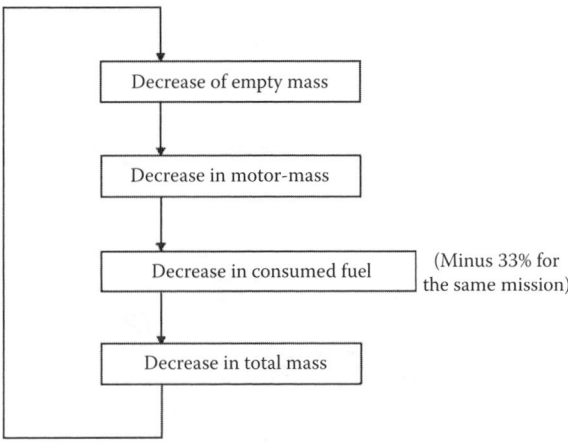

Figure 7.4   Cascading Effect in Mass Reduction

In practice: In reality, introduction of composites in the aircrafts is limited to certain parts of the structures. It is done case by case and in a progressive manner during the life of the aircraft (re-evaluation operation). One is then led to consider the different notions:

■ Notion of the exchange rate is the cost for a kilogram saved when one substitutes a classical metallic piece with a piece made mainly with composite. For the substitution light alloy—carbon/epoxy—this cost is on the order of $160 (1984) per kilogram when the piece is calculated in terms of rigidity (similar deformation for the same load). It is amortized over a period of at least one year for the gain in "paying passenger."
■ Notion of gain in paying passenger is the gain in terms of the number of passengers, of freight, or in fuel cost; for example, for a large carrier:
   ■ An aircraft of 150 tons, with 250 passengers consists of 60 tons of structure. A progressive introduction of 1600 kg of high performance composite materials leads to a gain of 16 more passengers along with their luggage.
■ A reduction of 1 kg mass leads to the reduction of fuel consumption of around 120 liters per year.

Why are the reductions of mass (average about 20%) not more spectacular?
Consider the example of a vertical stabilizer. The distribution of mass of a composite vertical stabilizer can be presented as follows:

■ Facings in carbon/epoxy: 30% of total mass
■ Honeycombs, adhesives: 35% of total mass
■ Attachments: 25% of total mass
■ Connections between carbon/epoxy components and attachments: overlayers of carbon/epoxy
■ Allowance for the aging of the carbon/epoxy: overdimensions of the facings (the stresses are magnified about 10% more for a subsonic aircraft and 13% for a supersonic aircraft)

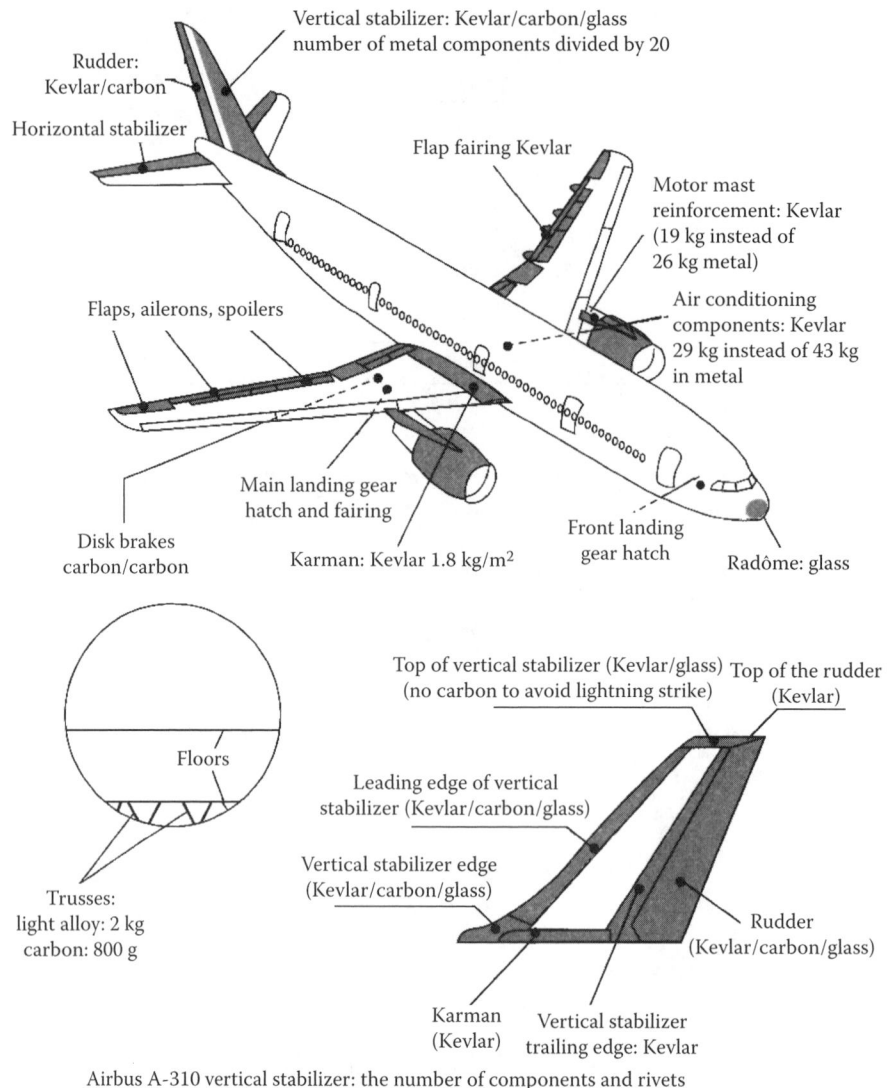

Airbus A-310 vertical stabilizer: the number of components and rivets
is divided by 20 in comparison with the classical solution

Figure 7.5   Composite Components in an Airbus A-310

In consequence, the global gain of mass in comparison with a classical metallic
construction for the vertical stabilizer is not more than an order of about 15%.

Example: European aircraft Airbus A310–300 (Figure 7.5).

- Total mass: 180 tons
- Mass of structure: 44.7 tons
- Mass of composites: 6.2 tons
- Mass of high performance composites: 1.1 tons
- Reduction of mass of structure: 1.4 tons

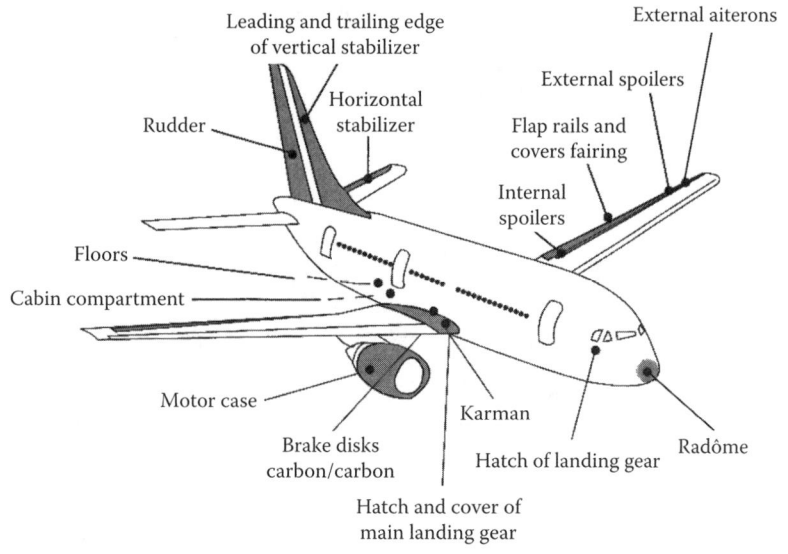

Figure 7.6a   Composite Components in an Airbus A-320

- Percentage of composites: 13.8% of mass of structure. A reduction of mass of the structure of 1 kg augments the range of the aircraft by 1 nautical mile.

Example: European aircraft Airbus A-320 (Figure 7.6a).

- Total mass: 72 tons
- Empty mass: 40 tons
- Mass of structure: 21 tons
- Mass of composite materials: 4.5 tons, corresponding to a reduction of mass of the structure of 1.1 tons. The percent of composite mass is 21.5% of the mass of the structure.
- A few other characteristics: Length: 37.6 m; breadth: 34 m; 150 to 176 passengers transported from 3,500 to 5,500 km; maximum cruising speed: 868 km/h

Example: European aircraft Airbus A-380.800 (Figure 7.6b).

- Total mass: 560 tons
- Empty mass: 240 tons
- Percentage of composites: 25% of the structural mass (weight of carbon/epoxy: 40 tons)
- A few other characteristics: Length: 72.7 m; Width: 79.6 m; Height: 24 m. Passenger weight: 55 tons (525 passengers) over a range of 14,800 km.

Example: Transport aircraft B-787.800 Boeing (USA) (Figure 7.6c).

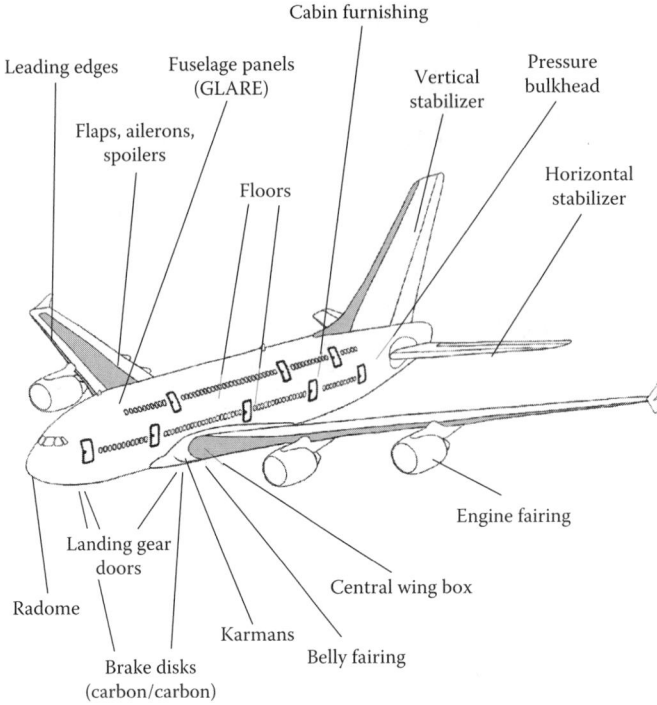

Figure 7.6b Composite Components in an Airbus A-380

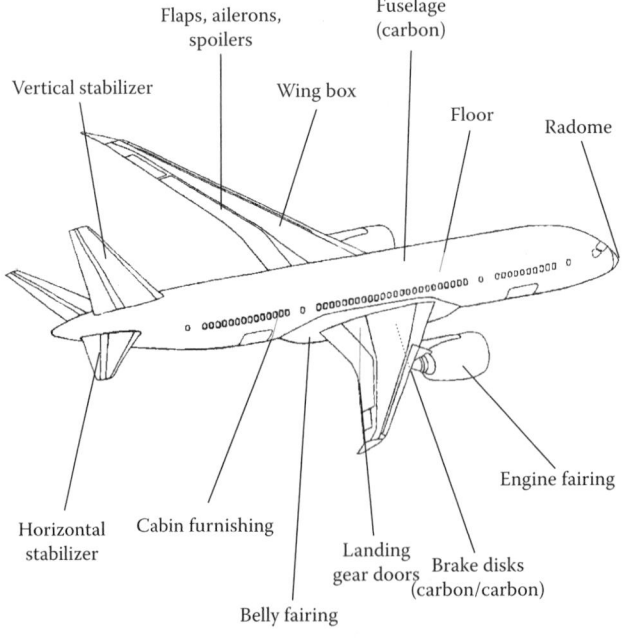

Figure 7.6c Composite Components in a Boeing B-787

■ Percentage of composite: 50% of structural mass. This aircraft materializes a real "technological jump" because the composite percentage is more than double in comparison with previous aircrafts. In addition appear new parts made of composite in a large civil transport aircraft such as:
  ■ The principal wing box (carbon/epoxy)
  ■ The fuselage (fiber placement using placing heads on rotating mandrel, with local fabric reinforcements around openings such as windows, doors, and for fastenings).
■ Some other characteristics: Length: 56 m; Width: 51 m; 217 passengers over a range of 15,700 km.

Example: Future supersonic aircraft *ATSF* (Figure 7.7), *Aerospatiale* (FRA) and *British Aerospace* (UK). Principal characteristics defined at the stage before the project

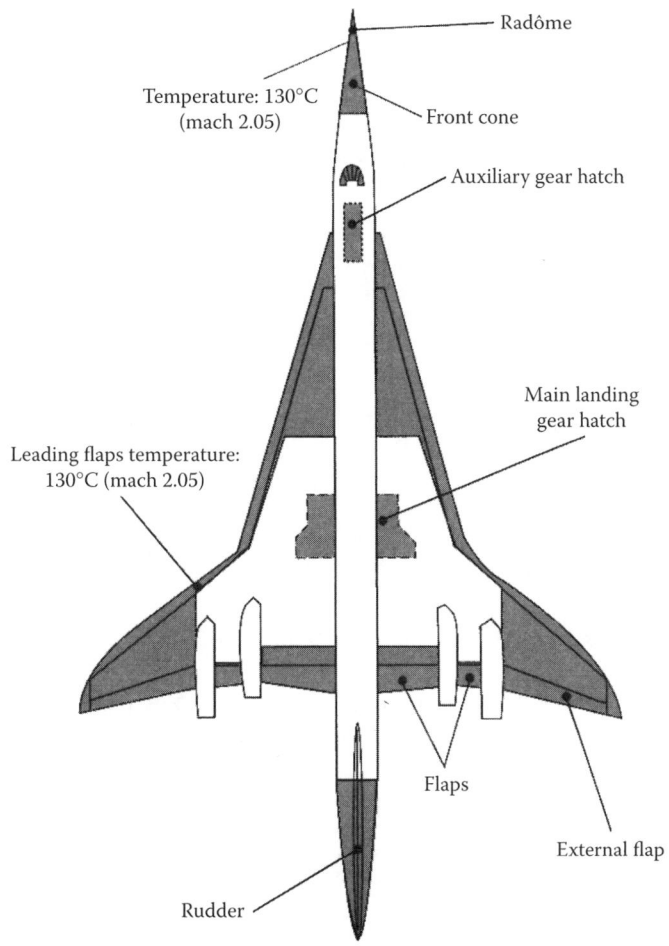

Figure 7.7 Composite Components in a Future Supersonic Aircraft

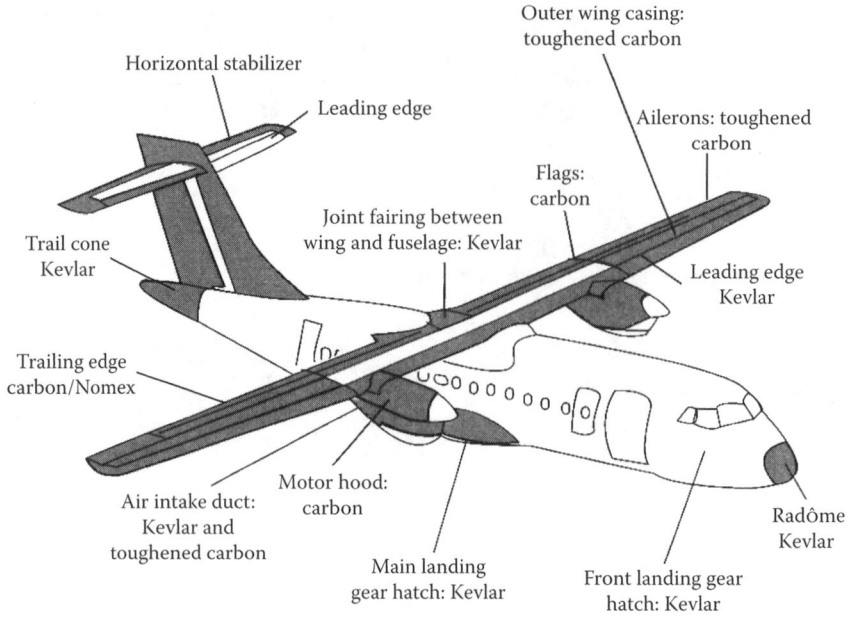

Figure 7.8 Composite Components in the Regional Transport Aircraft ATR 72

- Transport of 200 passengers over a distance of 12,000 km
- Cruising speed between Mach 2 (2,200 km/hr) and Mach 2.4 (2,600 km/hr)
- Economically viable for a single type of aircraft on the market (enlarged international cooperation)

7.1.6 Regional Jets

Example: Regional transport aircraft ATR 72, ATR (FRA–ITA) (Figure 7.8):

- Total mass: 20 tons
- Percentage of composite materials more than 25% of the mass of the structure
- Transports 66 passengers over a distance of 2,600 kilometers
- Interior equipment: Facings of panels for portholes and ceiling, baggage compartment, bulkheads, toilets, storing armors in glass-carbon/phenolic resins/NOMEX honeycomb; decoration by a film of "TEDLAR"

Example: Business aircraft Falcon 10 (Figure 7.9). Aircrafts (AMD–BA) (FRA).

- The principal wing box (primary structure) is constructed in ribbed panels of carbon/epoxy; it has been flying experimentally since 1985.
  Mass of wing box: 339 kg
  Reduction of mass in comparison with conventional metallic construction: 80 kg (20%)

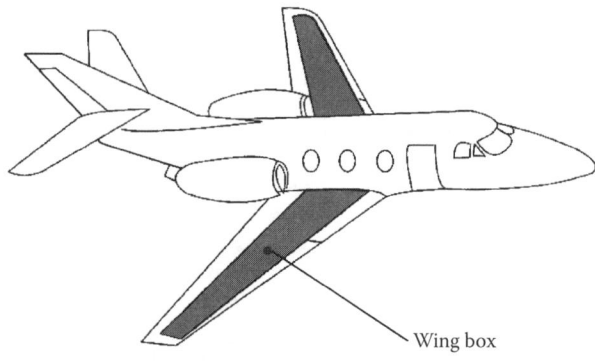

Figure 7.9 Business Aircraft Falcon 10

- The connection between the wing and the fuselage and the attachment between the landing gears and the wing box are made using metallic pieces.

7.1.7 Light Aircraft

These are the aircrafts for tourism and for the gliders. The new generation of these aircrafts is characterized by

- A large utilization of composites
- A renewal in aerodynamic solutions

Remarks: In addition to the problems to be resolved by the manufacturers (small manufacturers for the most part), comes the preparation of the dossiers of certification including composite primary structures.

- The useful reduction of masses, the range of flight, and the cruising speed, due to the utilization of composites, are amplified more clearly in these types of aircrafts.

Example: Back propeller aircraft. The principle of this is illustrated in Figure 7.10, with the advantages and drawbacks. The modification of the mass distribution, due to the displacement of the motor, requires a propeller shaft (foreseen to be in carbon/epoxy) and a wing shifted to the back. One can propose the "all composite" solution as follows:

- The solution "long shaft." Sup'air airplane Centrair (FRA); Figure 7.11
- The solution of shifted wing and additional duck wing: Beech Starship aircraft (USA); Figure 7.12:
  Eight to ten passengers: 650 km/hr with low fuel consumption
  Structure in carbon/epoxy
  Mass of the wings: 800 kg (reduction of 35% in comparison with a metallic solution)
  Mass of fuselage (structure): 240 kg
  Mass of the composite: about 70% of the mass of the structure

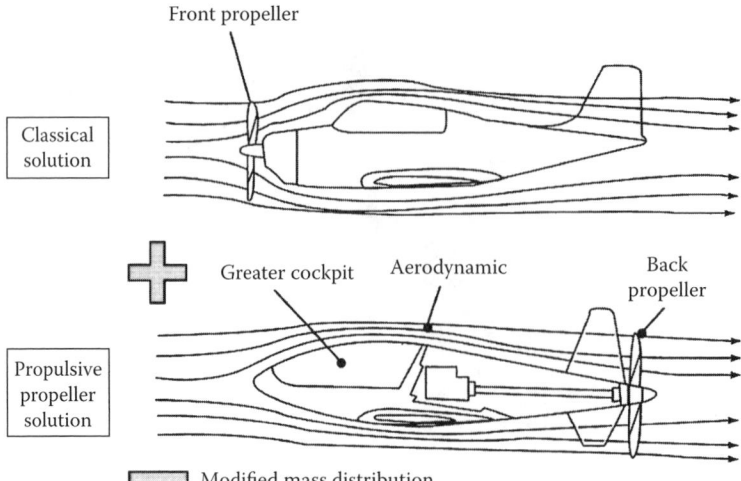

Figure 7.10 Propulsive Propeller Configuration

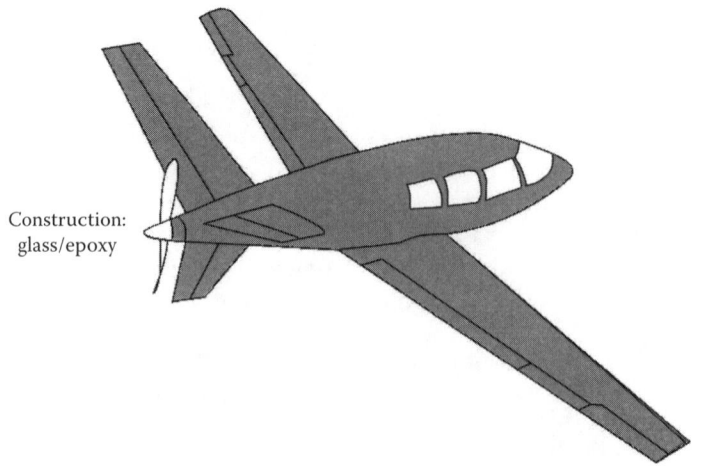

Figure 7.11 An All-composite Airplane

Example: The modern glider planes. These are made entirely of composites. Figure 7.13 shows a plane made of glass/epoxy:

■ Two-seater glider plane: Marianne Centrair (FRA):
   Mass: 440 kg
   Wings: 2 parts bonded
   Fuselage: 2 parts bonded

7.1.8 Fighter Aircraft

For this type of aircraft, there is a progressive replacement of metallic elements by composite elements. Beyond specific characteristics already mentioned previ-

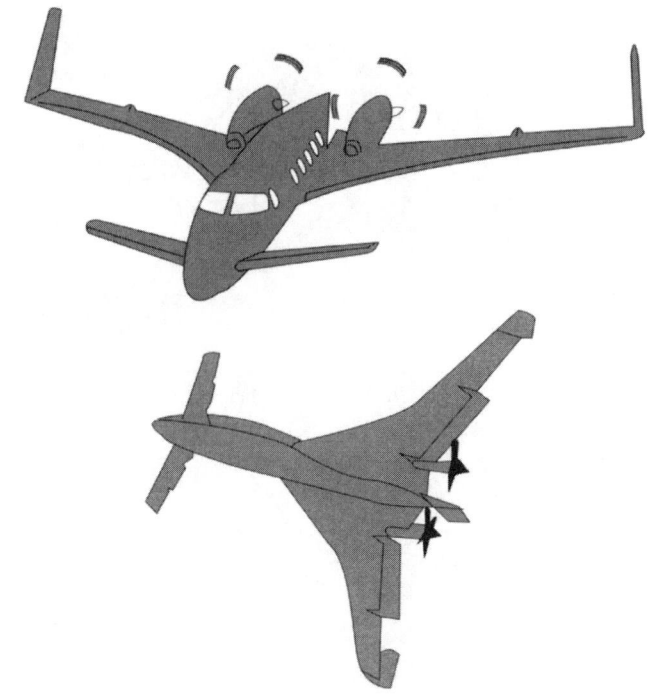

Figure 7.12 Beech Starship Aircraft

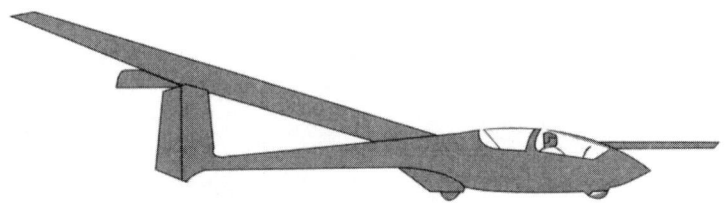

Figure 7.13 The Marianne Centrair Airplane

ously for the large aircrafts, here the composite components have to assure the necessary rigidity for the wing box to conserve the ability of command in a domain of flight larger than for the case of large civil aircrafts.

- For the flight with electrical commands, the use of composites allows for an evolution of the aerodynamic design for better maneuverability.
- In the near future, 25 to 40% of the structure of the fighter aircrafts will be made of composite materials.

Example: European airplane Alphajet (Figure 7.14).
Example: Airplane Mirage 2000 A.M.D.–B.A. (FRA; Figure 7.15). Mass of composite materials: 65 kg (On this aircraft there are boron/epoxy composite components.)

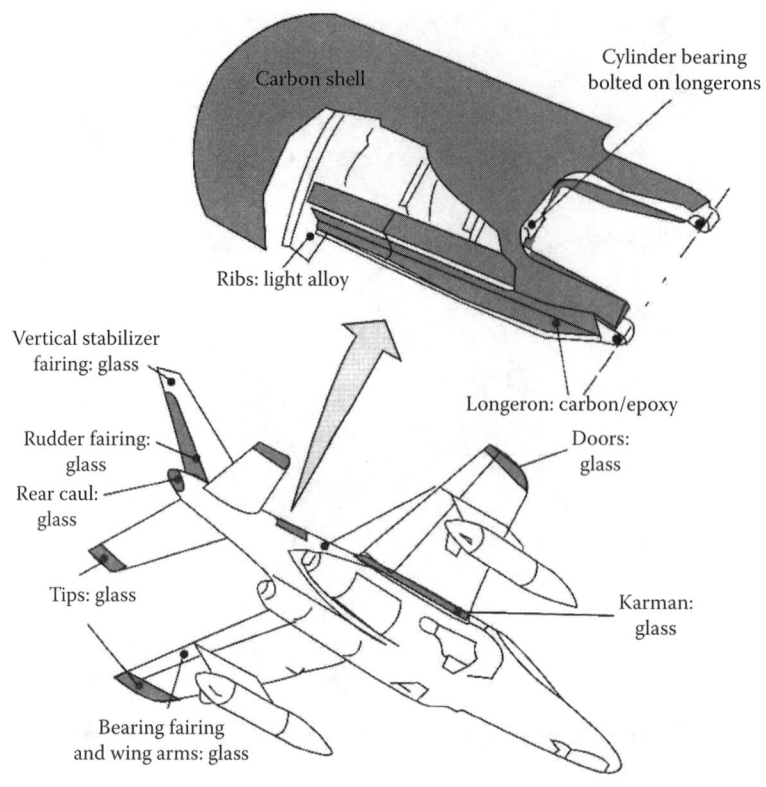

Figure 7.14 Alphajet Plane

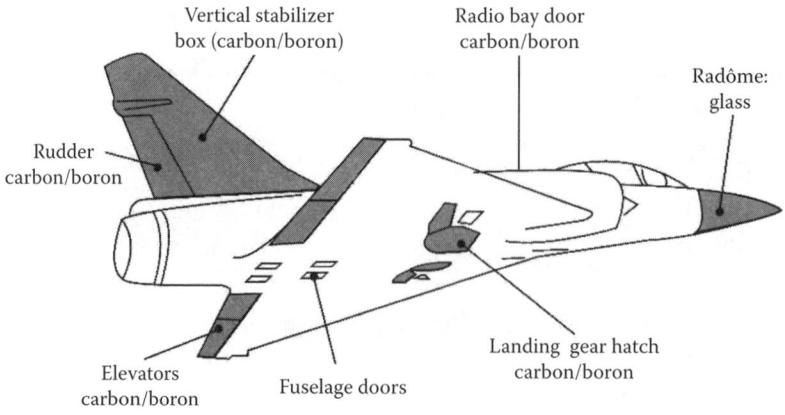

Figure 7.15 Mirage Airplane

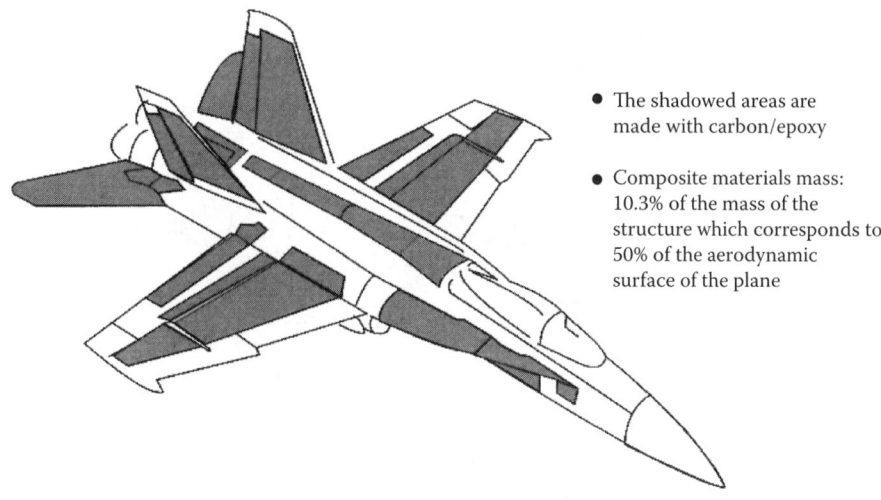

• The shadowed areas are made with carbon/epoxy

• Composite materials mass: 10.3% of the mass of the structure which corresponds to 50% of the aerodynamic surface of the plane

Figure 7.16 F-18 Hornet Airplane

■ Characteristics of boron: The diameter of the fiber varies between 0.1 mm and 0.2 mm depending on the demand of the customer. The radius of curvature, in fact, cannot be less than 4 or 5 mm. One finds
  ■ for the sheets: width is 1 m, 80 filaments/cm, length: 3.5 m.
  ■ for the fabrics: patented process;[8] the warp direction consists of textile filaments, the fill direction consists of boron filaments.

Example: Airplane F-18 Hornet, M.D. Douglas/Northrop (USA; Figure 7.16).
Example: Airplane X-29 Grumman (USA; Figure 7.17).
Example: Airplane Rafale A.M.D.–B.A. (FRA; Figure 7.18). Note that on this airplane there is a very large usage of high performance composites (carbon/epoxy and Kevlar/epoxy). Mass of composite materials is 1,110 kg, leading to a 25% reduction in the mass of the structure.[9] Figure 7.18 shows the main components using composites.

7.1.9 Architecture of Composite Parts in Aircraft

7.1.9.1 Sandwich Design

Sandwich with transverse honeycomb: Following the nature of the component, one uses two methods of fabrication:

■ **Multisteps** in which the facings of the sandwich piece are polymerized separately and then are placed on the core having an adhesive film. After that the assembly is polymerized following the method represented in Section 4.4.2, where eventually polymerization is done in an autoclave.[10]

---

[8] Patent Avions M. Dassault-Bréguet Aviation/Brochier.
[9] This is to compare with the number for Mirage 2000 airplane presented previously.
[10] See also Section 2.1.3.

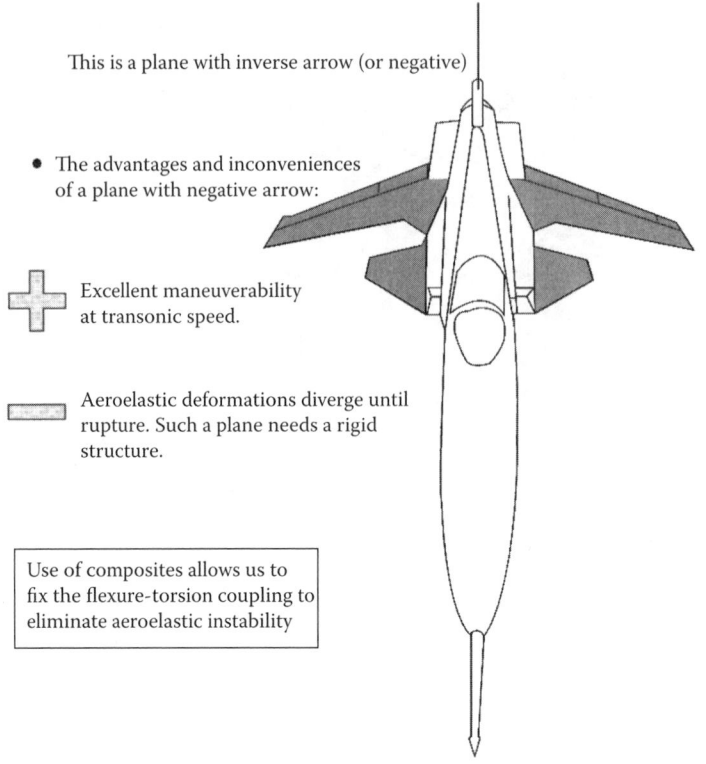

This is a plane with inverse arrow (or negative)

● The advantages and inconveniences
of a plane with negative arrow:

Excellent maneuverability
at transonic speed.

Aeroelastic deformations diverge until
rupture. Such a plane needs a rigid
structure.

Use of composites allows us to
fix the flexure-torsion coupling to
eliminate aeroelastic instability

Figure 7.17   X-29 Grumman Airplane

■ **Monostep** in which, after the honeycomb core is formed, the facings are
placed directly on this core. The assembly is polymerized using the same
method as for the multistep method.

Example: Wing box (Figure 7.19).

The honeycomb core assures the rigidity of the component. However, the
mass of the piece depends on the thickness of the core.

Example: Horizontal empennage of a fighting aircraft (Figure 7.20).

Remarks: One avoids drilling holes in boron/epoxy pieces as much as possible.
This operation is onerous and requires ultrasonic technique with diamond tools.

There is a problem with corrosion of metallic honeycomb. The corrosion is
due to the progressive condensation of water in the cavities. This is combined
with the mechanical and thermal stresses (fatigue) in the structure.

Remedies include

■ Covering the honeycomb core surface with a resin film
■ Introducing an inorganic inhibitor at the potential points of attack to prevent
the reaction with water

Sandwich construction for panels: When the thickness of the piece becomes
large (on the order of 150 mm), the facings are stiffened separately by using a
honeycomb core, following the arrangements shown in Figure 7.21. When the

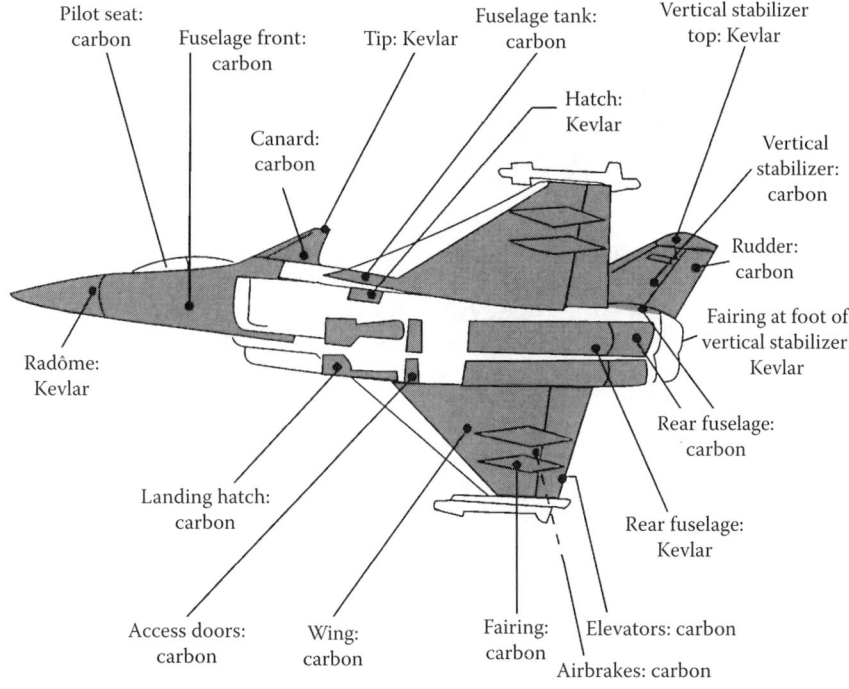

Figure 7.18   Rafale Airplane

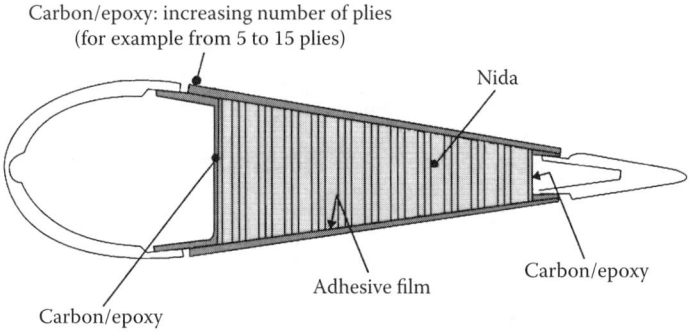

Figure 7.19   Wing Box

piece is of large size (long), the requirement of "nondeformation" can require the interposition of intermediate "ribs."

Each component (facings, ribs) is polymerized in the assembled state, following the technique of monostep described above.

Example: High lift devices flap (Figure 7.22)

Sandwich for the reinforcement of spars and ribs: One can increase the rigidity in flexure and in torsion by introducing honeycombs, as represented in Figure 7.23.

Stiffened panels: Currently one can find the ribbed panels in metallic construction. The ribs can be added, or monolithic, meaning that they are formed as

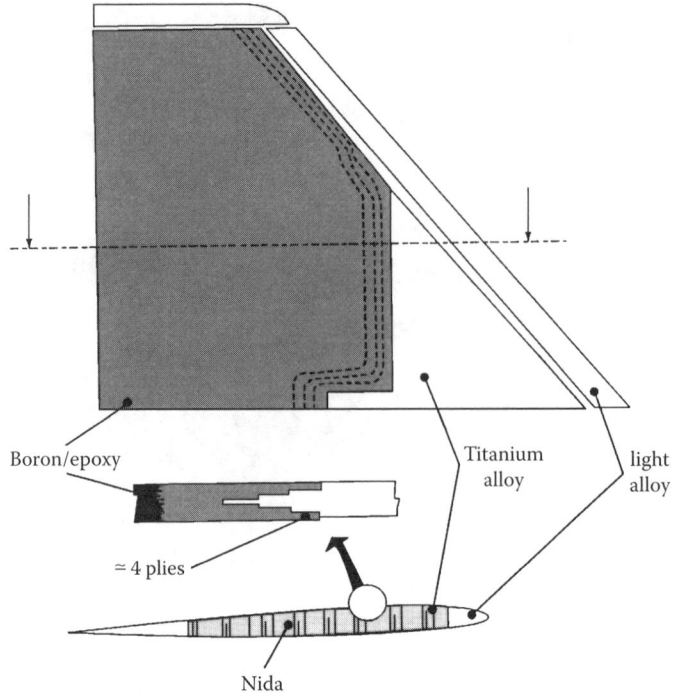

Figure 7.20   Horizontal Empennage of a Fighting Aircraft

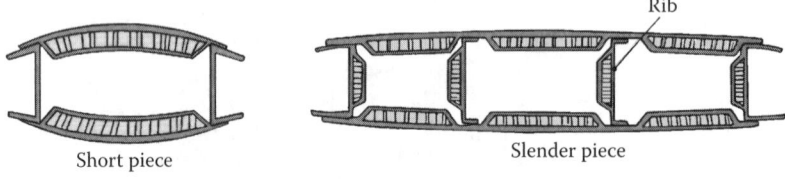

Figure 7.21   Sandwich Construction for Panels

part of the same piece when the panel is made. One observes in parallel here with the same technique in ribbed panels in composites.

- **Added stiffeners:** The composite stiffeners can have the forms shown in Figure 7.24.

Example: Wing box (Figure 7.25).

- **Monolithic stiffeners:** These are cured at the same time with the facings. These can support higher loads than the previous case, but with higher cost. The mode of fabrication is shown schematically in Figure 7.26 for the so-called "omega" ribs. One uses cores, which are made partly by thermoexpandable silicone. One also uses hollow silicone cores stiffened by compressed air, or cores that melt at temperature on the order of 170°C, which is slightly higher than the polymerization temperature of the piece.

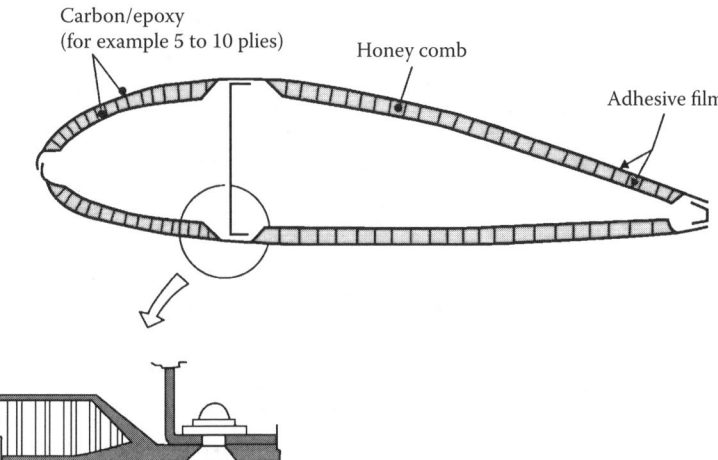

Figure 7.22 High Lift Devices Flap

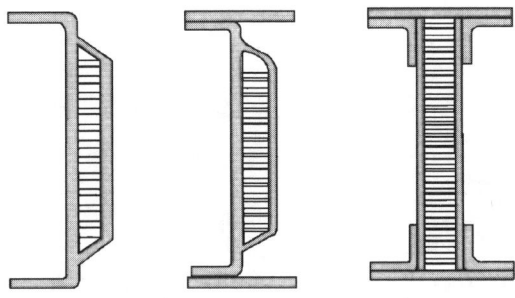

Figure 7.23 Sandwich for Spars and Ribs

Figure 7.24 Stiffeners Form

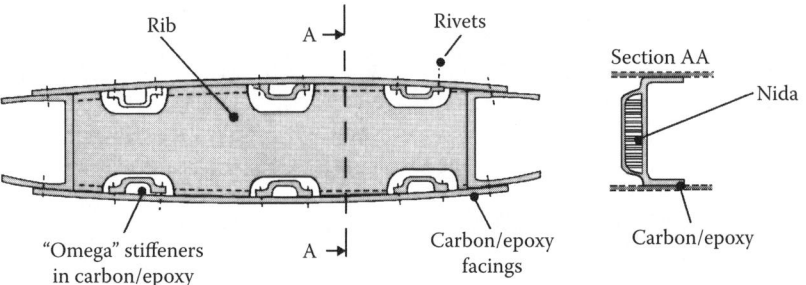

Figure 7.25 Wing Box

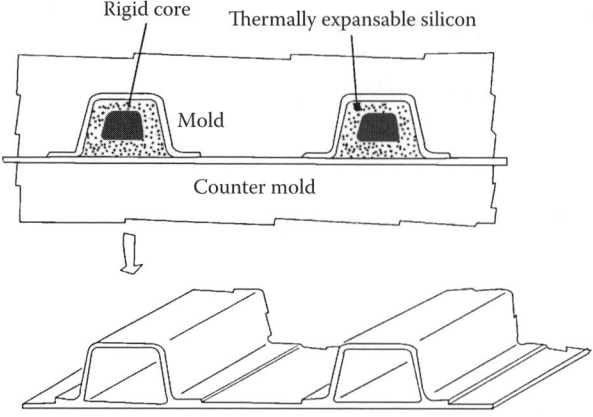

Figure 7.26 Monolithic Stiffeners

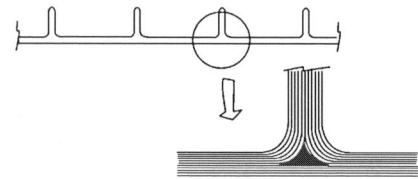

Figure 7.27 Flanged Plate

Example: Flanged plates (Figure 7.27)

Example: Stabilizer panel (European aircraft Airbus A-300; see Figure 7.28). The carbon/epoxy ribs are obtained by combining the autoclave pressure and the thermal dilatation of detachable metallic light alloy modules.[11] The steps of the process are shown schematically in Figure 7.28:

Example: Wing tip (Airplane ATR 72, ATR (FRA-ITA). This is a primary wing piece (its failure will bring about the failure of the airplane). It consists of two carbon/epoxy panels with monolithic ribs, two spars in carbon/epoxy, and 18 metallic flanges in light alloy folded sheets, as shown schematically in Figure 7.29.

- Mass of box: 260 kg (reduction of 65 kg as compared with a metallic solution).

Remarks: Lightning protection of such a structure requires specific precautions such as:

- Incorporation of a conducting fabric made of bronze wires on all external surfaces
- Installation of lightning conductors along the spars
- Protection of the connections

---

[11] See Section 1.6 for the coefficients of thermal expansion as compared between light alloy and carbon.

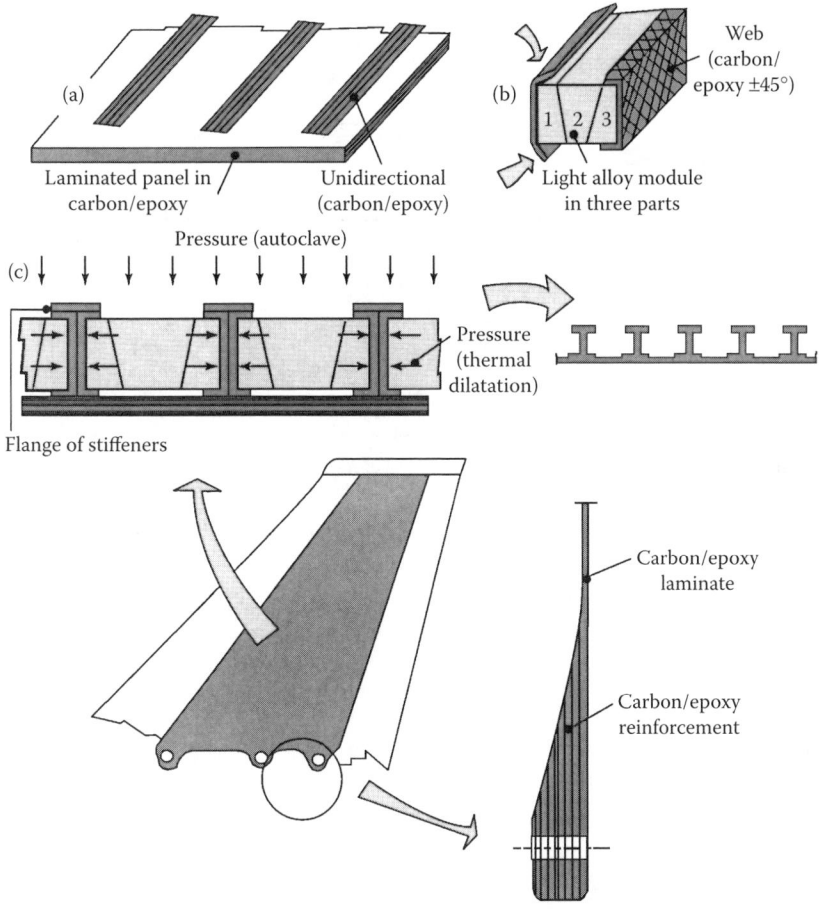

Figure 7.28 Stabilizer Panel

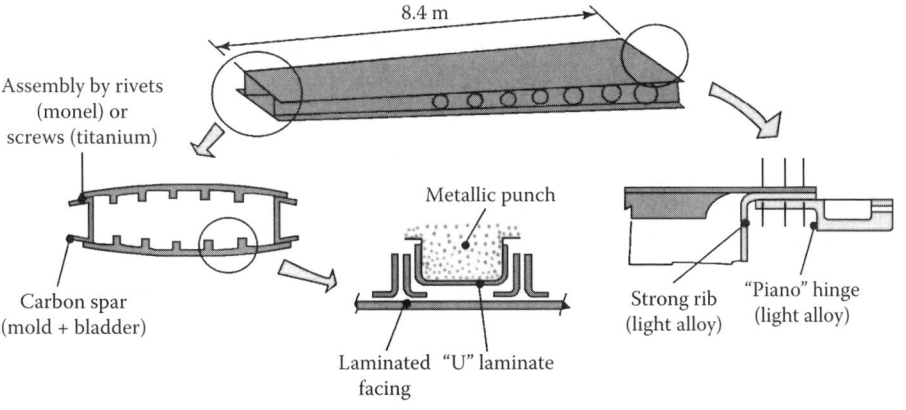

Figure 7.29 Wing Tip on ATR 72 Airplane

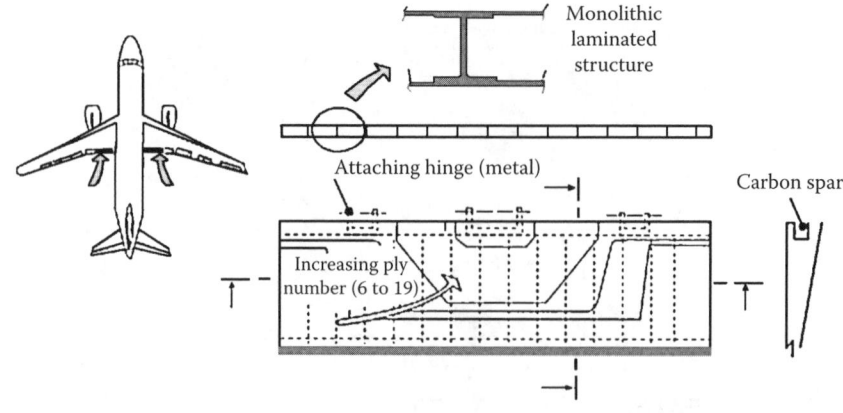

Figure 7.30   Air Brakes on Airbus A-320

Example: Air brakes (European airplane Airbus A-320; Figure 7.30).

7.1.10   Elements of Braking

Different from the brakes for ground vehicles, the brakes for aircrafts are characterized by successively distinct phases of isolated operations over time. These repeat themselves in almost identical conditions from one landing to the next. These are "**heat absorption**" brakes, which are activated for only a few seconds (about 20 seconds). The subsequent cooling of these brakes happens progressively afterwards. The heat coming from the transformation of kinetic energy is stored in the components participating in the friction phenomenon, which serve as "**heat sinks.**" These components must have the following characteristics:

- Being able to create a high braking moment which remains stable as the temperature increases
- Being able to support a very important "**thermal shock**," on the order of 10^6 joules per kg mass of the component
- Being refractory and retaining a good dimensional stability
- Being able to retain their mechanical properties at high temperature
- Having as low mass as possible

The corresponding brakes are of "disk" type. The materials that can be used to make these friction disks are compared in Figure 7.31. One can see the interest in using three-dimensional composite materials in carbon/carbon, which have the following characteristics:

- Their dynamic friction coefficient is stable with respect to the temperature, varying from 0.25 to 0.3.
- They resist thermal shock and are refractory until 1600°C.
- They retain their mechanical properties at high temperature.[12]
- They are light weight (specific mass of 1900 kg/m^3).

---

[12] See order of magnitude of the mechanical properties in Section 3.6.

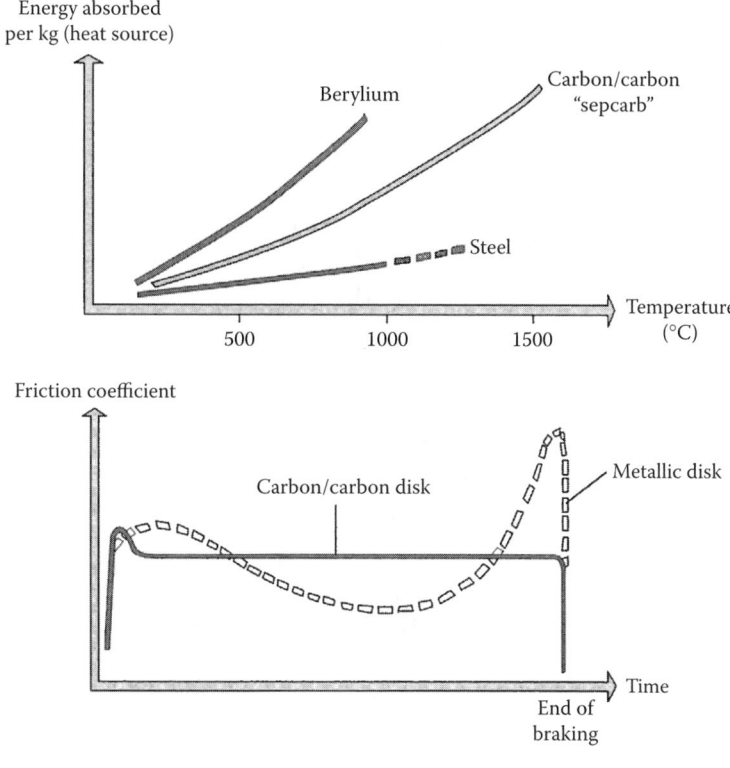

Figure 7.31   Airbrake Materials

Example: Disk brakes in carbon/carbon (Figure 7.32) "Aerolor" (FRA).

Example: Supersonic transport aircraft Concorde (FRA-U.K.). The gain in paying passenger due to this type of brakes represents six passengers and their luggage, or 600 kg.

Such types of brakes are also developed for rapid trains, as well as for racing cars and motorcycles.

7.1.11 The Future

One can cite as the principal objectives for the airplane manufacturers:

- Obtain with composite materials a minimum mass reduction of 20% as compared with the metallic solution using the new light alloys **aluminum/lithium**. This implies the improvement in the performance of fibers and matrices.
- Improvement on the manufacturing process. The graph in Figure 7.33 shows that significant efforts remain to be done to reduce the labor cost in the manufacturing of high performance composite components.

Apart from improvement such as automatic draping or electron beam curing or by x-ray, other more classical techniques described in Chapter 2 have actually been adapted for the fabrication of aircrafts or their components. One can mention

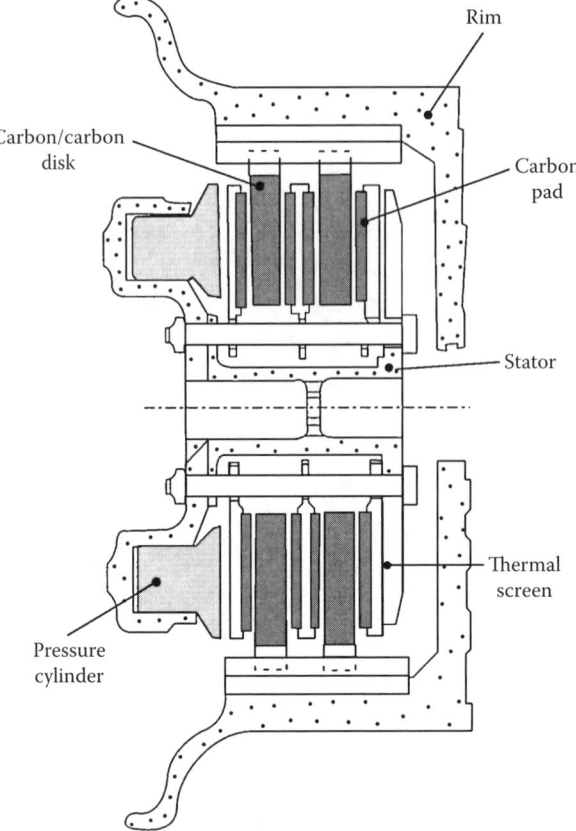

Case of take-off aborted on Airbus A340 (front landing).
Absorbed energy: 100 Mjoules; temperature: 2000°C.

Figure 7.32   Disk Brakes in Carbon/carbon

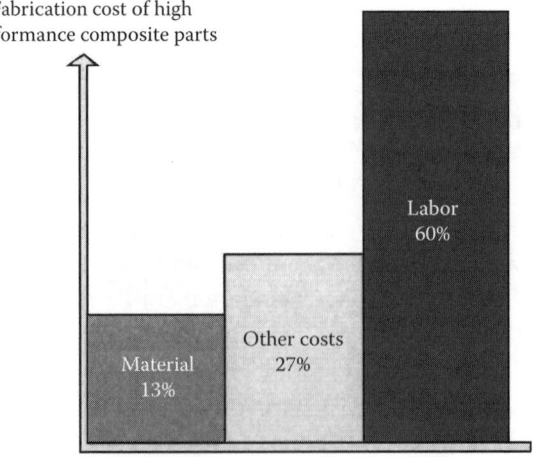

Figure 7.33   Different Costs in Composite Solution

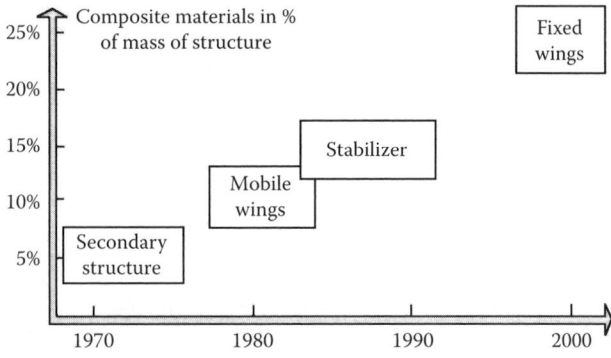

Figure 7.34 Evolution of Percent of Composite Materials in Aircraft Structure

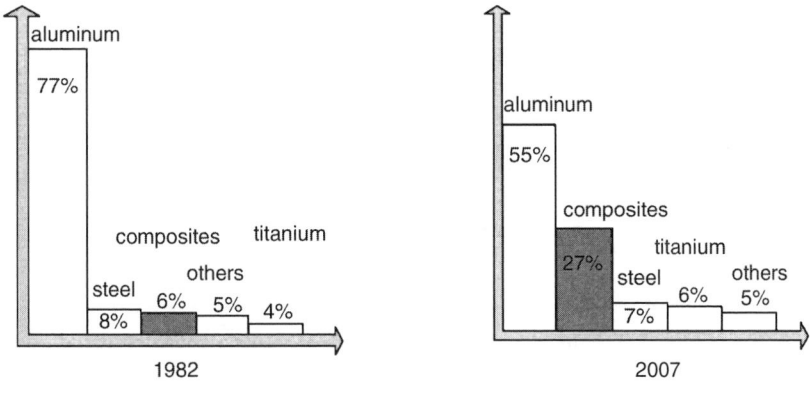

Figure 7.35 Relative Mass of Principal Materials Used in Aircraft Structures

- RTM process (see Section 2.3.1), for example, tail cone of airplane Airbus A-321.
- High pressure injection (see Section 2.1.5) with short fibers or with particles.
- Thermoforming (see Sections 2.3.1 and 2.2.3) for the small pieces.
- Compression molding (see Sections 2.3.1 and 2.1.2).

One can see in Figure 7.34 that about 20% of the mass of the structure of civil aircrafts should integrate composite materials. Figure 7.35 shows the evolution foreseen for the relative mass of the principal materials used for aircraft structures.

7.2 HELICOPTERS

7.2.1 The Situation

This category of aircraft is less advanced than the airplanes in terms of technological evolution towards "perfection" as shown in the graph in Figure 7.36.

In fact, taking into account the possibilities of specific mass reduction for this type of aircraft, one can see in the graph in Figure 7.37 the increasing trend in

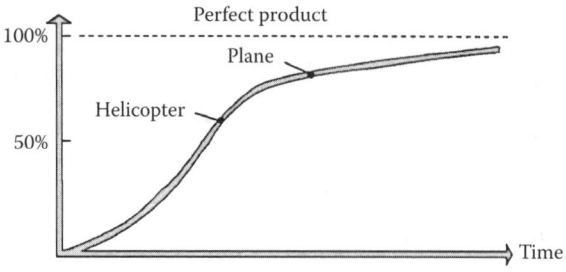

Figure 7.36 Degree of Technological Evolution

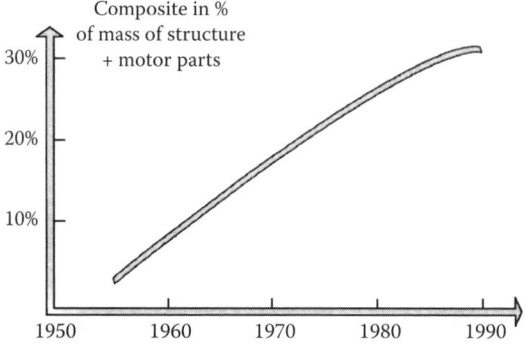

Figure 7.37 Evolution of Composites Mass on Helicopters

the strong integration of composites, with higher percentages as compared with the case of airplanes.

Example: Evolution of the mass of composite materials on the helicopters of Aerospatiale (FRA). In comparison with conventional metallic construction, one can mention the following mass reduction:

- 15% on the secondary structures
- Up to 50% in the working pieces, such as the elements of transmission of power and control

7.2.2 Composite Zones

Figure 7.38 shows the composite components in a helicopter.

7.2.3 Blades

The blades are the essential elements of the aircraft. They consist principally of the following:

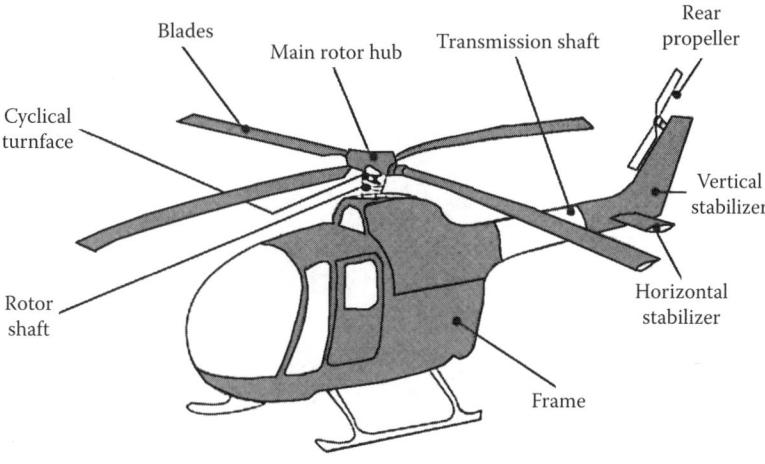

Figure 7.38   Composite Components in a Helicopter

- An envelope or a **box** that assures an aerodynamic profile and a stiffener for torsion (the blade does not twist under aerodynamic forces, at least for the actual generation of aircrafts).
- A spar which resists the centrifugal tension on the blade as well as the flexure caused by the lift and drag loads. It is made of glass/epoxy ("R" glass, more resistant and less sensitive to aging by humidity).
- A rear edge that stiffens the blade in flexure in the direction of the drag.
- A filler material (foam or honeycomb) that prevents the deformation of the profile.

Figure 7.39 shows the different parts of the blade.

7.2.3.1   Advantages

The list of advantages obtained with this type of design is impressive:

- The blade is molded (molding by assembly of two half shells under pressure). This solution allows one to obtain an **optimized profile** (variable chord and thickness, nonsymmetric profile, nonlinear twist).
- The stiffeners for fluttering and for torsion can be controlled thanks to judicious usage of composite materials.

7.2.3.2   Consequences

The payload is augmented. The mass reduction attains 400 kg for the aircraft Superpuma Aesrospatiale (FRA). The cruising speed is increased for the same power. The gain is 32 km/h at 1500 m altitude as compared with a previous helicopter.

- The cost of fabrication is reduced by **50%** in comparison with conventional metallic solution.

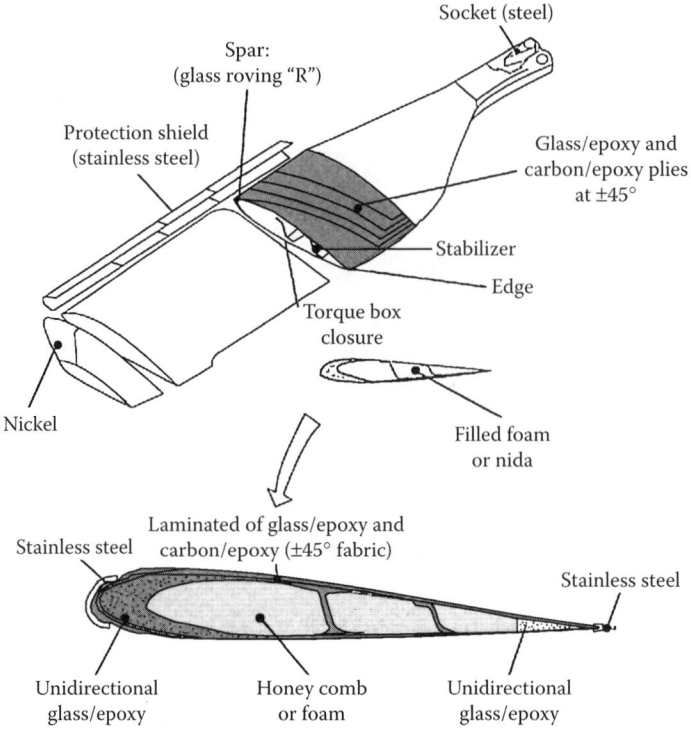

Figure 7.39 Helicopter Blade

- The cost of operation is reduced.
- The life of the blade is practically **unlimited.** None of the load in the range of flight of the aircraft can lead to the fatigue damage. It is quasi-indestructible, even when testing the specimens.
- Increased security: the blade has the **fail safe** character.[13] An impact (projectile, collision) causes a local deterioration which does not lead to the fall of the aircraft.
- The blade is repairable with a relatively simple process.[14]
- The blade is not sensitive to corrosion.

Remarks: The blade as conceived can be ultralight. However, light weight cannot be below a value that assures a minimum inertia that is indispensable for the good operation of the rotor.

7.2.4 Yoke Rotor

This is the mechanical assembly that allows

---

[13] See Section 7.1.4.
[14] See Section 4.4.4.

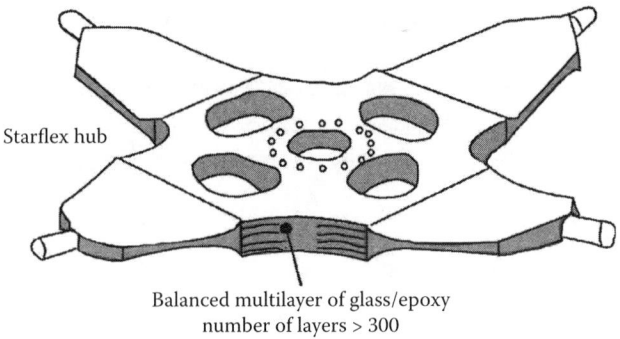

Starflex hub

Balanced multilayer of glass/epoxy
number of layers > 300

Figure 7.40   The Starflex Yoke

- The rotation of the blades
- The small amplitude angular displacements of the blades during rotation
- Pitch control, that is, the control of the aerodynamic incidence of the profiles of the blades

To assure all these functions, the previous classical metallic rotors were very complex. They consisted of many pieces—in particular the ball bearings—and numerous points of lubrication. In fact, the maintenance was very costly.

The modern rotors—first developed by Aerospatiale (FRA)—substitute for these classical articulations the degrees of freedom starting from the elastic deformation:

- Composites made of metal/elastomer
- Laminates

Example: The yoke "Starflex" Aerospatiale (FRA). As its name indicates, this yoke has the form of a star with flexible arms (see Figure 7.40) obtained by draping a large amount of balanced glass/epoxy fabric and molding under heat and pressure.

The different degrees of freedom necessary for the operation as cited above are made possible by the capabilities as shown in Figure 7.41:

- The elastic arm assures angular displacement called "lift fluttering."
- The elastic ball and socket allows for the rotation noted as pitch on the figure. This translates into a variation in the incidence of the profile.
- The elastomer bearing allows for fluttering of the blade in the plane of the figure, called "drag fluttering."

Consequences include the following:

- A spectacular decrease in the number of components. (One goes from 377 pieces for a classical metallic solution with 30 bearings to 70 pieces for a composite solution without any bearings.)
- There is a corresponding mass reduction of 40 kg.
- There is a reduced cost of fabrication.

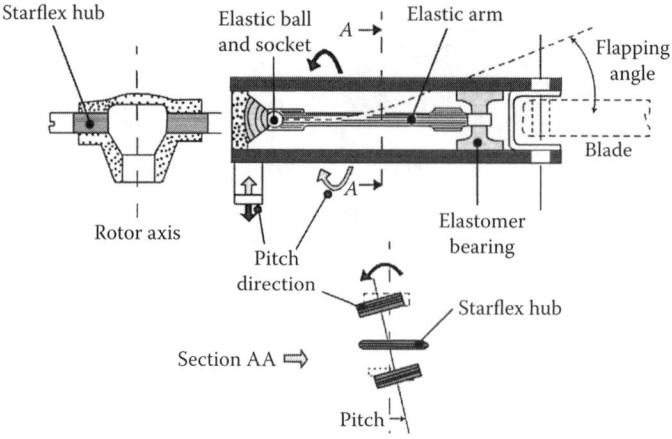

Figure 7.41   Details of the Starflex Yoke System

■ The maintenance is reduced in considerable proportion, lowering significantly the hourly cost of the flight.

■ There is increased security (more viability of the mechanical assembly).

7.2.4.1   Evolution of the Yoke Rotors

Example: The yoke "Triflex" Aerospatiale (FRA). The elastic deformation of the arms (see Figure 7.42) is sufficient to assure the displacements in lift, drag, and the variable incidence (pitch). The liaison arm/yoke and arm/joints of the blades are delicate works.

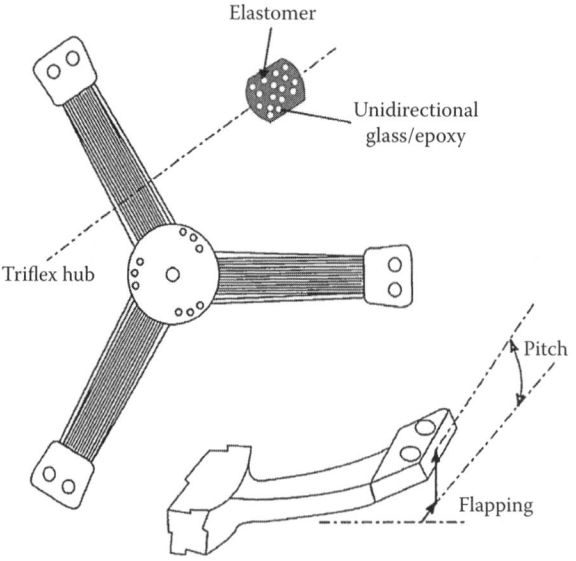

Figure 7.42   Triflex Yoke

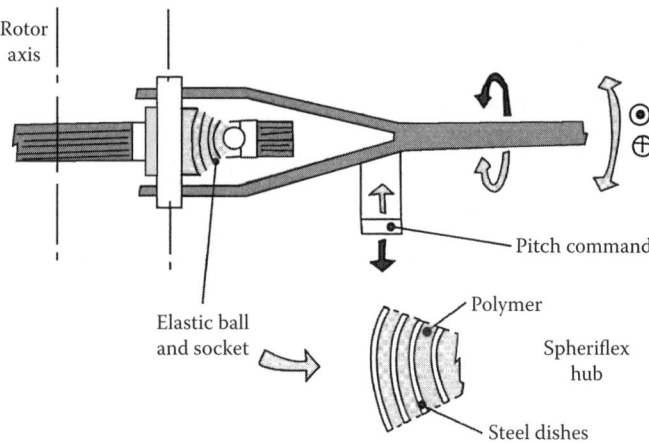

Figure 7.43   Spheriflex Yoke

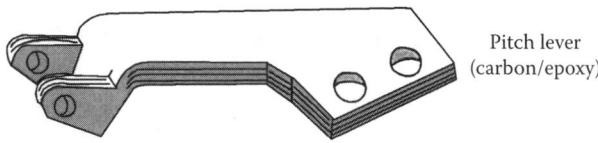

Figure 7.44   Pitch Lever

Example: The yoke "Spheriflex" Aerospatiale (FRA). This is so called because only the elastic ball-and-socket allows the various angular displacements: flapping and pitching. The foot of the blade is modified as a consequence (see Figure 7.43). The number of components becomes extremely reduced, with a minimum volume (less than the volume of the previous solutions).

7.2.5   Other Composite Working Components

Most of these are made of carbon/epoxy. The parts already in service or in development include

- The rotor mast
- The plate for the cyclical control of the pitch
- The struts for the control of the pitch
- The levers for the pitch (see Figure 7.44) where the composite design leads to a reduction of mass of 45% as compared with the metallic solution
- Lever winch gallows[15]
- The stabilizer

Example: Aircraft Dauphin Aerospatiale (FRA).
In the case of the military helicopters:

---
[15] See Section 1.5.

Stabilizer in light alloy	Stabilizer in carbon/epoxy
231 parts	88 parts
5900 rivets	0 rivets
Mass = 1	Mass = 0.78
Global cost = 1	Global cost = 0.66

- Utilization of composite materials reduces the "**radar signature**" of the helicopter.
- The damages by the projectiles on the blades, yokes, and command struts evolve more slowly in the composite parts and allow the aircraft to be able to return to its base (except for the case of a projectile shell break with a diameter larger than 20 mm).
- The "crash" resistance[16] is less for a composite structure as compared with conventional structure.

7.3 PROPELLER BLADES FOR AIRPLANES

The design using composites for the propeller blades for airplanes is analogous to that for the helicopter blades. They consist essentially of a torsion box of composite with a spar made of metal or composites.

Example: Propeller blade Hamilton Standard (USA) for the motor "14 SF" for the airplane ATR 42 (Figure 7.45).[17]

Example: Propeller blade Ratier-Figeac (FRA) for the motor of the plane TRAN-SALL (FRA–GER). The adoption of a spar in unidirectional glass and a torsion box in carbon leads to a particular weight reduction, when the diameters of the rotors become important, as indicated in Figure 7.46, which has the following characteristics:

- Diameter of the 4-bladed rotor: 5.5 m
- Mass of a composite blade: 51 kg

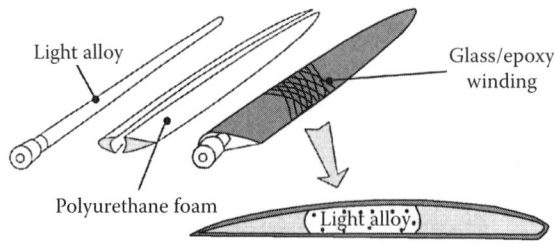

Figure 7.45 Propeller Blade, Hamilton Standard

- Mass reduction as compared with a metallic blade: 53 kg (mass of a metallic blade: 104 kg)

---

[16] See Section 7.1.4.
[17] See Section 7.1.6.

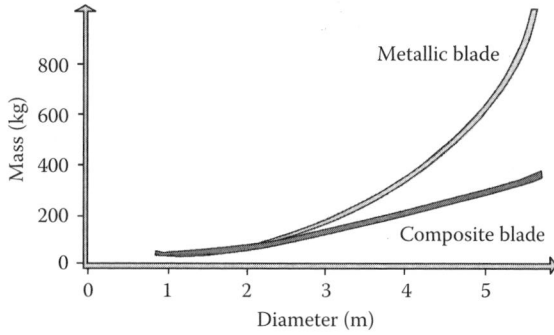

Figure 7.46   Mass Saving in Using Composites for Blades

■ Total mass reduction (2 × 4 blades): 430 kg

The centrifugal inertia force at the foot of the blade decreases from 105,000 daN to 30,000 daN. This is taken up by the glass fibers in the spar which are bonded to a steel piece having the form of a tulip and addition of circumferential hoop winding of rovings to allow for the "fail safe" design. If there is bonding rupture, the blade is retained on its base by the hoop windings.

The construction of the Ratier-Figeac propeller blade is shown schematically in Figure 7.47.

Example: Airplane with rocking rotors XV-15 BELL (USA; Figure 7.48), propellers Dowty Rotol Ltd. (U.K.), which has the following characteristics:

■ Rotor diameter: 7.6 m
■ Chord at foot of blade: 0.5 m
■ Mass of composites: 70% of the mass of the structure
■ Wing: stiffened panels in carbon/epoxy

The new generation of the propellers (called "**rapid**" or "**transonic**") is destined to propel commercial airplanes with a speed close to that obtained with the jets (Mach 0.8 to 0.85, or more than 850 km/h). Interest in such propellers rests in a propulsion efficiency higher than that of modern jets (with double flux) at high speeds, as shown in Figure 7.49.

For good aerodynamic and acoustic behavior, the propellers are characterized by low thickness, a large "chord" and a strong curvature which gives them the form of an oriental saber. The complexity of the geometry combined with important speeds of rotation (more than 4000 revolutions per minute) requires the composite construction.

Example: Propeller "Propfan" Hamilton Standard (USA; Figure 7.50). Made of two rotors in "counter-rotation," this is designed to propel the middle range airplanes. The method of fabrication is the same as the more classical propellers mentioned above: a spar made with light metal alloy forged and machined. A filling foam is molded around the spar to give the form of the torsion box in glass/epoxy.

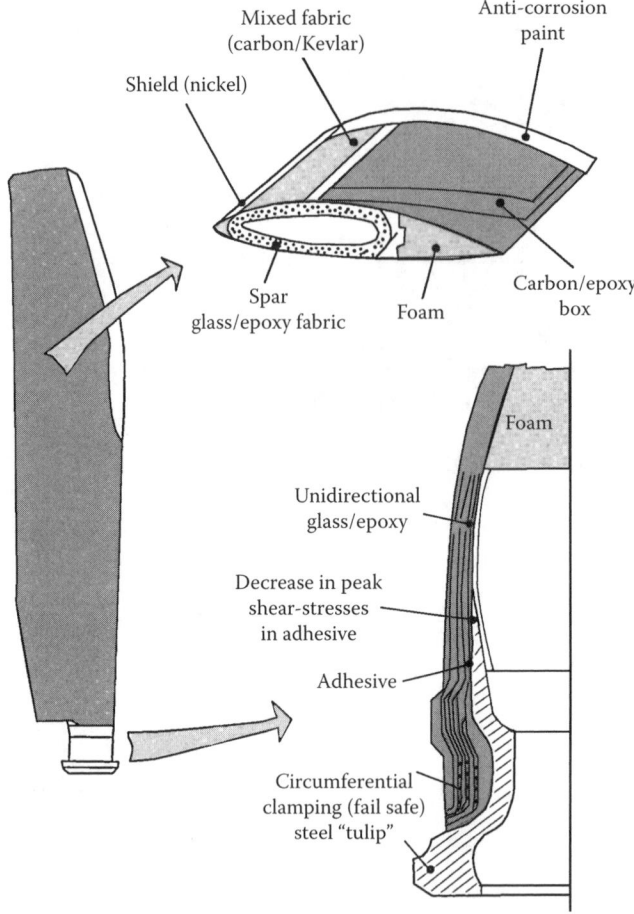

Figure 7.47   Ratier-Figeac Propeller Blade

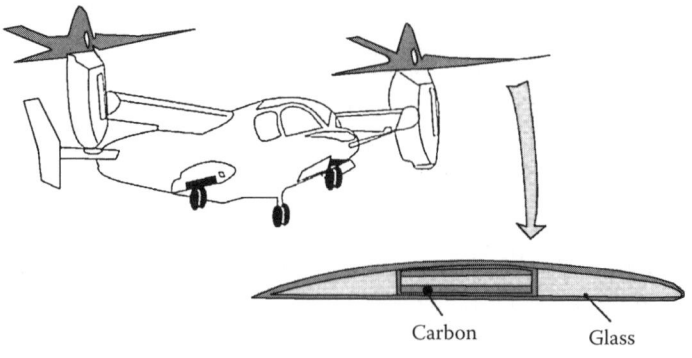

Figure 7.48   Rocking Rotor in Bell XV-15 Plane

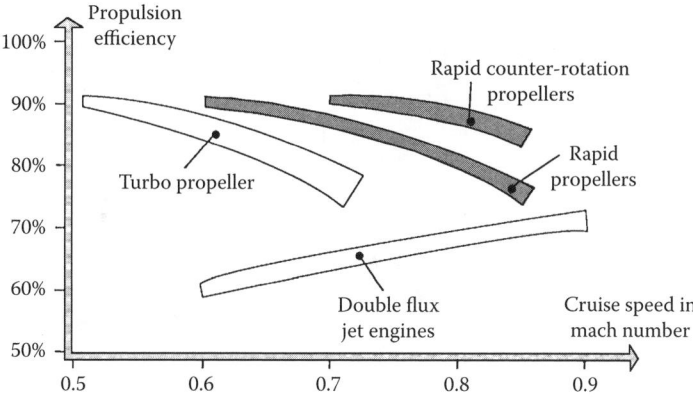

Figure 7.49   Propulsion Efficiency of Propellers

Figure 7.50   Propfan Propeller

Example: Transonic propeller "Charme" ONERA (FRA; Figure 7.51). The number of blades is high (12) to reduce the load in each blade. The blades are made of carbon/epoxy bonded to the titanium foot of the blades.

7.4 TURBINE BLADES IN COMPOSITES

Research to increase the efficiency of aeronautical motors leads to the increase of

- Air compression ratios, which reach the values on the order of 45.
- Temperatures of the gases in the turbines. The fixed or mobile blades in the release stages have to operate with a gas temperature on the order of 1500°C.

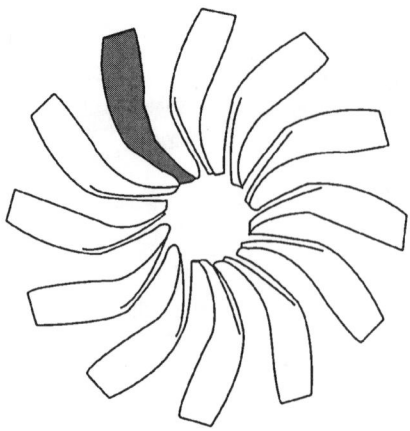

Figure 7.51   The Transonic Propeller "Charme"

Among the materials experimented with in view of increasing the operating temperature are the "oriented eutectics." These are composed of two distinct phases:

■ A superalloy phase with nickel base
■ A carbon phase

The solidification of this alloy is directed with a solidification front (interface between the liquid and the solid). The carbon phase is developed in the form of fibers with diameters on the order of microns and of infinite length, and parallel to the direction of displacement of the solidification front. One then obtains a unidirectional composite metal/metal.

The advantages and disadvantages of these types of materials can be summarized as follows:

■ **Pluses:**
It is possible to obtain directly, using solidification, pieces with the form of the turbine blades.
There can be a gain of 30°C on the maximum operating temperature of actual blades.
■ **Minuses:**
The growing speed of the composite is low: 2 cm/hour.
The proportion of fibers is low (5% to 10%) and cannot be controlled because it is constrained by the composition of the eutectic.
The material is sensitive to thermal fatigue due to the run–stop operating cycles of the turbine.
Fracture resistance at temperatures above 900°C is weaker than that for super alloys in actual service.

7.5 SPACE APPLICATIONS

There is no doubt that for launchers, space shuttles, and satellites the reduction of mass is most critical. Each kilogram reduction on the launcher for the European rocket Ariane E.S.A. (EU) gives a gain in payload of 30,000 US dollars.

7.5.1 Satellites

The structural part of the satellites is essentially constituted of an assembly of tubes and plates. Principally the structure has to

- Resist average and fluctuating accelerations of the launch, counted as number of times the acceleration of gravity (g = 9.81 m/sec^2), up to 5 × g continuously and 5 × g maximum amplitude in sinusoidal fluctuation, for frequencies up to 40 Hz. (In order to avoid resonance, the structure has to be very rigid. It is the **rigidity** that appears like the factor that controls the dimensions.)
- Be quasi-insensitive to temperature variations (such as in the case of precision optical instruments: telescope, high-resolution camera). Here, carbon is used for the tubular structure (very low coefficient of expansion, on the order of 10^{-7}).[18]

The primary structures of satellites can include sandwich plates, with the following characteristics:

- Light alloy honeycomb cores (not Nomex[19] because of gas emission in space).
- Facings (skins) made of laminates, without mirror symmetry for maximum lightness. The thickness of the skin is on the order of 0.1 mm. When demolding,[20] they are very deformed. These are then bonded onto the aluminum core. One then obtains a sandwich plate that is globally balanced.

Example: Camera V.H.R. (Visible high-resolution) SPOT (FRA), which is the upper part of the satellite shown in Figure 7.52.

In the case of the spatial structures, among the foreseen solutions for the construction of tubular space stations, one can experiment on tubes made of extruded carbon with junctions of half-shell made of carbon.

7.5.2 Pressure Vessels

The pressure vessels contain the combustibles, fuel or "powder," for the propulsion. These can be made by winding impregnated unidirectional ribbons over a mandrel with a particular form. The mandrel must be resistant to shrinkage after polymerization and should be designed to be extractable (see Figure 7.53).[21]

The **efficiency** of a filament-wound pressure vessel is defined as:

$$\text{efficiency ratio}_{(\text{meters})} = \frac{p(\text{burst pressure})}{\rho g(\text{specific weight})}$$

and has the units of length, for example:

- Efficiency of glass/epoxy: 25 km.
- Efficiency of Kevlar/epoxy: 35 km.

which explains the predominant use of Kevlar/epoxy in these applications.

---

[18] See Section 1.6.
[19] See Section 1.6.
[20] See Section 5.2.3.
[21] See Section 2.1.7.

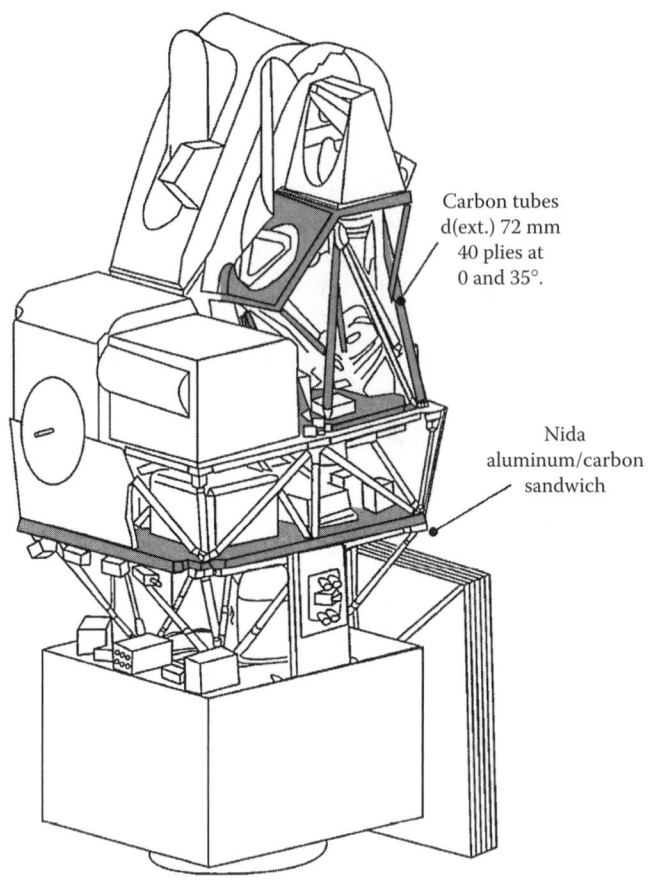

Figure 7.52   Camera V.H.R. Spot

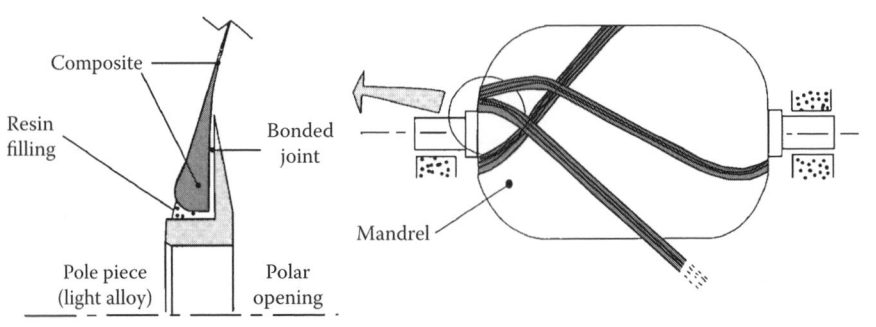

Figure 7.53   Filament-wound Pressure Vessel

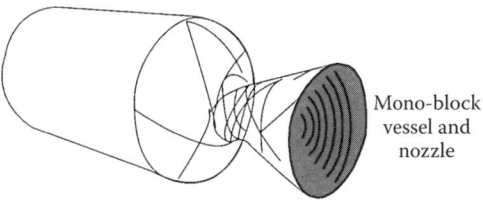

Mono-block
vessel and
nozzle

Figure 7.54   Filament Winding for Mono-block Part

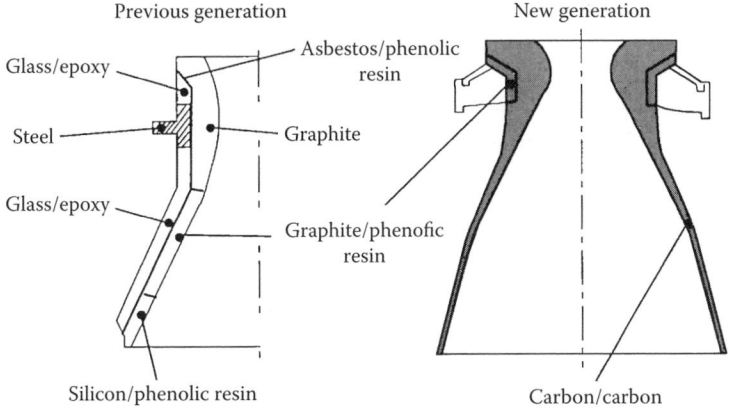

Previous generation   New generation

Glass/epoxy   Asbestos/phenolic
                                   resin

Steel   Graphite

Glass/epoxy

                             Graphite/phenofic
                                   resin

Silicon/phenolic resin   Carbon/carbon

Figure 7.55   Evolution of the Structure of Nozzles

For some applications, the principle of filament winding allows one to obtain vessels and tubes at the same winding session (see Figure 7.54).

7.5.3 Nozzles

The propulsion nozzles of solid propergol (powder) have operating temperatures up to 3000°C during several dozens of seconds, with pressures varying between a few bars and several dozens of bars.[22] The material making up the internal liner disappears progressively by decomposition, melting, vaporization, and sublimation. This is the phenomenon of **ablation**.

The materials that can play such a role have to possess

- ■ A strong resistance to ablation at a high operating temperature
- ■ A low specific mass
- ■ A strong resistance to mechanical and thermal shock

Figure 7.55 shows the evolution of the structure of these nozzles until the advent of the three-dimensional carbon/epoxy composite materials with the mechanical characteristics indicated in Section 3.6.

---

[22] 1 bar = 0.1 *MPa*.

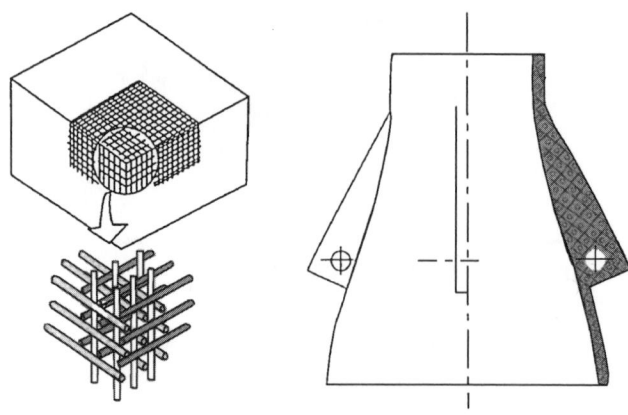

Figure 7.56   Sepcarb Material for Propulsion Nozzles

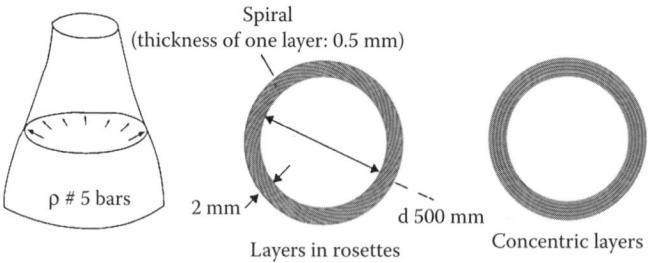

Figure 7.57   Nozzles in Rosette Form

Example: Material "Sepcarb" European company for propulsion (FRA; Figure 7.56).[23] The quantity of heat before ablation can reach 84×10^6 joules per kilogram of material. For example the motor for peak operation of the European launcher Ariane, with the divergent nozzle made of carbon/epoxy, has the following characteristics:

■ A mass reduction of 50% in comparison with previous nozzle constructions
■ A gain of the launch force of 10% thanks to higher elongation

Example: Divergent nozzle with "rosette" layering. Figure 7.57 shows the difference in constitution of this type of nozzle and a nozzle with classical concentric stratification, with a few orders of dimensional amplitude.

To compare with the concentric stratification, this design:

■ allows more convenient machining (more precise work of the lathe tool).
■ is more resistant to delamination.

---

[23] See Section 3.6.

7.5.4 Other Composite Components

7.5.4.1 For Thermal Protection

One can distinguish two modes on entrance into the atmosphere during the return of the space vehicles:

- Rapid entrance with strong incidence: This is the case of the ballistic missiles and manned capsules. The heat flux is very high (on the order of 10,000 kW/m^2) with relatively short time of entrance. One can use, depending on the particular case:
  - Heat sinks[24] in carbon/carbon or in beryllium (for case of the ballistic missiles).
  - Ablative materials (see above for the case of the nozzles) for the manned capsules.

- Slow entrance with weak incidence: This is the case of hypersonic planes or "space shuttles." The duration of the entrance is on the order of 2000 seconds. The heat fluxes are weaker but can attain hundreds of kilowatts per square meters of the structure at the beginning of the entrance (80 km altitude), for example:
  - 500 kW/m^2 at the leading edge
  - 100 to 200 kW/m^2 on the under part

The entrance temperatures reach 1700°C, or 2000°C at the nose of the shuttle. There are several types of thermal protection, depending on the zones of the equipment and the reutilization of the facing:

- Heat sinks[25] associated with insulation
- Reflective thermal barrier (lining of the vehicle reflects the heat flux it receives)
- Ablative facing (The transformation of the facing by fusion, vaporization, sublimation, chemical decomposition absorbs the heat, and the vaporized gases cool the remaining layer, decreasing also the convective thermal flux.)

The areal masses of these devices are related to the limiting admissible temperatures of the structure immediately below (see Figure 7.58).

Example: NASA space shuttle (USA), which has an empty mass of 70 tons.

Depending on the zones, one uses the linings made of composites of carbon/carbon or silicon/silicon and pieces of structure (horizontal members, cross members) in boron/aluminum. The useful temperature of the latter is 300°C for continuous use and up to 600°C for peak applications.

The under part is protected by composite "tiles" in silicon/silicon ceramic[26] that constitutes a reflective thermal barrier. The tiles are separated from the structure of light alloy or laminated boron/aluminum by a sandwich of felt and nonflammable

---

[24] See Section 7.1.10.

[25] See Section 3.7.

[26] See Sections 2.2.4 and 3.6.

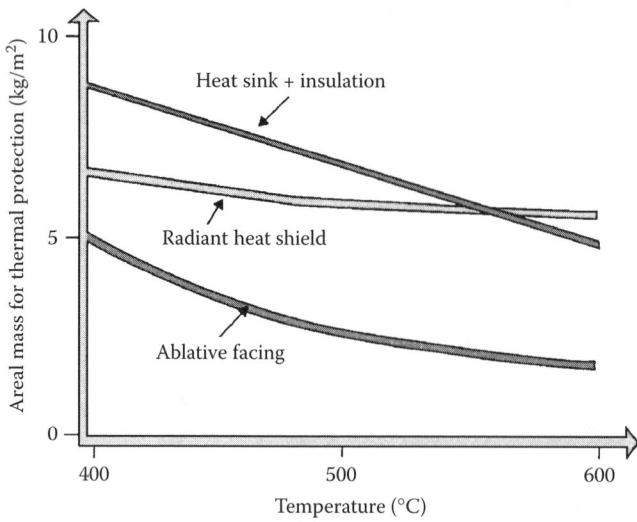

Figure 7.58   Areal Mass for Thermal Protection

nylon/silicon/"NOMEX" honeycomb. There are about 30,000 tiles. Their installation is shown in Figure 7.59.

Example: Space shuttle Hermes (EU), which has an empty mass of 8.5 tons. The tiles are replaced by thousands of pieces made of carbon/carbon, silicon/silicon (see Figure 7.60). The pieces have to be reused for thirty landings.

7.5.4.2   For Energy Storage

On board satellites and space stations, systems using the composite flywheels for the supply of electric power and for the control of attitude provide a mass reduction of 25% as compared with conventional storage methods using batteries and gyroscopic means (specific power on the order of 5 kW per kilogram of the device). In addition, in the Strategic Defense Initiative program (USA), the devices of flywheels can deliver high levels of specific powers on the order of 100 kW per kilogram of the device.

The peripheral speeds can attain 1400 m/sec (filament-wound flywheels in carbon), and the speeds of rotation from 40,000 to 60,000 rpm.

Example: Development of an energy storage module (USA; Figure 7.61).

■ Total mass: 200 kg (occupied volume: 0.15 m^3)
■ Specific energy: 230 kJ/kg (total energy # 46,000 kJ)
■ Peripheric speed: 1100 m/sec

Figure 7.62 shows different solutions for the construction of carbon/epoxy flywheels.

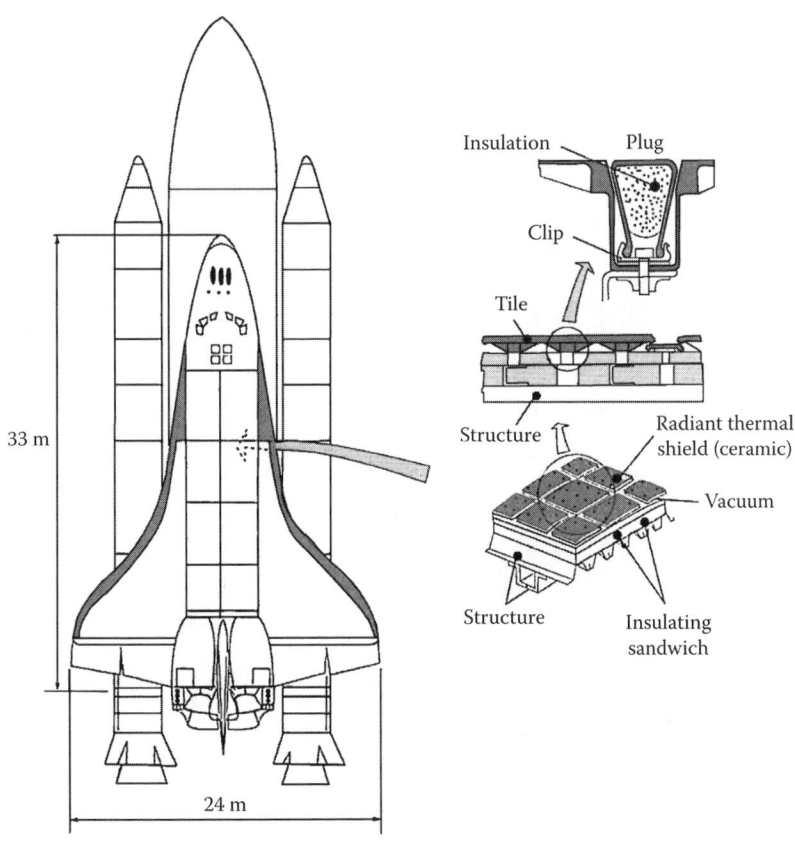

Figure 7.59 NASA Space Shuttle

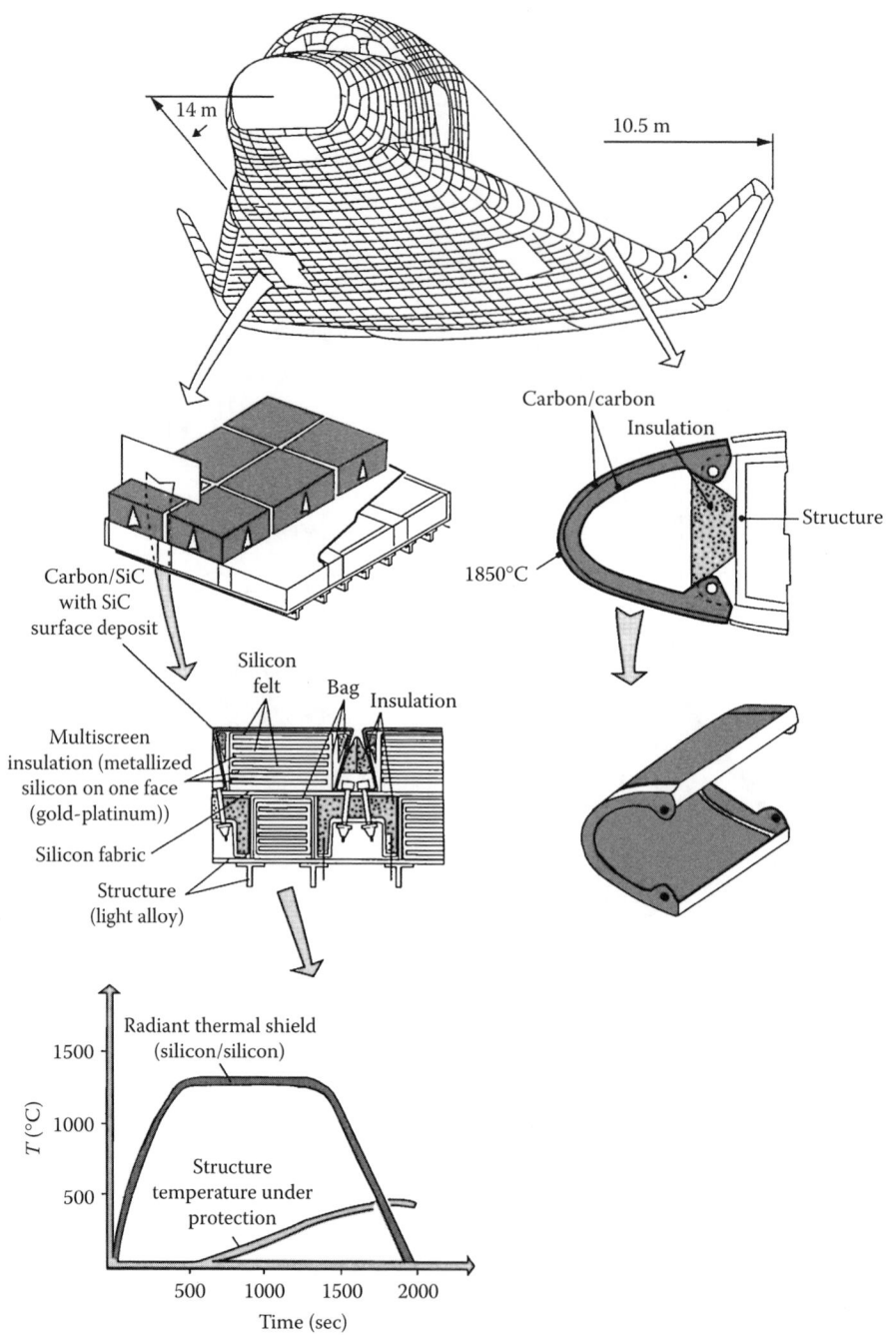

Figure 7.60   Space Shuttle Hermes

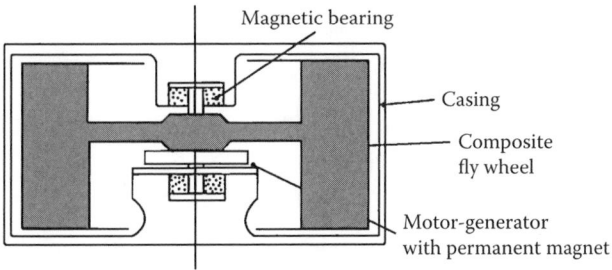

Figure 7.61   Flywheel Energy Storage

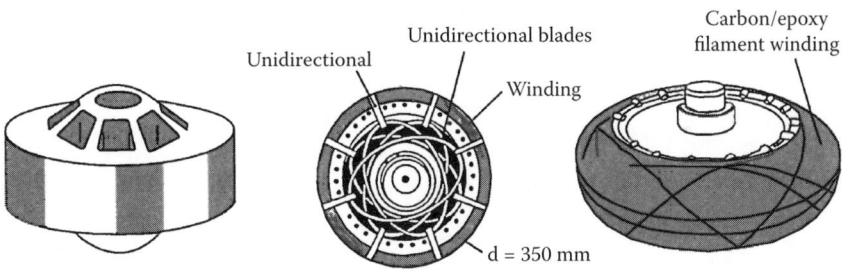

Figure 7.62   Different Flywheel Designs

8

COMPOSITE MATERIALS FOR OTHER APPLICATIONS

We have given in Chapter 1 an idea on the diversity of the products which can be made using composite materials.[1] In this chapter we examine a few of these products, which form a good part in the evolution of these materials, excluding the aerospace sector presented in the previous chapter.

8.1 COMPOSITE MATERIALS AND THE MANUFACTURING OF AUTOMOBILES

8.1.1 Introduction

Composite materials have been introduced progressively in automobiles, following polymer materials, a few of which have been used as matrices. It is interesting to examine the relative masses of different materials which are used in the construction of automobiles. This is shown in the graph in Figure 8.1. Even though the relative mass of polymer-based materials appears low, one needs to take into account that the specific mass of steel is about 4 times greater than that of polymers. This explains the higher percentage in terms of volume for the polymers. Among the polymers, the relative distribution can be shown as in Figure 8.2.

The materials called "plastics" include those so-called "reinforced plastics" for composite pieces that do not have very high performance.

8.1.2 Evaluation and Evolution

A few dates on the introduction of composite parts (fibers + matrix) include:

- The antiques as shown in Figure 8.3
- 1968: wheel rims in glass/epoxy in automobile S.M.Citroen (FRA)
- 1970: shock absorber shield made of glass/polyester in automobile R5 Renault (FRA)

---

[1] See Section 1.3.

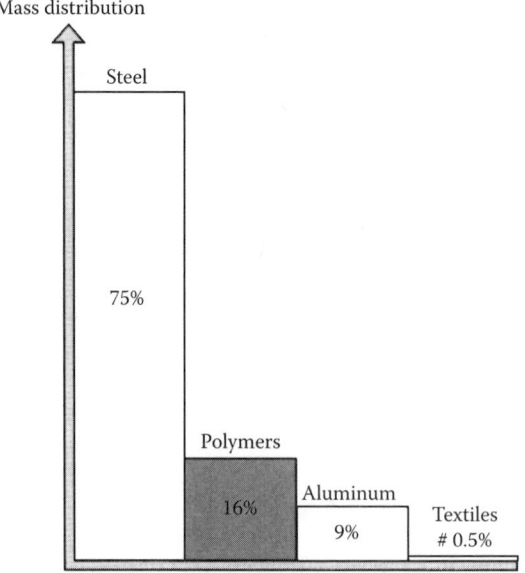

Figure 8.1   Use of Different Materials in Automobiles

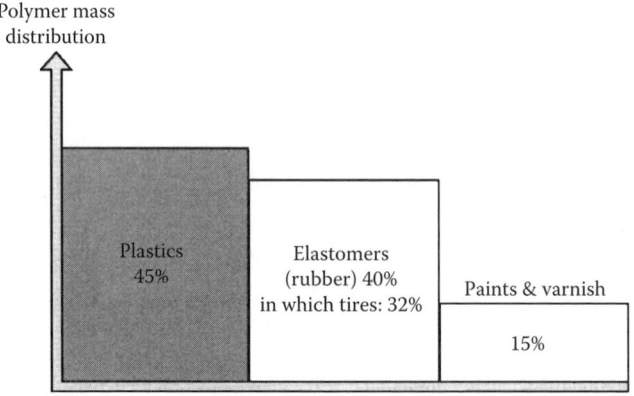

Figure 8.2   Mass Distribution among Polymer Materials

Consequences of the introduction of composite pieces in automobiles are now well-known. They allow a number of advantages. One can find several common points with aeronautic construction. There are also disadvantages that are more specific to automobiles.

- ■ Advantages include
  - ■ Lightening of the vehicles: A reduction of mass of 1 kg induces a final reduction of 1.5 kg, taking into account the consecutive lightening of the mechanical components.
  - ■ Cost reduction: This is due to the reduction of the number of pieces required for a certain component and to noise reduction and isolation.

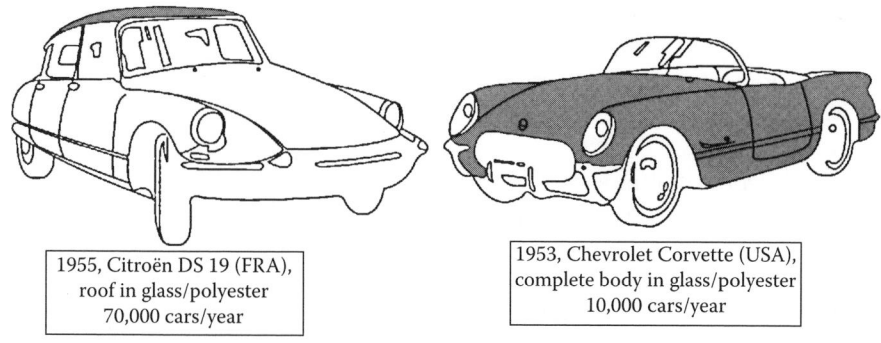

1955, Citroën DS 19 (FRA),
roof in glass/polyester
70,000 cars/year

1953, Chevrolet Corvette (USA),
complete body in glass/polyester
10,000 cars/year

Figure 8.3   Composite Pieces in Antique Cars

- The better corrosion resistance of the composite pieces.
- Significant disadvantages are
  - It is difficult, for fabrication in large volume, to obtain as good a surface finish as that of painted sheet metals.
  - For the car body, the painting process and the treatment of the surfaces require high temperature exposure.

How to Evaluate the Gains:

In theory: These are the experimental vehicles; Ford, Peugeot (1979). As compared with the metallic pieces, composite parts have obtained mass reduction of

- 20% to 30% on the pieces for the body
- 40% to 60% on the mechanical pieces

Example: Ford vehicle, which has a mass in metallic construction of 617 kg and a mass in composite construction of 300 kg for a global gain of 52%. It is convenient to consider this case as "technological prowess" far from the priority of economic constraints.

In practice: Over the past years, an increasing number of pieces made of glass fibers/organic matrices have been introduced. The following list contains pieces that are in actual service or in development.

- Components for the body
  - Motor cap
  - Hood cover
  - Hatchback door
  - Fenders
  - Roofs
  - Opening roof
  - Doors
  - Shock absorber
- Interior components
  - Seat frames

- Side panel and central consoles
- Holders
- Components under the hood
    - Headlight supports
    - Oil tanks
    - Direction columns
    - Cover for cylinder heads
    - Cover for distributor
    - Transmission shafts
    - Motor and gearbox parts
- Components for the structure
    - Chassis parts
    - Leaf springs
    - Floor elements

Figure 8.4 shows the importance of the volumes actually occupied by the composites in an automobile.

Example: Automobile BX Citroen (FRA)1983 with a total mass of 885 kg. Many of the molded pieces made of glass/resin composites as shown in Figure 8.5 are now commonly used by the automobile manufacturers. We note in particular the two elements below, the importance and large volume production of which (rate of production of more than 1000 pieces per day), indicate a significant penetration of composites in the manufacturing of automobiles.

- The hood is made of glass/polyester molded at high temperature in a press (20,000 kN) with the deposition of a gel coat during molding[2] to assure the quality of the surface. The following comparison is eloquent:
    - Conventional metallic construction (GS Citroen): 7 elements
    - Composite construction: 1 element
    - Mass gain: 7.8 kg or 46%.
- The rear window frame is made of injection molded glass/polyester (23,000 kN press). The mechanical characteristics obtained with this method (fracture resistance, modulus, impact resistance) are in the vicinity of those obtained by compression molding. One obtains a single piece that can support the rear glass piece, the aerodynamic details, the hinges, and the lower part of the trunk. The advantages over the classical construction are great:
    - Classical construction (GS Citroen): 27 elements
    - Composite construction: 7 elements
    - Mass reduction: 1.7 kg or 16%

Example: Automobile Alpine V6 Turbo, Renault (FRA), 1986 (Figure 8.6). The entire body in glass/polyester composites is not obtained by molding according to the technique used for the previous model A 310 (contact molding).[3] It is made by bonding around fifty elements in glass/polyester on a tubular chassis.

---

[2] See Section 2.1.1.
[3] See Section 2.1.1.

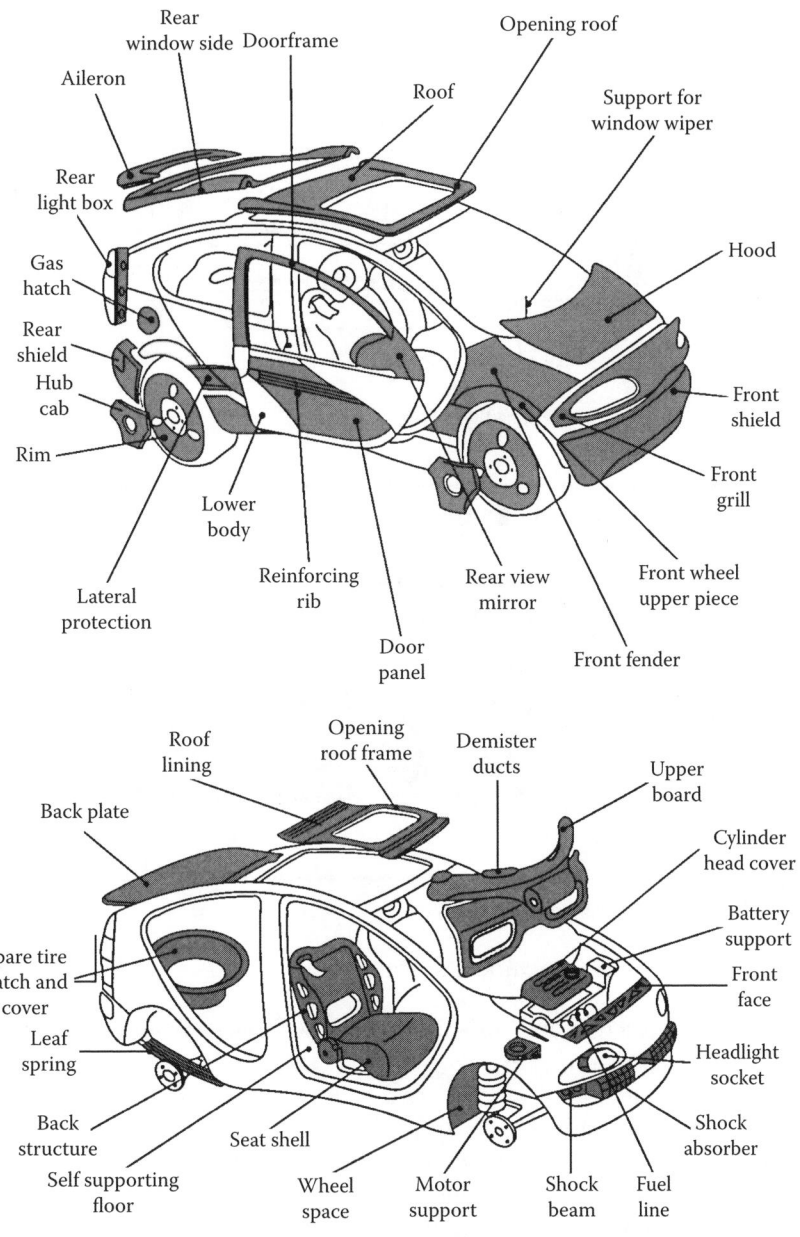

Figure 8.4 Composite Pieces in an Automobile

- The panels are made by molding using a press at low pressure and temperature (6 minutes at 45°C).
- Contouring is done using a high-speed water jet.[4]
- Structural bonding is done on a frame at 60°C. Robots control it. The classical mechanical nuts and bolts are replaced by 15 kg of adhesives.

---

[4] See Section 2.2.5.

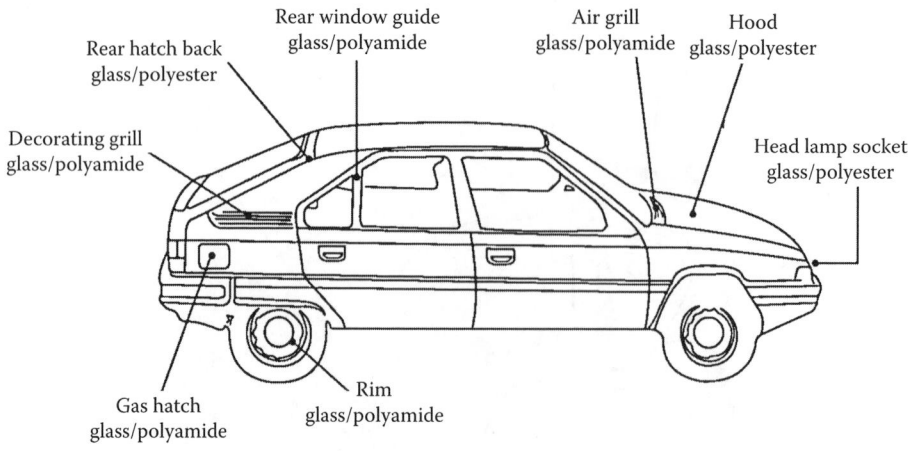

Rear hatch back
glass/polyester

Rear window guide
glass/polyamide

Air grill
glass/polyamide

Hood
glass/polyester

Decorating grill
glass/polyamide

Head lamp socket
glass/polyester

Gas hatch
glass/polyamide

Rim
glass/polyamide

Figure 8.5   Composite Pieces in BX Citroen

Figure 8.6   Automobile Alpine V6, Renault

Significant advantages include the following:

- There is reduction in fabrication time: 80 hours versus 120 hours for the construction of the previous model A 310.
- Excellent fatigue resistance is realized: (mileage > 300,000 km).
- There is good filter for noise from mechanical sources.
- The flexibility in the method of fabrication: The tooling in the press is interchangeable in order to produce small series of different pieces on the same press. This process is well adapted to a low rate of fabrication (10 cars per day).
- Mass reduction—as compared with the technique used in the previous model, which itself was using composites—is 100 kg.

For a cylinder size of 2500 cm^3 (power of 147 kW or 240 CV), it is one of the most rapid series of vehicles ever produced in France previously (250 km/h) with a remarkable ratio of quality/price as compared with other competing European vehicles (Germany in particular).

Example: Racing car "F.1" Ferrari (ITA) (Figure 8.7). This car body is a sandwich made of NOMEX honeycomb/carbon/epoxy. In addition, a crossing tube made of carbon/epoxy transmits to the chassis aerodynamic effects that act on the rear flap. This is attached to the chassis by light alloy parts, bonded to the composite part

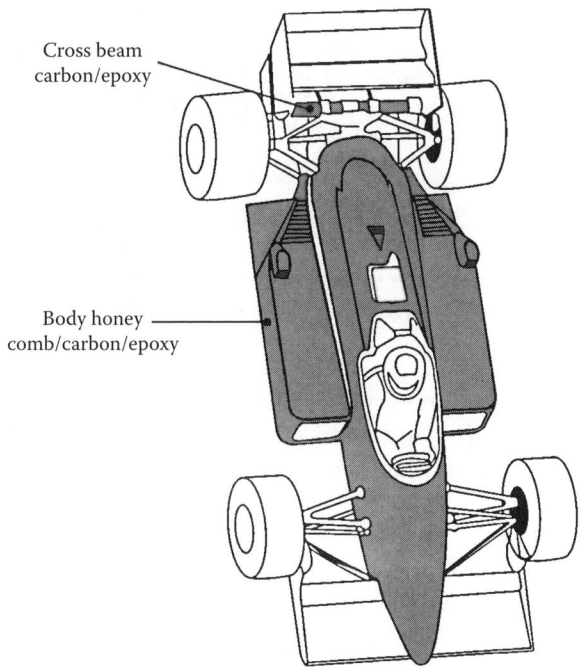

Cross beam
carbon/epoxy

Body honey
comb/carbon/epoxy

Figure 8.7 Ferrari F.1 Racing Car

with structural araldite epoxy adhesive. There is weight reduction compared with previous metallic solution, and one also sees very good fatigue resistance, which is important in regard to mechanical vibrations.

8.1.3 Research and Development

A number of working pieces—traditionally made of metallic alloys—of road vehicles have been designed and constructed in composite materials, and they have actually been tested and commercialized:

8.1.3.1 Chassis Components

Research and Development work has been concerned with the spars, floors, front structures, rear structures, and also the complete structure.

- Principal advantage: Reduction in the number of parts and thus in the cost.
- Secondary advantage: Mass reduction (beams for truck chassis in Kevlar/carbon/epoxy lead to a mass reduction of 38%—46 kg versus 74 kg for metal).

The problems involved are numerous:

- How to assemble the pieces.
- What will be the mechanical behavior when subjected to strong impacts?
- How is the rate of production to be augmented? The actual fabrication methods are too slow (decrease in the cycle time by using automation).

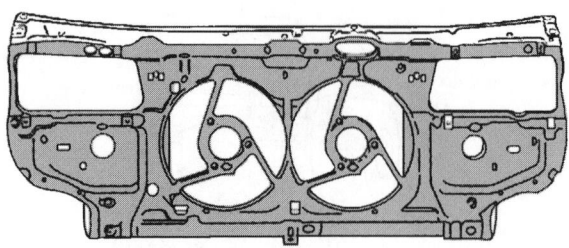

Figure 8.8 Front Face of 405 Peugeot

Example: The superior cross beam or "front face" of the automobile 405 Peugeot (FRA). This component (see Figure 8.8) is subjected to repeated load cycles in tension, flexure, and torsion. It also supports several dozens of components and equipment that form the front face of the vehicle. Characteristics include:

> Part molded in glass/polyester (V_f = 42%)
> Fabrication process: SMC[5]: Press 15,000 N
> Rate of production: 1200 pieces/day
> Machining/drilling (70 holes); installation of inserts (30) and components made by laser, numerical machining, and robots

8.1.3.2 Suspension Components

- Springs: One of the principal characteristics of the unidirectionals (namely glass/resin) is their capacity to accumulate elastic energy.[6] Herein lies the interest in making composite springs. In theory, a glass/resin spring is capable of storing 5 to 7 times more elastic energy than a steel spring of the same mass.

Other advantages include:

- The composite springs are "nonbreakable." Damage only translates into a minor modification of the behavior of the component.
- It is possible to integrate many functions in one particular system, leading to a reduction in the number of parts, an optimal occupation of space, and an improvement in road behavior.
- The mass reduction is important (see Figure 8.9)

The disadvantages: It is difficult to adapt the product to the requirements of the production. It is not sufficient to demonstrate the technical feasibility; one must optimize the three-criteria product-process-production rate (rates of production of several thousands of parts per day in the automotive industry, to be made using

---

[5] S.M.C. process: See Section 2.1.3 and 3.2.
[6] See Section 3.3.2, comparison of load-elongation diagrams for a metal and a unidirectional.

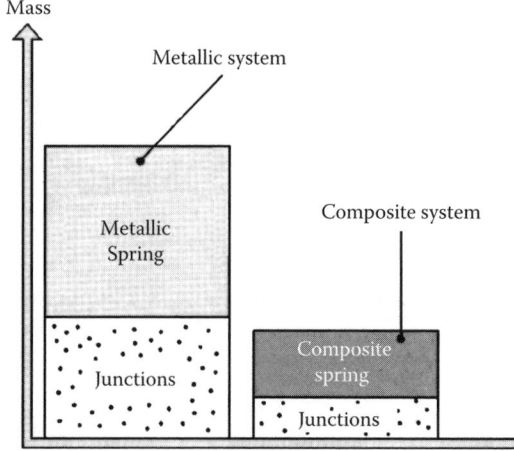

Figure 8.9 Comparison Between Metallic and Composite Springs

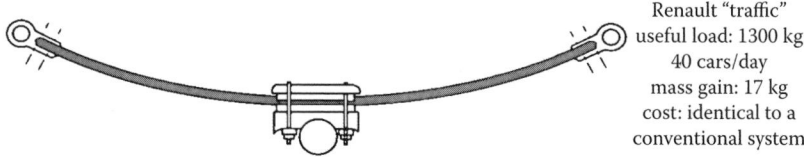

Renault "traffic"
useful load: 1300 kg
40 cars/day
mass gain: 17 kg
cost: identical to a
conventional system

Figure 8.10 Leaf Spring

a few processes, i.e., filament winding, compression molding, pultrusion, and pultrusion-forming).[7]

The current development and commercialization efforts deal with leaf springs and torsion beam springs.

Example: Single leaf spring (see Figure 8.10). A spring made of many metallic leaves is replaced by a single leaf spring made of composite in glass/epoxy. Many vehicles are sold with this type of spring, for example, Rover–GB; Nissan–JAP; General Motors–USA; Renault–FRA.

Example: Multifunctional system (Bertin–FRA). This prototype for the front suspension of the automobile combines the different functions of spring, rolling return, and wheel guide (see Figure 8.11).

Example: Stabilizing system. This is used for the connection between an automobile and a caravan (Bertin/Tunesi–FRA). The combined functions are shown schematically in Figure 8.12. The mass is divided by 4.5 in comparison with an "all metal" solution.

Example: The automobile suspension triangle has two parts (FRA) that are bonded to make a box (see Figure 8.13).

---

[7] See Chapter 2.

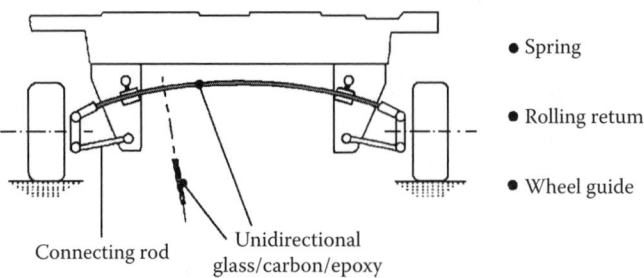

Figure 8.11   Combination of Functions

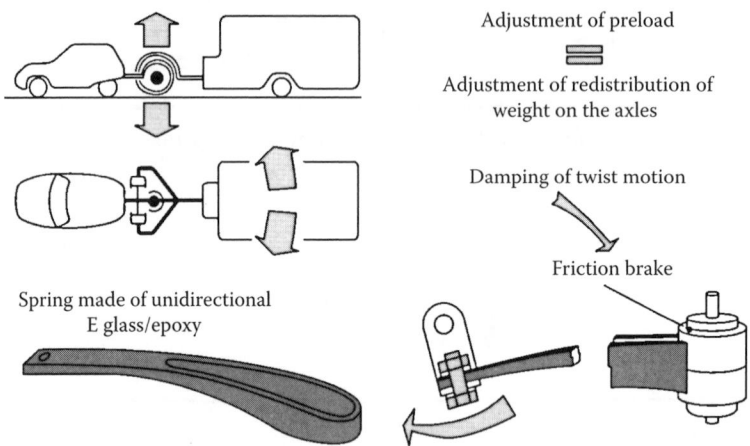

Figure 8.12   Stabilizing System

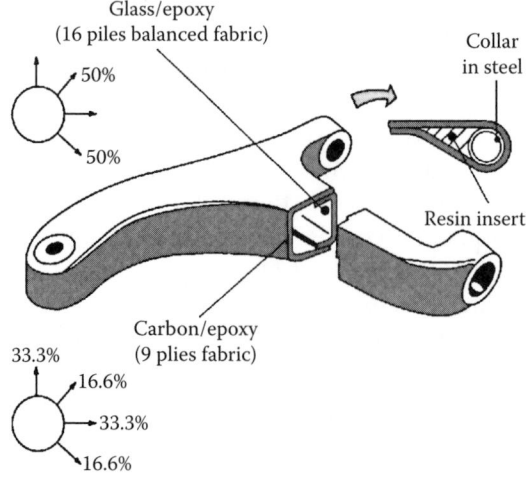

Figure 8.13   Composite Suspension Triangle

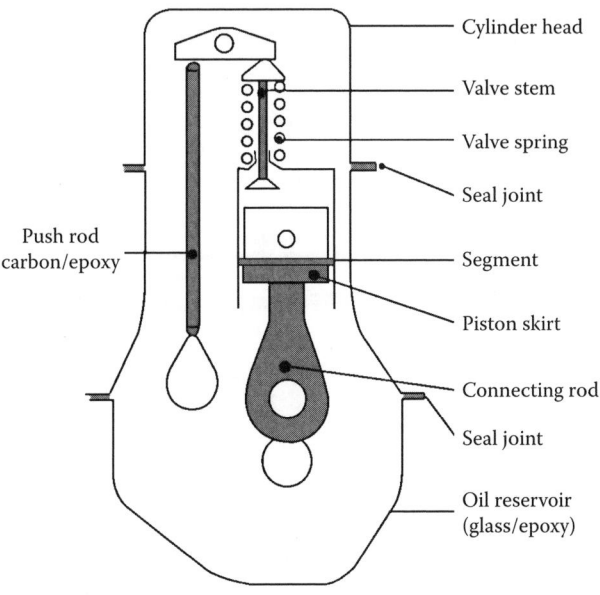

Cylinder head

Valve stem

Valve spring

Seal joint

Push rod
carbon/epoxy

Segment

Piston skirt

Connecting rod

Seal joint

Oil reservoir
(glass/epoxy)

Figure 8.14 Composite Mechanical Components

8.1.3.3 Mechanical Pieces

- Motor: The parts shown schematically in Figure 8.14 are in the experimental stage or in service in thermal motors. For pieces that have to operate at high temperatures, one should use the high temperature material system glass/polyamide (up to 300°C).[8] One can also mention the pump blades and the synchronizers in speed boxes made of glass/polyamide.
- Composite transmission shafts[9]: These are used in
  - Competition vehicles (rallies), allowing high speed of rotation with low inertia
  - Small and large trucks

Figure 8.15 shows how the low mass density associated with high rigidity in flexure allows the elimination of the intermediate bearing (this induces also a supplementary reduction in mass and cost).

For the transmission shafts equipped with supports, one obtains the following advantages:

- Reduction of mass of 30% to 60% in comparison with the transmission shafts with universal cardan joints
- Reduction in mechanical vibrations
- Decrease in acoustic vibration level (in particular the "peak")
- Good resistance against chemical agents
- Very good fatigue resistance

---

[8] The polyamide resin is said to be "thermally stable"; that is, it can maintain its mechanical properties at high temperatures (up to 500°C for one hour).

[9] See Application 18.1.4.

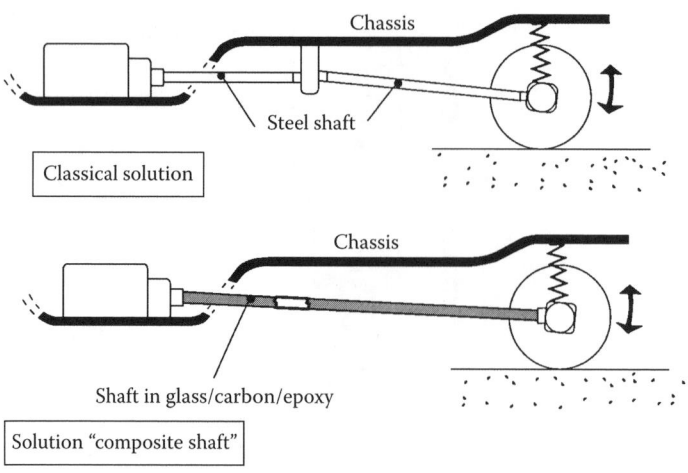

Figure 8.15 Composite Shafts

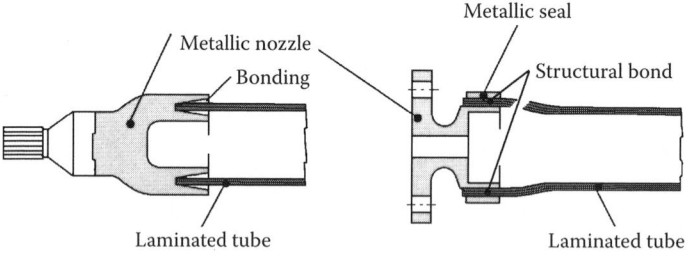

Figure 8.16 Composite-metal Shaft Bonding

■ Lateral transmission shafts are used for vehicles with front drive. They are used to eliminate the homokinetic joints that are actually used. They are made of a weak matrix material and wound fibers that allow the freedom of flexure for the transmission shaft.

8.2 COMPOSITES IN NAVAL CONSTRUCTION

8.2.1 Competition

8.2.1.1 Multishell Sail Boats

In the past years there has been a spectacular development in the sailboat competition, with significant research activities on the improvement of the qualities of the boats, and the design of sail boats called "**multishells**" with large dimensions, made of high performance composites, characterized by

■ Low mass leading to reduced "**water drafts**"
■ New and more performing "**riggings**"[10]

---

[10] There is a record speed of 34 knots with a class C catamaran (7.6 m in length).

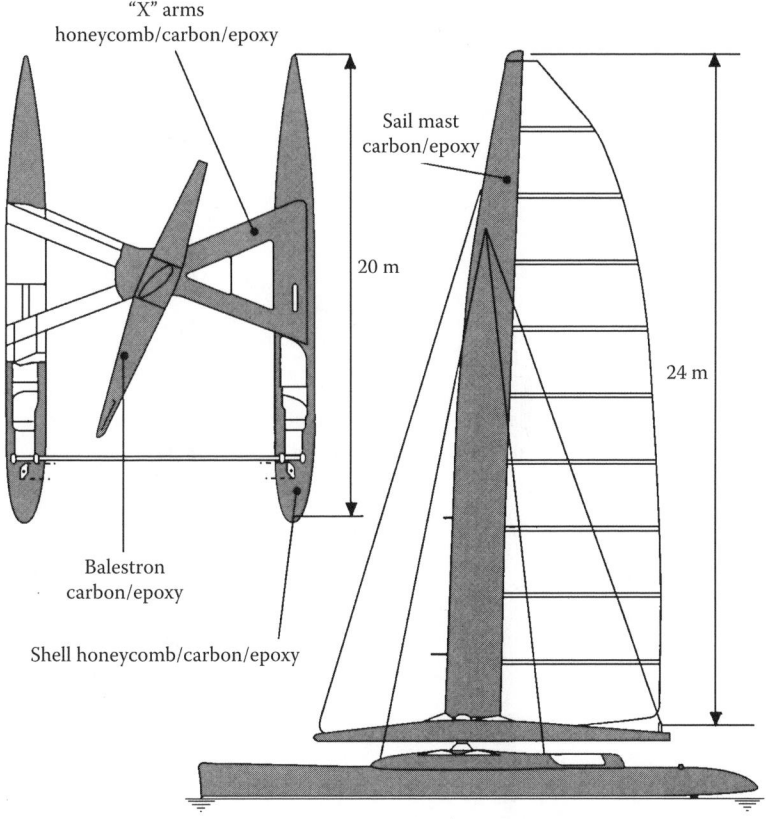

"X" arms
honeycomb/carbon/epoxy

Sail mast
carbon/epoxy

20 m

24 m

Balestron
carbon/epoxy

Shell honeycomb/carbon/epoxy

Figure 8.17   Elf Aquitaine Catamaran

■ Resistance against intense fatigue loadings, namely for the joint mechanisms between the shells

Example: Catamaran Elf Aquitaine (FRA) 1983 (see Figure 8.17). This is a large boat (20 m) in high performance composite materials. It has the following principal characteristics:

■ A mast-sail constituted of two half-shells in carbon/epoxy, 24 meters long
■ Connecting arms for shells with "x" shape that work in flexure to take up the difference in pitching between the two shells
■ A total fully equipped mass of 5.3 tons, corresponding to a mass reduction of 50% as compared with a construction in light alloy

Example: Competition skiff WM (FRA) 1995 (see Figure 8.18).
Example: Surf board 1995 (see Figure 8.19).

8.2.2   Ships

In the defense domain, there are composite boats of large dimensions (>50 m): escort-patrollers are expected to be 90 m in length.

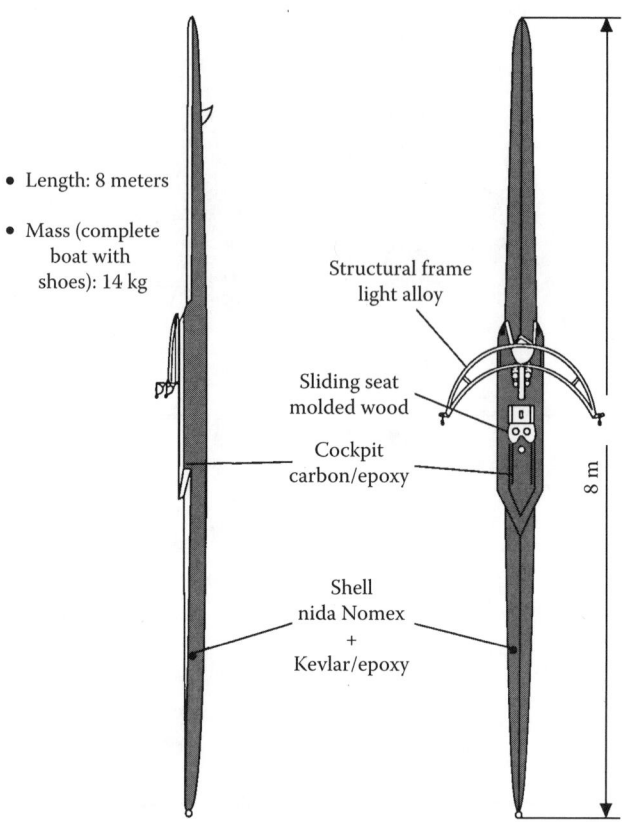

- Length: 8 meters

- Mass (complete boat with shoes): 14 kg

Structural frame light alloy

Sliding seat molded wood

Cockpit carbon/epoxy

Shell nida Nomex + Kevlar/epoxy

8 m

Figure 8.18   Competition Skiff

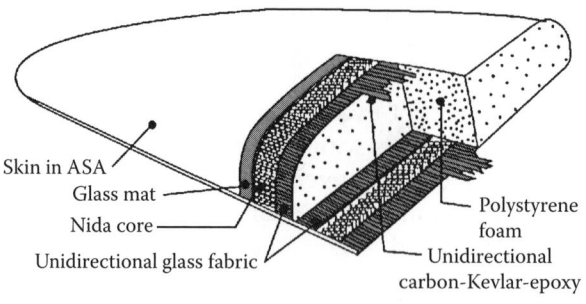

Skin in ASA

Glass mat

Nida core

Unidirectional glass fabric

Polystyrene foam

Unidirectional carbon-Kevlar-epoxy

Figure 8.19   Surf Board

Figure 8.20   Anti-mine Ocean Liner

Example: Anti-mine ocean liner BAMO (FRA). The Catamaran shell is 52 m in length, 15 m in width, and molded in 8 parts. It has 250 tons of glass/polyester composites in monolithic and sandwich construction with balsa core for the bridges and walls (see Figure 8.20).

8.3 SPORTS AND RECREATION

8.3.1 Skis

Initially made of monolithic wood, the ski has evolved toward composite solutions in which each phase — each in itself a composite material — fulfills a determined function. Figure 8.21 illustrates a transverse cross section of a ski of a previous generation: steel for the elasticity of the element, wood to dampen the vibration, interspersed aluminum for its better adherence to wood.

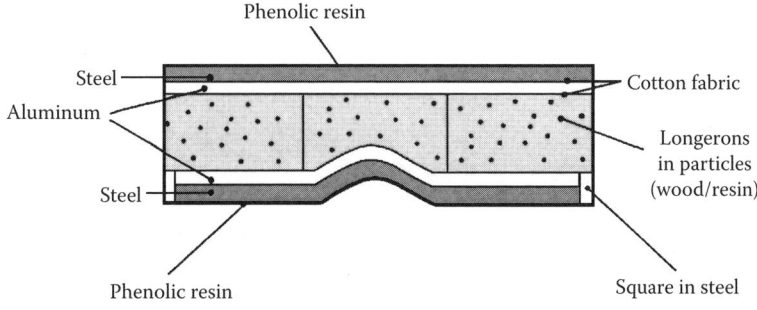

Figure 8.21   Cross Section of an Antique Ski

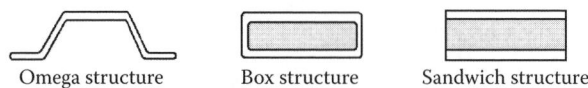

Omega structure   Box structure   Sandwich structure

Figure 8.22   Different Ski Cross Section Forms

Among the essential mechanical characteristics, the manufacturer has to master the following:

- Flexibility in flexure
- Ski stiffness in torsion (for turning)
- Elastic limit
- Fracture limit

One must also note the large diversities of the quality of

- The snow and the slopes (operating conditions)
- The skiers (different levels)

Taking into account the above specifications requires the manufacturers to provide a large variety of skis. The principal components include

- The **structure** is the part of the ski that assures the essential functions as mentioned previously. It may require a piece—or an assembly of several pieces—along the longitudinal direction of the ski, and the cross section area may take the form shown in Figure 8.22. In fact, other elements can be added to these requirements, which will allow one to control separately stiffness in flexure and stiffness in torsion. The ratio of these two stiffnesses will determine the behavior of the ski, particularly during turning. For this, one uses mainly the following materials:
  - Glass/epoxy
  - Carbon/epoxy
  - Kevlar/epoxy
  - Honeycombs
  - Zicral (AU 2 ZN)
  - Steel alloy

Example: Skis *Dynastar* (FRA): The real appearance of a section is complex, as shown in Figure 8.23, on which one can recognize the "omega," the box, and the sandwich structures described previously.

- The **filling** contributes little to the mechanical characteristics of the whole structure (10% to 15%), but it constitutes the big part in the total volume (70 to 80%). Its specific mass has to be as small as possible. The filling can be of
  - Wood, which is sensitive to humidity, with scatter in mechanical characteristics and specific mass depending on the lots
  - Polyurethane foam, with weak mechanical characteristics
  - Acrylic foam, which is very onerous

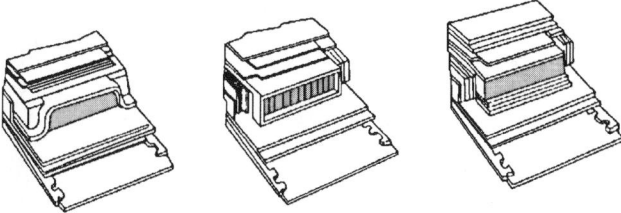

Figure 8.23   Section of Ski Dynastar

Figure 8.24   Composite Bicycle

- The **covers** and the **edges** are in glass/phenol or in glass/granix.
- The **upper edges** are in zicral, the lower edges are in steel.
- The **synthetic inner soles** have high specific mass.

8.3.2 Bicycles

Initially reserved only for competition, numerous variations with frames and wheels made of carbon/epoxy can now be found (see Figure 8.24).

8.4 OTHER APPLICATIONS

8.4.1 Wind Turbines

The renewed interest in wind turbines has been caused by the use of composites in the fabrication of the blades. These can be of large dimensions (lengths exceeding 24 meters).

   Example: Figure 8.25 gives an idea of the size of a wind turbine that develops a maximum power of 33 kW. The blades in glass/epoxy are made here by vacuum forming[11] using two half-shell assemblies by epoxy adhesive bonding.

---

[11] See Section 2.1.3.

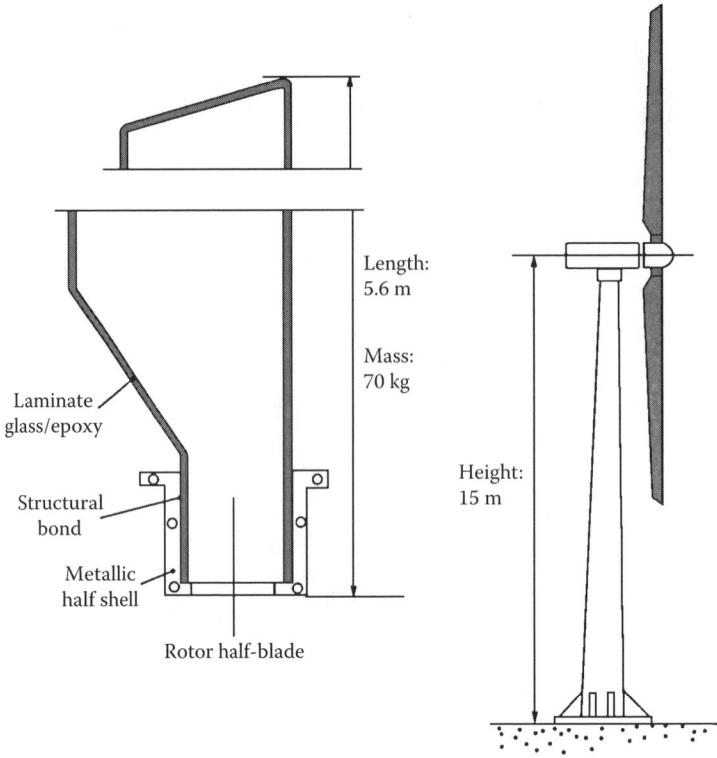

Length:
5.6 m

Mass:
70 kg

Laminate
glass/epoxy

Structural
bond

Metallic
half shell

Rotor half-blade

Height:
15 m

Figure 8.25   Wind Turbines

The blades can also be done by filament winding, a process well suited for the construction of torsion box (Figure 8.26). Nevertheless, for longer lengths, the flexure of the blade varies during winding (difference in flexural rigidity in one neutral plane from the other).

One can mention also the composite propeller blades for the cooling blowers, which borrow the technology of the aircraft propellers,[12] which have a speed of rotation larger than for the wind turbines.

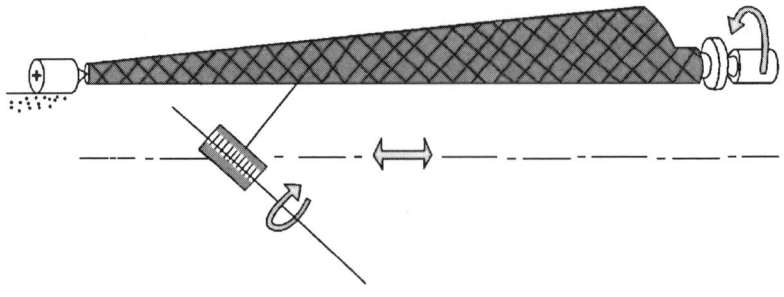

Figure 8.26   Filament Winding for Wind Turbine

[12] See Section 7.3.

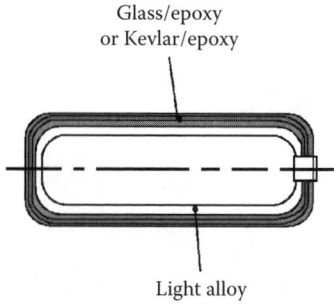

Glass/epoxy
or Kevlar/epoxy

Light alloy

Figure 8.27 Composite Gas Bottles

8.4.2 Compressed Gas Bottles

These are made using filament-wound glass/epoxy or Kevlar/epoxy (see Figure 8.27) reinforcing a thin envelope in light alloy which is used for sealing purposes.

- The pressure can reach 350 bars in operating conditions (rupture at more than 1,000 bars).
- The ratio (gas volume/bottle mass) is multiplied by 4 in comparison with the steel solution.
- Applications include
  - Very light breathing apparatus (for scuba diving)
  - Reservoirs for gaseous fuels
  - Accessories for missiles

8.4.3 Buggy Chassis

These are made using glass/epoxy. They are very resilient and can help reduce noise. They are very light with a reduced number of pieces in comparison with the metallic solution. They also offer very good fatigue endurance, as has already been described above for glass/epoxy.[13]

Other advantages include the possibility of integration of the "spring" function in the structure of the chassis and increase in the "critical" speed from which a proper mode of vibration can develop in the suspension (see Figure 8.28).

8.4.4 Tubes for Off-Shore Installations

These are used in deep waters. The weight of the metallic tubes—or **risers**—increases proportionally with the depth and can attain high values (one third of the limit stress for a depth of 1000 m). This gives rise to interest in using tubes made of glass/carbon/resin, which are three to four times lighter than tubes made of steel.

Example: Production tubes for a cable-held platform. The platform is held with cables toward the bottom (see Figure 8.29). A large number of production

---

[13] See Section 5.4.4.

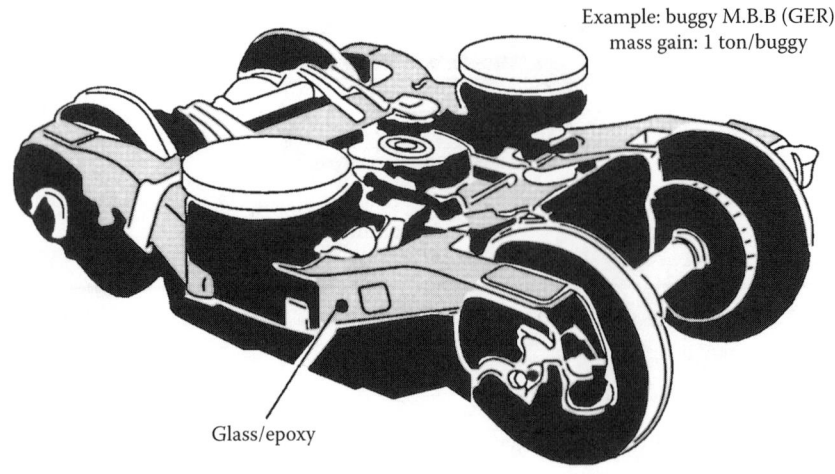

Example: buggy M.B.B (GER)
mass gain: 1 ton/buggy

Glass/epoxy

Figure 8.28   M.B.B. Buggy

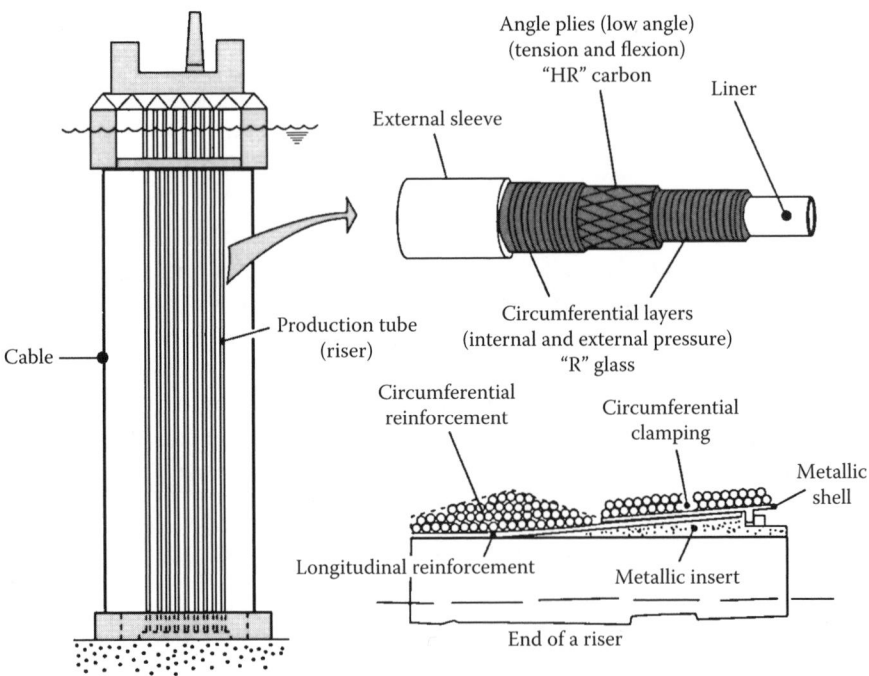

Angle plies (low angle)
(tension and flexion)
"HR" carbon

Liner

External sleeve

Cable

Production tube
(riser)

Circumferential layers
(internal and external pressure)
"R" glass

Circumferential
reinforcement

Circumferential
clamping

Metallic
shell

Longitudinal reinforcement

Metallic insert

End of a riser

Figure 8.29   Riser Tubes

tubes connect the underwater bottom to the platform. They are subjected to the static and dynamic loadings (due to underwater currents) as:

- Tension
- Flexure
- Circumferential extension and contraction due to external and internal pressures

Characteristics: Safety factor as compared with complete rupture: 2 to 3.

The micro cracks in the resin require internal and external sealings using elastomers.

8.4.5 Biomechanics Applications

The carbon/carbon composites (see Section 3.6) have the rare properties of not provoking fibrous outgrowth when in contact with the blood stream; they are so-called **thrombo resistant**. In addition, the following qualities also favor their implantation in the human body:

- Chemical resistance and inertness
- Mechanical and fatigue resistance
- Controllable flexibility due to the nature of composite materials
- Low specific mass
- Transparency to different rays
- Possible sterilization at very high temperatures

The principal applications (to be expanded) for the moment are

- Hip and knee implants (in development)
- Osteosynthesis plates
- Dental implants
- Implant apparatus

8.4.6 Telepherique Cabin

A substitution using composites on the classical solution for the telepherique cars made of metals gives, at equal mass, a notable augmentation of the useful payload.

Example: Company Ingenex/telepherique of Argentieres (FRA): Increasing capacity while keeping existing installations, that is cables—pylons—motorization.

- Previous metallic telepherique: useful payload 45 passengers
- New composite telepherique, carbon/Kevlar/epoxy (Figure 8.30): useful load 70 passengers (with the same mass as the previous construction)
- Augmentation of capacity: 55%

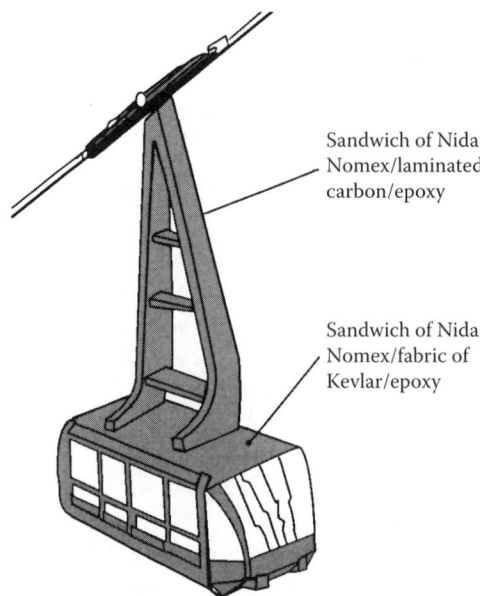

Sandwich of Nida Nomex/laminated carbon/epoxy

Sandwich of Nida Nomex/fabric of Kevlar/epoxy

Figure 8.30   Telepherique Cabin

- Cost comparison: renewal of all installation with the metallic solution ($11 million)
- Renewal of two telepherique cabins with the composite solution ($1.1 million): cost divided by 10

PART II

MECHANICAL BEHAVIOR
OF LAMINATED
MATERIALS

We have introduced in the previous part the anisotropic properties of composite materials from a qualitative point of view,[1] and we have indicated the characteristic elastic constants for the behavior of an anisotropic layer in its plane.

We have also mentioned the relations that allow one to predict the mechanical behavior of a fiber/matrix combination from the properties of the individual constituents.[2] In Chapter 5,[3] we have also given the elements necessary for the sizing of the laminates made of carbon/epoxy, Kevlar/epoxy, and glass/epoxy, in terms of strength and deformation.

This second part is dedicated to the justification and application of these properties and results. It requires a detailed study of the behavior of anisotropic composite layers and of the stacking that makes up the laminate. It is useful to note that the basics of the mechanics of continuous media—namely, the state of stress and strain at a point—already described in great details in many texts on mechanics of materials, are supposed to be known.

---

[1] See Section 3.1.
[2] See Section 3.3.1.
[3] See Section 5.2/5.3.

9

ANISOTROPIC ELASTIC MEDIA

9.1 REVIEW OF NOTATIONS

9.1.1 Continuum Mechanics

We consider the following classical notions and notations of the mechanics of continuous media:

- **State of stress at a point:** This is defined by a second order **tensor** with the symbol Σ. The 3 by 3 matrix associated with this tensor is symmetric. In this matrix, there are six distinct terms, which are denoted as σ_{ij}:

$$\sigma_{11};\ \sigma_{22};\ \sigma_{33};\ \sigma_{23};\ \sigma_{13};\ \sigma_{12}$$

- **State of strain at a point:** This is defined as a second order tensor $\boldsymbol{\varepsilon}$. The 3 by 3 matrix for this tensor is symmetric. It consists of six distinct terms denoted as ε_{ij}:

$$\varepsilon_{11};\ \varepsilon_{22};\ \varepsilon_{33};\ \varepsilon_{23};\ \varepsilon_{13};\ \varepsilon_{12}$$

- **Linear elastic medium:** The strains are linear and homogeneous functions of the stresses. The corresponding relations are:

$$\varepsilon_{ij} = \varphi_{ijk\ell} \times \sigma_{k\ell}{}^{1}$$

- **Homogeneous medium:** In this case, the matrix terms φ_{ijkl} characterizing the elastic behavior of the medium are not point functions. They are the same at all points in the considered medium.

---

[1] For example:

$\varepsilon_{11} = \varphi_{1111}\ \sigma_{11} + \varphi_{1112}\ \sigma_{12} + \varphi_{1113}\ \sigma_{13} + \varphi_{1121}\ \sigma_{21} + \varphi_{1122}\ \sigma_{22} + \varphi_{1123}\ \sigma_{23} + \varphi_{1131}\ \sigma_{31} + \varphi_{1132}\ \sigma_{32} + \varphi_{1133}\ \sigma_{33}.$

9.1.2 Number of Distinct $\varphi_{ijk\ell}$ Terms

The above stress–strain relation can be written in matrix form as:

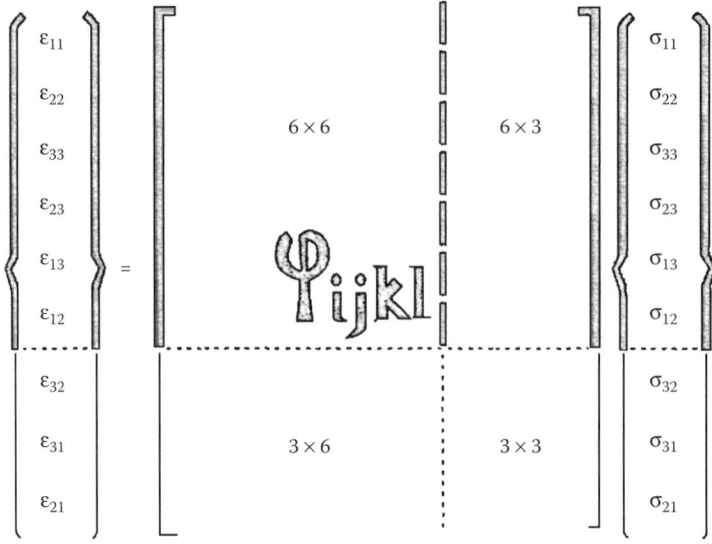

- Due to the symmetry of the stresses ($\sigma_{k\ell} = \sigma_{\ell k}$), the corresponding coefficients are the same, i.e., $\varphi_{ijk\ell} = \varphi_{ij\ell k}$.
- Due to the symmetry of the strains ($\varepsilon_{ij} = \varepsilon_{ji}$), the corresponding coefficients are the same, i.e., $\varphi_{ijk\ell} = \varphi_{jik\ell}$. In other words, the knowledge of only the coefficients of the 6 × 6 matrix written above is required.
- In addition, application of the theorem of virtual work on the stresses shows that the coefficients $\varphi_{ijk\ell}$ are symmetric, meaning: $\varphi_{ijk\ell} = \varphi_{k\ell ij}$.[2]

Therefore, the 6 × 6 matrix mentioned previously is symmetric. The number of distinct coefficients is:

$$\frac{6(6 + 1)}{2} = 21 \text{ coefficients}$$

---

[2] Consider two simple stress states:

- **State No. 1:** One single stress, $(\sigma_{k\ell})_1$, which causes the strain:

$$(\varepsilon_{ij})_1 = \varphi_{ijk\ell} \, (\sigma_{k\ell})_1$$

- **State No. 2:** One single stress, $(\sigma_{pq})_2$, which causes the strain:

$$(\varepsilon_{mn})_2 = \varphi_{mnpq} \, (\sigma_{pq})_2$$

One can write that the work of the stress in state No. 1 on the strain in state No. 2 is equal to the work of the stress in state No. 2 on the strain in state No. 1, as:

$$(\sigma_{k\ell})_1 \times (\varepsilon_{k\ell})_2 = (\sigma_{pq})_2 \times (\varepsilon_{pq})_1$$

which means: $(\sigma_{k\ell})_1 \times \varphi_{k\ell pq} \times (\sigma_{pq})_2 = (\sigma_{pq})_2 \times \varphi_{pqk\ell} \times (\sigma_{k\ell})_1$

from which one has: $\qquad \varphi_{k\ell pq} = \varphi_{pqk\ell}$

■ In summary:

$$
\boxed{
\begin{array}{l}
\text{stress reciprocity: } \varphi_{ijk\ell} = \varphi_{ij\ell k} \\[4pt]
\text{strain definition: } \varphi_{ijk\ell} = \varphi_{jik\ell} \\[4pt]
\text{symmetry: } \varphi_{ijk\ell} = \varphi_{k\ell ij} \\[4pt]
\text{There remain 21 distinct coefficients } \varphi_{ijk\ell}
\end{array}
}
\tag{9.1}
$$

The previous stress–strain relation can then be written as:

$$
\begin{Bmatrix}
\varepsilon_{11} \\
\varepsilon_{22} \\
\varepsilon_{33} \\
\varepsilon_{23} \\
\varepsilon_{13} \\
\varepsilon_{12}
\end{Bmatrix}
=
\begin{bmatrix}
\varphi_{1111} & \varphi_{1122} & \varphi_{1133} & 2\varphi_{1123} & 2\varphi_{1113} & 2\varphi_{1112} \\
\varphi_{2211} & \varphi_{2222} & \varphi_{2233} & 2\varphi_{2223} & 2\varphi_{2213} & 2\varphi_{2212} \\
\varphi_{3311} & \varphi_{3322} & \varphi_{3333} & 2\varphi_{3323} & 2\varphi_{3313} & 2\varphi_{3312} \\
\varphi_{2311} & \varphi_{2322} & \varphi_{2333} & 2\varphi_{2323} & 2\varphi_{2313} & 2\varphi_{2312} \\
\varphi_{1311} & \varphi_{1322} & \varphi_{1333} & 2\varphi_{1323} & 2\varphi_{1313} & 2\varphi_{1312} \\
\varphi_{1211} & \varphi_{1222} & \varphi_{1233} & 2\varphi_{1223} & 2\varphi_{1213} & 2\varphi_{1212}
\end{bmatrix}
\begin{Bmatrix}
\sigma_{11} \\
\sigma_{22} \\
\sigma_{33} \\
\sigma_{23} \\
\sigma_{13} \\
\sigma_{12}
\end{Bmatrix}
$$

The matrix above does not have the general symmetry as in the general form (9 × 9) presented previously. (Note the coefficients 2 in this matrix.) One can get around this inconvenience by doubling the terms ε_{23}, ε_{13}, ε_{12}, introducing the shear strains:

$$
\gamma_{23} = 2\varepsilon_{23}; \quad \gamma_{13} = 2\varepsilon_{13}; \quad \gamma_{12} = 2\varepsilon_{12}
$$

from which the stress–strain behavior can then be written in a symmetric form as:

$$
\begin{Bmatrix}
\varepsilon_{11} \\
\varepsilon_{22} \\
\varepsilon_{33} \\
2\varepsilon_{23} \;\; \gamma_{23} \\
2\varepsilon_{13} \;\; \gamma_{13} \\
2\varepsilon_{12} \;\; \gamma_{12}
\end{Bmatrix}
=
\begin{bmatrix}
\varphi_{1111} & \varphi_{1122} & \varphi_{1133} & 2\varphi_{1123} & 2\varphi_{1113} & 2\varphi_{1112} \\
\varphi_{2211} & \varphi_{2222} & \varphi_{2233} & 2\varphi_{2223} & 2\varphi_{2213} & 2\varphi_{2212} \\
\varphi_{3311} & \varphi_{3322} & \varphi_{3333} & 2\varphi_{3323} & 2\varphi_{3313} & 2\varphi_{3312} \\
2\varphi_{2311} & 2\varphi_{2322} & 2\varphi_{2333} & 4\varphi_{2323} & 4\varphi_{2313} & 4\varphi_{2312} \\
2\varphi_{1311} & 2\varphi_{1322} & 2\varphi_{1333} & 4\varphi_{1323} & 4\varphi_{1313} & 4\varphi_{1312} \\
2\varphi_{1211} & 2\varphi_{1222} & 2\varphi_{1233} & 4\varphi_{1223} & 4\varphi_{1213} & 4\varphi_{1212}
\end{bmatrix}
\begin{Bmatrix}
\sigma_{11} \\
\sigma_{22} \\
\sigma_{33} \\
\sigma_{23} \\
\sigma_{13} \\
\sigma_{12}
\end{Bmatrix}
\tag{9.2}
$$

9.2 ORTHOTROPIC MATERIALS

Definition: An orthotropic material is a homogeneous linear elastic material having two planes of symmetry at every point in terms of mechanical properties, these two planes being perpendicular to each other.

Then one can show that[3] the number of independent elastic constants is nine. The constitutive relation expressed in the so-called "**orthotropic**" axes, defined by three axes constructed on the two orthogonal planes and their intersection line, can be written in the following form, called the **engineering notation** because it utilizes the elastic modulus and Poisson ratios:

$$
\begin{Bmatrix} \varepsilon_{11} \\ \varepsilon_{22} \\ \varepsilon_{33} \\ \gamma_{23} \\ \gamma_{13} \\ \gamma_{12} \end{Bmatrix} = \begin{bmatrix} \dfrac{1}{E_1} & -\dfrac{v_{21}}{E_2} & -\dfrac{v_{31}}{E_3} & 0 & 0 & 0 \\ -\dfrac{v_{12}}{E_1} & \dfrac{1}{E_2} & -\dfrac{v_{32}}{E_3} & 0 & 0 & 0 \\ -\dfrac{v_{13}}{E_1} & -\dfrac{v_{23}}{E_2} & \dfrac{1}{E_3} & 0 & 0 & 0 \\ 0 & 0 & 0 & \dfrac{1}{G_{23}} & 0 & 0 \\ 0 & 0 & 0 & 0 & \dfrac{1}{G_{13}} & 0 \\ 0 & 0 & 0 & 0 & 0 & \dfrac{1}{G_{12}} \end{bmatrix} \begin{Bmatrix} \sigma_{11} \\ \sigma_{22} \\ \sigma_{33} \\ \sigma_{23} \\ \sigma_{13} \\ \sigma_{12} \end{Bmatrix}
\tag{9.3}
$$

where

E_1, E_2, E_3 are the longitudinal elastic moduli.
G_{23}, G_{13}, G_{12} are the shear moduli.
v_{12}, v_{13}, v_{23}, v_{21}, v_{31}, v_{32} are the Poisson ratios.

In addition, the symmetry of the stress–strain matrix above leads to the following relations:

$$
\frac{v_{21}}{E_2} = \frac{v_{12}}{E_1}; \quad \frac{v_{31}}{E_3} = \frac{v_{13}}{E_1}; \quad \frac{v_{32}}{E_3} = \frac{v_{23}}{E_2}
\tag{9.4}
$$

9.3 TRANSVERSELY ISOTROPIC MATERIALS

Definition: A transversely isotropic material is a homogeneous linear elastic material such that any plane including a preferred axis, is a plane of mechanical symmetry.

One can show that[4] the constitutive relation has five independent elastic constants. For the fiber/matrix composite shown in Figure 9.1 the preferred axis is ℓ. The fibers are distributed uniformly in the direction along ℓ. All directions perpendicular to the fibers characterize the transverse direction t.

The engineering stress–strain relation has the form:

---

[3] Proof is shown in Section 13.1.
[4] Proof is shown in Section 13.2.

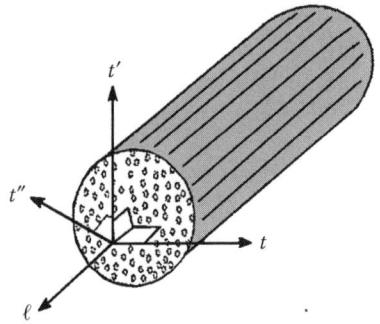

Figure 9.1   Transversely Isotropic Unidirectional

$$
\left\{
\begin{array}{c}
\varepsilon_{\ell\ell} \\[2mm]
\varepsilon_{tt} \\[2mm]
\varepsilon_{t't'} \\[2mm]
\gamma_{tt'} \\[2mm]
\gamma_{\ell t'} \\[2mm]
\gamma_{\ell t}
\end{array}
\right\}
=
\begin{bmatrix}
\dfrac{1}{E_\ell} & -\dfrac{v_{t\ell}}{E_t} & -\dfrac{v_{t\ell}}{E_t} & 0 & 0 & 0 \\[3mm]
-\dfrac{v_{\ell t}}{E_\ell} & \dfrac{1}{E_t} & -\dfrac{v_t}{E_t} & 0 & 0 & 0 \\[3mm]
-\dfrac{v_{\ell t}}{E_\ell} & -\dfrac{v_t}{E_t} & \dfrac{1}{E_t} & 0 & 0 & 0 \\[3mm]
0 & 0 & 0 & \dfrac{2(1+v_t)}{E_t} & 0 & 0 \\[3mm]
0 & 0 & 0 & 0 & \dfrac{1}{G_{\ell t}} & 0 \\[3mm]
0 & 0 & 0 & 0 & 0 & \dfrac{1}{G_{\ell t}}
\end{bmatrix}
\left\{
\begin{array}{c}
\sigma_{\ell\ell} \\[2mm]
\sigma_{tt} \\[2mm]
\sigma_{t't'} \\[2mm]
\tau_{tt'} \\[2mm]
\tau_{\ell t'} \\[2mm]
\tau_{\ell t}
\end{array}
\right\}
\qquad (9.5)
$$

Remarks: The independent elastic constants are

■ Young modulus along the ℓ direction: E_ℓ.
■ Young modulus along any transverse direction t: E_t.
■ Shear modulus in the plane ℓ, t: $G_{\ell t}$.
■ Poisson coefficients: $v_{\ell t}$ and v_t.

The symmetry of the coefficients of the constitutive relation leads to

$$
\boxed{\dfrac{v_{\ell t}}{E_\ell} = \dfrac{v_{t\ell}}{E_t}}
$$

One may also note that the shear modulus in the plane t, t' can be written as:

$$
\dfrac{E_t}{2(1+v_t)}
$$

This equation is classic for isotropic materials.

10

ELASTIC CONSTANTS OF UNIDIRECTIONAL COMPOSITES

In this chapter we examine a distinct combination of two materials (matrix and fiber), with simple geometry and loading conditions, in order to estimate the elastic properties of the equivalent material, i.e., of the composite.

10.1 LONGITUDINAL MODULUS E_ℓ

The two materials are shown schematically in Figure 10.1 where

  m stands for matrix.
  f stands for fiber.

- **Hypothesis:** The two materials are bonded together. More precisely, one makes the following assumptions:
- Both the matrix m and the fiber f have the same longitudinal strain ε_l.
- The interface between the two materials allows the z normal strains in the two materials to be different.

$$\varepsilon_z \neq \varepsilon_z$$
$$\textcircled{m} \quad \textcircled{f}$$

The state of stresses resulting from a force F can therefore be written as:

$$\Sigma \rightarrow \atop \textcircled{m} \begin{bmatrix} \sigma_\ell & 0 & 0 \\ 0 & 0 & 0 \\ 0 & 0 & 0 \end{bmatrix}_{\textcircled{m}} \qquad \Sigma \rightarrow \atop \textcircled{f} \begin{bmatrix} \sigma_\ell & 0 & 0 \\ 0 & 0 & 0 \\ 0 & 0 & 0 \end{bmatrix}_{\textcircled{f}}$$

and the corresponding state of strains:

$$\varepsilon \rightarrow \atop \textcircled{m} \begin{bmatrix} \varepsilon_\ell & 0 & 0 \\ 0 & \varepsilon_t & 0 \\ 0 & 0 & \varepsilon_z \end{bmatrix}_{\textcircled{m}} \qquad \varepsilon \rightarrow \atop \textcircled{f} \begin{bmatrix} \varepsilon_\ell & 0 & 0 \\ 0 & \varepsilon_t & 0 \\ 0 & 0 & \varepsilon_z \end{bmatrix}_{\textcircled{f}}$$

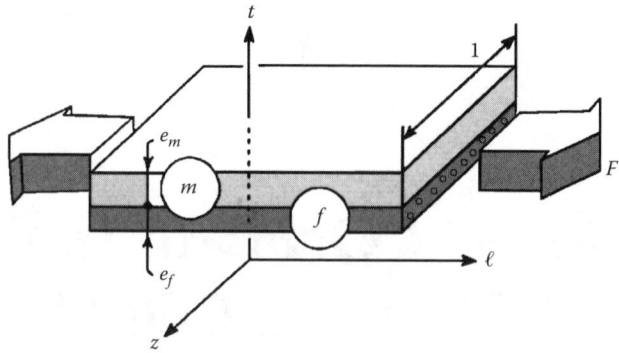

Figure 10.1 Longitudinal Modulus E_ℓ

Each material is assumed to be linear elastic and isotropic, with the following stress–strain relation:

$$\varepsilon = \frac{1+v}{E}\Sigma - \frac{v}{E}\ \text{trace}\ (\Sigma)\mathbf{I} \tag{10.1}$$

in which $\boldsymbol{\varepsilon}$ represents the strain tensor, Σ represents the stress tensor, and I the unity tensor. E and v are the elastic constants of the considered material.

For the composite $(m + f)$, one uses Equation 9.5 with restriction to the plane l,t. It reduces to:

$$\begin{Bmatrix} \varepsilon_\ell \\ \varepsilon_t \\ \gamma_{\ell t} \end{Bmatrix} = \begin{bmatrix} \dfrac{1}{E_\ell} & -\dfrac{v_{t\ell}}{E_t} & 0 \\ -\dfrac{v_{\ell t}}{E_\ell} & \dfrac{1}{E_t} & 0 \\ 0 & 0 & \dfrac{1}{G_{\ell t}} \end{bmatrix} \begin{Bmatrix} \sigma_\ell \\ \sigma_t \\ \tau_{\ell t} \end{Bmatrix}$$

The stress $\sigma_{\ell(m+f)}$ can be written as (see Figure 10.1 above):

$$\sigma_\ell = \frac{F}{S} = \frac{F}{(e_m + e_f) \times 1} = \sigma_\ell \frac{e_m}{e_m + e_f} + \sigma_\ell \frac{e_f}{e_m + e_f}$$

$$\boxed{m} + \boxed{f} \qquad\qquad \boxed{m} \qquad \boxed{f}$$

which can be written in terms of the volume fractions of the fiber and the matrix as[1]

$$\sigma_\ell = \sigma_\ell V_m + \sigma_\ell V_f$$

$$\boxed{m} + \boxed{f} \qquad \boxed{m} \quad \boxed{f}$$

---

[1] See Section 3.2.2.

Expressing the stresses in terms of the strains for each material yields

$$E_\ell \varepsilon_\ell = E_m \varepsilon_\ell V_m + E_f \varepsilon_\ell V_f$$

then:

$$\boxed{E_\ell = E_m V_m + E_f V_f} \tag{10.2}$$

Note: Among the real phenomena that are not taken into account in the expression of E_l is the lack of perfect straightness of the fibers in the matrix. Also, the modulus E_l depends on the sign of the stress (tension or compression). In rigorous consideration, the material is "**bimodulus.**"

Example: Unidirectional layers with 60% fiber volume fraction ($V_f = 0.60$) with epoxy matrix:

	Kevlar	"HR" Carbon	"HM" Carbon
E_ℓ tension (MPa)	85,000	134,000	180,000
E_ℓ compression (MPa)	80,300	134,000	160,000

10.2 POISSON COEFFICIENT

Considering again the loading defined in the previous paragraph, the transverse strain for the matrix m and fiber f can be written as:

$$\varepsilon_t = -\frac{v}{E}\sigma_\ell = -v\varepsilon_\ell$$

and for the composite ($m + f$):

$$\underset{\textstyle \widehat{m} + \widehat{f}}{\varepsilon_t} = \underset{\textstyle \widehat{m} + \widehat{f}}{-\frac{v_{\ell t}}{E_\ell} \times \sigma_\ell} = -v_{\ell t}\varepsilon_\ell$$

The strain in the transverse direction can also be written as:

$$\underset{\textstyle \widehat{m} + \widehat{f}}{\varepsilon_t} = \frac{\Delta(e_m + e_f)}{e_m + e_f} = \frac{\Delta e_m}{e_m}V_m + \frac{\Delta e_f}{e_f}V_f$$

$$\underset{\textstyle \widehat{m} + \widehat{f}}{\varepsilon_t} = \underset{\textstyle \widehat{m}}{\varepsilon_t V_m} + \underset{\textstyle \widehat{f}}{\varepsilon_t V_f}$$

Because ε_ℓ has the same value in m and f:

$$-v_{\ell t}\varepsilon_\ell = -v_m \varepsilon_\ell V_m - v_f \varepsilon_\ell V_f$$

$$\boxed{v_{\ell t} = v_m V_m + v_f V_f} \tag{10.3}$$

10.3 TRANSVERSE MODULUS E_t

To evaluate the modulus along the transverse direction E_t, the two materials are shown in the Figure 10.2. In addition, one uses the following simplifications:

- **Hypothesis:** At the interface between the two materials, assume the following:
    - Freedom of movement in the l direction allows for different strains in the two materials:

$$\varepsilon_\ell \neq \varepsilon_\ell$$
$$\overset{}{\underset{(m)}{}} \quad \overset{}{\underset{(f)}{}}$$

    - Freedom of movement in the z direction allows for different strains in the two materials:

$$\varepsilon_z \neq \varepsilon_z$$
$$\overset{}{\underset{(m)}{}} \quad \overset{}{\underset{(f)}{}}$$

Then, the state of stress created by a load F (see Figure 10.2), can be reduced for each material to the following:

$$\Sigma \rightarrow \begin{bmatrix} 0 & 0 & 0 \\ 0 & \sigma_t & 0 \\ 0 & 0 & 0 \end{bmatrix}$$

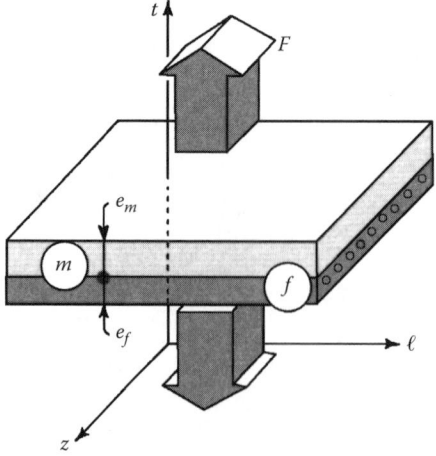

Figure 10.2   Transverse Modulus E_t

The strains can be written as:

$$\varepsilon \rightarrow \begin{bmatrix} \varepsilon_\ell & 0 & 0 \\ 0 & \varepsilon_t & 0 \\ 0 & 0 & \varepsilon_z \end{bmatrix} \textcircled{m}$$

or

$$\textcircled{f}$$

Then for the composite $(m + f)$, one has

$$\varepsilon_t = \frac{1}{E_t}\sigma_t$$

On the other hand, using direct calculation leads to (see Figure 10.2)

$$\varepsilon_t = \frac{\Delta(e_m + e_f)}{e_m + e_f} = \underset{\textcircled{m}}{\varepsilon_t V_m} + \underset{\textcircled{f}}{\varepsilon_t V_f}$$

then:

$$\frac{1}{E_t}\sigma_t = \frac{1}{E_m}\sigma_t V_m + \frac{1}{E_f}\sigma_t V_f$$

$$\boxed{\frac{1}{E_t} = \frac{V_m}{E_m} + \frac{V_f}{E_f} \quad \text{or} \quad E_t = E_m \left[\frac{1}{(1 - V_f) + \frac{E_m}{E_f}V_f} \right]} \tag{10.4}$$

Remarks:

- Due to the above simplifications that allow the possibility for relative sliding along the l and z directions at the interface, the transverse modulus E_t above may not be accurate.
- One finds in the technical literature many more complex formulae giving E_t. However, none can provide guaranteed good result.
- Taking into consideration the load applied (see Figure 10.2),: the modulus E_f that appears in Equation 10.4 is the modulus of elasticity of the fiber in a direction that is perpendicular to the fiber axis. This modulus can be very different from the modulus along the axis of the fiber, due to the anisotropy that exists in fibers.[2]

---

[2] This point was discussed in Paragraph 3.3.1.

10.4 SHEAR MODULUS $G_{\ell t}$

Load application that can be used to evaluate the shear modulus $G_{\ell t}$ is shown schematically in the Figure 10.3, both with the angular deformations that are produced. The state of stress, identical for both the matrix and fiber material, can be written as:

$$\Sigma \rightarrow \begin{bmatrix} 0 & \tau_{\ell t} & 0 \\ \tau_{\ell t} & 0 & 0 \\ 0 & 0 & 0 \end{bmatrix}$$

The corresponding strains can be written as:

$$\underset{\textcircled{m} \text{ or } \textcircled{f}}{\varepsilon} \rightarrow \begin{bmatrix} 0 & \varepsilon_{\ell t} & 0 \\ \varepsilon_{\ell t} & 0 & 0 \\ 0 & 0 & 0 \end{bmatrix}$$

Using the constitutive equation, one has

$$\varepsilon_{\ell t} = \frac{1+\nu}{E}\tau_{\ell t} = \frac{\tau_{\ell t}}{2G}$$

then:

$$\gamma_{\ell t} = \frac{\tau_{\ell t}}{G}$$

Also, from Figure 10.3, one has

$$\underset{\textcircled{m} + \textcircled{f}}{\gamma_{\ell t}} \; (e_m + e_f) = \underset{\textcircled{m}}{\gamma_{\ell t}e_m} + \underset{\textcircled{f}}{\gamma_{\ell t}e_f}$$

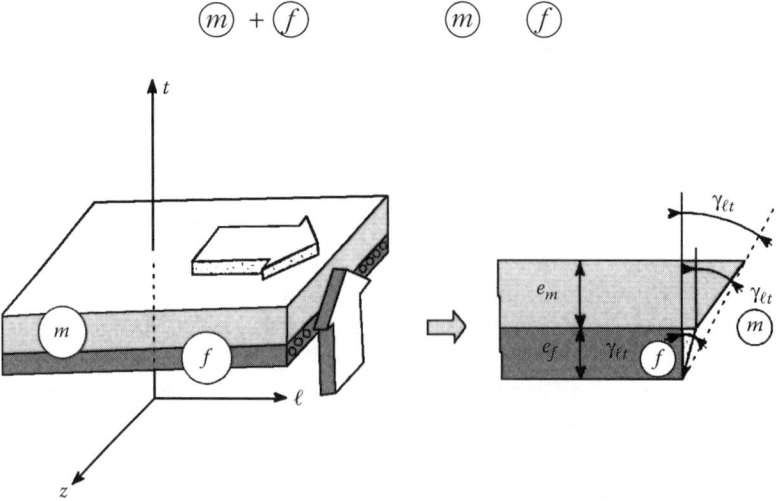

Figure 10.3   Shear modulus $G_{\ell t}$

which can be rewritten as:

$$\gamma_{\ell t} = \gamma_{\ell t} V_m + \gamma_{\ell t} V_f$$

$$\underbrace{(m)}_{} + \underbrace{(f)}_{} \quad \underbrace{(m)}_{} \quad \underbrace{(f)}_{}$$

$$\frac{\tau_{\ell t}}{G_{\ell t}} = \frac{\tau_{\ell t}}{G_m} V_m + \frac{\tau_{\ell t}}{G_f} V_f$$

$$\frac{1}{G_{\ell t}} = \frac{V_m}{G_m} + \frac{V_f}{G_f}$$

$$\boxed{G_{\ell t} = G_m \left[\frac{1}{(1 - V_f) + \frac{G_m}{G_f} V_f} \right]}$$
(10.5)[3]

10.5 THERMOELASTIC PROPERTIES

10.5.1 Isotropic Material: Recall

When the influence of temperature is taken into consideration, Hooke's law for the case of no temperature influence:

$$\varepsilon = \frac{1 + \nu}{E} \Sigma - \frac{\nu}{E} \text{trace}(\Sigma) \mathbf{I}$$

is replaced by the *Hooke–Duhamel* law:

$$\varepsilon = \frac{1 + \nu}{E} \Sigma - \frac{\nu}{E} \text{trace}(\Sigma) \mathbf{I} + \alpha \Delta T \mathbf{I}$$
(10.6)

where

ε = Strain tensor
Σ = Stress tensor
I = Unity tensor
E, ν = Elastic constants for the considered material
α = Coefficient of thermal expansion[4]
ΔT = Change in temperature with respect to a reference temperature at which the stresses and strains are nil

10.5.2 Case of Unidirectional Composite

The coefficient of thermal expansion of the matrix is usually much larger (more than ten times) than that of the fiber.[4] In Figure 10.4, one can imagine that even in the absence of mechanical loading, a change in temperature ΔT will produce a

---

[3] A few values of the shear modulus G_f are shown in Section 3.3.1.
[4] See Section 1.6, "Principal Physical Properties."

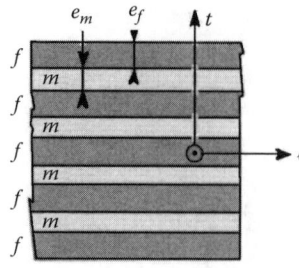

Figure 10.4   Unidirectional Composite

longitudinal strain in the composite. This longitudinal strain has a value that is intermediate between the strain of the fiber alone and that of the matrix alone. Then, in the composite one finds internal stresses along the direction l (along the direction t, the fiber and matrix can expand differently). One then has:

■ For the stresses:

$$\Sigma \rightarrow \begin{bmatrix} \sigma_\ell & 0 & 0 \\ 0 & 0 & 0 \\ 0 & 0 & 0 \end{bmatrix}_{(m)} \qquad \Sigma \rightarrow \begin{bmatrix} \sigma_\ell & 0 & 0 \\ 0 & 0 & 0 \\ 0 & 0 & 0 \end{bmatrix}_{(f)}$$

■ For the strains:

$$\varepsilon \rightarrow \begin{bmatrix} \varepsilon_\ell & 0 & 0 \\ 0 & \varepsilon_t & 0 \\ 0 & 0 & \varepsilon_z \end{bmatrix}_{(m)} \qquad \varepsilon \rightarrow \begin{bmatrix} \varepsilon_\ell & 0 & 0 \\ 0 & \varepsilon_t & 0 \\ 0 & 0 & \varepsilon_z \end{bmatrix}_{(f)}$$

10.5.2.1   Coefficient of Thermal Expansion along the Direction l

One has for the fiber and the matrix, respectively:

$$\varepsilon_\ell = \frac{\sigma_\ell}{E_m}_{(m)} + \alpha_m \Delta T = \varepsilon_\ell = \frac{\sigma_\ell}{E_f}_{(f)} + \alpha_f \Delta T$$

The external equilibrium can be written as (see Figure 10.4):

$$\sigma_\ell \times e_m + \sigma_\ell \times e_f = 0$$
$$_{(m)} _{(f)}$$

where, taking into account the equality of the strains:

$$\frac{\sigma_\ell}{E_m}\underset{(m)}{} + \alpha_m \Delta T = -\sigma_\ell \times \frac{e_m}{e_f} \times \frac{1}{E_f}\underset{(m)}{} + \alpha_f \Delta T$$

$$\sigma_\ell \underset{(m)}{} = \frac{(\alpha_f - \alpha_m)\Delta T}{\dfrac{1}{E_m} + \dfrac{e_m}{e_f} \times \dfrac{1}{E_f}} = \frac{(\alpha_f - \alpha_m)\Delta T}{\dfrac{1}{E_m} + \dfrac{V_m}{V_f} \times \dfrac{1}{E_f}}$$

V represents the volume fraction. The longitudinal strain can then be written as:

$$\varepsilon_\ell \underset{(m)}{} = \varepsilon_\ell \underset{(f)}{} = \frac{(\alpha_f E_f V_f + \alpha_m E_m V_m)(\Delta T)}{E_f V_f + E_m V_m}$$

It is also the longitudinal strain that is created only by the effect of temperature:

$$\varepsilon_\ell \underset{(m)\,+\,(f)}{} = \alpha_\ell \Delta T$$

where α_ℓ is the longitudinal coefficient of thermal expansion. One can then equate the above expressions to obtain:

$$\boxed{\alpha_\ell = \frac{\alpha_f E_f V_f + \alpha_m E_m V_m}{E_f V_f + E_m V_m}} \tag{10.7}$$

10.5.2.2 Coefficient of Thermal Expansion along the Transverse Direction t

The global thermal strain can be written as (see Figure 10.4):

$$\varepsilon_t \underset{(m)\,+\,(f)}{} = \frac{\Delta(e_m + e_f)}{e_m + e_f} = \varepsilon_t \underset{(m)}{} \frac{e_m}{e_m + e_f} + \varepsilon_t \underset{(f)}{} \frac{e_f}{e_m + e_f}$$

then:

$$\varepsilon_t \underset{(m)\,+\,(f)}{} = \varepsilon_t \underset{(m)}{} \times V_m + \varepsilon_t \underset{(f)}{} \times V_f$$

Using the Hooke and Duhamel law (Equation 10.6)[5]:

$$\varepsilon_t \underset{(m)\,+\,(f)}{} = \left(-\frac{\nu_m}{E_m}\sigma_\ell + \alpha_m \Delta T\right) V_m \underset{(m)}{} + \left(-\frac{\nu_f}{E_f}\sigma_\ell + \alpha_f \Delta T\right) V_f \underset{(f)}{}$$

---

[5] For the Poisson coefficients of common fibers, see Section 3.3.1.

Using the expressions for stresses obtained before, one obtains:

$$\underset{(m) + (f)}{\varepsilon_t} = \left\{ (\alpha_m V_m + \alpha_f V_f) + \frac{(V_f E_m - V_m E_f)}{E_m V_m + E_f V_f} V_m V_f (\alpha_f - \alpha_m) \right\} \Delta T$$

The quantity between the brackets represent the coefficient of thermal expansion along the transverse direction t, α_t, which can be written as:

$$\alpha_t = \alpha_m V_m + \alpha_f V_f + \frac{(V_f E_m - V_m E_f)}{\dfrac{E_m}{V_f} + \dfrac{E_f}{V_m}} \times (\alpha_f - \alpha_m) \tag{10.8}$$

10.5.3 Thermomechanical Behavior of a Unidirectional Layer

Under the combined effect of the stresses and temperature, the global thermomechanical strains of a unidirectional layer can be obtained using the following relation:

$$\left\{ \begin{array}{c} \varepsilon_\ell \\ \varepsilon_t \\ \gamma_{\ell t} \end{array} \right\} = \left[\begin{array}{ccc} \dfrac{1}{E_\ell} & -\dfrac{\nu_{t\ell}}{E_t} & 0 \\ -\dfrac{\nu_{\ell t}}{E_\ell} & \dfrac{1}{E_t} & 0 \\ 0 & 0 & \dfrac{1}{G_{\ell t}} \end{array} \right] \left\{ \begin{array}{c} \sigma_\ell \\ \sigma_t \\ \tau_{\ell t} \end{array} \right\} + \Delta T \left\{ \begin{array}{c} \alpha_\ell \\ \alpha_t \\ 0 \end{array} \right\} \tag{10.9}$$

in which the coefficients E_ℓ, E_t, $\nu_{\ell t}$, $G_{\ell t}$, α_ℓ and α_t have the values given by the Equations 10.2 to 10.8, respectively.

11

ELASTIC CONSTANTS OF A PLY ALONG AN ARBITRARY DIRECTION

To study the behavior of a laminate made up of many plies with different orientations, it is necessary to know the behavior of each of the plies in directions that are different from the principal material directions of the ply. We propose to determine the elastic constants for this ply behavior using relatively simple calculations.

11.1 COMPLIANCE COEFFICIENTS

The ply is already defined in Chapter 3.[1] Let ℓ, t and z be the orthotropic axes of a ply shown in the Figure 11.1.[2] For a thin laminate made up by a superposition of many plies, we assume that the stresses σ_{zz} are zero. It is then possible, for an orthotropic material, to write the stress–strain relation in the plane ℓ, t starting from Equation 9.3 or 9.5 in the form:

$$
\begin{Bmatrix} \varepsilon_\ell \\ \varepsilon_t \\ \gamma_{\ell t} \end{Bmatrix} = \begin{Bmatrix} \dfrac{1}{E_\ell} & -\dfrac{v_{t\ell}}{E_t} & 0 \\ -\dfrac{v_{\ell t}}{E_\ell} & \dfrac{1}{E_t} & 0 \\ 0 & 0 & \dfrac{1}{G_{\ell t}} \end{Bmatrix} \begin{Bmatrix} \sigma_\ell \\ \sigma_t \\ \tau_{\ell t} \end{Bmatrix}
\tag{11.1}
$$

Problem: How can one transform this relation expressed in the coordinates ℓ, t into a relation expressed in coordinates x, y inclined at an angle of θ with the ℓ, t coordinates (see Figure 11.1).[3]
First recall the following:

---

[1] See Section 3.2.
[2] The orthotropic axes 1,2,3 in Equation 9.3 are now called l, t, z, respectively.
[3] What follows is treated more globally and completely in Section 13.2.2.

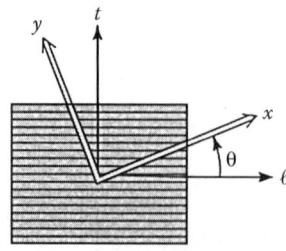

Figure 11.1   Orthotropic Axes and Arbitrary Direction in the Plane of a Ply

■ **Recall 1:** The stress $\vec{\sigma}$ acting on a surface with a normal vector $\vec{n}$ is given by

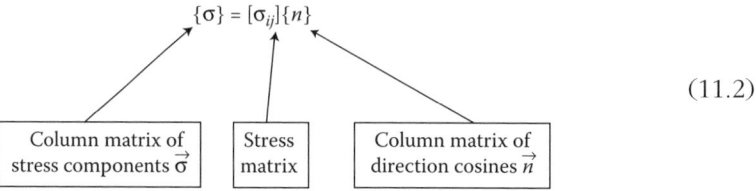

$$\{\sigma\} = [\sigma_{ij}]\{n\}$$

(11.2)

| Column matrix of stress components $\vec{\sigma}$ | Stress matrix | Column matrix of direction cosines $\vec{n}$ |

■ **Recall 2:** The coordinates of the same vector $\vec{V}$ in axes x,y as well as ℓ,t, such that $(\vec{x}.\vec{\ell}) = \theta$, are

$$\vec{V} = V_\ell \vec{\ell} + V_t \vec{t} = V_x \vec{x} + V_y \vec{y}$$

with the relation:

$$\begin{Bmatrix} V_x \\ V_y \end{Bmatrix} = \begin{Bmatrix} c & s \\ -s & c \end{Bmatrix} \begin{Bmatrix} V_\ell \\ V_t \end{Bmatrix} \quad \begin{pmatrix} c = \cos\theta \\ s = \sin\theta \end{pmatrix}$$

(11.3)

In axes ℓ,t the stress acting on the surface with a normal $\vec{x}$ can be expressed as follows, using Equation 11.2 above:

$$\{\sigma_{/x}\}_{\ell,t} = [\sigma_{ij}]_{\ell,t}\{x\}_{\ell,t} = [\sigma_{ij}]_{\ell,t}\begin{Bmatrix} c \\ s \end{Bmatrix}$$

where $\{\sigma_{/x}\}$ is the stress vector and $[\sigma_{ij}]$ is the stress matrix and in axes x,y, following Equation 11.3:

$$\{\sigma_{/x}\}_{x,y} = \begin{bmatrix} c & s \\ -s & c \end{bmatrix}[\sigma_{ij}]_{\ell,t}\begin{Bmatrix} c \\ s \end{Bmatrix}$$

In a similar manner, the stresses acting on the surface with the normal $\vec{y}$ are written in the x,y axes as:

$$\{\sigma_{/y}\}_{x,y} = \begin{bmatrix} c & s \\ -s & c \end{bmatrix} [\sigma_{ij}]_{\ell,t} \begin{Bmatrix} -s \\ c \end{Bmatrix}$$

Therefore, the matrix of stresses in the x,y axes is:

$$[\sigma_{ij}]_{x,y} = [\sigma_{/x}, \sigma_{/y}] = \begin{bmatrix} c & s \\ -s & c \end{bmatrix} [\sigma_{ij}]_{\ell,t} \begin{bmatrix} c & -s \\ s & c \end{bmatrix}$$

in setting:

$$[P] = \begin{bmatrix} c & s \\ -s & c \end{bmatrix}$$

and observing that the matrix $[P]$ is orthogonal ($^t[P] = [P]^{-1}$), one has[4]

$$[\sigma_{ij}]_{\ell,t} = {}^t[P][\sigma_{ij}]_{x,y}[P]$$

where $^t[P]$ is the transpose of matrix $[P]$.

This expression can be developed to become

$$\begin{bmatrix} \sigma_\ell & \tau_{\ell t} \\ \tau_{\ell t} & \sigma_t \end{bmatrix} = \begin{bmatrix} c & -s \\ s & c \end{bmatrix} \begin{bmatrix} \sigma_x & \tau_{xy} \\ \tau_{xy} & \sigma_y \end{bmatrix} \begin{bmatrix} c & s \\ -s & c \end{bmatrix}$$

One can also rearrange the equation to be

$$\begin{Bmatrix} \sigma_\ell \\ \sigma_t \\ \tau_{\ell t} \end{Bmatrix} = \begin{bmatrix} c^2 & s^2 & -2cs \\ s^2 & c^2 & 2cs \\ sc & -sc & (c^2 - s^2) \end{bmatrix} \begin{Bmatrix} \sigma_x \\ \sigma_y \\ \tau_{xy} \end{Bmatrix} \tag{11.4}$$

Then:

$$[\sigma]_{\ell,t} = [T][\sigma]_{x,y}$$

---

[4] One has: $\sigma_{x,y} = P\sigma_{\ell,t}\,{}^tP$; $\sigma_{\ell t}\,{}^tP = {}^tP\sigma_{xy}$; $P\sigma_{\ell,t} = \sigma_{x,y}P$; $\sigma_{\ell,t} = {}^tP\sigma_{x,y}P$.

with[5]

$$[T] = \begin{bmatrix} c^2 & s^2 & -2sc \\ s^2 & c^2 & 2sc \\ sc & -sc & (c^2 - s^2) \end{bmatrix}$$

In a similar manner, the strain components can be transformed as:

$$\begin{Bmatrix} \varepsilon_x \\ \varepsilon_y \\ \varepsilon_{xy} \end{Bmatrix} = \begin{bmatrix} c^2 & s^2 & 2cs \\ s^2 & c^2 & -2cs \\ -cs & cs & (c^2 - s^2) \end{bmatrix} \begin{Bmatrix} \varepsilon_\ell \\ \varepsilon_t \\ \varepsilon_{\ell t} \end{Bmatrix}$$

or:

$$\begin{Bmatrix} \varepsilon_x \\ \varepsilon_y \\ \gamma_{xy} \end{Bmatrix} = \begin{bmatrix} c^2 & s^2 & cs \\ s^2 & c^2 & -cs \\ -2cs & 2cs & (c^2 - s^2) \end{bmatrix} \begin{Bmatrix} \varepsilon_\ell \\ \varepsilon_t \\ \gamma_{\ell t} \end{Bmatrix}$$

then:

$$\begin{Bmatrix} \varepsilon \\ \gamma \end{Bmatrix}_{x,y} = [T'] \begin{Bmatrix} \varepsilon \\ \gamma \end{Bmatrix}_{\ell,t}$$

with:

$$[T'] = \begin{bmatrix} c^2 & s^2 & cs \\ s^2 & c^2 & -cs \\ -2cs & 2cs & (c^2 - s^2) \end{bmatrix}$$

The stress–strain Equation 11.1 can then be expressed in the axes x, y since we have written:

$$\begin{Bmatrix} \varepsilon \\ \gamma \end{Bmatrix}_{x,y} = [T'] \begin{Bmatrix} \varepsilon \\ \gamma \end{Bmatrix}_{\ell,t} \; ; \; \begin{Bmatrix} \varepsilon \\ \gamma \end{Bmatrix}_{\ell,t} = \begin{bmatrix} \dfrac{1}{E_\ell} & \dfrac{-v_{t\ell}}{E_t} & 0 \\ \dfrac{-v_{\ell t}}{E_\ell} & \dfrac{1}{E_t} & 0 \\ 0 & 0 & \dfrac{1}{G_{\ell t}} \end{bmatrix} \{\sigma\}_{\ell,t}; \; \{\sigma\}_{\ell,t} = [T]\{\sigma\}_{x,y};$$

---

[5] This $[T]$ matrix is readily established if one knows the relation that allows one to express the components of a tensor in one system in terms of the components of the same tensor in another system. Here this relation is: $\sigma_{IJ} = \cos_I^m \cos_J^n \sigma_{mn}$ with $\cos_I^m = \cos(\vec{m}, \vec{I})$; see Section 13.1.

where after substitution:

$$
\begin{Bmatrix} \varepsilon_x \\ \varepsilon_y \\ \gamma_{xy} \end{Bmatrix} = [T'] \underbrace{\begin{bmatrix} \dfrac{1}{E_\ell} & \dfrac{v_{t\ell}}{-E_t} & 0 \\ -\dfrac{v_{\ell t}}{E_\ell} & \dfrac{1}{E_t} & 0 \\ 0 & 0 & \dfrac{1}{G_{\ell t}} \end{bmatrix} [T]}_{\text{new matrix of elastic coefficients in } x,y \text{ axes}} \begin{Bmatrix} \sigma_x \\ \sigma_y \\ \tau_{xy} \end{Bmatrix}
$$

When all calculations are performed, one obtains the following constitutive relation, written in the coordinates x,y that make an angle θ with the axes ℓ,t. The elastic moduli and Poisson coefficients appear in these relations. One can also see the existence of the coupling coefficients η and $\mu,$[6] which demonstrates that a normal stress can produce a distortion.[7]

$$
\begin{Bmatrix} \varepsilon_x \\ \varepsilon_y \\ \gamma_{xy} \end{Bmatrix} = \begin{bmatrix} \dfrac{1}{E_x} & -\dfrac{v_{yx}}{E_y} & \dfrac{\eta_{xy}}{G_{xy}} \\ -\dfrac{v_{xy}}{E_x} & \dfrac{1}{E_y} & \dfrac{\mu_{xy}}{G_{xy}} \\ \dfrac{\eta_x}{E_x} & \dfrac{\mu_y}{E_y} & \dfrac{1}{G_{xy}} \end{bmatrix} \begin{bmatrix} \sigma_x \\ \sigma_y \\ \tau_{xy} \end{bmatrix}
$$

with:

$$
E_x(\theta) = \cfrac{1}{\dfrac{c^4}{E_\ell} + \dfrac{s^4}{E_t} + c^2 s^2 \left(\dfrac{1}{G_{\ell t}} - 2\dfrac{v_{t\ell}}{E_t} \right)}
$$

$$
E_y(\theta) = \cfrac{1}{\dfrac{s^4}{E_\ell} + \dfrac{c^4}{E_t} + c^2 s^2 \left(\dfrac{1}{G_{\ell t}} - 2\dfrac{v_{t\ell}}{E_t} \right)}
$$

(11.5)

$$
G_{xy}(\theta) = \cfrac{1}{4 c^2 s^2 \left(\dfrac{1}{E_\ell} + \dfrac{1}{E_t} + 2\dfrac{v_{t\ell}}{E_t} \right) + \dfrac{(c^2 - s^2)^2}{G_{\ell t}}}
$$

$$
\frac{v_{yx}}{E_y}(\theta) = \frac{v_{t\ell}}{E_t}(c^4 + s^4) - c^2 s^2 \left(\frac{1}{E_\ell} + \frac{1}{E_t} - \frac{1}{G_{\ell t}} \right)
$$

$$
\frac{\eta_{xy}}{G_{xy}}(\theta) = -2cs\left\{ \frac{c^2}{E_\ell} - \frac{s^2}{E_t} + (c^2 - s^2)\left(\frac{v_{t\ell}}{E_t} - \frac{1}{2G_{\ell t}} \right) \right\}
$$

$$
\frac{\mu_{xy}}{G_{xy}}(\theta) = -2cs\left\{ \frac{s^2}{E_\ell} - \frac{c^2}{E_t} - (c^2 - s^2)\left(\frac{v_{t\ell}}{E_t} - \frac{1}{2G_{\ell t}} \right) \right\}
$$

---

[6] Recall that the matrix of elastic coefficients is symmetric, meaning in particular: $\eta_{xy}/G_{xy} = \eta_x/E_x$ and $\mu_{xy}/G_{xy} = \mu_y/E_y$.

[7] See example described in Section 3.1.

11.2 STIFFNESS COEFFICIENTS

When one inverts Equation 11.1 written in the coordinate axes l,t of a ply, one obtains

$$
\begin{Bmatrix} \sigma_\ell \\ \sigma_t \\ \tau_{\ell t} \end{Bmatrix} = \begin{bmatrix} \dfrac{E_\ell}{(1 - v_{\ell t} v_{t l})} & \dfrac{v_{t \ell} E_\ell}{(1 - v_{\ell t} v_{t \ell})} & 0 \\ \dfrac{v_{\ell t} E_t}{(1 - v_{\ell t} v_{t \ell})} & \dfrac{E_t}{(1 - v_{\ell t} v_{t \ell})} & 0 \\ 0 & 0 & G_{\ell t} \end{bmatrix} \begin{Bmatrix} \varepsilon_\ell \\ \varepsilon_t \\ \gamma_{\ell t} \end{Bmatrix}
$$

where the "**stiffness**" coefficients appear, as opposed to the Equation 11.1 where the "**compliance**" coefficients appear. To simplify the writing, one can denote

$$
\begin{Bmatrix} \sigma_l \\ \sigma_t \\ \tau_{\ell t} \end{Bmatrix} = \begin{bmatrix} \bar{E}_\ell & v_{t \ell} \bar{E}_\ell & 0 \\ v_{\ell t} \bar{E}_t & \bar{E}_t & 0 \\ 0 & 0 & G_{\ell t} \end{bmatrix} \begin{Bmatrix} \varepsilon_\ell \\ \varepsilon_t \\ \gamma_{\ell t} \end{Bmatrix} \tag{11.6}
$$

An identical procedure can be followed to arrive at the stress–strain relation:

$$
\begin{Bmatrix} \sigma_x \\ \sigma_y \\ \tau_{xy} \end{Bmatrix} = \underbrace{\begin{bmatrix} c^2 & s^2 & 2cs \\ s^2 & c^2 & -2cs \\ -cs & cs & (c^2 - s^2) \end{bmatrix}}_{[T_1]} \begin{Bmatrix} \sigma_\ell \\ \sigma_t \\ \tau_{\ell t} \end{Bmatrix} :
$$

$$
\begin{Bmatrix} \varepsilon_\ell \\ \varepsilon_t \\ \gamma_{\ell t} \end{Bmatrix} = \underbrace{\begin{bmatrix} c^2 & s^2 & -cs \\ s^2 & c^2 & cs \\ 2cs & -2cs & (c^2 - s^2) \end{bmatrix}}_{[T_1']} \begin{Bmatrix} \varepsilon_x \\ \varepsilon_y \\ \gamma_{xy} \end{Bmatrix} \tag{11.7}
$$

Recall that axes x,y are derived from the axes ℓ,t by a rotation θ about the third axis z. Substituting Equations 11.7 into 11.6, one obtains

$$
\begin{Bmatrix} \sigma_x \\ \sigma_y \\ \tau_{xy} \end{Bmatrix} = [T_1] \begin{bmatrix} \bar{E}_\ell & v_{t \ell} \bar{E}_\ell & 0 \\ v_{\ell t} \bar{E}_t & \bar{E}_t & 0 \\ 0 & 0 & G_{\ell t} \end{bmatrix} [T_1'] \begin{Bmatrix} \varepsilon_x \\ \varepsilon_y \\ \gamma_{xy} \end{Bmatrix}
$$

which can be rewritten as:

$$
\begin{Bmatrix} \sigma_x \\ \sigma_y \\ \tau_{xy} \end{Bmatrix} = \begin{bmatrix} \bar{E}_{11} & \bar{E}_{12} & \bar{E}_{13} \\ \bar{E}_{21} & \bar{E}_{22} & \bar{E}_{23} \\ \bar{E}_{31} & \bar{E}_{32} & \bar{E}_{33} \end{bmatrix} \begin{Bmatrix} \varepsilon_x \\ \varepsilon_y \\ \gamma_{xy} \end{Bmatrix}
$$

Once the calculations are performed, one obtains the following expressions for the stiffness coefficients $\bar{E}_{ij}$, where $c = \cos\theta$ and $s = \sin\theta$.

$$\begin{Bmatrix} \sigma_x \\ \sigma_y \\ \tau_{xy} \end{Bmatrix} = \begin{bmatrix} \bar{E}_{11} & \bar{E}_{12} & \bar{E}_{13} \\ \bar{E}_{21} & \bar{E}_{22} & \bar{E}_{23} \\ \bar{E}_{31} & \bar{E}_{32} & \bar{E}_{33} \end{bmatrix} \begin{Bmatrix} \varepsilon_x \\ \varepsilon_y \\ \gamma_{xy} \end{Bmatrix}$$

with:

$$\bar{E}_{11}(\theta) = c^4 \bar{E}_\ell + s^4 \bar{E}_t + 2c^2 s^2 (v_{t\ell} \bar{E}_\ell + 2G_{\ell t})$$

$$\bar{E}_{22}(\theta) = s^4 \bar{E}_\ell + c^4 \bar{E}_t + 2c^2 s^2 (v_{t\ell} \bar{E}_\ell + 2G_{\ell t})$$

$$\bar{E}_{33}(\theta) = c^2 s^2 (\bar{E}_\ell + \bar{E}_t - 2v_{t\ell} \bar{E}_\ell) + (c^2 - s^2)^2 G_{\ell t} \qquad (11.8)$$

$$\bar{E}_{12}(\theta) = c^2 s^2 (\bar{E}_\ell + \bar{E}_t - 4G_{\ell t}) + (c^4 + s^4) v_{t\ell} \bar{E}_\ell$$

$$\bar{E}_{13}(\theta) = -cs\{c^2 \bar{E}_\ell - s^2 \bar{E}_t - (c^2 - s^2)(v_{t\ell} \bar{E}_\ell + 2G_{\ell t})\}$$

$$\bar{E}_{23}(\theta) = -cs\{s^2 \bar{E}_\ell - c^2 \bar{E}_t + (c^2 - s^2)(v_{t\ell} \bar{E}_\ell + 2G_{\ell t})\}$$

expressions in which:

$$\bar{E}_\ell = E_\ell / (1 - v_{\ell t} v_{t\ell}) : \bar{E}_t = E_t / (1 - v_{\ell t} v_{t\ell})$$

The rate of variation of stiffness coefficients $\bar{E}_{ij}$ as functions of the angle θ is represented in Figure 11.2 for a ply characterized by moduli E_ℓ and E_t with very different values, for example the case of unidirectional layers of fiber/resin.[8]

11.3 CASE OF THERMOMECHANICAL LOADING

11.3.1 Compliance Coefficients

When considering the temperature variations,[9] one must substitute the stress–strain Equation 11.1 with Equation 10.9:

$$\begin{Bmatrix} \varepsilon_\ell \\ \varepsilon_t \\ \gamma_{\ell t} \end{Bmatrix} = \begin{bmatrix} \dfrac{1}{E_\ell} & -\dfrac{v_{t\ell}}{E_t} & 0 \\ -\dfrac{v_{\ell t}}{E_\ell} & \dfrac{1}{E_t} & 0 \\ 0 & 0 & \dfrac{1}{G_{\ell t}} \end{bmatrix} \begin{Bmatrix} \sigma_\ell \\ \sigma_t \\ \tau_{\ell t} \end{Bmatrix} + \Delta T \begin{Bmatrix} \alpha_\ell \\ \alpha_t \\ 0 \end{Bmatrix}$$

---

[8] See characteristics of the fiber/resin unidirectionals in Paragraph 3.3.3.

[9] See Section 10.5.

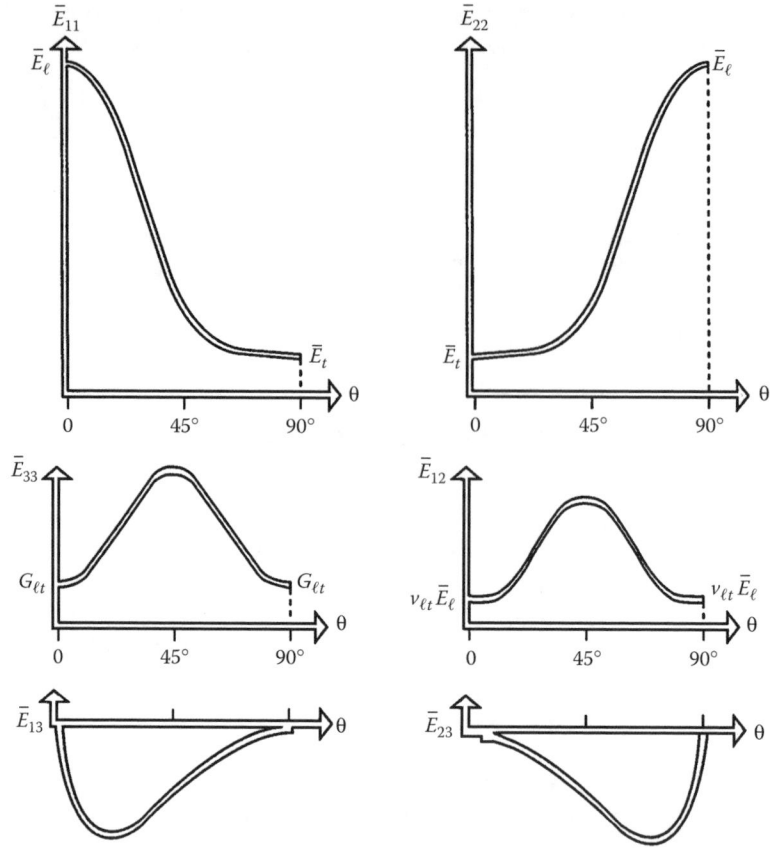

Figure 11.2   Variation of Stiffness Coefficients for a Misaligned Ply

in which α_ℓ and α_t are the coefficients of thermal expansion of the unidirectional layer along the longitudinal direction ℓ and transverse direction t, respectively. Following the same procedure as in Section 11.1 with the same notations, one can write

$$\left\{ \begin{matrix} \varepsilon \\ \gamma \end{matrix} \right\}_{x,y} = [T'] \left\{ \begin{matrix} \varepsilon \\ \gamma \end{matrix} \right\}_{\ell,t} \; ; \{\sigma\}_{\ell,t} = [T]\{\sigma\}_{x,y}$$

Then, upon substituting,

$$
\begin{Bmatrix} \varepsilon_x \\ \varepsilon_y \\ \gamma_{xy} \end{Bmatrix} = [T'] \begin{bmatrix} \dfrac{1}{E_\ell} & -\dfrac{v_{t\ell}}{E_t} & 0 \\ -\dfrac{v_{\ell t}}{E_\ell} & \dfrac{1}{E_t} & 0 \\ 0 & 0 & \dfrac{1}{G_{\ell t}} \end{bmatrix} [T] \begin{Bmatrix} \sigma_x \\ \sigma_y \\ \tau_{xy} \end{Bmatrix} + \Delta T [T'] \begin{Bmatrix} \alpha_\ell \\ \alpha_t \\ 0 \end{Bmatrix}
$$

One finds again in the first part of the second term a matrix of compliance coefficients, the terms of which are described in details in Equation 11.5. The second part of the second term is written as:

$$
\Delta T \begin{bmatrix} c^2 & s^2 & cs \\ s^2 & c^2 & -cs \\ -2cs & 2cs & (c^2 - s^2) \end{bmatrix} \begin{Bmatrix} \alpha_\ell \\ \alpha_t \\ 0 \end{Bmatrix} = \Delta T \begin{Bmatrix} c^2 \alpha_\ell + s^2 \alpha_t \\ s^2 \alpha_\ell + c^2 \alpha_t \\ 2cs(\alpha_t - \alpha_\ell) \end{Bmatrix}
$$

Therefore, the thermomechanical relation for a unidirectional layer written in the axes x,y, different from the ℓ,t coordinates, can be summarized as follows:

$$
\begin{Bmatrix} \varepsilon_x \\ \varepsilon_y \\ \gamma_{xy} \end{Bmatrix} = \begin{bmatrix} \dfrac{1}{E_x} & -\dfrac{v_{yx}}{E_y} & \dfrac{\eta_{xy}}{G_{xy}} \\ -\dfrac{v_{xy}}{E_x} & \dfrac{1}{E_y} & \dfrac{\mu_{xy}}{G_{xy}} \\ \dfrac{\eta_x}{E_x} & \dfrac{\mu_y}{E_y} & \dfrac{1}{G_{xy}} \end{bmatrix} \begin{Bmatrix} \sigma_x \\ \sigma_y \\ \tau_{xy} \end{Bmatrix} + \Delta T \begin{Bmatrix} \alpha_x \\ \alpha_y \\ \alpha_{xy} \end{Bmatrix}
\qquad (11.9)
$$

$E_x, E_y, G_{xy}, v_{xy}, v_{yx}, \eta_{xy}, \mu_{xy}$ are given by the relations [11.5]

$$
\alpha_x = c^2 \alpha_\ell + s^2 \alpha_t
$$

$$
\alpha_y = s^2 \alpha_\ell + c^2 \alpha_t
$$

$$
\alpha_{xy} = 2cs(\alpha_t - \alpha_\ell)
$$

$$
c = \cos\theta; \; s = \sin\theta
$$

11.3.2 Stiffness Coefficients

Inverting Equation 10.9 gives

$$
\begin{Bmatrix} \sigma_\ell \\ \sigma_t \\ \tau_{\ell t} \end{Bmatrix} = \begin{bmatrix} \dfrac{E_\ell}{(1-\nu_{\ell t}\nu_{t\ell})} & \dfrac{\nu_{t\ell}E_\ell}{(1-\nu_{\ell t}\nu_{t\ell})} & 0 \\ \dfrac{\nu_{\ell t}E_t}{(1-\nu_{\ell t}\nu_{t\ell})} & \dfrac{E_t}{(1-\nu_{\ell t}\nu_{t\ell})} & 0 \\ 0 & 0 & G_{\ell t} \end{bmatrix} \begin{Bmatrix} \varepsilon_\ell \\ \varepsilon_t \\ \gamma_{\ell t} \end{Bmatrix} \dots
$$

$$
\dots -\Delta T \begin{Bmatrix} \dfrac{E_\ell}{(1-\nu_{\ell t}\nu_{t\ell})}\alpha_\ell + \dfrac{\nu_{t\ell}E_\ell}{(1-\nu_{\ell t}\nu_{t\ell})}\alpha_t \\ \dfrac{\nu_{\ell t}E_t}{(1-\nu_{\ell t}\nu_{t\ell})}\alpha_l + \dfrac{E_t}{(1-\nu_{\ell t}\nu_{t\ell})}\alpha_t \\ 0 \end{Bmatrix}
$$

Following the procedure of Section 11.2, with the same notations, one can write:

$$
\{\sigma\}_{x,y} = [T_1]\{\sigma\}_{\ell,t}; \quad \begin{Bmatrix} \varepsilon \\ \gamma \end{Bmatrix}_{\ell,t} = [T'_1]\begin{Bmatrix} \varepsilon \\ \gamma \end{Bmatrix}_{x,y}
$$

where after substitution:

$$
\begin{Bmatrix} \sigma_x \\ \sigma_y \\ \tau_{xy} \end{Bmatrix} = [T_1]\begin{bmatrix} \bar{E}_\ell & \nu_{t\ell}\bar{E}_\ell & 0 \\ \nu_{\ell t}\bar{E}_t & \bar{E}_t & 0 \\ 0 & 0 & G_{\ell t} \end{bmatrix}[T'_1]\begin{Bmatrix} \varepsilon_x \\ \varepsilon_y \\ \gamma_{xy} \end{Bmatrix} - \Delta T[T_1]\begin{Bmatrix} \bar{E}_\ell\alpha_\ell + \nu_{t\ell}\bar{E}_\ell\alpha_t \\ \nu_{\ell t}\bar{E}_t\alpha_\ell + \bar{E}_t\alpha_t \\ 0 \end{Bmatrix}
$$

One finds again, in the first part of the second term, the matrix detailed in Equation 11.8. The second part of the second term can be developed as follows:

$$
-\Delta T \begin{bmatrix} c^2 & s^2 & 2cs \\ s^2 & c^2 & -2cs \\ -cs & cs & (c^2-s^2) \end{bmatrix}\begin{Bmatrix} \bar{E}_\ell\alpha_\ell + \nu_{t\ell}\bar{E}_\ell\alpha_t \\ \nu_{\ell t}\bar{E}_t\alpha_\ell + \bar{E}_t\alpha_t \\ 0 \end{Bmatrix} = \dots
$$

$$
\dots - \Delta T \begin{Bmatrix} c^2\bar{E}_\ell(\alpha_\ell + \nu_{t\ell}\alpha_t) + s^2\bar{E}_t(\nu_{\ell t}\alpha_\ell + \alpha_t) \\ s^2\bar{E}_\ell(\alpha_\ell + \nu_{t\ell}\alpha_t) + c^2\bar{E}_t(\nu_{\ell t}\alpha_\ell + \alpha_t) \\ cs[\bar{E}_t(\nu_{\ell t}\alpha_\ell + \alpha_t) - \bar{E}_\ell(\alpha_\ell + \nu_{t\ell}\alpha_t)] \end{Bmatrix}
$$

Therefore, the thermomechanical behavior of a unidirectional layer in the coordinate axes x,y can be written in the following form, in terms of the properties in the ℓ,t coordinates.

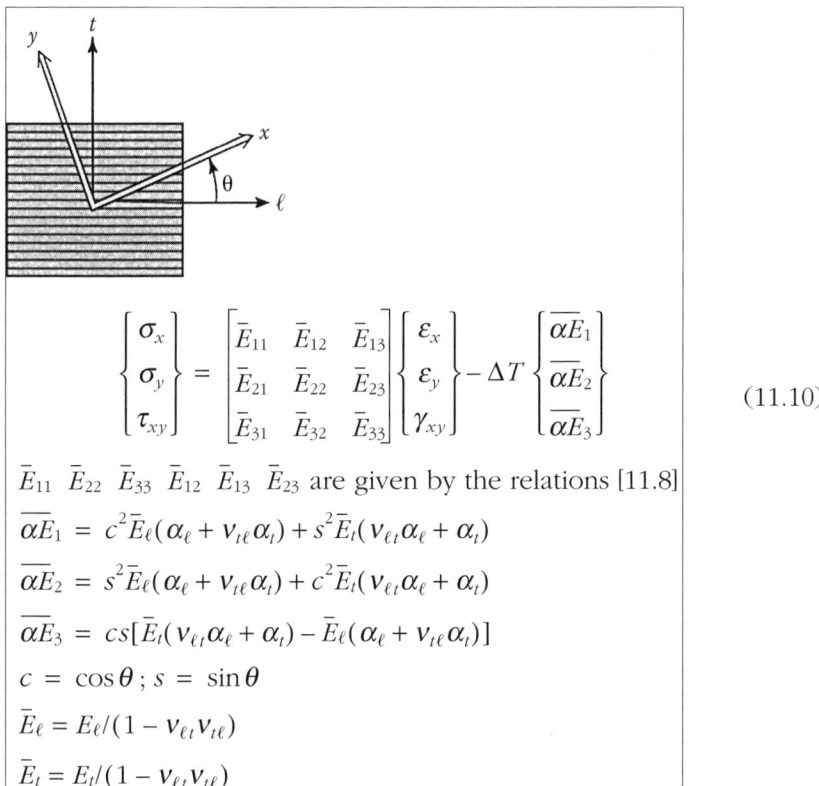

$$
\left\{\begin{array}{c} \sigma_x \\ \sigma_y \\ \tau_{xy} \end{array}\right\} = \left[\begin{array}{ccc} \bar{E}_{11} & \bar{E}_{12} & \bar{E}_{13} \\ \bar{E}_{21} & \bar{E}_{22} & \bar{E}_{23} \\ \bar{E}_{31} & \bar{E}_{32} & \bar{E}_{33} \end{array}\right] \left\{\begin{array}{c} \varepsilon_x \\ \varepsilon_y \\ \gamma_{xy} \end{array}\right\} - \Delta T \left\{\begin{array}{c} \overline{\alpha E_1} \\ \overline{\alpha E_2} \\ \overline{\alpha E_3} \end{array}\right\}
$$

(11.10)

$\bar{E}_{11}\ \bar{E}_{22}\ \bar{E}_{33}\ \bar{E}_{12}\ \bar{E}_{13}\ \bar{E}_{23}$ are given by the relations [11.8]

$$\overline{\alpha E_1} = c^2 \bar{E}_\ell (\alpha_\ell + v_{t\ell}\alpha_t) + s^2 \bar{E}_t (v_{\ell t}\alpha_\ell + \alpha_t)$$

$$\overline{\alpha E_2} = s^2 \bar{E}_\ell (\alpha_\ell + v_{t\ell}\alpha_t) + c^2 \bar{E}_t (v_{\ell t}\alpha_\ell + \alpha_t)$$

$$\overline{\alpha E_3} = cs[\bar{E}_t (v_{\ell t}\alpha_\ell + \alpha_t) - \bar{E}_\ell (\alpha_\ell + v_{t\ell}\alpha_t)]$$

$$c = \cos\theta\ ;\ s = \sin\theta$$

$$\bar{E}_\ell = E_\ell / (1 - v_{\ell t}v_{t\ell})$$

$$\bar{E}_t = E_t / (1 - v_{\ell t}v_{t\ell})$$

12

MECHANICAL BEHAVIOR OF THIN LAMINATED PLATES

The definition of a laminate was given in Chapter 5.[1] In the same chapter the practical calculation methods for the laminate was also described. We propose here to justify these methods, meaning to study the behavior of the laminate when it is subjected to a combination of loadings. This study is necessary if one wants to have correct design with strains or stresses within their admissible values.[2]

12.1 LAMINATE WITH MIDPLANE SYMMETRY

12.1.1 Membrane Behavior

We consider in the following a laminate with **midplane symmetry**.[3] The total thickness of the laminate is denoted as h. It consists of n plies. Ply number k has a thickness denoted as e_k (see Figure 12.1). Plane x,y is the plane of symmetry.

12.1.1.1 Loadings

The laminate is subjected to loadings in its plane. The stress resultants are denoted as N_x, N_y, $T_{xy} = T_{yx}$. These are the membrane stress resultants. They are defined as:

- N_x: Stress resultant in the x direction over a unit width along the y direction.

$$N_x = \int_{-h/2}^{h/2} \sigma_x dz = \sum_{k=1^{st}ply}^{n^{th}ply} (\sigma_x)_k \times e_k \qquad (12.1)$$

- N_y: Stress resultant along the y direction over a unit width along the x direction.

$$N_y = \int_{-h/2}^{h/2} \sigma_y \, dz = \sum_{k=1^{st}ply}^{n^{th}ply} (\sigma_y)_k \times e_k \qquad (12.2)$$

---

[1] See Section 5.2.
[2] The problem of buckling of the laminates is not the scope of this chapter. See Appendix 2.
[3] See Section 5.2.3.

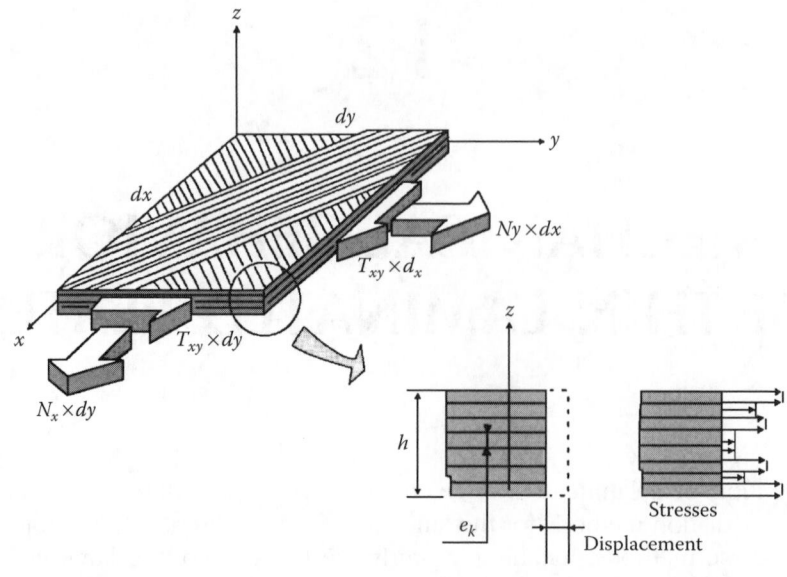

Figure 12.1 Definition of Laminate and Membrane Loading

■ T_{xy} (or T_{yx}): Membrane shear stress resultant over a unit width along the y direction (or respectively along the x direction):

$$T_{xy} = \int_{-b/2}^{b/2} \tau_{xy}\, dz = \sum_{k=1^{st}\text{ply}}^{n^{th}\text{ply}} (\tau_{xy})_k \times e_k \qquad (12.3)$$

12.1.1.2 Displacement Field

The elastic displacement at each point of the laminate is assumed to be two dimensional, in the x,y plane of the laminate. It has the components: u_o , v_o. The nonzero strains can be written as:

$$\varepsilon_{ox} = \partial u_0/\partial x$$
$$\varepsilon_{oy} = \partial v_0/\partial y$$
$$\gamma_{oxy} = \partial u_0/\partial y + \partial v_0/\partial x$$

It was shown in the previous chapter (Equation 11.8) that one can express, in a given coordinate system, the stresses in a ply as functions of the strains. Then the stress resultant N_x defined in Equation 12.1 can be written as follows:

$$N_x = \sum_{k=1^{st}\text{ply}}^{n^{th}\text{ply}} \{\bar{E}_{11}^k \varepsilon_{ox} + \bar{E}_{12}^k \varepsilon_{oy} + \bar{E}_{13}^k \gamma_{oxy}\} e_k$$

then:

$$N_x = A_{11}\varepsilon_{ox} + A_{12}\varepsilon_{oy} + A_{13}\gamma_{oxy}$$

with:

$$A_{11} = \sum_{k=1^{st}\text{ply}}^{n^{th}\text{ply}} \bar{E}_{11}^{k} e_{k}; \quad A_{12} = \sum_{k=1^{st}\text{ply}}^{n^{th}\text{ply}} \bar{E}_{12}^{k} e_{k}; \quad A_{13} = \sum_{k=1^{st}\text{ply}}^{n^{th}\text{ply}} \bar{E}_{13}^{k} e_{k}$$

In an analogous manner, one obtains for the Equation 12.2:

$$N_y = A_{21}\varepsilon_{ox} + A_{22}\varepsilon_{oy} + A_{23}\gamma_{oxy}$$

with:

$$A_{2j} = \sum_{k=1^{st}\text{ply}}^{n^{th}\text{ply}} \bar{E}_{2j}^{k} e_{k}$$

and for the shear stress resultant T_{xy} one can write, starting from Equation 12.3:

$$T_{xy} = A_{31}\varepsilon_{ox} + A_{32}\varepsilon_{oy} + A_{33}\gamma_{oxy}$$

with:

$$A_{3j} = \sum_{k=1^{st}\text{ply}}^{n^{th}\text{ply}} \bar{E}_{3j}^{k} e_{k}$$

Therefore, it is possible to express the stress resultants in the following matrix form:

$$\begin{Bmatrix} N_x \\ N_y \\ T_{xy} \end{Bmatrix} = \begin{bmatrix} A_{11} & A_{12} & A_{13} \\ A_{21} & A_{22} & A_{23} \\ A_{31} & A_{32} & A_{33} \end{bmatrix} \begin{Bmatrix} \varepsilon_{ox} \\ \varepsilon_{oy} \\ \gamma_{oxy} \end{Bmatrix}$$

with:

$$A_{ij} = \sum_{k=1^{st}\text{ply}}^{n^{th}\text{ply}} \bar{E}_{ij}^{k} e_{k} = A_{ji}$$

(12.4) [4]

Remarks:

- One observes from the above expressions that coefficients A_{ij} are independent of the stacking order of the plies.
- One can also see that the normal stress resultants N_x or N_y create angular distortions. This coupling will disappear if the laminate is **balanced**. This means that apart from the midplane symmetry, there are as many and identical plies that make with the x axis an angle $+\theta$ as those that make

---

[4] The developments of $\bar{E}_{ij}$ are given in Equation 11.8.

with the x axis an angle $-\theta$.[5] In effect, the coefficients $\bar{E}_{13}$ and $\bar{E}_{23}$ are antisymmetric in θ[6] and, therefore, cancel each other out for the pairs of plies at $\pm\theta$ when one calculates the terms A_{13} and A_{23}. The result is then:

$$A_{13} = A_{23} = 0$$

and the stress–strain relation for the laminate is reduced to

$$
\begin{Bmatrix} N_x \\ N_y \\ T_{xy} \end{Bmatrix} = \begin{bmatrix} A_{11} & A_{12} & 0 \\ A_{12} & A_{22} & 0 \\ 0 & 0 & A_{33} \end{bmatrix} \begin{Bmatrix} \varepsilon_{ox} \\ \varepsilon_{oy} \\ \gamma_{oxy} \end{Bmatrix}
\tag{12.5}
$$

■ It is possible to substitute the stress resultants N_x, N_y, T_{xy} with the global average stresses (which are fictitious):

$$
\begin{aligned}
\sigma_{ox} &= N_x/h \\
\sigma_{oy} &= N_y/h \\
\tau_{oxy} &= T_{xy}/h
\end{aligned}
\tag{12.6}
$$

One then deduces from Equation 12.4 the average membrane stress–strain behavior of a laminate as:

$$
\begin{Bmatrix} \sigma_{ox} \\ \sigma_{oy} \\ \tau_{oxy} \end{Bmatrix} = \frac{1}{h} \begin{bmatrix} A_{11} & A_{12} & A_{13} \\ A_{21} & A_{22} & A_{23} \\ A_{31} & A_{32} & A_{33} \end{bmatrix} \begin{Bmatrix} \varepsilon_{ox} \\ \varepsilon_{oy} \\ \gamma_{oxy} \end{Bmatrix}
\tag{12.7}
$$

■ One can also note that according to Equation 12.4 the terms of the matrix $\frac{1}{h}[A]$ above can be written as:

$$
\frac{1}{h} \times A_{ij} = \sum_{k=1^{\text{st}} \text{ply}}^{n^{\text{th}} \text{ply}} \bar{E}_{ij}^{k} \times \frac{e_k}{h}
$$

Then the ratios e_k/h can be rearranged to obtain the proportions of plies having the same orientation. In case where these proportions have already been fixed—and therefore their numerical values are known—it becomes possible to calculate the terms $\frac{1}{h}A_{ij}$ **without knowing the thickness**. For example, if the selected orientations are 0°, 90°, +45°, −45°, and by denoting $p^k(\%)$ as the percentages of the plies along the different orientations, one has:

$$
\frac{1}{h} \times A_{ij} = \bar{E}_{ij}^{0°} \times p^{0°} + \bar{E}_{ij}^{90°} \times p^{90°} + \bar{E}_{ij}^{+45°} \times p^{+45°} + \bar{E}_{ij}^{-45°} \times p^{-45°}
\tag{12.8}
$$

---

[5] See Figure 12.1 and figure in the Equation 11.8.

[6] The expressions developed for $\bar{E}_{ij}$ are given in Equation 11.8.

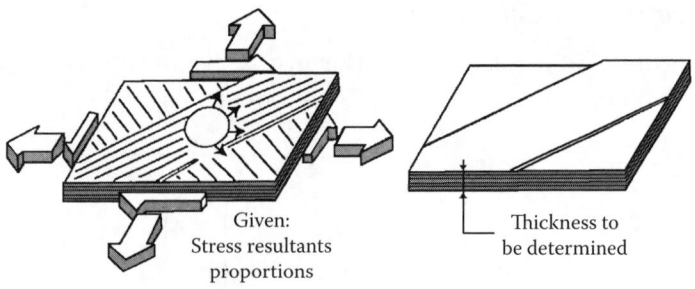

Given:
Stress resultants
proportions

Thickness to
be determined

Figure 12.2 Practical Determination of a Laminate Subject to Membrane Loading

12.1.2 Apparent Moduli of the Laminate

Inversion of Equation 12.7 above allows one to obtain what can be called as **apparent moduli** and coupling coefficients associated with the membrane behavior in the plane x,y. These coefficients appear in the following relation:

$$
\begin{Bmatrix} \varepsilon_{ox} \\ \varepsilon_{oy} \\ \gamma_{oxy} \end{Bmatrix} = h[A]^{-1} \begin{Bmatrix} \sigma_{ox} \\ \sigma_{oy} \\ \tau_{oxy} \end{Bmatrix} = \begin{bmatrix} \dfrac{1}{\overline{E}_x} & -\dfrac{\overline{v}_{yx}}{\overline{E}_y} & \dfrac{\overline{\eta}_{xy}}{\overline{G}_{xy}} \\ -\dfrac{\overline{v}_{xy}}{\overline{E}_x} & \dfrac{1}{\overline{E}_y} & \dfrac{\overline{\mu}_{xy}}{\overline{G}_{xy}} \\ \dfrac{\overline{\eta}_x}{\overline{E}_x} & \dfrac{\overline{\mu}_y}{\overline{E}_y} & \dfrac{1}{\overline{G}_{xy}} \end{bmatrix} \begin{Bmatrix} \sigma_{ox} \\ \sigma_{oy} \\ \tau_{oxy} \end{Bmatrix} \qquad (12.9)
$$

12.1.3 Consequence: Practical Determination of a Laminate Subject to Membrane Loading

Given:

- The stress resultants are given and denoted as: N_x, N_y, T_{xy}.
- Using the values of these stress resultants, one can estimate the ply proportions in the four orientations.[7] Assume in Figure 12.2 that the plies are identical (same material and same thickness).

The problem is to determine

- The apparent elastic moduli of the laminate and the associated coupling coefficients, in order to estimate strains under loading
- The minimum thickness for the laminate in order to avoid rupture of one of the plies in the laminate

---

[7] See Section 5.4.3.

12.1.3.1 Principle of Calculation

Apparent moduli of the laminate: The matrix $\frac{1}{h}[A]$ evaluated using Equation 12.8 can be inverted, and one obtains Equation 12.9 as:

$$
\begin{Bmatrix} \varepsilon_{ox} \\ \varepsilon_{oy} \\ \gamma_{oxy} \end{Bmatrix} = \begin{bmatrix} \dfrac{1}{\bar{E}_x} & -\dfrac{\bar{v}_{yx}}{\bar{E}_y} & \dfrac{\bar{\eta}_{xy}}{\bar{G}_{xy}} \\ -\dfrac{\bar{v}_{xy}}{\bar{E}_x} & \dfrac{1}{\bar{E}_y} & \dfrac{\bar{\mu}_{xy}}{\bar{G}_{xy}} \\ \dfrac{\bar{\eta}_x}{\bar{E}_x} & \dfrac{\bar{\mu}_y}{\bar{E}_y} & \dfrac{1}{\bar{G}_{xy}} \end{bmatrix} \begin{Bmatrix} \sigma_{ox} \\ \sigma_{oy} \\ \tau_{oxy} \end{Bmatrix}
$$

We have already determined the apparent moduli and the coupling coefficients of the laminate.

Nonrupture of the laminate: Let σ_ℓ, σ_t, and $\tau_{\ell t}$ be the stresses in the orthotropic axes ℓ, t of one of the plies making up the laminate that is subjected to the loadings N_x, N_y, T_{xy}. Let h be the thickness of the laminate (unknown at the moment) so that the rupture limit of the ply using the Hill–Tsai failure criterion is just reached.

One then has for this ply[8]:

$$
\frac{\sigma_\ell^2}{\underset{\text{rupture}}{\sigma_\ell^2}} + \frac{\sigma_t^2}{\underset{\text{rupture}}{\sigma_t^2}} - \frac{\sigma_\ell \sigma_t}{\underset{\text{rupture}}{\sigma_\ell^2}} + \frac{\tau_{\ell t}^2}{\underset{\text{rupture}}{\tau_{\ell t}^2}} = 1
$$

Multiplying the two parts of this equation with the square of thickness h:

$$
\frac{(\sigma_\ell h)^2}{\underset{\text{rupture}}{\sigma_\ell^2}} + \frac{(\sigma_t h)^2}{\underset{\text{rupture}}{\sigma_t^2}} - \frac{(\sigma_\ell h)(\sigma_t h)}{\underset{\text{rupture}}{\sigma_\ell^2}} + \frac{(\tau_{\ell t} h)^2}{\underset{\text{rupture}}{\tau_{\ell t}^2}} = h^2 \tag{12.10}
$$

To obtain the values $(\sigma_\ell h)$, $(\sigma_t h)$, $(\tau_{\ell t} h)$, one has to multiply with h the global stresses σ_{ox}, σ_{oy}, τ_{oxy} that are applied on the laminate, to become $(\sigma_{ox} h)$, $(\sigma_{oy} h)$, $(\tau_{oxy} h)$ which are the known stress resultants:

$$
N_x = (\sigma_{ox} h); \quad N_y = (\sigma_{oy} h); \quad T_{xy} = (\tau_{oxy} h)
$$

Then, for a ply, the calculation of the Hill–Tsai criterion can be done by substituting for the unknown global stresses the known stress resultants N_x, N_y, T_{xy}. This leads to the calculation of the thickness h so that the ply under consideration does not fracture.

In this way, each ply number k leads to a laminate thickness value denoted as h_k. The final thickness to be retained will be the one with the highest value.

---

[8] For the Hill–Tsai failure criterion, see Section 5.2.3 and detailed explanation in Chapter 14.

12.1.3.2 Calculation Procedure

1. **Complete calculation:** The ply proportions are given, the matrix $\frac{1}{h}[A]$ of the Equation 12.7 is known, and then—after inversion—we obtain the elastic moduli of the laminate (Equation 12.9).[9] Multiplying 12.9 with the thickness h (unknown) of the laminate:

$$
\begin{Bmatrix} h\varepsilon_{ox} \\ h\varepsilon_{oy} \\ h\gamma_{oxy} \end{Bmatrix} = \begin{bmatrix} \dfrac{1}{\overline{E}_x} & -\dfrac{\overline{v}_{yx}}{\overline{E}_y} & \dfrac{\overline{\eta}_{xy}}{\overline{G}_{xy}} \\ -\dfrac{\overline{v}_{xy}}{\overline{E}_x} & \dfrac{1}{\overline{E}_y} & \dfrac{\overline{\mu}_{xy}}{\overline{G}_{xy}} \\ \dfrac{\overline{\eta}_x}{\overline{E}_x} & \dfrac{\overline{\mu}_y}{\overline{E}_y} & \dfrac{1}{\overline{G}_{xy}} \end{bmatrix} \begin{Bmatrix} N_x \\ N_y \\ T_{xy} \end{Bmatrix}
$$

Then introducing a multiplication factor of h for the stresses in the ply—or the group of plies—corresponding to the orientation k (see Equation 11.8):

$$
\begin{Bmatrix} h\sigma_x \\ h\sigma_y \\ h\tau_{xy} \end{Bmatrix} = \begin{bmatrix} \overline{E}_{11} & \overline{E}_{12} & \overline{E}_{13} \\ \overline{E}_{21} & \overline{E}_{22} & \overline{E}_{23} \\ \overline{E}_{31} & \overline{E}_{32} & \overline{E}_{33} \end{bmatrix} \begin{Bmatrix} h\varepsilon_{ox} \\ h\varepsilon_{oy} \\ h\gamma_{oxy} \end{Bmatrix}
$$

ply $n°k$ ply $n°k$ laminate

and in the orthotropic coordinates of the ply (see Equation 11.4):

$$
\begin{Bmatrix} h\sigma_\ell \\ h\sigma_t \\ h\tau_{\ell t} \end{Bmatrix} = \begin{bmatrix} c^2 & s^2 & -2cs \\ s^2 & c^2 & 2cs \\ sc & -sc & (c^2-s^2) \end{bmatrix} \begin{Bmatrix} h\sigma_x \\ h\sigma_y \\ h\tau_{xy} \end{Bmatrix} \qquad c = \cos\theta; \ s = \sin\theta
$$

ply $n°k$ ply $n°k$ ply $n°k$

Saturation of the Hill–Tsai criterion leads then to Equation 12.10 where the above known stress resultants values appear in the numerator as:

$$
\frac{(h\sigma_\ell)^2}{\sigma_{\ell}^2} + \frac{(h\sigma_t)^2}{\sigma_{t}^2} - \frac{(h\sigma_\ell)(h\sigma_t)}{\sigma_{\ell}^2} + \frac{(h\tau_{\ell t}^2)}{\tau_{\ell t}^2} = h^2 \times 1
$$

rupture   rupture   rupture   rupture

After having written an analogous expression for each orientation k of the plies, one retains for the final value of the laminate thickness, the maximum value found for h.

---

[9] One can read directly these moduli in Tables 5.1 to 5.15 of Section 5.4.2 for balanced laminates of carbon, Kevlar, and glass/epoxy with $V_f = 60\%$ fiber volume fraction.

2. **Simplified calculation:** One can write more rapidly the Equation 12.10 if one knows at the beginning for each orientation the stresses due to a global uniaxial state of unit stress applied on the laminate: first $\sigma'_{ox} = 1$ (for example, 1 MPa), then $\sigma''_{oy} = 1$ MPa, then $\tau'''_{oxy} = 1$ MPa.

■ Assume first that the state of stress is given as:

$$\left| \begin{array}{l} \sigma'_{ox} = 1(\mathrm{MPa}) \\ \sigma'_{ox} = 0 \\ \tau'_{oxy} = 0 \end{array} \right.$$

Inverting the Equation 12.9 leads to

$$\left\{ \begin{array}{c} \varepsilon'_{ox} \\ \varepsilon'_{oy} \\ \gamma'_{oxy} \end{array} \right\} = \begin{bmatrix} \dfrac{1}{\bar{E}_x} & -\dfrac{\bar{v}_{yx}}{\bar{E}_y} & \dfrac{\bar{\eta}_{xy}}{\bar{G}_{xy}} \\ -\dfrac{\bar{v}_{xy}}{\bar{E}_x} & \dfrac{1}{\bar{E}_y} & \dfrac{\bar{\mu}_{xy}}{\bar{G}_{xy}} \\ \dfrac{\bar{\eta}_x}{\bar{E}_x} & \dfrac{\bar{\mu}_y}{\bar{E}_y} & \dfrac{1}{\bar{G}_{xy}} \end{bmatrix} \left\{ \begin{array}{c} 1 \ \mathrm{MPa} \\ 0 \\ 0 \end{array} \right\}$$

which can be considered as "**unitary strains**" of the laminate. These allow the calculation of the stresses in each ply by means of Equations 11.8 and then 11.4, successively, as:

$$\left\{ \begin{array}{c} \sigma'_x \\ \sigma'_y \\ \tau'_{xy} \end{array} \right\} = \begin{bmatrix} \bar{E}_{11} & \bar{E}_{12} & \bar{E}_{13} \\ \bar{E}_{21} & \bar{E}_{22} & \bar{E}_{23} \\ \bar{E}_{31} & \bar{E}_{32} & \bar{E}_{33} \end{bmatrix} \left\{ \begin{array}{c} \varepsilon'_{ox} \\ \varepsilon'_{oy} \\ \gamma'_{oxy} \end{array} \right\}$$

$$\text{ply } n°k \qquad\qquad \text{ply } n°k \qquad\qquad \text{laminate}$$

and in the orthotropic coordinates of the ply (Equation 11.4):

$$\left\{ \begin{array}{c} \sigma'_\ell \\ \sigma'_t \\ \tau'_{\ell t} \end{array} \right\} = \begin{bmatrix} c^2 & s^2 & -2cs \\ s^2 & c^2 & 2cs \\ sc & -sc & (c^2 - s^2) \end{bmatrix} \left\{ \begin{array}{c} \sigma'_x \\ \sigma'_y \\ \tau'_{xy} \end{array} \right\} \qquad \begin{array}{l} c = \cos\theta \\ s = \sin\theta \end{array}$$

$$\text{ply } n°k \qquad\qquad \text{ply } n°k \qquad\qquad \text{ply } n°k$$

■ Consider then the state of stresses:

$$\left| \begin{array}{l} \sigma''_{ox} = 0 \\ \sigma''_{oy} = 1 \ (\mathrm{MPa}) \\ \tau''_{oxy} = 0 \end{array} \right.$$

Following the same procedure, one can calculate σ''_ℓ, σ''_t, and $\tau''_{\ell t}$ in the orthotropic axes of each ply for a global stress on the laminate that is reduced to $\sigma''_{oy} = 1$ MPa.

■ Finally consider the state of stresses:

$$\left|\begin{array}{l} \sigma_{ox}''' = 0 \\[6pt] \sigma_{oy}''' = 0 \\[6pt] \tau_{oxy}''' = 1 \ (\text{MPa}) \end{array}\right.$$

Following the same procedure, one obtains σ_ℓ'', σ_t'', and $\tau_{\ell t}''$ in the orthotropic axes of each ply for a global stress applied on the laminate, and that is reduced to $\tau_{oxy}'' = 1$ MPa.[10]

It is then easy to determine by simple rule of proportion (or multiplication)[11] the quantities $(\sigma_\ell b)$, $(\sigma_t b)$, and $(\tau_{\ell t} b)$ in each ply, corresponding to loadings that are no longer unitary, but equal successively to

$$N_x = (\sigma_{ox} b)$$

then:

$$N_y = (\sigma_{oy} b)$$

then:

$$T_{xy} = (\tau_{oxy} b)$$

Subsequently, the **principle of superposition** allows one to determine $(\sigma_\ell b)_{\text{total}}$ $(\sigma_t b)_{\text{total}}$ and $(\tau_{\ell t} b)_{\text{total}}$ when one applies **simultaneously** N_x, N_y, and T_{xy}. From these it is possible to write the modified Hill–Tsai expression in the form of Equation 12.10, which will provide the thickness for the laminate needed to avoid the fracture of the ply under consideration.

If b_k is the laminate thickness obtained from the ply number k, after having gone over all the plies, one will retain for the final thickness b the thickness of highest value found as:

$$b = \sup \{b_k\}^{12}$$

Remark: The principle of calculation is conserved when the plies have different thicknesses with any orientations. It then becomes indispensable to program the procedure, or to use existing computer programs. Then one can propose a complete composition for the laminate and verify that the solution is satisfactory regarding the criterion mentioned previously (deformation and fracture). This is

---

[10] This calculation can be easily programmed on a computer: *cf.* Application 18.2.2 "Program for Calculation of a Laminate." One will find in Appendix 1 at the end of the book the values σ_ℓ, σ_t, $\tau_{\ell t}$ obtained for the particular case of a carbon/epoxy laminate with ply orientations of 0°, 90°, +45°, −45°. These values are given in Plates 1 to 12.

[11] For example, one has the following:

$$\sigma_{ox}' = 1 \ \text{MPa} \rightarrow \sigma_\ell', \ \sigma_t', \ \tau_{\ell t}'$$

$$\sigma_{ox}(\text{MPa}) \rightarrow \sigma_\ell, \ \sigma_t, \ \tau_{\ell t}$$

then: $\dfrac{\sigma_{ox}}{\sigma_{ox}'} = \dfrac{\sigma_\ell}{\sigma_\ell'} \Rightarrow \sigma_\ell = \sigma_\ell' \times \dfrac{\sigma_{ox}}{1}$, and $b\sigma_\ell = \dfrac{\sigma_\ell'}{1} \times N_x$

[12] This method to determine the thickness is illustrated by an example: See Application 18.1.6.

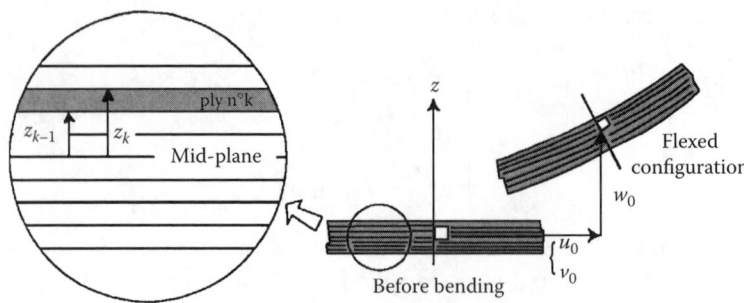

Figure 12.3   Bending of the Laminate

facilitated by using the user friendly aspect of the program, allowing rapid return of the solution.

12.1.4   Flexure Behavior

In the previous paragraph, we have limited discussion to loadings consisting of N_x, N_y, and T_{xy} applying in the midplane of the laminate. We will now examine the cases that can cause deformation outside of the plane of the laminate. The laminate considered is—as before—supposed to have **midplane symmetry**.

12.1.4.1   Displacement Fields

- **Hypothesis:** Assume that a line perpendicular to the midplane of laminate before deformation (see Figure 12.3) remains perpendicular to the midplane surface after deformation.
- **Consequence:** If one denotes as before u_o and v_o the components of the displacement in the midplane and w_o as the displacement out of the plane (see Figure 12.3), the displacement of any point at a position z in the laminate (in the nondeformed configuration) can be written as

$$\left\{ \begin{aligned} u &= u_o - z\frac{\partial w_0}{\partial x} \\[2mm] v &= v_o - z\frac{\partial w_0}{\partial y} \\[2mm] w &= w_o \end{aligned} \right. \tag{12.11}$$

One can then deduce the nonzero strains:

$$\left\{ \begin{aligned} \varepsilon_x &= \varepsilon_{ox} - z\frac{\partial^2 w_0}{\partial x^2} \\[2mm] \varepsilon_y &= \varepsilon_{oy} - z\frac{\partial^2 w_0}{\partial y^2} \\[2mm] \gamma_{xy} &= \gamma_{oxy} - z \times 2\frac{\partial^2 w_0}{\partial x \partial y} \end{aligned} \right. \tag{12.12}$$

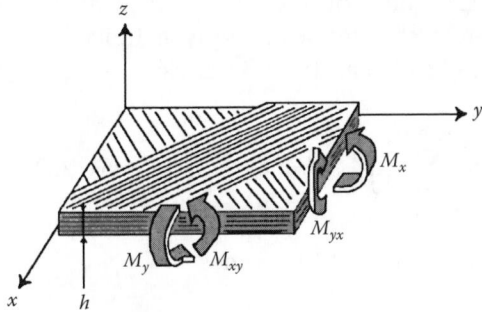

Figure 12.4   Moment Resultants

12.1.4.2 Loadings

In addition to the membrane stress resultants N_x, N_y, T_{xy} in the previous paragraphs, one can add the **moment resultants** along the x and y directions (see Figure 12.4).

As in the case of the membrane stress resultants, the moment resultants serve to synthesize the cohesive forces that appear by sectioning, following classical method that is common for all structures (beams, plates, etc.). One can interpret these as the unit moments of the cohesive forces.[13] They are written as:

- M_y: Moment resultant along the y axis, due to the stresses σ_x, over a unit width along the y direction.

$$M_y = \int_{-b/2}^{b/2} \sigma_x z \, dz \qquad (12.13)$$

- M_x: Moment resultant along the x direction, due to the stress σ_y, over a unit width along the x direction.

$$M_x = -\int_{-b/2}^{b/2} \sigma_y z \, dz \qquad (12.14)$$

- M_{xy}: (or $-M_{yx}$): Twisting moment along the x axis (or y axis), due to the shear stress τ_{xy} over a unit width along the y direction (or x direction):

$$M_{xy} = -\int_{-b/2}^{b/2} \tau_{xy} z \, dz \qquad (12.15)$$

---

[13] The expression of M_y can be written in integral form as:

$$M_y = \left[\int_{-b/2}^{b/2} z\vec{z} \wedge \sigma_x \vec{x} dz \right] \cdot \vec{y} = \int_{-b/2}^{b/2} \sigma_x z \, dz$$

also:

$$M_x = \left[\int_{-b/2}^{b/2} z\vec{z} \wedge \sigma_y \vec{y} dz \right] \cdot \vec{x} = -\int_{-b/2}^{b/2} \sigma_y z \, dz$$

Finally:

$$M_{xy} = \left[\int_{-b/2}^{b/2} z\vec{z} \wedge \tau_{xy} \vec{y} dz \right] \cdot \vec{x} = -\int_{-b/2}^{b/2} \tau_{xy} z \, dz$$

Taking Equation 11.8 into consideration, which allows one to express, in a certain coordinate system, the stresses in a ply as functions of strains, the moment resultant M_y (Equation 12.13) can be written as:

$$M_y = \sum_{k=1^{st}\text{ply}}^{n^{th}\text{ply}} \left\{ \int_{z_{k-1}}^{z_k} (\bar{E}_{11}^k \, \varepsilon_x + \bar{E}_{12}^k \, \varepsilon_y + \bar{E}_{13}^k \, \gamma_{xy}) z \; dz \right\}$$

which, when using Equation 12.12 becomes

$$M_y = \sum_{k=1^{st}\text{ply}}^{n^{th}\text{ply}} \left\{ \int_{z_{k-1}}^{z_k} \left\{ \bar{E}_{11}^k \left(z\varepsilon_{ox} - z^2\frac{\partial^2 w_o}{\partial x^2} \right) + \bar{E}_{12}^k \left(z\varepsilon_{oy} - z^2\frac{\partial^2 w_o}{\partial y^2} \right) \cdots \right.\right.$$

$$\left.\left. \cdots + \bar{E}_{13}^k \left(z\gamma_{oxy} - z^2 2\frac{\partial^2 w_o}{\partial x \partial y} \right) \right\} dz \right\}$$

Due to midplane symmetry, every integral of the form:

$$\int_{z_{k-1}}^{z_k} \bar{E}_{1j} z \; dz$$

in the above expression is accompanied by an integral of the form:

$$\int_{-z_k}^{-z_{k-1}} \bar{E}_{1j} z \; dz$$

that is opposite in sign. Integrals of this type disappear and there remains

$$M_y = \sum_{k=1^{st}\text{ply}}^{n^{th}\text{ply}} -\left\{ \bar{E}_{11}^k \frac{(z_k^3 - z_{k-1}^3)}{3}\frac{\partial^2 w_o}{\partial x^2} + \bar{E}_{12}^k \frac{(z_k^3 - z_{k-1}^3)}{3}\frac{\partial^2 w_o}{\partial y^2} \cdots \right.$$

$$\left. \cdots + \bar{E}_{13}^k \frac{(z_k^3 - z_{k-1}^3)}{3} 2\frac{\partial^2 w_o}{\partial x \partial y} \right\}$$

which can be written as:

$$M_y = -C_{11}\frac{\partial^2 w_o}{\partial x^2} - C_{12}\frac{\partial^2 w_o}{\partial y^2} - C_{13}2\frac{\partial^2 w_o}{\partial x \partial y}$$

with

$$C_{1j} = \sum_{k=1^{st}\text{ply}}^{n^{th}\text{ply}} \bar{E}_{1j}^k \frac{(z_k^3 - z_{k-1}^3)}{3}$$

Proceeding in an analogous manner with M_x and M_{xy} (Equations 12.14 and 12.15), one obtains the following matrix form:

$$\begin{Bmatrix} M_y \\ -M_x \\ -M_{xy} \end{Bmatrix} = \begin{bmatrix} C_{11} & C_{12} & C_{13} \\ C_{21} & C_{22} & C_{23} \\ C_{31} & C_{32} & C_{33} \end{bmatrix} \begin{Bmatrix} -\dfrac{\partial^2 w_o}{\partial x^2} \\ -\dfrac{\partial^2 w_o}{\partial y^2} \\ -2\dfrac{\partial^2 w_o}{\partial x \partial y} \end{Bmatrix}$$

with :

$$C_{ij} = \sum_{k=1^{st}\text{ply}}^{n^{th}\text{ply}} \bar{E}_{ij}^k \frac{(z_k^3 - z_{k-1}^3)}{3}$$

(12.16)

Remarks:

- One can observe that in Equation 12.16 the coefficients C_{ij} **depend** on the stacking sequence of the plies.
- Does a laminated plate bend under membrane loadings? Using the displacement field due to flexure to express, for example, the stress resultant N_x (Equation 12.11), one has

$$N_x = \sum_{k=1^{st}\text{ply}}^{n^{th}\text{ply}} \left\{ \int_{z_{k-1}}^{z_k} \left\{ \bar{E}_{11}^k \left(\varepsilon_{ox} - z\frac{\partial^2 w_o}{\partial x^2} \right) + \bar{E}_{12}^k \left(\varepsilon_{oy} - z\frac{\partial^2 w_o}{\partial y^2} \right) \cdots \right. \right.$$

$$\left. \left. \cdots + \bar{E}_{13}^k \left(\gamma_{oxy} - z \times 2\frac{\partial^2 w_o}{\partial x \partial y} \right) \right\} dz \right\}$$

Making use of the remark mentioned above, the midplane symmetry causes the disappearance of integrals of the type:

$$\int_{z_{k-1}}^{z_k} \bar{E}_{1j}^k z \; dz$$

As a consequence, one finds again the Equation 12.4 as:

$$N_1 = A_{11}\varepsilon_{ox} + A_{12}\varepsilon_{oy} + A_{13}\gamma_{oxy}$$

As a result of the midplane symmetry, the membrane behavior is independent of the flexural behavior.

- Even in the case of balanced laminate (as many plies oriented at angle θ as the number of plies oriented at an angle $-\theta$), terms C_{13} and C_{23} in Equation 12.16 are not zero. This modifies the deformed bending configuration compared with the isotropic case (see Figure 12.5).

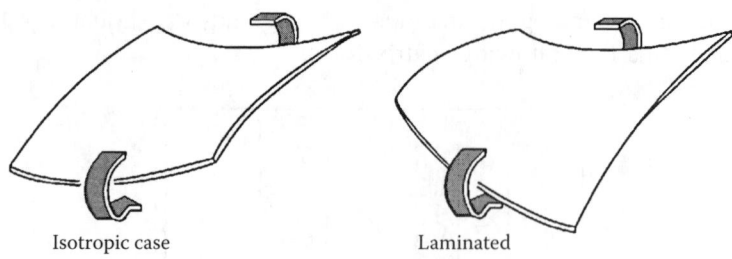

Isotropic case   Laminated

Figure 12.5   Bending Configurations of a Plate

■ The terms C_{13} and C_{23} disappear only in the following cases:

a) The plies are oriented uniquely in the 0° and 90°. Then the product $\cos\theta \times \sin\theta$ is zero and[14]:

$$\bar{E}_{13}^k = \bar{E}_{23}^k = 0 \quad \forall k$$

b) The laminate [0/90/45/−45] is constituted mainly with balanced fabric layers (in each fabric layer, the fibers along the warp and fill directions are, by first approximation[15] at the same z location), or mats layers.

■ The stresses in the different plies are obtained from the Equations 11.8. For example, in the ply number k, one has:

$$\sigma_x = \bar{E}_{11}^k \; \varepsilon_x + \bar{E}_{12}^k \; \varepsilon_y + \bar{E}_{13}^k \; \gamma_{xy}$$

and taking into consideration Equations 12.12 for the strains:

$$\sigma_x = [\bar{E}_{11}^k \; \varepsilon_{ox} + \bar{E}_{12}^k \; \varepsilon_{oy} + \bar{E}_{13}^k \; \gamma_{oxy}] - z\left[\bar{E}_{11}^k \frac{\partial^2 w_o}{\partial x^2} \cdots \right.$$
$$\left. \cdots + \bar{E}_{12}^k \frac{\partial^2 w_o}{\partial y^2} + \bar{E}_{13}^k \times 2\frac{\partial^2 w_o}{\partial x \partial y}\right]$$

one can resume by:

$$\sigma_x = \sigma_{x_{\text{membrane}}} + \sigma_{x_{\text{flexure}}}$$

Along the thickness of the laminate, the stress σ_x can be considered as the sum of two parts: a constant part and a linearly varying part, as seen in Figure 12.6. One can also observe analogous forms for the stresses σ_y and τ_{xy}.

---

[14] See Equations 11.8

[15] See Section 5.2.3.5, Particular Cases of Balanced Fabrics.

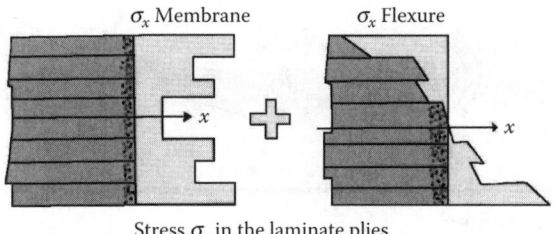

Stress σ_x in the laminate plies

Figure 12.6   Total Normal Stress in a Laminate

12.1.5 Consequence: Practical Determination for a Laminate Subject to Flexure

Given:

- The moment resultants are known.
- Using these resultants, one is led to estimate proportions of plies along the four orientations (or more, eventually)[16] and to predict the stacking sequence.

Principle for the calculation:

- **Nonrupture of laminate:** Following a procedure analogous to that described in Section 12.1.3, it is possible to calculate the stresses σ_ℓ, σ_t, $\tau_{\ell t}$ along the orthotropic axes of each of the plies. This allows the control of their integrity using the Hill–Tsai failure criterion. This requires the use of a computer program which can allow the adjustment of the composition of the laminate.
- **Flexure deformation:** The determination of the deformed configuration of the laminate under flexure poses the same problem as with the isotropic plates: outside of a few cases of academic interest, it is necessary to use a computer program based on the finite element method.[17]

12.1.6 Simplified Calculation for Flexure

It is possible, for a first estimate, to perform simplified calculations by considering that the moment M_y is related uniquely to the curvature $\frac{\partial^2 w_o}{\partial x^2}$ and the moment M_x to the curvature $\frac{\partial^2 w_o}{\partial y^2}$. One then can determine experimentally:

1. The apparent failure stresses in flexure

An experiment on a sample can provide the value for the moment at failure, denoted by $M_{rupture}$ on Figure 12.7 (per unit width of the sample). Analogy with

---

[16] See Section 5.2.

[17] These elements are constituted on the basis presented above and can include the effects which were not taken into account previously: in particular, the transverse shear stresses in flexure due to the transverse shear stress resultants (consult this subject in Chapter 17).

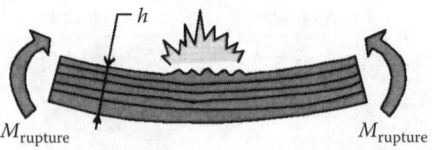

Figure 12.7   Bending Failure

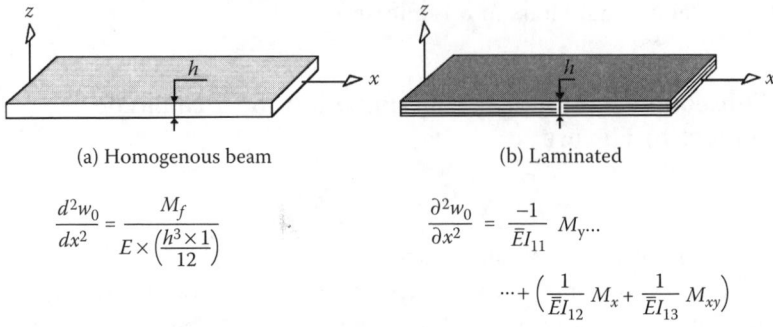

$$\frac{d^2w_0}{dx^2} = \frac{M_f}{E \times \left(\frac{h^3 \times 1}{12}\right)}$$

$$\frac{\partial^2w_0}{\partial x^2} = \frac{-1}{\overline{EI}_{11}}\,M_y\ldots$$

$$\ldots + \left(\frac{1}{\overline{EI}_{12}}\,M_x + \frac{1}{\overline{EI}_{13}}\,M_{xy}\right)$$

Figure 12.8   Homogeneous and Laminated Beams

the flexure of beams leads to:

$$\|\sigma_{\text{rupture}}\| = \frac{M_{\text{rupture}} \times h/2}{h^3/12} \quad \text{then: } \|\sigma_{\text{rupture}}\| = M_{\text{rupture}} \times \frac{6}{h^2}$$

2. Apparent flexure moduli

These are obtained starting from the comparison of relations between the "composite" and "homogeneous." One recalls on Figure 12.8(a) the relation between the moment and curvature for a homogeneous beam with unit width, obtained by integration of the local behavior[18]:

$$\varepsilon_x = \frac{\sigma_x}{E} \rightarrow \frac{h^3 \times 1}{12} \times \frac{d^2w_o}{dx^2} = -\frac{M_f}{E}$$

If one notes that Equation 12.16, recalled as:

$$\begin{Bmatrix} M_y \\ -M_x \\ -M_{xy} \end{Bmatrix} = [C] \begin{Bmatrix} -\partial^2 w_o/\partial x^2 \\ -\partial^2 w_o/\partial y^2 \\ -2 \times \partial^2 w_o/\partial x\,dy \end{Bmatrix}$$

---

[18] Recall that $\varepsilon_x = \dfrac{\partial u}{\partial x}$ with $u = -z\dfrac{dw_o}{dx}$; then $z^2\dfrac{d^2w_o}{dx^2} = -z\dfrac{\sigma_x}{E}$ which can be integrated into the thickness.

can be inverted, and noting:

$$[C]^{-1} = \begin{bmatrix} 1/\overline{EI}_{11} & 1/\overline{EI}_{12} & 1/\overline{EI}_{13} \\ 1/\overline{EI}_{21} & 1/\overline{EI}_{22} & 1/\overline{EI}_{23} \\ 1/\overline{EI}_{31} & 1/\overline{EI}_{32} & 1/\overline{EI}_{33} \end{bmatrix}$$

one obtains:

$$\frac{\partial^2 w_o}{\partial x^2} = \frac{-1}{\overline{EI}_{11}} \times M_y + \frac{1}{\overline{EI}_{12}} \times M_x + \frac{1}{\overline{EI}_{13}} \times M_{xy}$$

The identification of the behavior noted in Figure 12.8(a), on the one hand, with only the first term of the moment M_y in the equation in Figure 12.8(b), on the other hand, gives:

$$\overline{EI}_{11} \equiv E \frac{b^3 \times 1}{12}$$

leads to an approximate equation of an equivalent modulus E that one can interpret as the **flexure modulus** along the x direction of the homogeneous material:

$$\boxed{E_{\text{flexure} \atop (\text{along } x)} = \frac{12}{b^3} \times \overline{EI}_{11}}$$

Note: When the plies of the laminate are oriented uniquely along the 0° and 90° directions, or when the laminate [0/90/45/−45] is constituted uniquely of balanced fabrics and of mats, excluding the unidirectional layers, then one has in the matrix [C]:

$$C_{13} = C_{23} = 0$$

then:

$$\overline{EI}_{11} = C_{11} - \frac{C_{12}^2}{C_{22}}$$

12.1.7 Case of Thermomechanical Loading

12.1.7.1 Membrane Behavior

When one considers variation in temperature, which is assumed to be **identical** in all plies of the laminate, the stresses are given by the modified Equations 11.10. Following the procedure of Section 12.1.1, with the same hypotheses and notations, the stress resultant N_x (Equation 12.1) becomes

$$N_x = \sum_{k=1^{\text{st}}\text{ply}}^{n^{\text{th}}\text{ply}} \{ \overline{E}_{11}^k \varepsilon_{ox} + \overline{E}_{12}^k \varepsilon_{oy} + \overline{E}_{13}^k \gamma_{oxy} \} e_k - \Delta T \sum_{k=1^{\text{st}}\text{ply}}^{n^{\text{th}}\text{ply}} \overline{\alpha E}_1^k \times e_k$$

then:

$$N_x = A_{11}\varepsilon_{ox} + A_{12}\varepsilon_{oy} + A_{13}\gamma_{oxy} - \Delta T \langle \alpha E b \rangle_x$$

with:

$$A_{1j} = \sum_{k=1^{st}\text{ply}}^{n^{th}\text{ply}} \bar{E}_{1j}^k \; e_k \; ; \langle \alpha Eb \rangle_x = \sum_{k=1^{st}\text{ply}}^{n^{th}\text{ply}} \overline{\alpha E}_1^k \; e_k$$

Following the same procedure for N_y and T_{xy}, the stress resultants are expressed as:

$$\begin{Bmatrix} N_x \\ N_y \\ T_{xy} \end{Bmatrix} = \begin{bmatrix} A_{11} & A_{12} & A_{13} \\ A_{21} & A_{22} & A_{23} \\ A_{31} & A_{32} & A_{33} \end{bmatrix} \begin{Bmatrix} \varepsilon_{ox} \\ \varepsilon_{oy} \\ \gamma_{oxy} \end{Bmatrix} - \Delta T \begin{Bmatrix} \langle \alpha Eb \rangle_x \\ \langle \alpha Eb \rangle_y \\ \langle \alpha Eb \rangle_{xy} \end{Bmatrix}$$

with :

$$A_{ij} = \sum_{k=1^{st}\text{ply}}^{n^{th}\text{ply}} \bar{E}_{ij}^k \times e_k = A_{ji} \qquad \textbf{cf. [11.8]}$$

$$\left. \begin{aligned} \langle \alpha Eb \rangle_x &= \sum_{k=1^{st}\text{ply}}^{n^{th}\text{ply}} \overline{\alpha E}_1^k \times e_k \\ \langle \alpha Eb \rangle_y &= \sum_{k=1^{st}\text{ply}}^{n^{th}\text{ply}} \overline{\alpha E}_2^k \times e_k \\ \langle \alpha Eb \rangle_{xy} &= \sum_{k=1^{st}\text{ply}}^{n^{th}\text{ply}} \overline{\alpha E}_3^k \times e_k \end{aligned} \right\} \qquad \textbf{cf. [11.10]}$$

(12.17)

Inversion of the above relation allows one to show the apparent moduli of the laminate (see Paragraph 12.1.2) and thermal membrane strains:

$$\begin{Bmatrix} \varepsilon_{ox} \\ \varepsilon_{oy} \\ \gamma_{oxy} \end{Bmatrix} = b[A]^{-1} \begin{Bmatrix} \sigma_{ox} \\ \sigma_{oy} \\ \tau_{oxy} \end{Bmatrix} + \Delta T [A]^{-1} \begin{Bmatrix} \langle \alpha Eb \rangle_x \\ \langle \alpha Eb \rangle_y \\ \langle \alpha Eb \rangle_{xy} \end{Bmatrix}$$

or with Equation 12.9:

$$\begin{Bmatrix} \varepsilon_{ox} \\ \varepsilon_{oy} \\ \gamma_{oxy} \end{Bmatrix} = \begin{bmatrix} \dfrac{1}{\bar{E}_x} & -\dfrac{\bar{\nu}_{yx}}{\bar{E}_y} & \dfrac{\bar{\eta}_{xy}}{\bar{G}_{xy}} \\ -\dfrac{\bar{\nu}_{xy}}{\bar{E}_x} & \dfrac{1}{\bar{E}_y} & \dfrac{\bar{\mu}_{xy}}{\bar{G}_{xy}} \\ \dfrac{\bar{\eta}_x}{\bar{E}_x} & \dfrac{\bar{\mu}_y}{\bar{E}_y} & \dfrac{1}{\bar{G}_{xy}} \end{bmatrix} \begin{Bmatrix} \sigma_{ox} \\ \sigma_{oy} \\ \tau_{oxy} \end{Bmatrix} + \Delta T [A]^{-1} \begin{Bmatrix} \langle \alpha Eb \rangle_x \\ \langle \alpha Eb \rangle_y \\ \langle \alpha Eb \rangle_{xy} \end{Bmatrix}$$

which can be rewritten as:

$$
\begin{Bmatrix} \varepsilon_{ox} \\[2ex] \varepsilon_{oy} \\[2ex] \gamma_{oxy} \end{Bmatrix} = h[A]^{-1} \begin{Bmatrix} \sigma_{ox} \\[2ex] \sigma_{oy} \\[2ex] \tau_{oxy} \end{Bmatrix} + \Delta T \times h[A]^{-1} \begin{Bmatrix} \frac{1}{h}\langle \alpha E h\rangle_x \\[2ex] \frac{1}{h}\langle \alpha E h\rangle_y \\[2ex] \frac{1}{h}\langle \alpha E h\rangle_{xy} \end{Bmatrix}
$$

Remarks:

- Evaluation of terms $(1/h)\langle \alpha E h\rangle_x$, $(1/h)\langle \alpha E h\rangle_y$, and $(1/h)\langle \alpha E h\rangle_{xy}$ only requires the knowledge of the proportions of plies along the different orientations and not their thicknesses.[19]
- The matrix $h[A]^{-1}$, already mentioned in Section 12.1.2, contains the global moduli of the laminate. One can then write (see Equation 12.9):

$$
\begin{Bmatrix} \varepsilon_{ox} \\[2ex] \varepsilon_{oy} \\[2ex] \gamma_{oxy} \end{Bmatrix} = \begin{bmatrix} \frac{1}{\overline{E}_x} & -\frac{\overline{v}_{yx}}{\overline{E}_y} & \frac{\overline{\eta}_{xy}}{\overline{G}_{xy}} \\[2ex] -\frac{\overline{v}_{xy}}{\overline{E}_x} & \frac{1}{\overline{E}_y} & \frac{\overline{\mu}_{xy}}{\overline{G}_{xy}} \\[2ex] \frac{\overline{\eta}_x}{\overline{E}_x} & \frac{\overline{\mu}_y}{\overline{E}_y} & \frac{1}{\overline{G}_{xy}} \end{bmatrix} \begin{Bmatrix} \sigma_{ox} \\[2ex] \sigma_{oy} \\[2ex] \tau_{oxy} \end{Bmatrix} + \Delta T \begin{bmatrix} \frac{1}{\overline{E}_x} & \frac{\overline{v}_{yx}}{\overline{E}_y} & \frac{\overline{\eta}_{xy}}{\overline{G}_{xy}} \\[2ex] -\frac{\overline{v}_{xy}}{\overline{E}_x} & \frac{1}{\overline{E}_y} & \frac{\overline{\mu}_{xy}}{\overline{G}_{xy}} \\[2ex] \frac{\overline{\eta}_x}{\overline{E}_x} & \frac{\overline{\mu}_y}{\overline{E}_y} & \frac{1}{\overline{G}_{xy}} \end{bmatrix} \begin{Bmatrix} \frac{1}{h}\langle \alpha E h\rangle_x \\[2ex] \frac{1}{h}\langle \alpha E h\rangle_y \\[2ex] \frac{1}{h}\langle \alpha E h\rangle_{xy} \end{Bmatrix}
$$

The last term of the above equation allows one to show the global expansion coefficients of the laminate, which are denoted as α_{ox}, α_{oy}, and α_{oxy}, as below:

$$
\begin{Bmatrix} \alpha_{ox} \\[2ex] \alpha_{oy} \\[2ex] \alpha_{oxy} \end{Bmatrix} = \begin{bmatrix} \frac{1}{\overline{E}_x} & -\frac{\overline{v}_{yx}}{\overline{E}_y} & \frac{\overline{\eta}_{xy}}{\overline{G}_{xy}} \\[2ex] -\frac{\overline{v}_{xy}}{\overline{E}_x} & \frac{1}{\overline{E}_y} & \frac{\overline{\mu}_{xy}}{\overline{G}_{xy}} \\[2ex] \frac{\overline{\eta}_x}{\overline{E}_x} & \frac{\overline{\mu}_y}{\overline{E}_y} & \frac{1}{\overline{G}_{xy}} \end{bmatrix} \begin{Bmatrix} \frac{1}{h}\langle \alpha E h\rangle_x \\[2ex] \frac{1}{h}\langle \alpha E h\rangle_y \\[2ex] \frac{1}{h}\langle \alpha E h\rangle_{xy} \end{Bmatrix} \qquad (12.18)
$$

In summary, the membrane thermomechanical behavior of a laminate with midplane symmetry can be written as:

$$
\begin{Bmatrix} \varepsilon_{ox} \\[2ex] \varepsilon_{oy} \\[2ex] \gamma_{oxy} \end{Bmatrix} = \begin{bmatrix} \frac{1}{\overline{E}_x} & -\frac{\overline{v}_{yx}}{\overline{E}_y} & \frac{\overline{\eta}_{xy}}{\overline{G}_{xy}} \\[2ex] -\frac{\overline{v}_{xy}}{\overline{E}_x} & \frac{1}{\overline{E}_y} & \frac{\overline{\mu}_{xy}}{\overline{G}_{xy}} \\[2ex] \frac{\overline{\eta}_x}{\overline{E}_x} & \frac{\overline{\mu}_y}{\overline{E}_y} & \frac{1}{\overline{G}_{xy}} \end{bmatrix} \begin{Bmatrix} \sigma_{ox} \\[2ex] \sigma_{oy} \\[2ex] \tau_{oxy} \end{Bmatrix} + \Delta T \begin{Bmatrix} \alpha_{ox} \\[2ex] \alpha_{oy} \\[2ex] \alpha_{oxy} \end{Bmatrix} \qquad (12.19)
$$

---

[19] See Application 18.2 "Residual Thermal Stresses Due to Curing of the Laminate."

This is an equation in which α_{ox}, α_{oy}, and α_{oxy} are given by Equations 12.17 and 12.18.[20]

12.1.7.2 Flexure Behavior

Following the procedure in Section 12.1.4 with the same notations, the moment resultant M_y (Equation 12.13) becomes, using the modified Equations 11.10:

$$M_y = \sum_{k=1^{st}\,ply}^{n^{th}\,ply} \left\{ \int_{z_{k-1}}^{z_k} (\bar{E}_{11}\,\varepsilon_x + \bar{E}_{12}^{\,k}\,\varepsilon_y + \bar{E}_{13}^{\,k}\,\gamma_{xy})z\,dz \right\}$$

$$-\Delta T \sum_{k=1^{st}\,ply}^{n^{th}\,ply} \left(\int_{z_{k-1}}^{z_k} \overline{\alpha E}_1^{\,k} \times z\,dz \right)$$

The plate is assumed to have midplane symmetry, then each integral of the form $\int_{z_{k-1}}^{z_k} \overline{\alpha E}_1 z\,dz$ is associated with another integral such that $\int_{-z_k}^{-z_{k-1}} \overline{\alpha E}_1 z\,dz$ is equal and opposite in sign. There remains the following expression, with the notations of Section 12.1.4:

$$M_y = -C_{11}\frac{\partial^2 w_0}{\partial x^2} - C_{12}\frac{\partial^2 w_0}{\partial y^2} - C_{13} \times 2\frac{\partial^2 w_0}{\partial x \partial y}$$

Due to the midplane symmetry, the behavior in flexure 12.16 is not modified when the laminate is subjected to thermomechanical loading.

Remark:

In the preceding discussion, it is assumed that the temperature field is uniform across the thickness of the laminate.

12.2 LAMINATE WITHOUT MIDPLANE SYMMETRY

12.2.1 Coupled Membrane–Flexure Behavior

If one considers again the calculations of Section 12.1.4 without midplane symmetry, one can see the presence of new integrals as:

$$\int_{z_{k-1}}^{z_k} \bar{E}_{ij}^{\,k}\, z\,dz = \bar{E}_{ij}^{\,k}\left(\frac{z_k^2 - z_{k-1}^2}{2}\right)$$

for the ply k. When the summation over all plies is taken, these integrals lead to nonzero terms with the form:

$$B_{ij} = \sum_{k=1^{st}\,ply}^{n^{th}\,ply} \bar{E}_{ij}^{\,k}\left(\frac{z_k^2 - z_{k-1}^2}{2}\right)$$

---

[20] One indicates in Tables 5.4, 5.9, and 5.14 of Section 5.4 the values of expansion coefficients of the laminates made of carbon, Kevlar, and glass/epoxy with $V_f = 60\%$ fiber volume fraction.

Then one has for the development of M_y (see Section 12.1.4):

$$M_y = -C_{11}\frac{\partial^2 w_0}{\partial x^2} - C_{12}\frac{\partial^2 w_0}{\partial y^2} - C_{13} \times 2\frac{\partial^2 w_0}{\partial x \partial y} + B_{11}\varepsilon_{ox} + B_{12}\varepsilon_{oy} + B_{13}\gamma_{oxy}$$

In this expression appears the coupling between bending and membrane behavior.

In a similar manner, the stress resultant N_x which was developed in Section 12.1.4 is rewritten as:

$$N_x = A_{11}\varepsilon_{ox} + A_{12}\varepsilon_{oy} + A_{13}\gamma_{oxy} - B_{11}\frac{\partial^2 w_0}{\partial x^2} - B_{12}\frac{\partial^2 w_0}{\partial y^2} - B_{13} \times 2\frac{\partial^2 w_0}{\partial x \partial y}$$

where one can find the coupling as mentioned previously.

Developing along the same manner the resultants M_x, M_{xy}, N_y, and T_{xy}, one can regroup the obtained relations. Therefore, the global relation for the behavior can be written as:

$$\left\{\begin{array}{c} N_x \\ N_y \\ T_{xy} \\ \hline M_y \\ -M_x \\ -M_{xy} \end{array}\right\} = \left[\begin{array}{c|c} A & B \\ \hline B & C \end{array}\right] \left\{\begin{array}{c} \varepsilon_{ox} \\ \varepsilon_{oy} \\ \gamma_{oxy} \\ \hline -\partial^2 w_0/\partial x^2 \\ -\partial^2 w_0/\partial y^2 \\ -2\partial^2 w_0/\partial x \partial y \end{array}\right\}$$

$$\text{with :}$$

$$A_{ij} = \sum_{k=1^{\text{st}}\text{ply}}^{n^{\text{th}}\text{ply}} \bar{E}_{ij}^k e_k; \quad B_{ij} = \sum_{k=1^{\text{st}}\text{ply}}^{n^{\text{th}}\text{ply}} \bar{E}_{ij}^k \left(\frac{z_k^2 - z_{k-1}^2}{2}\right)$$

$$C_{ij} = \sum_{k=1^{\text{st}}\text{ply}}^{n^{\text{th}}\text{ply}} \bar{E}_{ij}^k \left(\frac{z_k^3 - z_{k-1}^3}{3}\right)$$

(12.20)

12.2.2 Case of Thermomechanical Loading

Using the expression developed for moment resultant M_y as shown in Section 12.1.7.2, one can find the following form of integrals for each ply k:

$$\int_{z_{k-1}}^{z_k} \overline{\alpha E_1}^k \times z\, dz = \overline{\alpha E_1}^k \left(\frac{z_k^2 - z_{k-1}^2}{2}\right)$$

after summing over all plies of the laminate, it appears a nonzero term of the form:

$$\langle \alpha E b^2 \rangle_x = \sum_{k=1^{\text{st}}\text{ply}}^{n^{\text{th}}\text{ply}} \overline{\alpha E_1}^k \left(\frac{z_k^2 - z_{k-1}^2}{2}\right)$$

A similar development for other resultants lead to the following relation for thermomechanical behavior:

$$
\left\{
\begin{array}{c}
N_x \\
N_y \\
T_{xy} \\
\hline
M_y \\
-M_x \\
-M_{xy}
\end{array}
\right\}
=
\left[
\begin{array}{c|c}
 & \\
A & B \\
 & \\
\hline
 & \\
B & C \\
 &
\end{array}
\right]
\left\{
\begin{array}{c}
\varepsilon_{ox} \\
\varepsilon_{oy} \\
\gamma_{oxy} \\
\hline
-\partial^2 w_0/\partial x^2 \\
-\partial^2 w_0/\partial y^2 \\
-2\partial^2 w_0/\partial x\partial y
\end{array}
\right\}
- \Delta T
\left\{
\begin{array}{c}
\langle \alpha E b \rangle_x \\
\langle \alpha E b \rangle_y \\
\langle \alpha E b \rangle_{xy} \\
\hline
\langle \alpha E b^2 \rangle_x \\
\langle \alpha E b^2 \rangle_y \\
\langle \alpha E b^2 \rangle_{xy}
\end{array}
\right\}
$$

with:
$$
A_{ij} = \sum_k \bar{E}_{ij}^k \, e_k; \quad B_{ij} = \sum_k \bar{E}_{ij}^k \left(\frac{z_k^2 - z_{k-1}^2}{2} \right)
$$

$$
C_{ij} = \sum_k \bar{E}_{ij}^k \left(\frac{z_k^3 - z_{k-1}^3}{3} \right)
$$

(12.21)

$$
\langle \alpha E b \rangle_x = \sum_k \overline{\alpha E_1}^k \, e_k; \quad \langle \alpha E b \rangle_y = \sum_k \overline{\alpha E_2}^k \, e_k; \quad \langle \alpha E b \rangle_{xy} = \sum_k \overline{\alpha E_3}^k \, e_k
$$

$$
\langle \alpha E b^2 \rangle_x = \sum_k \overline{\alpha E_1}^k \frac{(z_k^2 - z_{k-1}^2)}{2}; \quad \langle \alpha E b^2 \rangle_y = \sum_k \overline{\alpha E_2}^k \frac{(z_k^2 - z_{k-1}^2)}{2};
$$

$$
\langle \alpha E b^2 \rangle_{xy} = \sum_k \overline{\alpha E_3}^k \frac{(z_k^2 - z_{k-1}^2)}{2}
$$

PART III

JUSTIFICATIONS, COMPOSITE BEAMS, AND THICK PLATES

We regroup in Part III elements that are less utilized than those in the previous parts. Nevertheless they are of fundamental interest for a better understanding of the principles for calculation of composite components. In the first two chapters, we focused on anisotropic properties and fracture strength of orthotropic materials, and then more particularly on transversely isotropic ones. The following two chapters allow us to consider that composite components in the form of beams can be "homogenized." This means that their study is analogous to the study of homogeneous beams that are common in the literature. Finally, the last chapter in this part describes with a similar procedure the behavior of thick composite plates subject to transverse loadings.

13

ELASTIC COEFFICIENTS

The definition of a linear elastic anisotropic medium was given in Chapter 9. We have also given, without justification, the behavior relations characterizing the particular case of orthotropic materials. Now we propose to examine more closely the elastic constants which appear in stress–strain relations for these materials. In the case of transversely isotropic materials, we will study also the manner in which the constants evolve.

13.1 ELASTIC COEFFICIENTS IN AN ORTHOTROPIC MATERIAL

Recall: Consider the relation for elastic behavior written in Paragraph 9.1.1 in the form:

$$\varepsilon_{mn} = \varphi_{mnpq} \times \sigma_{pq}$$

Recall that the components φ_{mnpq} of a tensor expressed in the coordinate system 1,2,3 are written as Φ_{IJKL} in a coordinate system I,II,II using the relation:

$$\boxed{\Phi_{IJKL} = \cos_I^m \cos_J^n \cos_K^p \cos_L^q \varphi_{mnpq}} \tag{13.1}$$

in which:

$$\cos_I^m = \cos(\vec{m}, \vec{I})$$

By definition,[1] for mechanical behavior, an orthotropic medium has at any point two orthogonal planes of symmetry. Consider here two coordinate systems 1,2,3 and I,II,III, constructed on these planes and their intersection. One plane can be obtained from the other by a 180° rotation about the 3 axis as shown in Figure 13.1. One can deduce

$$[\cos_I^m] = \begin{bmatrix} -1 & 0 & 0 \\ 0 & -1 & 0 \\ 0 & 0 & 1 \end{bmatrix}$$

---

[1] See Section 9.2.

259

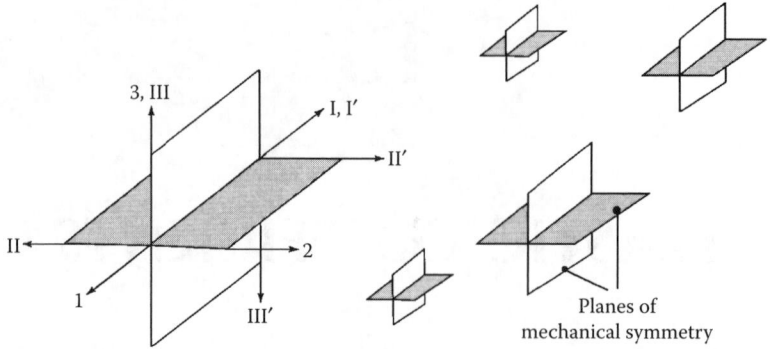

Figure 13.1 Orthotropic Medium

Application of Relation 13.1 above leads to

$$\Phi_{\text{I I I I}} = \varphi_{1111}; \quad \Phi_{\text{I I II II}} = \varphi_{1122}; \quad \Phi_{\text{I I III III}} = \varphi_{1133}$$

$$\Phi_{\text{II II II II}} = \varphi_{2222}; \quad \Phi_{\text{II II III III}} = \varphi_{2233}; \quad \Phi_{\text{III III III III}} = \varphi_{3333}$$

$$\Phi_{\text{II III II III}} = \varphi_{2323}; \quad \Phi_{\text{I III I III}} = \varphi_{1313}; \quad \Phi_{\text{I II I II}} = \varphi_{1212}$$

and:

$$\Phi_{\text{I I II III}} = -\varphi_{1123};$$

However, because the mechanical properties in the coordinates 1,2,3 and I,II,III are identical, one has

$$\Phi_{\text{I I II III}} = \varphi_{1123}$$

from this:

$$\Phi_{\text{I I II III}} = \varphi_{1123} = -\varphi_{1123} = 0$$

In an analogous manner:

$$\varphi_{\text{II II II III}} = 0; \quad \Phi_{\text{III III II III}} = 0$$

$$\Phi_{\text{I I I III}} = 0; \quad \Phi_{\text{II II I III}} = 0; \quad \Phi_{\text{III III I III}} = 0$$

$$\Phi_{\text{II III I II}} = 0; \quad \Phi_{\text{I III I II}} = 0$$

finally:

$$\Phi_{\text{I I I II}} = \varphi_{1112}; \quad \Phi_{\text{II II I II}} = \varphi_{2212}; \quad \Phi_{\text{III III I II}} = \varphi_{3312}$$

$$\Phi_{\text{II III I III}} = \varphi_{2313}$$

Until now, we have taken into account the symmetry with respect to plane 1,3. Consider now the coordinates 1,2,3 and I′, II′, III′ (see Figure 13.1), which can be obtained from each other by a 180° rotation about the 2 axis (symmetry with respect to plane 1,2). One has

$$[\cos_I^m] = \begin{bmatrix} -1 & 0 & 0 \\ 0 & 1 & 0 \\ 0 & 0 & -1 \end{bmatrix}$$

The same procedure as above will lead to

$$\Phi_{I'I'I'II'} = -\varphi_{1112} = \varphi_{1112} = 0; \quad \Phi_{II'II'I'II'} = -\varphi_{2212} = \varphi_{2212} = 0$$

$$\Phi_{III'III'I'II'} = -\varphi_{3312} = \varphi_{3312} = 0; \quad \Phi_{II'III'I'III'} = -\varphi_{2313} = \varphi_{2313} = 0$$

Considering the symmetry of the coefficients φ_{mnpq} indicated in Relation 9.1,[2] we have written here the only nonzero terms. For the mechanical behavior, one obtains by simplification of Equation 9.2:

$$\begin{Bmatrix} \varepsilon_{11} \\ \varepsilon_{22} \\ \varepsilon_{33} \\ \gamma_{23} \\ \gamma_{13} \\ \gamma_{12} \end{Bmatrix} = \begin{bmatrix} \varphi_{1111} & \varphi_{1122} & \varphi_{1133} & 0 & 0 & 0 \\ \varphi_{2211} & \varphi_{2222} & \varphi_{2233} & 0 & 0 & 0 \\ \varphi_{3311} & \varphi_{3322} & \varphi_{3333} & 0 & 0 & 0 \\ 0 & 0 & 0 & 4\varphi_{2323} & 0 & 0 \\ 0 & 0 & 0 & 0 & 4\varphi_{1313} & 0 \\ 0 & 0 & 0 & 0 & 0 & 4\varphi_{1212} \end{bmatrix} \begin{Bmatrix} \sigma_{11} \\ \sigma_{22} \\ \sigma_{33} \\ \tau_{23} \\ \tau_{13} \\ \tau_{12} \end{Bmatrix} \quad (13.2)$$

There remain then only **nine** distinct elastic coefficients, which can be written in the form of Young's moduli and Poisson ratios as:

$$\begin{Bmatrix} \varepsilon_{11} \\ \varepsilon_{22} \\ \varepsilon_{33} \\ \gamma_{23} \\ \gamma_{13} \\ \gamma_{12} \end{Bmatrix} = \begin{bmatrix} \frac{1}{E_1} & \frac{-v_{21}}{E_2} & \frac{-v_{31}}{E_3} & 0 & 0 & 0 \\ \frac{-v_{12}}{E_1} & \frac{1}{E_2} & \frac{-v_{32}}{E_3} & 0 & 0 & 0 \\ \frac{-v_{13}}{E_1} & \frac{-v_{23}}{E_2} & \frac{1}{E_3} & 0 & 0 & 0 \\ 0 & 0 & 0 & \frac{1}{G_{23}} & 0 & 0 \\ 0 & 0 & 0 & 0 & \frac{1}{G_{13}} & 0 \\ 0 & 0 & 0 & 0 & 0 & \frac{1}{G_{12}} \end{bmatrix} \begin{Bmatrix} \sigma_{11} \\ \sigma_{22} \\ \sigma_{33} \\ \tau_{23} \\ \tau_{13} \\ \tau_{12} \end{Bmatrix} \quad (13.3)$$

---

[2] Recall the symmetry relations: $\varphi_{ijkl} = \varphi_{ijlk}; \; \varphi_{ijkl} = \varphi_{jikl}; \; \varphi_{ijkl} = \varphi_{klij}$.

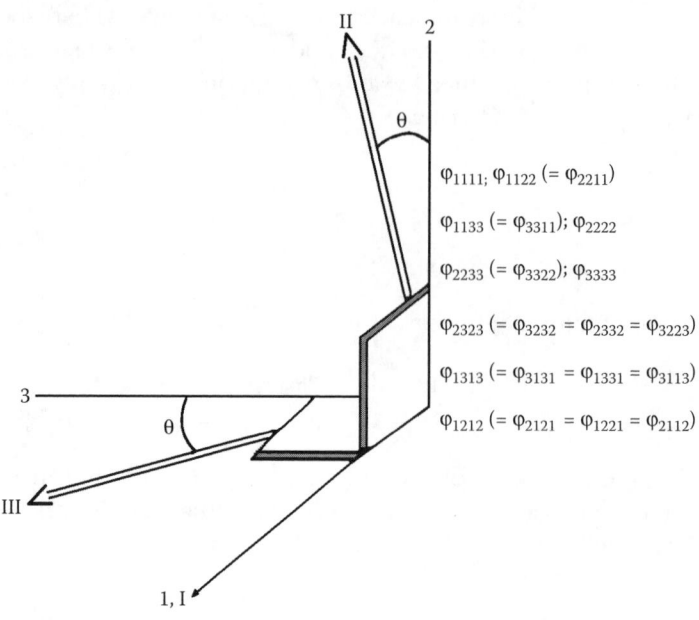

Figure 13.2 Transversely Isotropic Material

13.2 ELASTIC COEFFICIENTS FOR A TRANSVERSELY ISOTROPIC MATERIAL

Recall: By definition,[3] a transversely isotropic material (Figure 13.2) is such that any plane including a preferred axis is a plane of mechanical symmetry. We have already noted that this is a particular case of orthotropic materials. Therefore, the only nonzero elastic constants are shown in Figure 13.2.[4]

The preferred direction is axis 1 in Figure 13.2. Considering that the coordinates 1,2,3 and I,II,III can be obtained from each other by a rotation of θ, one then has

$$[\cos_I^m] = \begin{bmatrix} 1 & 0 & 0 \\ 0 & c & -s \\ 0 & s & c \end{bmatrix} \text{ with } \begin{aligned} c &= \cos\theta \\ s &= \sin\theta \end{aligned}$$

From the definition of material, the matrix of elastic coefficients has to remain invariant in this rotation. The Relation 13.1 allows one to write

- $\Phi_{I I I I} = \varphi_{1111}$

- $\Phi_{I I I I I I} = \varphi_{1122}c^2 + \varphi_{1133}s^2 = \varphi_{1122}$

---

[3] See Section 9.3.

[4] Using the symmetries in Equation 9.1 and also using the discussion of the previous paragraph.

then:

$$\varphi_{1122}(c^2 - 1) + \varphi_{1133}s^2 = 0$$

$$\boxed{\varphi_{1122} = \varphi_{1133}}$$

- $\Phi_{\text{II II II II}} = \varphi_{2222}c^4 + \varphi_{2233}s^2c^2 + \varphi_{2323}s^2c^2 + \varphi_{2332}s^2c^2 \cdots$

  $$\cdots + \varphi_{3223}s^2c^2 + \varphi_{3232}s^2c^2 + \varphi_{3322}s^2c^2 + \varphi_{3333}s^4$$

and:

$$\Phi_{\text{II II II II}} = \varphi_{2222}$$

Then, taking into account the symmetries, we obtain the relation:

$$\varphi_{2222}(c^4 - 1) + \varphi_{3333}s^4 + 2s^2c^2(\varphi_{2233} + 2\varphi_{2323}) = 0 \qquad \textbf{(a)}$$

- $\Phi_{\text{III III III III}} = \varphi_{2222}s^4 + \varphi_{2233}s^2c^2 + \varphi_{2323}s^2c^2 + \varphi_{2332}s^2c^2 \cdots$

  $$\cdots + \varphi_{3232}s^2c^2 + \varphi_{3232}s^2c^2 + \varphi_{3322}s^2c^2 + \varphi_{3333}s^4$$

and:

$$\Phi_{\text{III III III III}} = \varphi_{3333}$$

then taking in account symmetry, we have:

$$\varphi_{2222}s^4 + \varphi_{3333}(c^4 - 1) + 2s^2c^2(\varphi_{2233} + 2\varphi_{2323}) = 0 \qquad \textbf{(b)}$$

Examining member by member the difference of relations shown in (a) and (b) above, one obtains:

$$\boxed{\varphi_{2222} = \varphi_{3333}}$$

Replacing in (a):

$$\varphi_{2222}(c^4 + s^4 - 1) + 2s^2c^2(\varphi_{2233} + 2\varphi_{2323}) = 0$$
$$-2s^2c^2\varphi_{2222} + 2s^2c^2(\varphi_{2233} + 2\varphi_{2323}) = 0$$

$$\boxed{2\varphi_{2323} = \varphi_{2222} - \varphi_{2233}}$$

- $\Phi_{\text{I III I III}} = \varphi_{1212}s^2 + \varphi_{1313}c^2 = \varphi_{1313}$

then

$$\varphi_{1212}s^2 + \varphi_{1313}(c^2 - 1) = 0$$

$$\boxed{\varphi_{1212} = \varphi_{1313}}$$

We have written four relations for the nine coefficients; there remain five distinct elastic coefficients. Equation 13.2 is reduced to

$$
\begin{Bmatrix} \varepsilon_{11} \\ \varepsilon_{22} \\ \varepsilon_{33} \\ \gamma_{23} \\ \gamma_{13} \\ \gamma_{12} \end{Bmatrix}
=
\begin{bmatrix}
\varphi_{1111} & \varphi_{1122} & \varphi_{1122} & 0 & 0 & 0 \\
\varphi_{2211} & \varphi_{2222} & \varphi_{2233} & 0 & 0 & 0 \\
\varphi_{2211} & \varphi_{3322} & \varphi_{2222} & 0 & 0 & 0 \\
0 & 0 & 0 & 2(\varphi_{2222} - \varphi_{2233}) & 0 & 0 \\
0 & 0 & 0 & 0 & 4\varphi_{1212} & 0 \\
0 & 0 & 0 & 0 & 0 & 4\varphi_{1212}
\end{bmatrix}
\begin{Bmatrix} \sigma_{11} \\ \sigma_{22} \\ \sigma_{33} \\ \tau_{23} \\ \tau_{13} \\ \tau_{12} \end{Bmatrix}
\quad (13.4)
$$

or in the form of Young's moduli and Poisson ratios:

$$
\begin{Bmatrix} \varepsilon_{11} \\ \varepsilon_{22} \\ \varepsilon_{33} \\ \gamma_{23} \\ \gamma_{13} \\ \gamma_{12} \end{Bmatrix}
=
\begin{bmatrix}
\dfrac{1}{E_1} & \dfrac{-\nu_{21}}{E_2} & \dfrac{-\nu_{21}}{E_2} & 0 & 0 & 0 \\[2mm]
\dfrac{-\nu_{12}}{E_1} & \dfrac{1}{E_2} & \dfrac{-\nu}{E_2} & 0 & 0 & 0 \\[2mm]
\dfrac{-\nu_{12}}{E_1} & \dfrac{-\nu}{E_2} & \dfrac{1}{E_2} & 0 & 0 & 0 \\[2mm]
0 & 0 & 0 & \dfrac{2(1+\nu)}{E_2} & 0 & 0 \\[2mm]
0 & 0 & 0 & 0 & \dfrac{1}{G_{12}} & 0 \\[2mm]
0 & 0 & 0 & 0 & 0 & \dfrac{1}{G_{12}}
\end{bmatrix}
\begin{Bmatrix} \sigma_{11} \\ \sigma_{22} \\ \sigma_{33} \\ \tau_{23} \\ \tau_{13} \\ \tau_{12} \end{Bmatrix}
\quad (13.5)
$$

13.2.1 Rotation about an Orthotropic Transverse Axis

13.2.1.1 Problem

How can one transform the elastic coefficients of the previous constitutive equation when writing them in coordinate axes x,y,z other than the orthotropic axes ℓ,t,z?[5] The coordinate axes x,y,z are obtained from the orthotropic axes by a rotation θ about the z axis, as shown in Figure 13.3.

Recall Relation 13.1, which allows the calculation of components Φ_{IJKL} in the coordinate axes x,y,z as functions of the components φ_{mnpq} in the coordinate axes ℓ,t,z to be:

$$\boxed{\begin{array}{ll} \Phi_{IJKL} = \cos_I^m \cos_J^n \cos_K^p \cos_L^q \times \varphi_{mnpq} \\ \text{(axes } x, y, z) \qquad \qquad \text{(axes } \ell, t, z) \end{array}}$$

---

[5] The orthotropic axes 1,2,3 of Equation 13.5 are therefore denoted as ℓ,t,z.

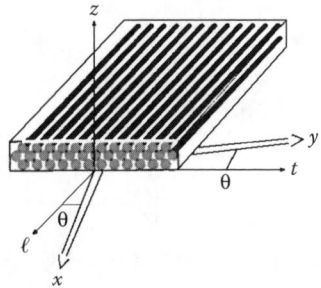

Figure 13.3   Rotation above an Orthotropic Transverse Axis

with (see Figure 13.3):

$$[\cos_l^m] = \begin{bmatrix} \cos(\ell, x) & \cos(\ell, y) & \cos(\ell, z) \\ \cos(t, x) & \cos(t, y) & \cos(t, z) \\ \cos(z, x) & \cos(z, y) & \cos(z, z) \end{bmatrix} = \begin{bmatrix} c & -s & 0 \\ s & c & 0 \\ 0 & 0 & 1 \end{bmatrix}$$

Noting that the only nonzero coefficients φ_{mnpq} are those that appear in Equation 13.4, one obtains:

- $\Phi_{IIII} = c^4 \varphi_{1111} + c^2 s^2 \varphi_{1122} + c^2 s^2 \varphi_{1212} + c^2 s^2 \varphi_{1221} \cdots$
  $$\cdots + c^2 s^2 \varphi_{2112} + c^2 s^2 \varphi_{2121} + c^2 s^2 \varphi_{2211} + s^4 \varphi_{2222}$$
- $\Phi_{IIII} = c^4 \varphi_{1111} + s^4 \varphi_{2222} + 2c^2 s^2 (\varphi_{1122} + 2\varphi_{1212})$

Expressing this coefficient as a function of the "technical" constants which appear in Equation 13.5, one obtains:

$$\Phi_{IIII} = \frac{c^4}{E_\ell} + \frac{s^4}{E_t} + s^2 c^2 \left(\frac{1}{G_{\ell t}} - 2\frac{\nu_{t\ell}}{E_t} \right)$$

- $\Phi_{IIIItt} = c^2 s^2 \varphi_{1111} + c^4 \varphi_{1122} - c^2 s^2 \varphi_{1212} - c^2 s^2 \varphi_{1221} \cdots$
  $$\cdots - s^2 c^2 \varphi_{2112} - s^2 c^2 \varphi_{2121} + s^4 \varphi_{2211} + s^2 c^2 \varphi_{2222}$$
- $\Phi_{IIIItt} = (c^4 + s^4) \varphi_{1122} + c^2 s^2 (\varphi_{1111} + \varphi_{2222} - 4c^2 s^2 \varphi_{1212})$

or in the "technical" form:

$$\Phi_{IIItt} = -\frac{\nu_{t\ell}}{E_t}(c^4 + s^4) + c^2 s^2 \left(\frac{1}{E_\ell} + \frac{1}{E_t} - \frac{1}{G_{\ell t}} \right)$$

- $\Phi_{IIIIIIIII} = c^2 \varphi_{1133} + s^2 \varphi_{2233}$ and as $\varphi_{1133} = \varphi_{1122}$[6]

  $\Phi_{IIIIIIIII} = c^2 \varphi_{1122} + s^2 \varphi_{2233}$

---

[6] Because this is a transversely isotropic material; see Equations 9.2 and 13.4.

or in the "technical" form:

$$\Phi_{\text{II I III III}} = -\left(c^2\frac{\nu_{t\ell}}{E_t} + s^2\frac{\nu}{E_t}\right)$$

- $\Phi_{\text{II I III}} = 0$
- $\Phi_{\text{II I I III}} = 0$
- $\Phi_{\text{I I I II}} = -c^3s\varphi_{1111} + c^3s\varphi_{1122} + c^3s\varphi_{1212} - cs^3\varphi_{1221} + sc^3\varphi_{2112}\ldots$

$$\ldots - s^3c\varphi_{2121} - s^3c\varphi_{2211} + s^3c\varphi_{2222}$$

$$\Phi_{\text{I I I II}} = -sc\{c^2\varphi_{1111} - s^2\varphi_{2222} - (c^2 - s^2)(\varphi_{1122} + 2\varphi_{1212})\}$$

or in the "technical" form:

$$\Phi_{\text{I I I II}} = -cs\left\{\frac{c^2}{E_\ell} - \frac{s^2}{E_t} + (c^2 - s^2)\left(\frac{\nu_{t\ell}}{E_t} - \frac{1}{2G_{\ell t}}\right)\right\}$$

- $\Phi_{\text{II II II II}} = s^4\varphi_{1111} + s^2c^2\varphi_{1122} + s^2c^2\varphi_{1212} + s^2c^2\varphi_{1221}\ldots$

$$\ldots + s^2c^2\varphi_{2112} + s^2c^2\varphi_{2121} + s^2c^2\varphi_{2211} + c^4\varphi_{2222}$$

$$\Phi_{\text{II II II II}} = s^4\varphi_{1111} + c^4\varphi_{2222} + s^2c^2(4\varphi_{1212} + 2\varphi_{1122})$$

or in "technical" form:

$$\Phi_{\text{II II II II}} = \frac{s^4}{E_\ell} + \frac{c^4}{E_t} + s^2c^2\left(\frac{1}{G_{\ell t}} - 2\frac{\nu_{t\ell}}{E_t}\right)$$

- $\Phi_{\text{II II III III}} = s^2\varphi_{1133} + c^2\varphi_{2233}$ and as $\varphi_{1133} = \varphi_{1122}$[7]

$$\Phi_{\text{II II III III}} = s^2\varphi_{1122} + c^2\varphi_{2233}$$

or in "technical" form:

$$\Phi_{\text{II II III III}} = -\left(s^2\frac{\nu_{t\ell}}{E_t} + c^2\frac{\nu}{E_t}\right)$$

- $\Phi_{\text{II II II III}} = 0$
- $\Phi_{\text{II II I III}} = 0$
- $\Phi_{\text{II II I II}} = -s^3c\varphi_{1111} + s^3c\varphi_{1122} - sc^3\varphi_{1212} + s^3c\varphi_{1221}\ldots$

$$\ldots(-sc^3\varphi_{2112} + s^3c\varphi_{2121} - sc^3\varphi_{2211} + c^3s\varphi_{2222})$$

$$\Phi_{\text{II II I II}} = -sc\{s^2\varphi_{1111} + c^2\varphi_{2222} + (c^2 - s^2)(\varphi_{1122} + 2\varphi_{1212})\}$$

---

[7] See Equations 9.2 and 13.4.

or in "technical" form:

$$\Phi_{\text{II II I II}} = -cs\left\{\frac{s^2}{E_\ell} - \frac{c^2}{E_t} - (c^2 - s^2)\left(\frac{v_{t\ell}}{E_t} - \frac{1}{2G_{\ell t}}\right)\right\}$$

- $\Phi_{\text{III III III III}} = \varphi_{3333}$

in "technical" form:

$$\Phi_{\text{III III III III}} = \frac{1}{E_t}$$

- $\Phi_{\text{III III II III}} = 0$
- $\Phi_{\text{III III I III}} = 0$
- $\Phi_{\text{III III II II}} = -sc\varphi_{3311} + sc\varphi_{3322}$ and as $\varphi_{3311} = \varphi_{1122}$[8]

  $\Phi_{\text{III III I II}} = -sc\varphi_{1122} + sc\varphi_{2233}$

in "technical" form:

$$\Phi_{\text{III III I II}} = -sc\left(\frac{v - v_{t\ell}}{E_t}\right)$$

- $\Phi_{\text{II III II III}} = s^2\varphi_{1313} + c^2\varphi_{2323}$

we know[8] that for a transversely isotropic material, one has:

$$\varphi_{1313} = \varphi_{1212} \quad \text{and} \quad 2\varphi_{2323} = \varphi_{2222} - \varphi_{2233}$$

then:

$$\Phi_{\text{II III II III}} = s^2\varphi_{1212} + c^2\left(\frac{\varphi_{2222} - \varphi_{2233}}{2}\right)$$

in "technical" form:

$$\Phi_{\text{II III II III}} = \frac{s^2}{4G_{\ell t}} + \frac{c^2(1 + v)}{2E_t}$$

- $\Phi_{\text{II III I III}} = -sc\varphi_{1313} + sc\varphi_{2323}$ is still:[8]

  $\Phi_{\text{II III I III}} = -sc\left(\varphi_{2121} - \frac{1}{2}(\varphi_{2222} - \varphi_{2233})\right)$

---

[8] See Equations 9.2 and 13.4.

or in "technical" form:

$$\Phi_{\text{II III I III}} = -sc\left(\frac{1}{4G_{\ell t}} - \frac{(1+\nu)}{2E_t}\right)$$

- $\Phi_{\text{III II I III}} = 0$
- $\Phi_{\text{I III I III}} = c^2\varphi_{1313} + s^2\varphi_{2323}$ is still:

$$\Phi_{\text{I III I III}} = c^2\varphi_{1212} + s^2\frac{(\varphi_{2222} - \varphi_{2233})}{2}$$

or in "technical" form:

$$\Phi_{\text{I III I III}} = \frac{c^2}{4G_{\ell t}} + s^2\frac{(1+\nu)}{2E_t}$$

- $\Phi_{\text{I III I II}} = 0$
- $\Phi_{\text{I II I II}} = s^2c^2\varphi_{1111} - s^2c^2\varphi_{1122} + c^4\varphi_{1212} - s^2c^2\varphi_{1221}\cdots$

$$\cdots - s^2c^2\varphi_{2112} + s^4\varphi_{2121} - s^2c^2\varphi_{2211} + s^2c^2\varphi_{2222}$$

$$\Phi_{\text{I II I II}} = s^2c^2(\varphi_{1111} + \varphi_{2222} - 2\varphi_{1122}) + (c^2 - s^2)^2\varphi_{1212}$$

or in "technical" form:

$$\Phi_{\text{I II I II}} = s^2c^2\left\{\frac{1}{E_\ell} + \frac{1}{E_t} + 2\frac{\nu_{t\ell}}{E_t}\right\} + (c^2 - s^2)^2\frac{1}{4G_{\ell t}}$$

All the nonzero coefficients Φ_{IJKL} found above allow one to write the constitutive relation in the form[9]:

$$
\begin{Bmatrix} \varepsilon_{xx} \\ \varepsilon_{yy} \\ \varepsilon_{zz} \\ \gamma_{yz} \\ \gamma_{xz} \\ \gamma_{xy} \end{Bmatrix}
=
\begin{bmatrix}
\Phi_{\text{I I I I}} & \Phi_{\text{I I II II}} & \Phi_{\text{I I III III}} & 0 & 0 & 2\Phi_{\text{I I I II}} \\
\Phi_{\text{II II I I}} & \Phi_{\text{II II II II}} & \Phi_{\text{II II III III}} & 0 & 0 & 2\Phi_{\text{II II I II}} \\
\Phi_{\text{III III I I}} & \Phi_{\text{III III II II}} & \Phi_{\text{III III III III}} & 0 & 0 & 2\Phi_{\text{III III I II}} \\
0 & 0 & 0 & 4\Phi_{\text{II III II III}} & 4\Phi_{\text{II III I III}} & 0 \\
0 & 0 & 0 & 4\Phi_{\text{I III II III}} & 4\Phi_{\text{I III I III}} & 0 \\
2\Phi_{\text{I II I I}} & 2\Phi_{\text{I II II II}} & 2\Phi_{\text{I II III III}} & 0 & 0 & 4\Phi_{\text{I II I II}}
\end{bmatrix}
\begin{Bmatrix} \sigma_{xx} \\ \sigma_{yy} \\ \sigma_{zz} \\ \tau_{yz} \\ \tau_{xz} \\ \tau_{xy} \end{Bmatrix}
$$

$$(13.6)$$

---

[9] This is deduced from the general Equation 9.2.

13.2.1.2 Technical Form

In analogy with the technical form of Equation 13.5, which was written in orthotropic axes, one can write the constitutive equation in terms of equivalent moduli and Poisson coefficients, as:

$$
\begin{Bmatrix} \varepsilon_{xx} \\ \varepsilon_{yy} \\ \varepsilon_{zz} \\ \gamma_{yz} \\ \gamma_{xz} \\ \gamma_{xy} \end{Bmatrix} = \begin{bmatrix} \dfrac{1}{E_x} & \dfrac{-v_{yx}}{E_y} & \dfrac{-v_{zx}}{E_z} & 0 & 0 & \dfrac{\eta_{xy}}{G_{xy}} \\ \dfrac{-v_{xy}}{E_x} & \dfrac{1}{E_y} & \dfrac{-v_{zy}}{E_z} & 0 & 0 & \dfrac{\mu_{xy}}{G_{xy}} \\ \dfrac{-v_{xz}}{E_x} & \dfrac{-v_{yz}}{E_y} & \dfrac{1}{E_z} & 0 & 0 & \dfrac{\zeta_{xy}}{G_{xy}} \\ 0 & 0 & 0 & \dfrac{1}{G_{yz}} & \dfrac{\xi_{xz}}{G_{xz}} & 0 \\ 0 & 0 & 0 & \dfrac{\xi_{yz}}{G_{yz}} & \dfrac{1}{G_{xz}} & 0 \\ \dfrac{\eta_x}{E_x} & \dfrac{\mu_y}{E_y} & \dfrac{\zeta_z}{E_z} & 0 & 0 & \dfrac{1}{G_{xy}} \end{bmatrix} \begin{Bmatrix} \sigma_{xx} \\ \sigma_{yy} \\ \sigma_{zz} \\ \tau_{yz} \\ \tau_{xz} \\ \tau_{xy} \end{Bmatrix}
$$

(13.7)

In this equation, there are coupling terms characterized by the coefficients η_{xy}, μ_{xy}, ζ_{xy}, and ξ_{xy}, which are not similar to the Poisson coefficients.

The values of elastic constants in Relation 13.7 are deduced immediately from the technical forms obtained above for the coefficients Φ_{IJKL}. These constants are detailed below. One obtains subsequently the elastic modulus and Poisson coefficients in the x, y, z coordinates.

$\dfrac{1}{E_x} = \dfrac{c^4}{E_\ell} + \dfrac{s^4}{E_t} + s^2 c^2 \left(\dfrac{1}{G_{\ell t}} - 2\dfrac{v_{t\ell}}{E_t} \right)$	$\Rightarrow\ E_x(\theta) = \dfrac{1}{\dfrac{c^4}{E_\ell} + \dfrac{s^4}{E_t} + s^2 c^2 \left(\dfrac{1}{G_{\ell t}} - \dfrac{2 v_{t\ell}}{E_t} \right)}$
$\dfrac{1}{E_y} = \dfrac{s^4}{E_\ell} + \dfrac{c^4}{E_t} + s^2 c^2 \left(\dfrac{1}{G_{\ell t}} - 2\dfrac{v_{t\ell}}{E_t} \right)$	$\Rightarrow\ E_y(\theta) = \dfrac{1}{\dfrac{s^4}{E_\ell} + \dfrac{c^4}{E_t} + s^2 c^2 \left(\dfrac{1}{G_{\ell t}} - \dfrac{2 v_{t\ell}}{E_t} \right)}$
$\dfrac{1}{E_z} = \dfrac{1}{E_t}$	$\Rightarrow\ E_z(\theta) = E_t (\forall \theta)$
$-\dfrac{v_{yx}}{E_y} = -\dfrac{v_{t\ell}}{E_t}(c^4 + s^4) \cdots$ $\cdots + c^2 s^2 \left(\dfrac{1}{E_\ell} + \dfrac{1}{E_t} - \dfrac{1}{G_{\ell t}} \right)$	$\Rightarrow\ \dfrac{v_{yx}}{E_y}(\theta) = \dfrac{v_{t\ell}}{E_t}(c^4 + s^4) \cdots$ $\cdots - c^2 s^2 \left(\dfrac{1}{E_\ell} + \dfrac{1}{E_t} - \dfrac{1}{G_{\ell t}} \right)$
$-\dfrac{v_{zx}}{E_z} = -\left(c^2 \dfrac{v_{t\ell}}{E_t} + s^2 \dfrac{v}{E_t} \right)$	$\Rightarrow\ v_{zx}(\theta) = c^2 v_{t\ell} + s^2 v$
$-\dfrac{v_{zy}}{E_z} = -\left(s^2 \dfrac{v_{t\ell}}{E_t} + c^2 \dfrac{v}{E_t} \right)$	$\Rightarrow\ v_{zy}(\theta) = s^2 v_{t\ell} + c^2 v$
$\dfrac{1}{G_{yz}} = c^2 \dfrac{2(1+v)}{E_t} + \dfrac{s^2}{G_{\ell t}}$	$\Rightarrow\ G_{yz}(\theta) = \dfrac{1}{c^2 \dfrac{2(1+v)}{E_t} + \dfrac{s^2}{G_{\ell t}}}$

$$\frac{1}{G_{xz}} = s^2\frac{2(1+v)}{E_t} + \frac{c^2}{G_{\ell t}} \qquad \Rightarrow \qquad G_{xz}(\theta) = \frac{1}{s^2\frac{2(1+v)}{E_t} + \frac{c^2}{G_{\ell t}}}$$

$$\frac{1}{G_{xy}} = 4c^2s^2\left(\frac{1}{E_\ell} + \frac{1}{E_t} + 2\frac{v_{t\ell}}{E_t}\right) + \frac{(c^2-s^2)^2}{G_{\ell t}} \qquad \Rightarrow \qquad G_{xy}(\theta) = \frac{1}{4c^2s^2\left(\frac{1}{E_\ell} + \frac{1}{E_t} + 2\frac{v_{t\ell}}{E_t}\right) + \frac{(c^2-s^2)^2}{G_{\ell t}}}$$

$$\frac{\eta_{xy}}{G_{xy}} = -2cs\left\{\frac{c^2}{E_\ell} - \frac{s^2}{E_t}\cdots \right. \qquad\qquad \frac{\mu_{xy}}{G_{xy}} = -2cs\left\{\frac{s^2}{E_\ell} - \frac{c^2}{E_t}\cdots \right.$$

;

$$\left. \cdots + (c^2-s^2)\left(\frac{v_{t\ell}}{E_t} - \frac{1}{2G_{\ell t}}\right)\right\} \qquad\qquad \left. \cdots - (c^2-s^2)\left(\frac{v_{t\ell}}{E_t} - \frac{1}{2G_{\ell t}}\right)\right\}$$

$$\frac{\zeta_{xy}}{G_{xy}} = -2cs\frac{(v-v_{t\ell})}{E_t} \qquad ; \qquad \frac{\zeta_{xz}}{G_{xz}} = -cs\left(\frac{1}{G_{\ell t}} - \frac{2(1+v)}{E_t}\right)$$

(13.8)

13.3 CASE OF A PLY

One can observe from Equation 13.7 that the stress–strain relations in the plane x,y appear decoupled in the case when $\sigma_{zz} = 0$. We suppose that this applies for the plies making a thin laminate. Each ply will be characterized in its plane by the following relations which are extracted from relations 13.5[10] and 13.7:

■ In the orthotropic axes ℓ,t:

$$\begin{Bmatrix} \varepsilon_\ell \\ \varepsilon_t \\ \gamma_{\ell t} \end{Bmatrix} = \begin{bmatrix} \frac{1}{E_\ell} & \frac{-v_{t\ell}}{E_t} & 0 \\ \frac{-v_{\ell t}}{E_\ell} & \frac{1}{E_t} & 0 \\ 0 & 0 & \frac{1}{G_{\ell t}} \end{bmatrix} \begin{Bmatrix} \sigma_\ell \\ \sigma_t \\ \tau_{\ell t} \end{Bmatrix}$$

(13.9)

■ In the x,y axes, making an angle θ with the orthotropic axes:

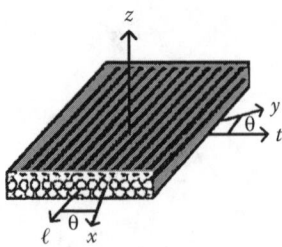

---

[10] The orthotropic axes of Equation 13.5 are denoted as l,t,z for a ply (see Section 3.3.1).

$$
\begin{Bmatrix} \varepsilon_{xx} \\ \varepsilon_{yy} \\ \gamma_{xy} \end{Bmatrix} = \begin{bmatrix} \dfrac{1}{E_x} & -\dfrac{v_{yx}}{E_y} & \dfrac{\eta_{xy}}{G_{xy}} \\[2mm] \dfrac{-v_{xy}}{E_x} & \dfrac{1}{E_y} & \dfrac{\mu_{xy}}{G_{xy}} \\[2mm] \dfrac{\eta_x}{E_x} & \dfrac{\mu_y}{E_y} & \dfrac{1}{G_{xy}} \end{bmatrix} \begin{Bmatrix} \sigma_{xx} \\ \sigma_{yy} \\ \tau_{xy} \end{Bmatrix} \tag{13.10}
$$

For the constants, the values were shown in detail in 13.8.

14

THE HILL–TSAI FAILURE
CRITERION

There are many failure criteria for orthotropic materials. The most commonly used for design calculations is the so-called "Hill–Tsai" criterion.[1] This criterion can be interpreted as analogous to the Von Mises criterion which is applicable to isotropic material in elastic deformation. We will review at the beginning the principal aspects of the Von Mises criterion.

14.1 ISOTROPIC MATERIAL: VON MISES CRITERION

The material is elastic and isotropic. In Figure 14.1, one denotes by I,II,III the principal directions of the stress tensor Σ for a given point. The corresponding matrix is

$$\begin{bmatrix} \sigma_{\mathrm{I}} & 0 & 0 \\ 0 & \sigma_{\mathrm{II}} & 0 \\ 0 & 0 & \sigma_{\mathrm{III}} \end{bmatrix}$$

The general form of the deformation energy dW for an elementary volume dV surrounding the point considered can be written as:

$$dW_{\text{total}} = \frac{1}{2}\sum_i \sum_j \sigma_{ij}\varepsilon_{ij}\ dV$$

which can be reduced to

$$dW_{\text{total}} = \frac{1}{2}(\sigma_{\mathrm{I}}\varepsilon_{\mathrm{I}} + \sigma_{\mathrm{II}}\varepsilon_{\mathrm{II}} + \sigma_{\mathrm{III}}\varepsilon_{\mathrm{III}})\ dV$$

---

[1] See failure of composite materials in Section 5.3.2.

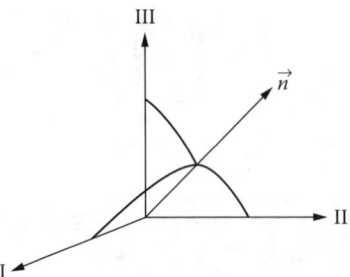

Figure 14.1   Principal Directions for the Stress Tensor

ε_I ε_{II} ε_{III} are the principal strains that one can express as functions of stresses using the constitutive Equation 10.1 as:

$$\varepsilon = \frac{1 + v}{E}\Sigma - \frac{v}{E}\text{trace}(\Sigma)I$$

This leads to:

$$\left(\frac{dW}{dV}\right)_{\text{total}} = \frac{1}{2}\left\{\frac{1 + v}{E}(\sigma_I^2 + \sigma_{II}^2 + \sigma_{III}^2) - \frac{v}{E}(\sigma_I + \sigma_{II} + \sigma_{III})^2\right\}$$

(Note that (dW/dV) represents energy per unit volume.)

The total elastic deformation above is due to the dilatation and distortion of the material. The Von Mises criterion postulates that the material resists to a state of isotropic (or spherical) stress, but will plastify when the distortion energy per unit volume reaches a critical value. One notes that:

$$\left(\frac{dW}{dV}\right)_{\text{distortion}} = \left(\frac{dW}{dV}\right)_{\text{total}} - \left(\frac{dW}{dV}\right)_{\text{spherical stress}}$$

The isotropic portion of the stress state here is written as: $\frac{\sigma_I + \sigma_{II} + \sigma_{III}}{3}$. It creates a state of isotropic dilatation (Equation 10.1):

$$\varepsilon = \frac{1 + v}{E}\left(\frac{\sigma_I + \sigma_{II} + \sigma_{III}}{3}\right) - \frac{v}{E}(\sigma_I + \sigma_{II} + \sigma_{III})$$

then:

$$\left(\frac{dW}{dV}\right)_{\text{spherical stress}} = \frac{1}{2}\left\{3\times\left(\frac{\sigma_I + \sigma_{II} + \sigma_{III}}{3}\right)\times\varepsilon\right\}$$

$$\left(\frac{dW}{dV}\right)_{\text{spherical stress}} = \frac{1}{2}\left\{\frac{1 + v(\sigma_I + \sigma_{II} + \sigma_{III})^2}{E}\frac{}{3} - \frac{v}{E}(\sigma_I + \sigma_{II} + \sigma_{III})^2\right\}$$

One obtains then by replacing:

$$\left(\frac{dW}{dV}\right)_{\text{distortion}} = \frac{1}{2}\left\{\frac{1+\nu}{E}(\sigma_I^2 + \sigma_{II}^2 + \sigma_{III}^2) - \frac{\nu}{E}(\sigma_I + \sigma_{II} + \sigma_{III})^2 \cdots \right.$$

$$\left. \cdots - \frac{1+\nu}{E}\frac{(\sigma_I + \sigma_{II} + \sigma_{III})^2}{3} + \frac{\nu}{E}(\sigma_I + \sigma_{II} + \sigma_{III})^2 \right\}$$

then:
$$\left(\frac{dW}{dV}\right)_{\text{distortion}} = \frac{1}{4G}\left\{(\sigma_I^2 + \sigma_{II}^2 + \sigma_{III}^2) - \frac{(\sigma_I + \sigma_{II} + \sigma_{III})^2}{3}\right\} \qquad (14.1)$$

One can rewrite as following the quantity in brackets:

$$\frac{2}{3}\left\{\sigma_I^2 + \sigma_{II}^2 + \sigma_{III}^2 - \sigma_I\sigma_{II} - \sigma_{II}\sigma_{III} - \sigma_{III}\sigma_I\right\}$$

$$\frac{2}{3}\left\{(\sigma_I + \sigma_{II} + \sigma_{III})^2 - 3(\sigma_I\sigma_{II} + \sigma_{II}\sigma_{III} + \sigma_{III}\sigma_I)\right\}$$

$$\boxed{\left(\frac{dW}{dV}\right)_{\text{distortion}} = \frac{1}{6G}\left\{(\sigma_I + \sigma_{II} + \sigma_{III})^2 \cdots \right.}$$
$$\boxed{\left. \cdots - 3(\sigma_I\sigma_{II} + \sigma_{II}\sigma_{III} + \sigma_{III}\sigma_I)\right\}} \qquad (14.2)$$

Remarks: If one denotes as $\vec{n}$ the direction making the same angle with each of the principal directions (following Figure 14.1), one observes on the face with the normal $\vec{n}$, a stress $\vec{\sigma}$ such that: $\vec{\sigma} = \Sigma(\vec{n})$

that is:

$$\{\sigma\} = \left\{\begin{array}{c} \sigma_I/\sqrt{3} \\ \sigma_{II}/\sqrt{3} \\ \sigma_{III}/\sqrt{3} \end{array}\right\}$$

which can be decomposed as:

■ A **normal stress:** $\sigma_n = \vec{\sigma} \cdot \vec{n}$

then:
$$\sigma_n = \frac{\sigma_I + \sigma_{II} + \sigma_{III}}{3}$$

The above consists of the average or isotropic part of the stress tensor.[2]

■ A **shear** stress:

$$\tau = \sqrt{\sigma^2 - \sigma_n^2}$$

---

[2] Recall the expression $\sigma_I + \sigma_{II} + \sigma_{III}$ that constitutes the first scalar invariant of the stress tensor.

then:

$$\tau^2 = \frac{1}{3}\left\{\sigma_I^2 + \sigma_{II}^2 + \sigma_{III}^2 - \left(\frac{\sigma_I + \sigma_{II} + \sigma_{III}}{3}\right)^2\right\}$$

which can be compared with Equation 14.1. Thus,

$$\left(\frac{dW}{dV}\right)_{distortion} = \frac{1}{2G}\left(\frac{3}{2}\tau^2\right)$$

The shear stress τ also appears as the shear characteristic of the distortion energy.

■ One recognizes in Equation 14.2 the presence of the first and second scalar invariants of the stress tensor independent of the coordinate system. In coordinate axes other than the principal directions, the second invariant can be written as:

$$(\sigma_{11}\sigma_{22} - \tau_{12}^2) + (\sigma_{22}\sigma_{33} - \tau_{23}^2) + (\sigma_{33}\sigma_{11} - \tau_{31}^2)$$

One then has for any coordinate system:

$$\left(\frac{dW}{dV}\right)_{distortion} = \frac{1}{6G}\{(\sigma_{11} + \sigma_{22} + \sigma_{33})^2 \cdots$$

$$\cdots - 3((\sigma_{11}\sigma_{22} - \tau_{12}^2) + (\sigma_{22}\sigma_{33} - \tau_{23}^2) + (\sigma_{33}\sigma_{11} - \tau_{31}^2))\}$$

then:

$$\left(\frac{dW}{dV}\right)_{distortion} = \frac{1}{12G}\{(\sigma_{11} - \sigma_{22})^2 + (\sigma_{22} - \sigma_{33})^2 \cdots$$

$$\cdots + (\sigma_{33} - \sigma_{11})^2 + 6(\tau_{12}^2 + \tau_{23}^2 + \tau_{31}^2)\}$$

The elastic domain (where the distortion energy is below a certain critical value) can then be characterized by the condition:

$$a\{(\sigma_{11} - \sigma_{22})^2 + (\sigma_{22} - \sigma_{33})^2 + (\sigma_{33} - \sigma_{11})^2 \cdots$$

$$\cdots + 6(\tau_{12}^2 + \tau_{23}^2 + \tau_{31}^2)\} < 1 \qquad (14.3)$$

To determine the constant, a uniaxial test is sufficient; in effect if one denotes by σ_e the elastic limit obtained from a tension–compression test, one has:

$$a \times 2\sigma_e^2 = 1$$

then:

$$a = 1/2\sigma_e^2$$

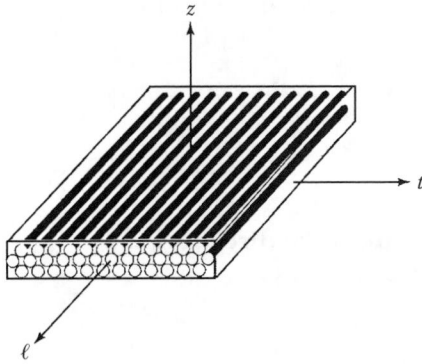

Figure 14.2   Principal Axes for an Orthotropic Ply

14.2 ORTHOTROPIC MATERIAL: HILL–TSAI CRITERION

14.2.1 Preliminary Remarks

A parallel with the Von Mises criterion can be seen with the following remarks:

- For an orthotropic material, the principal directions for the stresses do not coincide with the orthotropic directions, unlike the isotropic case.
- A uniaxial test is not enough to determine all the terms of the equation for the criterion because the mechanical behavior changes with the direction of loading.
- For the fiber/resin composites, the elastic limit corresponds with the rupture limit.
- The rupture strengths are very different when loading is applied along the *l* direction or along the *t* direction.
- The rupture strengths are different in tension as compared with in compression.

One can then write in the orthotropic coordinates ℓ,t,z — shown in Figure 14.2 — an expression similar to Equation 14.3, as:

$$a(\sigma_\ell - \sigma_t)^2 + b(\sigma_t - \sigma_z)^2 + c(\sigma_z - \sigma_\ell)^2 + d\tau_{\ell z}^2 + e\tau_{tz}^2 + f\tau_{\ell t}^2 \leq 1 \qquad (14.4)$$

14.2.2 Case of a Transversely Isotropic Material

In the following, we will limit ourselves, for the purpose of simplification, to the case of a transversely isotropic material.[3] The constants a, b, c, d, e, f in Equation 14.4 above will be determined using the results of the following tests:

- **Test along the longitudinal direction ℓ:**

$$a + c = \frac{1}{\sigma_{\ell \, \text{rupture}}^2}$$

---

[3] For an orthotropic material, the procedure is identical.

■ **Test along the transverse direction t:**

$$a + b = \frac{1}{\sigma_{t\ \text{rupture}}^2}$$

■ **Test along the transverse direction z: due to transverse isotropy:**

$$b + c = \frac{1}{\sigma_{t\ \text{rupture}}^2}$$

then:

$$a = c = \frac{1}{2\sigma_{\ell\ \text{rupture}}^2}$$

$$b = \frac{1}{\sigma_{t\ \text{rupture}}^2} - \frac{1}{2\sigma_{\ell\ \text{rupture}}^2}$$

■ **Shear tests:**

$$\blacksquare\ \tau_{\ell t} \rightarrow f = \frac{1}{\tau_{\ell t\ \text{rupture}}^2}$$

$$\blacksquare\ \tau_{tz} \rightarrow e = \frac{1}{\tau_{tz\ \text{rupture}}^2}$$

$$\blacksquare\ \tau_{\ell z} \rightarrow d = \frac{1}{\tau_{\ell t\ \text{rupture}}^2}$$

due to transverse isotropy.
Replacing in Equation 14.4:

$$\frac{1}{2\sigma_{\ell\ \text{rupture}}^2}\{(\sigma_\ell - \sigma_t)^2 + (\sigma_\ell - \sigma_z)^2\}\cdots$$

$$\cdots - \left(\frac{1}{2\sigma_{\ell\ \text{rupture}}^2} - \frac{1}{\sigma_{t\ \text{rupture}}^2}\right)(\sigma_t - \sigma_z)^2 + \frac{1}{\tau_{\ell t\ \text{rupture}}^2}(\tau_{\ell t}^2 + \tau_{\ell z}^2) + \frac{\tau_{tz}^2}{\tau_{tz\ \text{rupture}}^2} \leq 1$$

and in developing[4]:

$$
\frac{\sigma_\ell^2}{\sigma_{\ell\,\text{rupture}}^2} + \frac{\sigma_t^2 + \sigma_z^2}{\sigma_{t\,\text{rupture}}^2} - \frac{\sigma_\ell}{\sigma_{\ell\,\text{rupture}}^2}(\sigma_t + \sigma_z) + \sigma_z\sigma_t\left(\frac{1}{\sigma_{\ell\,\text{rupture}}^2} - \frac{2}{\sigma_{t\,\text{rupture}}^2}\right)\cdots
$$

$$
\cdots + \frac{\tau_{\ell t}^2 + \tau_{\ell z}^2}{\tau_{\ell t\,\text{rupture}}^2} + \frac{\tau_{tz}^2}{\tau_{tz\,\text{rupture}}^2} \le 1 \qquad (14.5)
$$

Remark: For the case of a "three-dimensional" orthotropic material, an analogous reasoning to the previous presentation leads to a more general criterion, which can be written as:

$$
\frac{\sigma_\ell^2}{\sigma_{\ell\,\text{rupt.}}^2} + \frac{\sigma_t^2}{\sigma_{t\,\text{rupt.}}^2} + \frac{\sigma_z^2}{\sigma_{z\,\text{rupt.}}^2} - \left(\frac{1}{\sigma_{\ell\,\text{rupt.}}^2} + \frac{1}{\sigma_{t\,\text{rupt.}}^2} - \frac{1}{\sigma_{z\,\text{rupt.}}^2}\right)\sigma_\ell\sigma_t \cdots
$$

$$
\cdots - \left(\frac{1}{\sigma_{t\,\text{rupt.}}^2} + \frac{1}{\sigma_{z\,\text{rupt.}}^2} - \frac{1}{\sigma_{\ell\,\text{rupt.}}^2}\right)\sigma_t\sigma_z - \left(\frac{1}{\sigma_{z\,\text{rupt.}}^2} + \frac{1}{\sigma_{\ell\,\text{rupt.}}^2} - \frac{1}{\sigma_{t\,\text{rupt.}}^2}\right)\sigma_z\sigma_\ell \cdots
$$

$$
\cdots + \frac{\tau_{\ell t}^2}{\tau_{\ell t\,\text{rupt.}}^2} + \frac{\tau_{tz}^2}{\tau_{tz\,\text{rupt.}}^2} + \frac{\tau_{z\ell}^2}{\tau_{z\ell\,\text{rupt.}}^2} \le 1
$$

14.2.3 Case of a Unidirectional Ply Under In-Plane Loading

When the stress state is plane-stress, in the plane defined by the axes ℓ, t (see Figure 14.2), one has

$$
\sigma_z = \tau_{\ell z} = \tau_{tz} = 0
$$

Equation 14.5 is simplified, and one obtains what is called "the Hill–Tsai criterion" for a ply subject to stresses within its plane:

$$
\frac{\sigma_\ell^2}{\sigma_{\ell\,\text{rupture}}^2} + \frac{\sigma_t^2}{\sigma_{t\,\text{rupture}}^2} - \frac{\sigma_\ell\sigma_t}{\sigma_{\ell\,\text{rupture}}^2} + \frac{\tau_{\ell t}^2}{\tau_{\ell t\,\text{rupture}}^2} < 1 \qquad (14.6)
$$

Remarks:

■ The rupture strengths of the "fiber/matrix" plies are different in tension and in compression.[5] Do not forget to place in the denominator of each of the first three terms of Equation 14.6 the values of the rupture strengths

---

[4] Attention, this is not valid for a fabric that is not transversely isotropic! (see Application 18.2.10).

[5] See values in Section 3.3.3.

corresponding to the type of loadings in the numerators (tension or compression).

■ **Safety factor:** Let $\alpha^2 < 1$ the Hill–Tsai expression found for a state of stress σ_ℓ, σ_t, $\tau_{\ell t}$. One can then increase the load by means of a multiplication coefficient k to reach the limit as:

$$\frac{(k\sigma_\ell)^2}{\underset{\text{rupture}}{\sigma_\ell^2}} + \frac{(k\sigma_t)^2}{\underset{\text{rupture}}{\sigma_t^2}} - \frac{(k\sigma_\ell)(k\sigma_t)}{\underset{\text{rupture}}{\sigma_\ell^2}} + \frac{(k\tau_{\ell t})^2}{\underset{\text{rupture}}{\tau_{\ell t}^2}} = k^2 \alpha^2 = 1$$

The margin of safety can then be defined as the expression:

$$\frac{(k\sigma_\ell) - \sigma_\ell}{\sigma_\ell} = k - 1$$

which can also be written as:

$$\boxed{\text{safety factor} = \frac{1}{\alpha} - 1}$$

14.3 VARIATION OF RESISTANCE OF A UNIDIRECTIONAL PLY WITH RESPECT TO THE DIRECTION OF LOADING

14.3.1 Tension and Compression Resistance

We propose to evaluate the maximum stress σ_x that one can apply on a ply in the direction x in Figure 14.3. The stresses σ_ℓ, σ_t, $\tau_{\ell t}$ in the orthotropic axes are given by Equation 11.4 as:

$$\begin{Bmatrix} \sigma_\ell \\ \sigma_t \\ \tau_{\ell t} \end{Bmatrix} = \begin{bmatrix} c^2 & s^2 & -2cs \\ s^2 & c^2 & 2cs \\ cs & -cs & (c^2 - s^2) \end{bmatrix} \begin{Bmatrix} \sigma_x \\ 0 \\ 0 \end{Bmatrix}$$

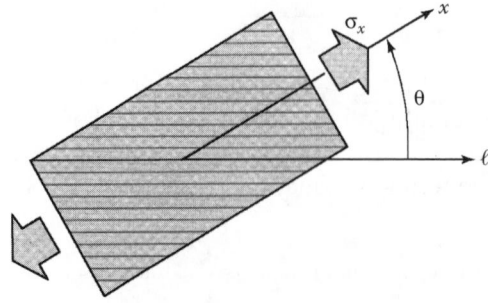

Figure 14.3　Direction of Loading Distinct from Orthotropic Axes

where one recalls that $c = \cos\theta$ and $s = \sin\theta$. Thus,

$$\sigma_\ell = c^2\sigma_x$$
$$\sigma_t = s^2\sigma_x$$
$$\tau_{\ell t} = cs\sigma_x$$

Replacing in the expression of the Hill–Tsai criterion of Equation 14.6, we have

$$\sigma_x^2\left\{\frac{c^4}{\sigma_\ell^2}_{\text{rupt.}} + \frac{s^4}{\sigma_t^2}_{\text{rupt.}} - \frac{c^2s^2}{\sigma_\ell^2}_{\text{rupt.}} + \frac{c^2s^2}{\tau_{\ell t}^2}_{\text{rupt.}}\right\} \leq 1$$

then:

$$\sigma_{x_{\text{rupture}}} = \frac{1}{\sqrt{\dfrac{c^4}{\sigma_{\ell\,\text{rupt.}}^2} + \dfrac{s^4}{\sigma_{t\,\text{rupt.}}^2} + c^2s^2\left(\dfrac{1}{\tau_{\ell t\,\text{rupt.}}^2} - \dfrac{1}{\sigma_{\ell\,\text{rupt.}}^2}\right)}}$$

Remarks:

■ If σ_x is in tension, then $\sigma_{\ell\,\text{rupture}}$ and $\sigma_{t\,\text{rupture}}$ are the limit stresses in tension (tensile strengths). In effect, when $\theta = 0$:

$$\sigma_{x\,\text{rupture}} = \sigma_{\ell\,\text{rupture}}$$

and when $\theta = 90°$:

$$\sigma_{x\,\text{rupture}} = \sigma_{t\,\text{rupture}}$$

■ The evolution of the $\sigma_{x\,\text{rupture}}$, when θ varies, was discussed in Section 3.3.2.

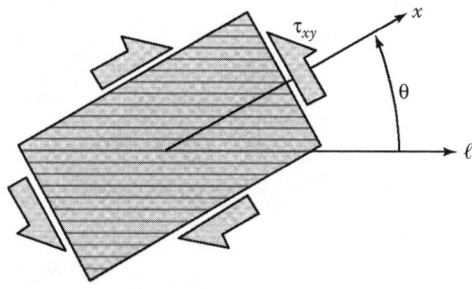

Figure 14.4 Pure Shear in x,y Axes

14.3.2 Shear Strength

For a state of pure shear represented in Figure 14.4, one will have in an analogous manner:

$$\begin{Bmatrix} \sigma_\ell \\ \sigma_t \\ \tau_{\ell t} \end{Bmatrix} = \begin{bmatrix} c^2 & s^2 & -2cs \\ s^2 & c^2 & 2cs \\ cs & -cs & (c^2 - s^2) \end{bmatrix} \begin{Bmatrix} 0 \\ 0 \\ \tau_{xy} \end{Bmatrix}$$

$$\begin{aligned} \sigma_\ell &= -2cs\,\tau_{xy} \\ \sigma_t &= 2cs\,\tau_{xy} \\ \tau_{\ell t} &= (c^2 - s^2)\,\tau_{xy} \end{aligned} \qquad (14.7)$$

Using this in the Hill–Tsai criterion in Equation 14.6:

$$\tau_{xy}^2 \left\{ \frac{4c^2 s^2}{\sigma_{\ell\,\text{rupture}}^2} + \frac{4c^2 s^2}{\sigma_{t\,\text{rupture}}^2} + \frac{4c^2 s^2}{\sigma_{\ell\,\text{rupture}}^2} + \frac{(c^2 - s^2)^2}{\tau_{\ell t\,\text{rupture}}^2} \right\} \le 1$$

then:

$$\tau_{xy\,\text{rupture}} = \frac{1}{\sqrt{4c^2 s^2 \left(\dfrac{2}{\sigma_{\ell\,\text{rupture}}^2} + \dfrac{1}{\sigma_{t\,\text{rupture}}^2} \right) + \dfrac{(c^2 - s^2)^2}{\tau_{\ell t\,\text{rupture}}^2}}}$$

Remarks: Here, taking into account Figure 14.4 ($\tau_{xy} > 0$) and Equations 14.7, $\sigma_{\ell\,\text{rupture}}$ will be the limit stress in compression, and $\sigma_{t\,\text{rupture}}$ the limit stress in tension for $0° \le \theta \le 90°$.

15

COMPOSITE BEAMS IN FLEXURE

Due to their slenderness, a number of composite elements (mechanical components or structural pieces) can be considered as beams. A few typical examples are shown schematically in Figure 15.1. The study of the behavior under loading of these elements (evaluation of stresses and displacements) becomes a very complex problem when one gets into three-dimensional aspects. In this chapter, we propose a monodimensional approach to the problem in an original method. It consists of the definition of displacements corresponding to the traditional stress and moment resultants for the applied loads. This leads to a **homogenized** formulation for the flexure—and for torsion. This means that the equilibrium and behavior relations are formally identical to those that characterize the behavior of classical homogeneous beams. Utilization of these relations for the calculation of stresses and displacements then leads to expressions that are analogous to the common beams.

We will limit ourselves to the composite beams with constant characteristics (geometry, materials) in any cross section, made of different materials—which we call **phases**—that are assumed to be perfectly bonded to each other.

To clarify the procedure and for better simplicity in the calculations, we will limit ourselves in this chapter to the case of composite beams with isotropic phases. The extension to the transversely isotropic materials is immediate. When the phases are orthotropic, with eventually orthotropic directions that are changing from one point to another in the section, the study will be analogous, with a much more involved formulation.[1]

15.1 FLEXURE OF SYMMETRIC BEAMS WITH ISOTROPIC PHASES

In the following, D symbolizes the domain occupied by the cross section, in the y,z plane. The **external** frontier is denoted as ∂D. One distinguishes also (see Figure 15.2) the internal frontiers which limit the phases, denoted by l_{ij} for two contiguous phases i and j. The area of the phase i is denoted as S_i; its moduli of elasticity are denoted by E_i and G_i. The elastic displacement at any point of the beam has the components: $u_x\,(x,y,z)$; $u_y\,(x,y,z)$; $u_z\,(x,y,z)$.

The beam is bending in the plane of symmetry x,y under external loads which are also symmetric with respect to this plane.

---

[1] The only restrictive condition lies in the fact that one of the orthotropic directions is supposed to remain parallel to the longitudinal axis of the beam. cf. bibliography.

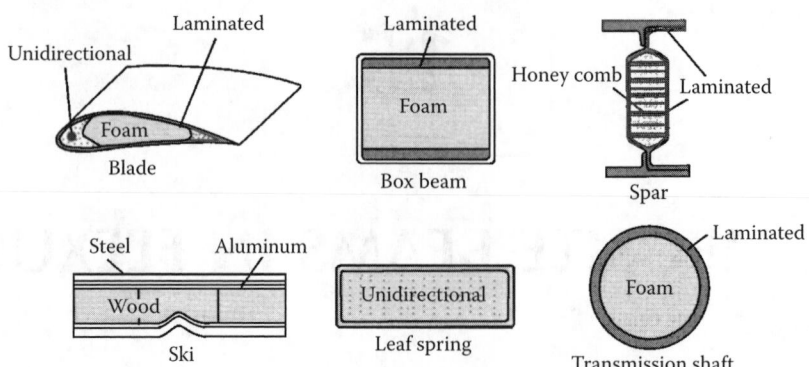

Figure 15.1 Composite Beams

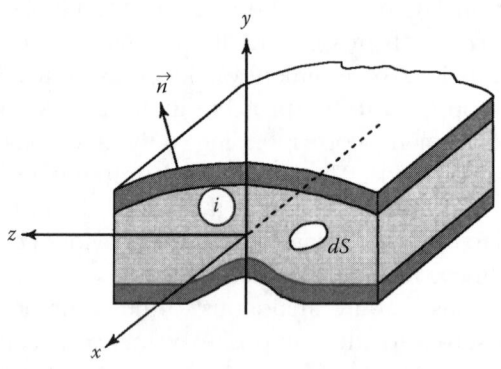

Figure 15.2 Composite Beam, with Plane of Symmetry

15.1.1 Degrees of Freedom

15.1.1.1 Equivalent Stiffness

One writes in condensed form the following integrals taken over the total cross section[2]:

$$\langle ES \rangle = \int_D E_i dS \quad \text{or} \quad = \sum_i E_i S_i$$
number of phases

$$\langle EI_z \rangle = \int_D E_i y^2 dS \quad \text{or} \quad = \sum_i E_i I_{zi} \tag{15.1}$$
number of phases

$$\langle GS \rangle = \int_D G_i dS \quad \text{or} \quad = \sum_i G_i S_i$$
number of phases

---

[2] I_{zi} is the quadratic moment of the Phase i with respect with z axis.

15.1.1.2 Longitudinal Displacement

By definition, longitudinal displacement is denoted by $u(x)$ and written as:

$$u(x) = \frac{1}{\langle ES \rangle} \int_D E_i u_x(x, y, z) \, dS$$

then appears of a mean displacement $u(x)$ and an incremental displacement Δu_x as:

$$u_x(x, y, z) = u(x) + \Delta u_x(x, y, z)$$

where one notes that:

$$\int_D E_i \Delta u_x \, dS = 0 \tag{15.2}$$

15.1.1.3 Rotations of the Sections

By definition, this is the fictitious rotation—or equivalent—given by the following expression:

$$\theta_z(x) = \frac{-1}{\langle EI_z \rangle} \int_D E_i u_x(x, y, z) \times y \, dS$$

Or, with the above:

$$\theta_z(x) = \frac{-1}{\langle EI_z \rangle} \left\{ u(x) \int_D E_i y \, dS + \int_D E_i \Delta u_x(x, y, z) y \, dS \right\}$$

15.1.1.4 Elastic Center

Origin 0 of the coordinate y is chosen such that the following integral is zero:

$$\int_D E_i y \, dS = 0$$

We call **elastic center** the corresponding point 0 that is located as in the expression above. Then Δu_x takes the form:

$$\Delta u_x(x, y, z) = -y \theta_z(x) + \eta_x(x, y, z)$$

with[3]:

$$\int_D E_i \eta_x y \, dS \quad \text{and} \quad \int_D E_i \eta_x \, dS = 0 \quad (*)$$

The displacement u_x (x,y,z) can then take the form:

$$u_x(x, y, z) = u(x) - y \theta_z(x) + \eta_x(x, y, z).$$

---

[3] In what follows, the second property is the consequence of Equation 15.2.

15.1.1.5 Transverse Displacement along y Direction

By definition, this is $v(x)$ that is given by the following expression:

$$v(x) = \frac{1}{\langle GS \rangle} \int_D G_i u_y(x, y, z) dS$$

It follows from this definition that:

$$u_y(x, y, z) = v(x) + \eta_y(x, y, z)$$

where one notes that:

$$\int_D G_i \eta_y \, dS = 0.$$

15.1.1.6 Transverse Displacement along z Direction

By definition, this is $w(x)$ given by

$$w(x) = \frac{1}{\langle GS \rangle} \int_D G_i u_z(x, y, z) dS$$

It follows from this definition and from the existence of the plane of symmetry x,y of the beam a zero average transverse displacement, as: $w(x) = 0$.

$$u_z(x, y, z) = 0 + \eta_z(x, y, z), \qquad \text{with} \qquad \int_D G_i \eta_z \, dS = 0.$$

In summary, we obtain the following elastic displacement field:

$$\begin{cases} u_x = u(x) - y\theta_z(x) + \eta_x(x, y, z) \\ u_y = v(x) + \quad\quad\quad\quad \eta_y(x, y, z) \\ u_z = \quad\quad\quad\quad\quad\quad \eta_z(x, y, z) \end{cases} \tag{15.3}$$

The origin of the axes is the elastic center such that:

$$\int_D E_i y \, dS = 0 \tag{15.4}$$

The three-dimensional incremental displacements η_x, η_y, η_z with respect to the unidimensional approximation u, v, θ_z verify the following:

$$\int_D E_i \eta_x \, dS = \int_D E_i y \eta_x \, dS = 0$$

$$\int_D G_i \eta_y \, dS = 0 \tag{15.5}$$

$$\int_D G_i \eta_z \, dS = 0$$

Remarks:

- η_x represents the **longitudinal distortion** of a cross section, that is, the quantity that this section displaces **out of the plane** which characterizes it if it moves truly as a rigid plane body.
- η_y and η_z represent the displacements that characterize the variations of the form of the cross section in its initial plane.

15.1.2 Perfect Bonding between the Phases

15.1.2.1 Displacements

The bonding is assumed to be perfect. Then the displacements are continuous when crossing through the interface between two phases in contact. For two phases in contact i and j, one has:

$$u_x \left(i \right) = u_x \left(j \right)$$
$$u_y \left(i \right) = u_y \left(j \right)$$
$$u_z \left(i \right) = u_z \left(j \right)$$

15.1.2.2 Strains

For the phases i and j in Figure 15.3, in the plane of an elemental interface with a normal vector of $\vec{n}$, the relations between the strain tensors ε are:

$$\vec{x} \cdot \varepsilon(\vec{x})_{\left(i\right)} = \vec{x} \cdot \varepsilon(\vec{x})_{\left(j\right)}$$

$$\vec{t} \cdot \varepsilon(\vec{x})_{\left(i\right)} = \vec{t} \cdot \varepsilon(\vec{x})_{\left(j\right)}$$

$$\vec{t} \cdot \varepsilon(\vec{t})_{\left(i\right)} = \vec{t} \cdot \varepsilon(\vec{t})_{\left(j\right)}$$

which can also be written as:

$$\varepsilon_{xx} = \varepsilon_{xx}$$
$$\left(i\right) \quad \left(j\right)$$
$$-\varepsilon_{xy} n_z + \varepsilon_{xz} n_y = -\varepsilon_{xy} n_z + \varepsilon_{xz} n_y$$
$$\left(i\right) \quad \left(i\right) \quad\quad \left(j\right) \quad \left(j\right)$$
$$\varepsilon_{yy} n_z^2 - 2\varepsilon_{yz} n_y n_z + \varepsilon_{zz} n_y^2 = \varepsilon_{yy} n_z^2 - 2\varepsilon_{yz} n_y n_z + \varepsilon_{zz} n_y^2$$
$$\left(i\right) \quad\quad \left(i\right) \quad\quad \left(i\right) \quad\quad \left(j\right) \quad\quad \left(j\right) \quad\quad \left(j\right)$$

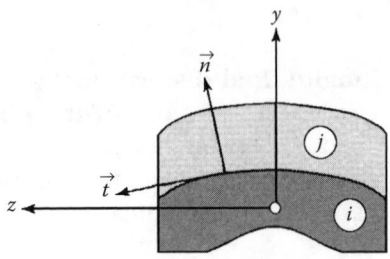

Figure 15.3   Interface between Two Phases

15.1.2.3 Stresses

The stress vector $\vec{\sigma} = \Sigma(\vec{n})$, where Σ represents the stress tensor, remains continuous across an element of the interface with normal $\vec{n}$ as:

$$
\underset{\widehat{i}}{\tau_{xy}n_y} + \underset{\widehat{i}}{\tau_{xz}n_z} = \underset{\widehat{j}}{\tau_{xy}n_y} + \underset{\widehat{j}}{\tau_{xz}n_z}
$$

$$
\underset{\widehat{i}}{\sigma_{yy}n_y} + \underset{\widehat{i}}{\tau_{yz}n_z} = \underset{\widehat{j}}{\sigma_{yy}n_y} + \underset{\widehat{j}}{\tau_{yz}n_z} \tag{15.6}
$$

$$
\underset{\widehat{i}}{\tau_{yz}n_y} + \underset{\widehat{i}}{\sigma_{zz}n_z} = \underset{\widehat{j}}{\tau_{yz}n_y} + \underset{\widehat{j}}{\sigma_{zz}n_z}
$$

15.1.3 Equilibrium Relations

Starting from the local equilibrium, in the absence of body forces, we have

$$
\frac{\partial \sigma_{ij}}{\partial x_j} = 0
$$

By integrating over the cross section, we have successively:

a)
$$
\frac{d}{dx}\int_D \sigma_{xx} dS + \int_D \left(\frac{\partial \tau_{xy}}{\partial y} + \frac{\partial \tau_{xz}}{\partial z}\right) dS = 0
$$

where **the normal stress resultant** N_x appears as:

$$
N_x = \int_D \sigma_{xx} dS \, .
$$

Then, transforming the second integral to an integral over the frontier ∂D of D[4]:

$$
\frac{dN_x}{dx} + \int_{\partial D}(\tau_{xy}n_y + \tau_{xz}n_z)d\Gamma = 0
$$

---

[4] Note that equality $\int_D(\frac{\partial \tau_{xy}}{\partial y} + \frac{\partial \tau_{xz}}{\partial z})dS = \int_{\partial D}(\tau_{xy}n_y + \tau_{xz}n_z)d\Gamma$ is made possible due to the continuity of the expression $(\tau_{xy}\,n_y + \tau_{xz}\,n_z)$ across the interfaces between the different phases (see Equation 15.6).

in which n_y and n_z are the cosines of the outward normal $\vec{n}$, and $d\Gamma$ represents element of frontier ∂D. If one assumes the absence of shear stresses applied over the lateral surface of the beam, then $\tau_{xy}\,n_y + \tau_{xz}\,n_z = 0$ along the external frontier ∂D. Then for longitudinal equilibrium we have[5]

b)

$$\frac{dN_x}{dx} = 0$$

$$\frac{d}{dx}\int_D \tau_{xy}dS + \int_D \left(\frac{\partial \sigma_{yy}}{\partial y} + \frac{\partial \tau_{yz}}{\partial z}\right)dS = 0$$

where one recognizes the **shear stress resultant:**

$$T_y = \int_D \tau_{xy}dS.$$

Then transforming the second integral into an integral over the external frontier ∂D of the domain D of the cross section[6]:

$$\frac{\partial T_y}{\partial x} + \int_{\partial D} (\sigma_{yy}n_y + \tau_{yz}n_z)d\Gamma = 0$$

if one remarks that:

$$\int_{\partial D} (\sigma_{yy}n_y + \tau_{yz}n_z)d\Gamma = \int_{\partial D} \vec{y}\cdot\Sigma(\vec{n})d\Gamma = \vec{y}\cdot\int_{\partial D} \vec{\sigma}d\Gamma = p_y\,(\text{N/m})$$

which is the transverse density of loading on the lateral surface of the beam, transverse equilibrium can be written as:

c)

$$\frac{dT_y}{dx} + p_y = 0$$

$$\frac{d}{dx}\int_D -y\sigma_{xx}dS + \int_D -y\left(\frac{\partial \tau_{xy}}{\partial y} + \frac{\partial \tau_{xz}}{\partial z}\right)dS = 0$$

---

[5] We have neglected the body forces which appear in the local equation of equilibrium in the form of a function f_x. If these exist (inertia forces, centrifugal forces, or vibration inertia, for example), one obtains for the equilibrium: $\frac{dN_x}{dx} + p_x = 0$ in which $p_x = \int_D f_x dS$ represents the longitudinal load density.

[6] Note that the equality $\int_D (\frac{\partial \sigma_{yy}}{\partial y} + \frac{\partial \tau_{yz}}{\partial z})dS = \int_{\partial D}(\sigma_{yy}n_y + \tau_{yz}n_z)d\Gamma$ is made possible due to the continuity of the expression $(\sigma_{yy}\,n_y + \tau_{yz}\,n_z)$ across the frontier lines between different phases (see Equation 15.6).

where appears the **moment resultant:**

$$M_z = \int_D -y\sigma_{xx}dS.$$

Then transforming the second integral[7]:

$$\frac{dM_z}{dx} + \int_{\partial D} -y(\tau_{xy}n_y + \tau_{xz}n_z)d\Gamma + \int_D \tau_{xy}dS = 0$$

where one notes that:

$$\int_{\partial D} -y(\tau_{xy}n_y + \tau_{xz}n_z)d\Gamma = \int_{\partial D} -y\vec{x}\cdot\Sigma(\vec{n})d\Gamma = \int_{\partial D} -y(\vec{\sigma}\cdot\vec{x})d\Gamma = \mu_{z(\text{mN/m})}$$

which can be called a density moment on the beam. Then one obtains the equilibrium relation:

$$\frac{dM_z}{dx} + T_y + \mu_z = 0$$

The case where a density moment could exist in statics is practically nil, we therefore assume that $\mu_z = 0$. In summary, one obtains for the equations of equilibrium:

$$\left|\begin{array}{l} \dfrac{dN_x}{dx} = 0 \\[2mm] \dfrac{dT_y}{dx} + p_y = 0 \\[2mm] \dfrac{dM_z}{dx} + T_y = 0 \end{array}\right. \tag{15.7}$$

15.1.4 Constitutive Relations

Taking into account the isotropic nature of the different phases, the constitutive relation can be written in tensor form for Phase i as:

$$\varepsilon = \frac{1 + v_i}{E_i}\Sigma - \frac{v_i}{E_i}\text{tr}(\Sigma)I \qquad (I = \text{unity tensor})$$

---

[7] Analogous note for the continuity of the expression $(\tau_{xy}n_y + \tau_{xz}n_z)$ across the lines of internal interfaces (Equation 15.6).

One deduces, in integrating over the domain occupied by the cross section of the beam:

a)
$$\int_D \varepsilon_{xx} E_i \, dS = \int_D \sigma_{xx} \, dS - \int_D v_i (\sigma_{yy} + \sigma_{zz}) \, dS$$

Taking into account the form of the displacements in Equation 15.3, one can write

$$\int_D \varepsilon_{xx} E_i \, dS = \int_D \frac{\partial u_x}{\partial x} E_i \, dS = -\frac{d\theta_z}{dx} \int_D' y E_i \, dS + \frac{du}{dx} \int_D E_i \, dS + \frac{\partial}{\partial x} \int_D \not{E_i} \eta_x \, dS$$

which leads, with the notation in Equation 15.1, to the relation:

$$N_x = \langle ES \rangle \frac{du}{dx} + \int_D v_i (\sigma_{yy} + \sigma_{zz}) \, dS \qquad (15.8)$$

b)
$$\int_D -y \varepsilon_{xx} E_i \, dS = \int_D -y \sigma_{xx} \, dS + \int_D v_i \, y (\sigma_{yy} + \sigma_{zz}) \, dS$$

Taking into account the form of the displacements in Equation 15.3 one can write

$$\int_D -y \varepsilon_{xx} E_i \, dS = \frac{d\theta_z}{dx} \int_D E_i y^2 \, dS - \frac{du}{dx} \int_D' E_i y \, dS - \frac{\partial}{\partial x} \int_D \not{E_i} y \eta_x \, dS$$

This leads, with the notation in Equation 15.1, to the relation:

$$M_z = \langle EI_z \rangle \frac{d\theta_z}{dx} - \int_D v_i y (\sigma_{yy} + \sigma_{zz}) \, dS$$

c)
$$\int_D 2 \varepsilon_{xy} G_i \, dS = \int_D \tau_{xy} \, dS \qquad (15.9)$$

Taking into account the form of the displacements in Equation 15.3, one can write

$$\int_D 2 \varepsilon_{xy} G_i \, dS = \int_D \left(\frac{\partial u_x}{\partial y} + \frac{\partial u_y}{\partial x} \right) G_i \, dS = -\theta_z \int_D G_i \, dS \cdots$$

$$\cdots + \int_D G_i \frac{\partial \eta_x}{\partial y} \, dS + \frac{dv}{dx} \int_D G_i \, dS + \frac{\partial}{\partial x} \int_{D} \eta_{\not{y}} G_i \, dS$$

which gives with the notation in Equation 15.1, the relation:

$$T_y = \langle GS \rangle \left(\frac{dv}{dx} - \theta_z \right) + \int_D G_i \frac{\partial \eta_x}{\partial y} \, dS \qquad (15.10)$$

15.1.5 Technical Formulation

15.1.5.1 Simplifications

We extend to the composite beams the simplifications made for the homogeneous beams as:

1. σ_{yy} and $\sigma_{zz} \ll \sigma_{xx}$ at almost all points of the cross section.[8]
2. We neglect the variation of warping $\{\eta_x, \eta_y, \eta_z\}$ between two neighboring infinitely near sections in order to calculate the flexure stresses, that are

$$\sigma_{xx}, \tau_{xy}, \text{ and } \tau_{xz}.[9]$$

15.1.5.2 Expression for the Normal Stresses

With the previous simplifications, one extracts from the constitutive relation:

$$\varepsilon_{xx} = \frac{\sigma_{xx}}{E_i} - \frac{v_i}{E_i}(\sigma_{yy} + \sigma_{zz})$$

the following simplified form:

$$\frac{\sigma_{xx}}{E_i} \# \frac{\partial u_x}{\partial x} = -y\frac{d\theta_z}{dx} + \frac{du}{dx} + \frac{\partial \eta_x'}{\partial x}$$

Then with $M_z \# <EI_z> \dfrac{d\theta_z}{dx}$ (Equation 15.9) and $N_x \# <ES> \dfrac{du}{dx}$ (Equation 15.8):

$$\boxed{\sigma_{xx} = -E_i\frac{M_z}{\langle EI_z\rangle}y + E_i\frac{N_x}{\langle ES\rangle}}$$

$$\underbrace{\qquad\qquad}_{\text{bending}}\quad\underbrace{\qquad}_{\text{extension}}$$

(15.11)

Remark: The continuity $(\varepsilon_{xx})_i = (\varepsilon_{xx})_j[10]$ at the interface between the Phases i and j leads to

$$\frac{(\sigma_{xx})_i}{E_i} = \frac{(\sigma_{xx})_j}{E_j}$$

---

[8] This hypothesis is better verified by the fact that the Poisson coefficients of the different phases have similar values.

[9] This hypothesis is known in the literature for the homogeneous beams as the generalized "Navier–Bernoulli" hypothesis.

[10] See Section 15.1.2.

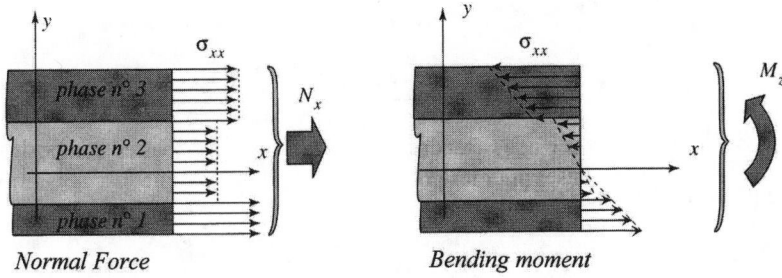

Figure 15.4 Normal and Bending Stresses

which shows the **discontinuity of normal stresses** due to the difference in the longitudinal moduli, as illustrated in Figure 15.4.

15.1.5.3 Shear Stress Expression

15.1.5.3.1 Characterization of Warping

Starting from the local equilibrium described by the relation:

$$\frac{\partial \sigma_{ij}}{\partial x_j} = 0$$

we study the flexure shear stress in the plane of the cross section, noted as:

$$\vec{\tau} = \tau_{xy}\vec{y} + \tau_{xy}\vec{z}$$

Taking into consideration Equations 15.11 and 15.7, and the simplification 2, at the beginning of Section 15.1.5.1, one can write

$$\frac{\partial \tau_{xy}}{\partial y} + \frac{\partial \tau_{xz}}{\partial z} = -\frac{\partial \sigma_{xx}}{\partial x} = \frac{E_i}{\langle EI_z \rangle}\frac{dM_z}{dx}y - \frac{E_i}{\langle ES \rangle}\frac{dN_x}{dx} = -\frac{E_i}{\langle EI_z \rangle}T_y \times y$$

and with the displacement field in Equation 15.3:

$$G_i\left(\frac{\partial^2 \eta_x}{\partial y^2} + \frac{\partial^2 \eta_x}{\partial z^2}\right) = -T_y\frac{E_i}{\langle EI_z \rangle} \times y$$

Putting η_x in the form:

$$\eta_x = \frac{T_y}{\langle GS \rangle} \times g(y,z) \tag{15.12}$$

leads to

$$\nabla^2 g = -\frac{E_i \langle GS \rangle}{G_i \langle EI_z \rangle} \times y \tag{15.13}$$

and Equation 15.10 becomes

$$T_y = \langle GS \rangle \left(\frac{dv}{dx} - \theta_z \right) + \int_D G_i \frac{T_y}{\langle GS \rangle} \frac{\partial g}{\partial y} dS$$

$$T_y \left(1 - \frac{1}{\langle GS \rangle} \int_D G_i \frac{\partial g}{\partial y} dS \right) = \langle GS \rangle \left(\frac{dv}{dx} - \theta_z \right)$$

or:

$$\boxed{T_y = \frac{\langle GS \rangle}{k} \left(\frac{dv}{dx} - \theta_z \right)} \tag{15.14}$$

In the above relation appears a k coefficient which is analogous to the shear coefficient for homogeneous beams.

15.1.5.3.2 External Limit Condition

We have supposed that the lateral surface of the beam is free from shear. This gives, along the external contour ∂D of the cross section, the relation:

$$\vec{\tau} \cdot \vec{n} = \tau_{xy} n_y + \tau_{xz} n_z = 0,$$

and, with the displacement field in Equation 15.3 and the simplifications described above:

$$\left(\frac{dv}{dx} - \theta_z \right) n_y + \overrightarrow{\text{grad}} \, \eta_x \cdot \vec{n} = 0$$

Introducing the function $g(y,z)$ (Equation 15.12) and with 15.14, one obtains

$$\overrightarrow{\text{grad}} \, g \cdot \vec{n} = \frac{\partial g}{\partial n} = -k n_y$$

Substituting the function $g(y,z)$ with the function $g_o(y,z)$ such that:

$$g_o(y,z) = g(y,z) + k \times y \tag{15.15}$$

one verifies that g_o is solution of the problem:

$$
\begin{cases}
\nabla^2 g_o = -\dfrac{E_i \langle GS \rangle}{G_i \langle EI_z \rangle} \times y \text{ in domain D} \\[3mm]
\dfrac{\partial g_o}{\partial n} = 0 \text{ on the boundary } \partial D
\end{cases}
$$

We denote $g_o(y,z)$ as the **longitudinal warping function** for the cross section.

15.1.5.3.3 Conditions at the Interfaces

The continuity conditions already described in Section 15.1.2 lead for the warping function, at the interface between two phases i and j,

$$g_{oi} = g_{oj}$$
$$\tau_{xyi} n_y + \tau_{xzi} n_z = \tau_{xyj} n_y + \tau_{xzj} n_z$$

then:

$$G_i \frac{\partial g_{oi}}{\partial n} = G_j \frac{\partial g_{oj}}{\partial n}$$

15.1.5.3.4 Uniqueness of the Solution

This is given by Equation 15.5 which is interpreted here as:

$$\int_D E_i g_o \, dS = 0$$

15.1.5.3.5 Form of the Shear Stresses

One can easily verify the following expressions:

$$\tau_{xy} = G_i \frac{T_y}{\langle GS \rangle} \frac{\partial g_o}{\partial y}$$

$$\tau_{xz} = G_i \frac{T_y}{\langle GS \rangle} \frac{\partial g_o}{\partial z}$$

then again:

$$\vec{\tau} = G_i \frac{T_y}{\langle GS \rangle} \overrightarrow{\text{grad}} \, g_o$$

15.1.5.3.6 Shear Coefficient for the Section

The shear coefficient for the section is obtained starting from the Equation 15.5:

$$\int_D E_i \eta_x \, dS = 0$$

This necessitates the knowledge of the warping function g_o. Then the above relation can be rewritten as:

$$\int_D E_i \frac{T_y}{\langle GS \rangle}(g_o - k \times y) \times y\, dS = 0$$

which leads to:

$$k = \frac{1}{\langle EI_z \rangle} \int_D E_i g_o y\, dS$$

In summary, in the absence of body forces (inertia forces for example), the bending of a composite beam in its plane of symmetry can be characterized by a homogenized formulation—equivalent to a classical homogeneous beam solution—in the following manner:

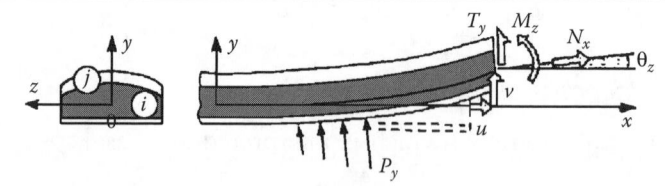

(15.16)

• **Elastic center 0:** it is such that $$\int_D E_i\, y\, dS = 0$$
• **Equivalent stiffnesses:** $$\langle ES \rangle = \sum_i E_i S_i; \quad \langle EI_z \rangle = \sum_i E_i I_{zi}; \quad \frac{\langle GS \rangle}{k} = \sum_i G_i S_i \times \frac{1}{k}$$
• **Equilibrium relations:** (stress resultants calculated at elastic center) $$\frac{dN_x}{dx} = 0; \quad \frac{dT_y}{dx} + p_y = 0; \quad \frac{dM_z}{dx} + T_y = 0$$
• **Constitutive relations:** $$N_x = \langle ES \rangle \frac{du}{dx}; \quad T_y = \frac{\langle GS \rangle}{k}\left(\frac{dv}{dx} - \theta_z\right); \quad M_z = \langle EI_z \rangle \frac{d\theta_z}{dx}$$

(Cont.)

Stresses:

normal stresses $\quad \sigma_{xx} = -E_i \dfrac{M_z}{\langle EI_z \rangle} y + E_i \dfrac{N_x}{\langle ES \rangle}$

shear stresses $\quad \left. \begin{array}{l} \tau_{xy} = G_i \dfrac{T_y}{\langle GS \rangle} \dfrac{\partial g_o}{\partial y} \\[4mm] \tau_{xz} = G_i \dfrac{T_y}{\langle GS \rangle} \dfrac{\partial g_o}{\partial z} \end{array} \right\} \vec{\tau} = \dfrac{G_i}{\langle GS \rangle} T_y \overrightarrow{\mathrm{grad}} g_o$

- longitudinal warping function $g_o(y, z)$: it is the solution to the problem

$$\begin{cases} \dfrac{\partial^2 g_o}{\partial y^2} + \dfrac{\partial^2 g_o}{\partial z^2} = -\dfrac{E_i \langle GS \rangle}{G_i \langle EI_z \rangle} y \text{ in domain D of the section.} \\[5mm] \dfrac{\partial g_o}{\partial n} = 0 \text{ on the boundary } \partial D \end{cases} \qquad (15.16)$$

with internal continuity:

$$\left. \begin{array}{l} g_{oi} = g_{oj} \\[3mm] G_i \dfrac{\partial g_{oi}}{\partial n} = G_j \dfrac{\partial g_{oj}}{\partial n} \end{array} \right\} \text{ along internal boundaries } \ell_{ij}$$

and the uniqueness condition:

$$\int_D E_i g_o \, dS = 0$$

Shear coefficient k: it is given by the formula:

$$k = \dfrac{1}{\langle EI_z \rangle} \int_D E_i g_o y \, dS$$

15.1.6 Energy Interpretation

15.1.6.1 Energy Due to Normal Stresses σ_{xx}

Denoting by dW_σ as the deformation energy of an elementary portion of a beam with length dx due to the application of normal stresses σ_{xx}, one has

$$dW_\sigma = \frac{1}{2} \int \sigma_{xx} \varepsilon_{xx} \, dV = \left\{ \frac{1}{2} \int_D \frac{\sigma_{xx}^2}{E_i} \, dS \right\} dx$$

Taking into account Equation 15.11 for the normal stresses:

$$\frac{dW_\sigma}{dx} = \frac{1}{2}\int_D \frac{1}{E_i}\left[-\frac{E_i}{\langle EI_z\rangle}M_z y + \frac{E_i}{\langle ES\rangle}N_x\right]^2 dS$$

$$= \frac{1}{2}\int_D E_i \frac{M_z^2}{\langle EI_z\rangle^2}y^2 dS + \frac{1}{2}\int_D E_i \frac{N_x^2}{\langle ES\rangle^2}dS\cdots$$

$$\cdots + \int_D E_i \frac{M_z N_x}{\langle EI_z\rangle\langle ES\rangle}y\, dS$$

(the above expression simplifies due to the definition of the elastic center 0 in 15.16); therefore:

$$\frac{dW_\sigma}{dx} = \frac{1}{2}\frac{M_z^2}{\langle EI_z\rangle} + \frac{1}{2}\frac{N_x^2}{\langle ES\rangle}$$

15.1.6.2 Energy Due to Shear Stresses $\vec{\tau}$

Denoting dW_τ as the deformation energy of an elementary portion of a beam with length dx due to the application of shear stresses $\vec{\tau}$, one has:

$$dW_\tau = \frac{1}{2}\int_D 2(\tau_{xy}\varepsilon_{xy} + \tau_{xz}\varepsilon_{xz})dV = \frac{1}{2}\left\{\int_D \frac{1}{G_i}(\tau_{xy}^2 + \tau_{xz}^2)dS\right\}dx$$

then, taking into account the form of the shear stresses in 15.16:

$$\frac{dW_\tau}{dx} = \frac{1}{2}\int_D G_i \frac{T_y^2}{\langle GS\rangle^2}\left\{\left(\frac{\partial g_o}{\partial y}\right)^2 + \left(\frac{\partial g_o}{\partial y}\right)^2\right\}dS$$

$$\frac{dW_\tau}{dx} = \frac{1}{2}\frac{T_y^2}{\langle GS\rangle^2}\int_D G_i\left\{\frac{\partial}{\partial y}\left(g_o\frac{\partial g_o}{\partial y}\right) + \frac{\partial}{\partial z}\left(g_o\frac{\partial g_o}{\partial z}\right) - g_o\cdot\nabla^2 g_o\right\}dS$$

with the value from 15.16 of the Laplacian of the warping function g_o[11]:

$$\frac{dW_\tau}{dx} = \frac{1}{2}\frac{T_y^2}{\langle GS\rangle^2}\left\{\int_D G_i g_o \frac{E_i\langle GS\rangle}{G_i\langle EI_z\rangle}y\,dS + \int_{\partial D} G_i g_o \frac{\partial g_o}{\partial n}d\Gamma\right\}$$

---

[11] The equality $\int_D G_i\{\frac{\partial}{\partial y}(g_o\frac{\partial g_o}{\partial y}) + \frac{\partial}{\partial z}(g_o\frac{\partial g_o}{\partial y})\}dS = \int_{\partial D}G_i g_o\frac{\partial g_o}{\partial n}d\Gamma$ is made possible due the continuity of the quantities $G_i g_o\frac{\partial g_o}{\partial n}$ at the interfaces l_{ij} (See Section 15.1.5.3.3, "Conditions at the Interfaces").

One knows in the above the expression of the shear coefficient k for the section (see 15.16). Then:

$$\frac{dW_\tau}{dx} = \frac{1}{2}k\frac{T_y^2}{\langle GS \rangle}$$

In summary, the strain energy density can be written as:

$$\boxed{\frac{dW}{dx} = \frac{1}{2}\frac{N_x^2}{\langle ES \rangle} + \frac{1}{2}\frac{M_z^2}{\langle EI_z \rangle} + \frac{1}{2}k\frac{T_y^2}{\langle GS \rangle}} \qquad (15.17)$$

Remarks:

- Note the analogy between this expression and that for the strain energy of a classical homogeneous beam, which should be written here as:

$$\frac{dW}{dx} = \frac{1}{2}\frac{N_x^2}{ES} + \frac{1}{2}\frac{M_z^2}{EI_z} + \frac{1}{2}k\frac{T_y^2}{GS}$$

- As a practical consequence of this homogenization, it becomes possible to determine the **equivalent characteristics** which are necessary for the entry of data into a computer program utilizing finite elements of classical homogeneous beams. The problem then comes to the numerical evaluation of the following values:

  Equivalent moduli: $E_{\text{equivalent}}$, $G_{\text{equivalent}}$, (or $\nu_{\text{equivalent}}$)

  Geometric characteristics: $S_{\text{equivalent}}$, $I_{z\ \text{equivalent}}$, and k

By taking $S_{\text{equivalent}} = S$ (real area of the cross section), one can easily verify that:

$$E_{\text{equivalent}} = \frac{\langle ES \rangle}{S}; \qquad G_{\text{equivalent}} = \frac{\langle GS \rangle}{S}$$

$$I_{z\ \text{equivalent}} = \frac{\langle EI_z \rangle}{E_{\text{equivalent}}}; \qquad \nu_{\text{equivalent}} = \frac{1}{2}\frac{\langle ES \rangle}{\langle GS \rangle} - 1$$

15.1.7 Extension to the Dynamic Case

The equilibrium relations of Section 15.1.3 were written in the absence of body forces. In vibratory motions, these body forces exist in the form of inertia forces. One then has

$$\frac{\partial \sigma_{ij}}{\partial x_j} - \rho \ddot{u}_i = 0$$

Following the main steps of the calculations in Section 15.1.3, for a beam under free vibration,[12] one obtains:

a)
$$\frac{\partial N_x}{\partial x} = \frac{\partial^2}{\partial t^2} \int_D \rho_i u_x \, dS$$

which leads, with Equation 15.3, to the following expression:

$$\frac{\partial N_x}{\partial x} = \langle \rho S \rangle \frac{\partial^2 u}{\partial t^2} - y_G \langle \rho S \rangle \frac{\partial^2 \theta_z}{\partial t^2}$$

in which we denote

$$\langle \rho S \rangle = \int_D \rho_i \, dS \quad \text{and} \quad y_G \langle \rho S \rangle = \int_D \rho_i y \, dS.$$

y_G appears here as the ordinate of the mass center (center of gravity) of the section. One has neglected the secondary coupling due to η_x:

b)
$$\frac{\partial T_y}{\partial x} = \frac{\partial^2}{\partial t^2} \int_D \rho_i u_y \, dS$$

with Equation 15.3 and neglecting the secondary coupling due to η_x:

$$\frac{\partial T_y}{\partial x} = \langle \rho S \rangle \frac{\partial^2 v}{\partial t^2}$$

c)
$$\frac{\partial M_z}{\partial x} + T_y = \frac{\partial^2}{\partial t^2} \int_D -y \rho_i u_x \, dS$$

with Equation 15.3, posing $\langle \rho I_z \rangle = \int_D \rho_i y^2 \, dS$, and neglecting the secondary coupling due to η_x:

$$\frac{\partial M_z}{\partial x} + T_y = \langle \rho I_z \rangle \frac{\partial^2 \theta_z}{\partial t^2} - y_G \langle \rho S \rangle \frac{\partial^2 u}{\partial t^2}$$

The above relations are to be joined with the constitutive relations in 15.16. However one must note these constitutive relations were written in the absence of body forces. Nevertheless, we will consider them to be valid, with the condition that the concerned frequencies are not too high. Generally this case corresponds to the mechanical frequencies, and one denotes this as the "*quasi static*" domain.

---

[12] One removes all the forces and moments on the beam except inertial forces and moments.

In summary, in dynamic regime, one has to replace the equilibrium relations and the constitutive behavior which appear in 15.16 with the following relations:

• Governing equations (stress resultants calculated at elastic center):

$$\frac{\partial N_x}{\partial x} = \langle \rho S \rangle \frac{\partial^2 u}{\partial t^2} - y_G \langle \rho S \rangle \frac{\partial^2 \theta_z}{\partial t^2}$$

$$\frac{\partial T_x}{\partial x} = \langle \rho S \rangle \frac{\partial^2 v}{\partial t^2}$$

$$\frac{\partial M_z}{\partial x} + T_y = \langle \rho I_z \rangle \frac{\partial^2 \theta_z}{\partial t^2} - y_G \langle \rho S \rangle \frac{\partial^2 u}{\partial t^2}$$

(15.18)

with

$$\langle \rho S \rangle = \sum_i \rho_i S_i; \quad \langle \rho I_z \rangle = \sum_i \rho_i I_{zi}; \quad y_G \langle \rho S \rangle = \int_D \rho_i \, y \, dS$$

• Constitutive relations:

$$N_x = \langle ES \rangle \frac{\partial u}{\partial x}; \quad T_y = \frac{\langle GS \rangle}{k}\left(\frac{\partial v}{\partial x} - \theta_z\right); \quad M_z = \langle EI_z \rangle \frac{\partial \theta_z}{\partial x}$$

Remark: We note in the above relations a nonclassical coupling between the longitudinal oscillations $u(x,t)$ and flexural oscillations. This coupling disappears if the elastic center is mixed with the center of gravity.[13]

15.2 CASE OF ANY CROSS SECTION (ASYMMETRIC)

Now, the cross section of the beam does not present any particular symmetry (see Figure 15.5). One can consider for this general case the procedure adopted in the previous paragraph for beams with symmetry. One notes the supplementary equivalent stiffness:

$$\langle EI_y \rangle = \int_D E_i z^2 dS = \sum_{\substack{i \\ \text{number of phases}}} E_i I_{yi}$$

It also appears an equivalent rotation $\theta_y(x)$ defined by the expression:

$$\theta_y(x) = \frac{1}{\langle EI_y \rangle} \int_D E_i u_x(x,y,z) \times z \, dS$$

Then, from the definitions of θ_y, u, and θ_z (Section 15.1.1):

$$\theta_y(x) = \frac{1}{\langle EI_y \rangle} \int_D E_i\{u - y\theta_z + \eta_{ox}\} \times z \, dS$$

[13] See Chapter 18, Application 18.3.9.

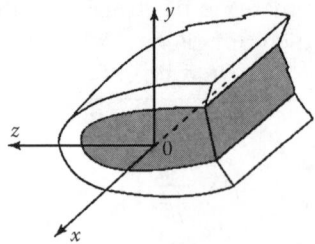

Figure 15.5   Composite Beam with any Cross Section

this expression simplifies if one chooses the origin of the coordinate z such that:

$$\int_D E_i z\, dS = 0; \quad \int_D E_i yz\, dS = 0$$

These relations, joined with the condition already established in the previous paragraph: $\int_D E_i y\, ds = 0$, allow one to define the position of the **elastic center** 0 of the section, as well as the orientation of the axes y and z that we will call **the principal axes** of the section. Then, in summary:

elastic center:	$\int_D E_i y\, dS = 0$
	$\int_D E_i z\, dS = 0$
principal axes:	$\int_D E_i yz\, dS = 0$

One can then rewrite the contribution η_{ox} to the longitudinal displacement u_x, which appeared above, in the form:

$$\eta_{ox}(x, y, z) = z \times \theta_y + \eta_x(x, y, z)$$

and one verifies that, by definition of the degree of freedom θ_y:

$$\int_D E_i \eta_x z\, dS = 0$$

The displacement $u_x\,(x, y, z)$ then takes the form:

$$u_x(x, y, z) = u(x) - y\theta_z(x) + z\theta_y(x) + \eta_x(x, y, z)$$

In addition, due to the disappearance of symmetry in the section, the transverse displacement $w(x)$ (Section 15.1.1.6) is not zero.

We then obtain for the elastic displacement field:

$$
\begin{vmatrix}
u_x = u(x) - y\theta_z + z\theta_y + \eta_x \\
u_y = v(x) + \eta_y \\
u_z = w(x) + \eta_z
\end{vmatrix}
$$

The three-dimensional incremental displacements η_x, η_y, η_z verify (see Section 15.1.1.6):

$$
\int_D E_i \eta_x \, dS = \int_D E_i \eta_x y \, dS = \int_D E_i \eta_x z \, dS = 0
$$

$$
\int_D G_i \eta_y \, dS = 0
$$

$$
\int_D G_i \eta_z \, dS = 0
$$

Starting from the above and following the same procedure as in the previous paragraph, successively for the bending in the plane x,y, with identical results, then in the plane x,z, we obtain results summarized in the following relations:

- **degree of freedom:**

  along x: $u(x)$

  along y: $v(x); \theta_y(x)$

  along z: $w(x); \theta_z(x)$

- **elastic center:** it is such that:

$$
\int_D E_i y \, dS = \int_D E_i z \, dS = 0
$$

- **principal axes:** they are such that:

$$
\int_D E_i yz \, dS = 0
$$

- **equivalent stiffnesses:**

$$
\langle ES \rangle = \sum_i E_i S_i
$$

$$
\langle EI_z \rangle = \sum_i E_i I_{zi}; \quad \langle EI_y \rangle = \sum_i E_i I_{yi}
$$

$$
\frac{\langle GS \rangle}{k_y} = \sum_i G_i S_i \times \frac{1}{k_y}; \quad \frac{\langle GS \rangle}{k_z} = \sum_i G_i S_i \times \frac{1}{k_z} \quad \text{(Continued)}
$$

(Cont.)

• **equilibrium relations:** (stress resultants calculated at elastic center)

$$\frac{dN_x}{dx} = 0$$

$$\frac{dT_y}{dx} + p_y = 0; \quad \frac{dT_z}{dx} + p_z = 0$$

$$\frac{dM_z}{dx} + T_y = 0; \quad \frac{dM_y}{dx} - T_z = 0$$

• **constitutive relations:**

$$N_x = \langle ES \rangle \frac{du}{dx}$$

$$T_y = \frac{\langle GS \rangle}{k_y}\left(\frac{dv}{dx} - \theta_z\right); \quad T_z = \frac{\langle GS \rangle}{k_z}\left(\frac{dw}{dx} + \theta_y\right) \qquad (15.19)[14]$$

$$M_z = \langle EI_z \rangle \frac{d\theta_z}{dx}; \quad M_y = \langle EI_y \rangle \frac{d\theta_y}{dx}$$

• **normal stresses:**

$$\sigma_{xx} = -E_i \frac{M_z}{\langle EI_z \rangle} y + E_i \frac{M_y}{\langle EI_y \rangle} z + E_i \frac{N_x}{\langle ES \rangle}$$

• **shear stresses:**

$$\tau_{xy} = \frac{G_i}{\langle GS \rangle}\left(T_y \frac{\partial g_o}{\partial y} + T_z \frac{\partial h_o}{\partial y}\right)$$

$$\tau_{xz} = \frac{G_i}{\langle GS \rangle}\left(T_y \frac{\partial g_o}{\partial z} + T_z \frac{\partial h_o}{\partial z}\right)$$

then

$$\vec{\tau} = \frac{G_i}{\langle GS \rangle}\left(T_y \overrightarrow{\text{grad }} g_o + T_z \overrightarrow{\text{grad }} h_o\right) \quad \text{(Continued)}$$

Remark: As already mentioned in Section 15.1.6 for a beam with a plane of symmetry, it is possible to evaluate the equivalent characteristics that one has to introduce in data form in order to utilize computer programs for the calculation by

---

[14] In fact, in place of the constitutive relation $T_y = \frac{\langle GS \rangle}{k_y}(\frac{dv}{dx} - \theta_z)$ it comes to a form such as: $k_y T_y + k_{yz} T_z = \langle GS \rangle(\frac{dv}{dz} - \theta_z)$ where appears a coupling coefficient k_{yz}. This means that a unique shear resultant T_z leads to flexure in the x,y plane. This secondary effect has been neglected here. Analogous remark yields for the constitutive relation $T_z = \frac{\langle GS \rangle}{k_z}(\frac{dw}{dx} + \theta_y)$.

- **Longitudinal warping functions:**

function $g_o(y,z)$: It is the solution to the problem

$$\begin{cases} \dfrac{\partial^2 g_o}{\partial y^2} + \dfrac{\partial^2 g_o}{\partial z^2} = -\dfrac{E_i}{G_i}\dfrac{\langle GS \rangle}{\langle EI_z \rangle}y \text{ in domain D of the section} \\ \\ \dfrac{\partial g_o}{\partial n} = 0 \text{ on the boundary } \partial D \end{cases}$$

with internal continuity:

$$\left. \begin{array}{l} g_{oi} = g_{oj} \\ \\ G_i\dfrac{\partial g_{oi}}{\partial n} = G_j\dfrac{\partial g_{oj}}{\partial n} \end{array} \right\} \text{along internal boundaries } \ell_{ij}$$

and the uniqueness condition: $\displaystyle\int_D E_i g_o \, dS = 0$

function $h_o(y,z)$: It is the solution to the problem:

$$\begin{cases} \dfrac{\partial^2 h_o}{\partial y^2} + \dfrac{\partial^2 h_o}{\partial z^2} = -\dfrac{E_i}{G_i}\dfrac{\langle GS \rangle}{\langle EI_y \rangle}z \text{ in domain D of the section} \\ \\ \dfrac{\partial h_o}{\partial n} = 0 \text{ on the boundary } \partial D \end{cases}$$

with internal continuity:

$$\left. \begin{array}{l} h_{oi} = h_{oj} \\ \\ G_i\dfrac{\partial h_{oi}}{\partial n} = G_j\dfrac{\partial h_{oj}}{\partial n} \end{array} \right\} \text{along internal boundaries } \ell_{ij}$$

and the uniqueness condition: $\displaystyle\int_D E_i h_o \, dS = 0$

- **Shear coefficients**[1]

coefficient k_y: it is given by the formula: $\quad k_y = \dfrac{1}{\langle EI_z \rangle}\displaystyle\int_D E_i g_o y \, dS$

coefficient k_z: it is given by the formula: $\quad k_z = \dfrac{1}{\langle EI_y \rangle}\displaystyle\int_D E_i h_o z \, dS$

- **Strain energy**

$$\dfrac{dW}{dx} = \dfrac{1}{2}\dfrac{N_x^2}{\langle ES \rangle} + \dfrac{1}{2}\dfrac{M_z^2}{\langle EI_z \rangle} + \dfrac{1}{2}\dfrac{M_y^2}{\langle EI_z \rangle} + \dfrac{1}{2}k_y\dfrac{T_y^2}{\langle GS \rangle} + \dfrac{1}{2}k_z\dfrac{T_z^2}{\langle GS \rangle}$$

[1] See Chapter 18, Applications 18.3.5, 18.3.6.

finite elements for the classical beams.[15] The characteristics $E_{equivalent}$, $G_{equivalent}$, I_z equivalent, $I_{y \, equivalent}$, can be obtained right away. On the contrary, the calculation of the shear coefficients k_y and k_z is not direct. At first it is necessary to know the values of the functions g_o and h_o, solutions in the domain occupied by the cross section of Poisson type problem, as one can take note in the preceding table. The nature of these problems makes it possible for each of the functions g_o and h_o to write an equivalent functional which allows the calculation of the function considered by means of discretization of the cross section using finite elements.

---

[15] It is convenient to note that a computer program based on elements of homogeneous beams cannot provide correct values for the stresses in a cross section, because these stresses are of particular formulation for composite beams (see 15.16).

16

COMPOSITE BEAMS IN TORSION

As in the previous chapter, we consider here the composite beams made of isotropic phases. Extension to transversely isotropic phases is straightforward. The study of orthotropic phases with one principal direction parallel to the axis of the beam, the other two principal directions in the plane of a cross section, does not present fundamental difficulties.

16.1 UNIFORM TORSION

We will keep the conventions and notations of the previous chapter. On Figure 16.1, 0 is the **elastic center**, x,y, and z are the **principal axes**. The beam is slender and uniformly twisted, this means that every cross section is subjected to a pure and constant torsion moment, along the x axis, denoted as M_x.

Then, under the application of this moment, each line in the beam, initially parallel to the x axis, becomes a **helicoid curve**, including (in the absence of symmetry in the cross section) the line which, initially, was coinciding with the elastic x axis itself. The only line which remains rectilinear is cutting the plane of all sections at a point which will be called **torsion center** and denoted as C, with coordinates y_C and z_C in the principal axes (see Figure 16.1).

16.1.1 Torsional Degree of Freedom

By definition, this is the rotation of each section about the x axis, denoted as θ_x.[1] The torsional moment M_x being constant, the angle θ_x evolves along the x axis in such a manner that, for any pair of cross sections spaced with a distance dx, one can observe a same increment of rotation $d\theta_x$; then:

$$\boxed{\dfrac{d\theta_x}{dx} = \text{constant}}$$

---

[1] Here it is not necessary to define the rotation θ_x by means of an integral of displacements, as in the previous chapter relating to flexure. In effect, we will see in the following that the displacement field associated with this pure rotation of the sections leads to the exact solution of the problem in the elastic domain (at least for the case of uniform warping).

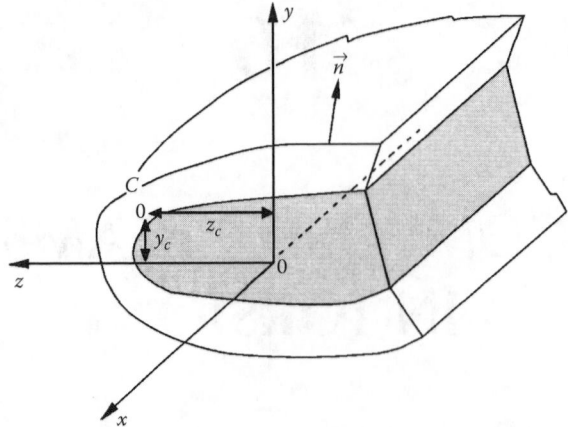

Figure 16.1   Elastic Center (O), Torsion Center (C), and Principal Axes

From this it comes that the angle of rotation of the sections varies linearly along the longitudinal axis x. As a consequence, we assume *a priori* the components of the displacement field u_x, u_y, u_z, to be written as:

$$
\begin{aligned}
u_x &= \frac{d\theta_x}{dx} \times \varphi(y, z)\\
u_y &= -(z - z_c)\theta_x\\
u_z &= (y - y_c)\theta_x
\end{aligned}
$$

(16.1)

where the function denoted as $\varphi(y,z)$ is characteristic of the cross section shape and of the materials that constitute the section. This is called the **warping function** for torsion.

16.1.2 Constitutive Relation

With the displacement field in Equation 16.1 the only nonzero strains are written as:

$$
\gamma_{xy} = \frac{d\theta_x}{dx}\left(\frac{\partial\varphi}{\partial y} - (z - z_c)\right)
$$

$$
\gamma_{xz} = \frac{d\theta_x}{dx}\left(\frac{\partial\varphi}{\partial z} + (y - y_c)\right)
$$

The only nonzero stresses are then the shear stresses τ_{xy} and τ_{xz}. The torsional moment can be deduced by integration over the domain of the straight section as:

$$
M_x = \int_D (y\tau_{xz} - z\tau_{xy})dS = \frac{d\theta_x}{dx}\int_D G_i\left\{y\left(\frac{\partial\varphi}{\partial z} - y_c\right)\cdots\right.
$$

$$
\left.\cdots - z\left(\frac{\partial\varphi}{\partial y} + z_c\right) + y^2 + z^2\right\}dS
$$

Substituting to the function $\varphi(y,z)$ the function $\Phi(y,z)$ such that:

$$\Phi(y, z) = \varphi(y, z) + yz_c - zy_c$$ (16.2)

it becomes:

$$M_x = \frac{d\theta_x}{dx} \times \int_D G_i \left(y\frac{\partial \Phi}{\partial z} - z\frac{\partial \Phi}{\partial y} + y^2 + z^2 \right) dS$$

In this expression, it is possible to define an **equivalent stiffness in torsion** with the form:

$$\langle GJ \rangle = \int_D G_i \left(y\frac{\partial \Phi}{\partial z} - z\frac{\partial \Phi}{\partial y} + y^2 + z^2 \right) dS$$ (16.3)

One obtains then for the constitutive relation:

$$M_x = \langle GJ \rangle \frac{\partial \theta_x}{\partial x}$$

16.1.3 Determination of the Function $\Phi(y, z)$

16.1.3.1 Local Equilibrium

Local equilibrium is written as:

$$\frac{\partial \tau_{xy}}{\partial y} + \frac{\partial \tau_{xz}}{\partial z} = 0$$

then with the displacement field in Equation 16.1:

$$\nabla^2 \varphi = 0$$

and with form 16.2 of the function Φ:

$$\nabla^2 \Phi = 0$$

16.1.3.2 Conditions at the External Boundary

The lateral surface being free of stresses, one can write along the external boundary ∂D:

$$\vec{\tau} \cdot \vec{n} = 0.$$

With the displacement field in Equation 16.1:

$$\left\{ \frac{\partial \phi}{\partial y} - (z - z_c) \right\} n_y + \left\{ \frac{\partial \phi}{\partial z} + (y - y_c) \right\} n_z = 0$$

then again:

$$\frac{\partial \Phi}{\partial y} n_y + \frac{\partial \Phi}{\partial z} n_z = z n_y - y n_z$$

16.1.3.3 Conditions at the Internal Boundaries

The continuity conditions of Section 15.1.2 are verified for u_y and u_z. At an interfacial line l_{ij} between two phases i and j, the continuity of u_x leads to

$$\Phi_i = \Phi_j$$

The continuity relations in Equation 15.6 lead to the continuity of $(\tau_{xy} n_y + \tau_{xz} n_z)$ when crossing the lines l_{ij}, that produces the continuity of

$$G_i \left(\frac{\partial \Phi_i}{\partial y} - z \right) n_y + G_i \left(\frac{\partial \Phi_i}{\partial z} + y \right) n_z$$

16.1.3.4 Uniqueness of the Function Φ

If one superimposes torsion and bending, by using the degrees of freedom for flexure defined in the previous chapter, the displacement component u_x becomes:

$$u_x = u - y \theta_z + z \theta_y + \frac{d\theta_x}{dx} \varphi + \eta_x$$

The longitudinal displacement $u(x)$ has to respond to its definition (Section 15.1.1), meaning

$$u = \frac{1}{\langle ES \rangle} \int_D E_i u_x \, dS$$

$$u = \frac{1}{\langle ES \rangle} \left\{ u \int_D E_i \, dS - \theta_z \int_D E_i y \, dS + \theta_y \int_D E_i z \, dS \cdots \right.$$

$$\left. \cdots + \frac{d\theta_x}{dx} \int_D E_i \varphi \, dS + \int_D E_i \eta_x \, dS \right\}$$

This requires that:

$$\int_D E_i \varphi \, dS = 0$$

Then, taking into account the form in Equation 16.2 of Φ and the properties of the elastic center:

$$\int_D E_i \Phi \, dS = 0$$

In summary, the function $\Phi(y,z)$ is the solution of the problem:

$$\begin{cases} \nabla^2\Phi = 0 & \text{in domain } D \text{ of the section} \\ \dfrac{\partial\Phi}{\partial n} = zn_y - yn_z & \text{on the external boundary } \partial D \end{cases}$$

with the internal continuity:

$$\left.\begin{array}{l} \Phi_i = \Phi_j \\ G_i\left(\dfrac{\partial\Phi_i}{\partial n} - (zn_y + yn_z)\right) = G_j\left(\dfrac{\partial\Phi_j}{\partial n} - (zn_y + yn_z)\right) \end{array}\right\} \begin{array}{l} \text{along the internal} \\ \text{boundaries } \ell_{ij} \end{array}$$

and the condition of uniqueness:

$$\int_D E_i\Phi\, dS = 0$$

16.1.4 Energy Interpretation

The strain energy of an elementary segment of a beam with thickness dx is written as:

$$dW = \frac{1}{2}\int 2(\tau_{xy}\varepsilon_{xy} + \tau_{xz}\varepsilon_{xz})dV = \left\{\frac{1}{2}\int_D G_i(\gamma_{xy}^2 + \gamma_{xz}^2)\,dS\right\}dx$$

then, taking into account the displacement field in Equation 16.1:

$$\frac{dW}{dx} = \frac{1}{2}\left(\frac{d\theta_x}{dx}\right)^2\int_D G_i\left\{\left(\frac{\partial\Phi}{\partial y} - z\right)^2 + \left(\frac{\partial\Phi}{\partial z} + y\right)^2\right\}dS$$

which can be rewritten as[2]:

$$\frac{dW}{dx} = \frac{1}{2}\left(\frac{d\theta_x}{dx}\right)^2\left\{\int_D G_i\left\{y\frac{\partial\Phi}{\partial z} - z\frac{\partial\Phi}{\partial y} + y^2 + z^2\right\}dS - \int_D G_i\Phi\cancel{\nabla^2\Phi}\,dS\ldots\right.$$

$$\left.\ldots + \int_{\partial D} G_i\Phi\left\{\left(\frac{\partial\Phi}{\partial y} - z\right)n_y + \left(\frac{\partial\Phi}{\partial z} + y\right)n_z\right\}d\Gamma\right\}$$

---

[2] In effect, one has, for example:

$$\left(\frac{\partial\Phi}{\partial y}\right)^2 - z\frac{\partial\Phi}{\partial y} = \frac{\partial\Phi}{\partial y}\left(\frac{\partial\Phi}{\partial y} - z\right) = \frac{\partial}{\partial y}\left\{\Phi\left(\frac{\partial\Phi}{\partial y} - z\right)\right\} - \Phi\frac{\partial^2\Phi}{\partial y^2}$$

where we note the presence of the stiffness in torsion $\langle GJ \rangle$ defined by Equation 16.3. Thus,

$$\frac{dW}{dx} = \frac{1}{2}\langle GJ \rangle \left(\frac{d\theta_x}{dx}\right)^2 \quad \text{or} \quad = \frac{1}{2}\frac{M_x^2}{\langle GJ \rangle}$$

16.2 LOCATION OF THE TORSION CENTER

Consider the cantilever beam that is clamped at its left end as shown schematically in Figure 16.2, and more particularly the segment limited by the cross sections denoted by D_0 and D_1. In the section D_1,0 is the elastic center and C is the torsion center the position of which we wish to determine.

With this objective, we will apply on the cross section D_1 the two following successive loadings:

- **Loading No. 1:** One applies on the torsion center C of the cross section D_1 a force $\vec{F}$ situated in the plane of the section.
- **Loading No. 2:** One applies on the same cross section D_1 a torsional moment denoted as M_x (see Figure 16.2).

When one applies these two loads successively, the final state is independent of the order of the application. As a consequence for the external forces acting on the isolated segment $(D_0 D_1)$, the work corresponding to loading No. 1 on the displacements created by loading No. 2 is equal to the work corresponding to loading No. 2 on the displacements created by loading No. 1. This can be written in the following form:

$$W_{\text{(loading 1} \times \text{displacement 2)}} = W_{\text{(loading 2} \times \text{displacement 1)}}$$

Now we evaluate these works:

a) W (loading 1 × displacement 2)

- On D_0: $\vec{F}$ creates the bending moments M_z and M_y, thus a normal stress distribution given in the principal axes by Equation 15.19 as:

$$(\sigma_{xx})_1 = -E_i \frac{M_z}{\langle EI_z \rangle} \times y + E_i \frac{M_y}{\langle EI_y \rangle} \times z$$

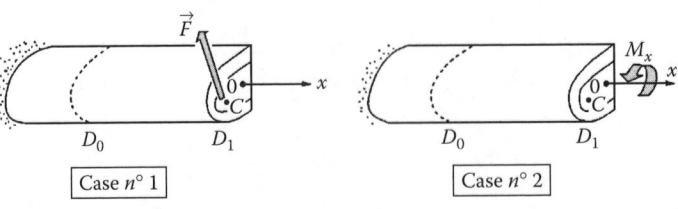

Case n° 1   Case n° 2

Figure 16.2 Cantilever Beam with Two Successive Loadings

Then, taking into account the displacement field in Equation 16.1, the work done on D_0 is

$$\int_D (\sigma_{xx})_1 \times (u_x)_2 \, dS = \int_D \left\{ -E_i \frac{M_z}{\langle EI_z \rangle} \times y + E_i \frac{M_y}{\langle EI_y \rangle} \times z \right\} \frac{d\theta_x}{dx} \varphi \, dS$$

$$= \frac{d\theta_x}{dx} \int_D \left\{ -E_i \frac{M_z}{\langle EI_z \rangle} \times y + E_i \frac{M_y}{\langle EI_y \rangle} \times z \right\} (\Phi - yz_c + zy_c) dS$$

■ On D_1: The torsion center C does not move in the plane of the cross section during torsion. The work done by the force $\vec{F}$ in the displacement field of torsion is nil.

b) W (loading 2 × displacement 1)

Force $\vec{F}$ as applied to the torsion center C does not lead to the rotation of the cross sections around the longitudinal axis x. From this the torsional moment M_x does not work on the bending displacement field due to $\vec{F}$.

The equality of the two works is then written as:

$$\frac{d\theta_x}{dx} \int_D \left\{ -E_i \frac{M_z}{\langle EI_z \rangle} \times y + E_i \frac{M_y}{\langle EI_y \rangle} \times z \right\} (\Phi - yz_c + zy_c) dS = 0$$

then:

$$\frac{M_z}{\langle EI_z \rangle} \int_D (E_i y \; \Phi - E_i y^2 z_c + E_i yzy_c) dS \cdots$$

$$\cdots + \frac{M_y}{\langle EI_y \rangle} \int_D (E_i z \; \Phi + E_i z^2 y_c - E_i yzz_c) ds = 0$$

This relation has to be verified when the force applied at C varies in magnitude and direction in the plane of the section. One can deduce from there that the relation is valid no matter what the values of M_z and M_y are. Both the above integrals are then nil. One extracts from this property the coordinates of the torsion center:

$$y_c = -\frac{1}{\langle EI_y \rangle} \int_D E_i z \Phi \, dS$$

$$z_c = \frac{1}{\langle EI_z \rangle} \int_D E_i y \; \Phi \, dS$$

In summary, the uniform torsion of a cylindrical composite beam made of perfectly bonded isotropic phases can be characterized by a homogenized formulation—equivalent to that of a classical homogeneous beam—in the following manner:

- **degree of freedom:** *about x axis:* θ_x

- **elastic center 0:** *it is such that* $\int_D E_i y \, dS = \int_D E_i z \, dS = 0$

- **principal axes:** *they are such that* $\int_D E_i yz \, dS = 0$

- **equivalent stiffnesses:**

$$\langle EI_z \rangle = \sum_i E_i I_{zi} \quad ; \quad \langle EI_y \rangle = \sum_i E_i I_{yi}$$

$$\langle GJ \rangle = \int_D G_i \left(y \frac{\partial \Phi}{\partial z} - z \frac{\partial \Phi}{\partial y} + y^2 + z^2 \right) dS$$

- **torsion center:** *coordinates in principal axes:*

$$y_c = -\frac{1}{\langle EI_y \rangle} \int_D E_i z \, \Phi \, dS$$

$$z_c = \frac{1}{\langle EI_z \rangle} \int_D E_i y \, \Phi \, dS$$

- **equilibrium relation:** $\dfrac{dM_x}{dx} = 0 \quad (M_x = constant)$

- **constitutive relation:** $M_x = \langle GJ \rangle \dfrac{d\theta_x}{dx}$

- **shear stresses** $\tau_{xy} = G_i \dfrac{d\theta_x}{dx} \left(\dfrac{\partial \Phi}{\partial y} - z \right)$

$$\tau_{xz} = G_i \frac{d\theta_x}{dx} \left(\frac{\partial \Phi}{\partial z} + y \right)$$

- **function $\Phi(y, z)$:** *it is the solution to the problem:*

$$\begin{cases} \dfrac{\partial^2 \Phi}{\partial y^2} + \dfrac{\partial^2 \Phi}{\partial z^2} = 0 & \text{in domain D of the section.} \\[2mm] \dfrac{\partial \Phi}{\partial n} = zn_y - yn_z & \text{on the external boundary } \partial D. \end{cases}$$

with internal continuity:

$$\left. \begin{aligned} \Phi_i &= \Phi_j \\ G_i \left(\frac{\partial \Phi_i}{\partial n} - zn_y + yn_z \right) &= G_j \left(\frac{\partial \Phi_j}{\partial n} - zn_y + yn_z \right) \end{aligned} \right\} \begin{aligned} &\text{along internal} \\ &\text{boundaries } \ell_{ij} \end{aligned}$$

and the uniqueness condition: $\int_D E_i \Phi \, dS = 0$

- **strain energy density**

$$\frac{dW}{dx} = \frac{1}{2} \frac{M_x^2}{\langle GJ \rangle}$$

(16.4)

Remarks:

■ A finite element computer program for the classical homogeneous beams is usable[3] with the condition that the equivalent rigidity in torsion $\langle GJ \rangle$ can be available. This requires the numerical calculation of the function Φ.[4] The latter is the solution of a Laplace type problem, which can be noted in relations 16.4. An equivalent functional is possible to define, which leads to the calculation of Φ by the finite element method, by means of discretization of the cross section.

■ **Flexion-torsion coupling:** When, due to the loads applied on the beam, there exists simultaneously bending and torsion of the beam, the approach of the previous chapter is always valid. Keeping the definitions in Sections 15.1.1 and 15.2 for the degrees of freedom u, v, θ_x, θ_y, one arrives at the following displacement field:

$$\begin{cases} u_x = u - y\theta_z + z\theta_y + \varphi\dfrac{d\theta_x}{dx} + \eta_x \\[2mm] u_y = v - z\theta_x + \eta_y \\[2mm] u_z = w + y\theta_x + \eta_z \end{cases}$$

The torsion being uniform, the equilibrium relations in 15.19 become more restrictive and can be reduced to:

$$\boxed{\begin{aligned} &\frac{dN_x}{dx} = 0; \quad \frac{dT_y}{dx} = 0; \quad \frac{dT_z}{dx} = 0 \\[2mm] &\frac{dM_x}{dx} = 0; \quad \frac{dM_z}{dx} + T_y = 0; \quad \frac{dM_y}{dx} - T_z = 0 \end{aligned}}$$

(16.5)

---

[3] Except if the considered application requires the calculation of the shear stresses in a cross section.

[4] One has to solve an analogous problem for the homogeneous beams, when one desires to calculate the torsional Saint-Venant stiffness:

$$J = \int_D \left(y\frac{\partial \Phi}{\partial z} - z\frac{\partial \Phi}{\partial y} + y^2 + z^2 \right) dS.$$

Taking into account six degrees of freedom also leads to six constitutive relations. One finds[5]:

$$N_x = \langle ES \rangle \frac{du}{dx}$$

$$T_y = \frac{\langle GS \rangle}{k_y} \left(\frac{dv}{dx} - \theta_z - z_c \frac{d\theta_x}{dx} \right)(*)$$

$$T_z = \frac{\langle GS \rangle}{k_z} \left(\frac{dw}{dx} + \theta_y + y_c \frac{d\theta_x}{dx} \right)(*)$$

$$M_x = \langle GJ \rangle \frac{d\theta_x}{dx} - z_c T_y + y_c T_z$$

$$M_y = \langle EI_y \rangle \frac{d\theta_y}{dx}$$

$$M_z = \langle EI_z \rangle \frac{d\theta_z}{dx}$$

(16.6)

---

[5] In each of the relations marked with (∗), there appears a supplementary coupling term connected to the existence of a third coefficient denoted as k_{yz}. The complete form is then:

$$k_y T_y + k_{yz} T_z = \langle GS \rangle \left(\frac{dv}{dx} - \theta_z - z_c \frac{d\theta_x}{dx} \right)$$

$$k_{yz} T_y + k_z T_z = \langle GS \rangle \left(\frac{dw}{dx} + \theta_y + y_c \frac{d\theta_x}{dx} \right)$$

This secondary coupling has been neglected in the indicated form.

17

FLEXURE OF THICK COMPOSITE PLATES

The mechanical behavior of a laminated plate as studied in Chapter 12 involves the definition of stress resultants N_x, N_y, T_{xy} and moment resultants M_x, M_y, M_{xy}. These resultants are obtained from the membrane stresses σ_x, σ_y, τ_{xy}. The other stress components, σ_z, τ_{xz}, τ_{yz} have not been taken into account until now.

In this chapter we note how these stresses exist and have influence on the mechanical behavior of the laminate. We will also examine the configurations of plates for which the influence of these stresses are significant (for example, plates with relatively high thicknesses, justifying the title given for this chapter). This study is based on the previous definition of the displacement parameters based on integral displacement forms, and constitutes an approach analogous (and also original) to that used in Chapter 15 for the description of the bending of composite beams.

17.1 PRELIMINARY REMARKS

17.1.1 Transverse Normal Stress σ_z

The plate is situated as in Chapter 12 from which the name of transverse normal stress σ_z. Such stress appears due to the application of a transverse load (concentrated or distributed) which will cause bending of the plate.

- A very local concentration of load in a very small zone cannot be examined with the theory of plates, which is not able to provide spatial distribution of the stresses in the neighborhood of the point of load application. This phenomenon is complex even in three-dimensional numerical modeling. Therefore, what will be presented will not be valid in the immediate surroundings of a very local transverse load (for example, an insert).
- A distributed load gives rise to the stresses σ_z with small amplitude as compared with the stresses σ_x and σ_y. This is the reason why σ_z is often neglected.

17.1.2 Transverse Shear Stresses τ_{xz} and τ_{yz}

Due to the assumption of perfect bonding between the plies, the stress vector remains continuous across an interfacial element with normal vector $\vec{n} = \vec{z}$, between two consecutive plies of the laminate. Then τ_{xz} and τ_{yz} remain continuous at the interfaces between plies (see Section 15.1.2). In addition, the upper face and lower face of the laminate are assumed to be free of tangential forces. The thickness of the laminate is denoted as h. One then has:

$$\tau_{xz} = \tau_{yz} = 0 \quad \text{for } z = \pm h/2$$

Assume stress and moment resultants to be constant in a given zone of the laminate:

$$N_x, N_y, T_{xy}, M_y, M_x, M_{xy} \text{ constants } \forall(x, y)$$

Then, by inversion of Equation 12.20, for example, one notes that the following global strain

$$\varepsilon_{ox}, \varepsilon_{oy}, \gamma_{oxy}, \partial^2 w_0/\partial x^2, \partial^2 w_0/\partial y^2, 2\partial^2 w_0/\partial x\partial y$$

are constant in the zone under consideration. The local strains of Equation 12.12 then depend only on the coordinate z of the laminate. This is the same for the membrane stresses σ_x, σ_y, τ_{xy}.

With the above consideration, local equilibrium can be written as (in the absence of body forces):

$$\frac{\partial \sigma_x'}{\partial x} + \frac{\partial \tau_{xy}'}{\partial y} + \frac{\partial \tau_{xz}}{\partial z} = 0$$

$$\frac{\partial \tau_{xy}'}{\partial x} + \frac{\partial \sigma_y'}{\partial y} + \frac{\partial \tau_{yz}}{\partial z} = 0 \tag{17.1}$$

The transverse shear stresses then appear to be constant across the thickness of a ply. Being continuous at the interface and nil at the location $z = \pm h/2$, they are nil in all the thickness of the laminate.

From this, these stresses do not play an important role in all cases: they do not always exist, their existence being related to variable stresses and moment resultants. When they exist and depending on the composition of the laminate, they can have influence on the deformation in bending, and on the interlaminar adhesion (between layers).

We will assume that these stresses exist, associated with the hypotheses of the following paragraph.

17.1.3 Hypotheses

- The plate has midplane symmetry.
- The plies are orthotropic, the orthotropic axes coinciding with the x, y, z axes of the laminate.[1]
- The stress σ_z is negligible.

---

[1] For example, this is the case for a laminate made of layers of balanced fabric at 0°, 90°, or 45°, −45°, for unidirectional layers at 0° and 90°, or for mats. Instead of this hypothesis, one can also adopt the less restrictive hypothesis of a balanced laminate. In this case the following calculations are much more involved, without appreciable gain on the enlargement of the field of applications examined in Section 17.6.3.

Remarks:

■ For each ply having orthotropic axes x, y, z, the constitutive Equation 13.3 can be written as, taking into account the simplification $\sigma_z \neq 0$:

$$
\begin{Bmatrix} \varepsilon_x \\ \varepsilon_y \\ \gamma_{xy} \\ \gamma_{xz} \\ \gamma_{yz} \end{Bmatrix} = \begin{bmatrix} \dfrac{1}{E_x} & -\dfrac{v_{yx}}{E_y} & 0 & 0 & 0 \\ -\dfrac{v_{xy}}{E_x} & \dfrac{1}{E_y} & 0 & 0 & 0 \\ 0 & 0 & \dfrac{1}{G_{xy}} & 0 & 0 \\ 0 & 0 & 0 & \dfrac{1}{G_{xz}} & 0 \\ 0 & 0 & 0 & 0 & \dfrac{1}{G_{yz}} \end{bmatrix} \begin{Bmatrix} \sigma_x \\ \sigma_y \\ \tau_{xy} \\ \tau_{xz} \\ \tau_{yz} \end{Bmatrix}
$$

or under inverse form

$$
\begin{Bmatrix} \sigma_x \\ \sigma_y \\ \tau_{xy} \\ \tau_{xz} \\ \tau_{yz} \end{Bmatrix} = \begin{bmatrix} \bar{E}_{11} & \bar{E}_{12} & 0 & 0 & 0 \\ \bar{E}_{21} & \bar{E}_{22} & 0 & 0 & 0 \\ 0 & 0 & \bar{E}_{33} = G_{xy} & 0 & 0 \\ 0 & 0 & 0 & \bar{E}_{44} = G_{xz} & 0 \\ 0 & 0 & 0 & 0 & \bar{E}_{55} = G_{yz} \end{bmatrix} \begin{Bmatrix} \varepsilon_x \\ \varepsilon_y \\ \gamma_{xy} \\ \gamma_{xz} \\ \gamma_{yz} \end{Bmatrix} \qquad (17.2)
$$

where:

$$
\bar{E}_{11} = \frac{E_x}{1 - v_{xy} v_{yx}}; \quad \bar{E}_{12} = \frac{v_{yx} E_x}{1 - v_{xy} v_{yx}}; \quad \bar{E}_{22} = \frac{E_y}{1 - v_{xy} v_{yx}}
$$

■ The transverse shear is at the origin of distortions as illustrated in Figure 17.1 for the shear stress τ_{yz}.

As a consequence, the displacements due to flexion discussed in Section 12.2.1 can be adapted as shown in Figure 17.2.

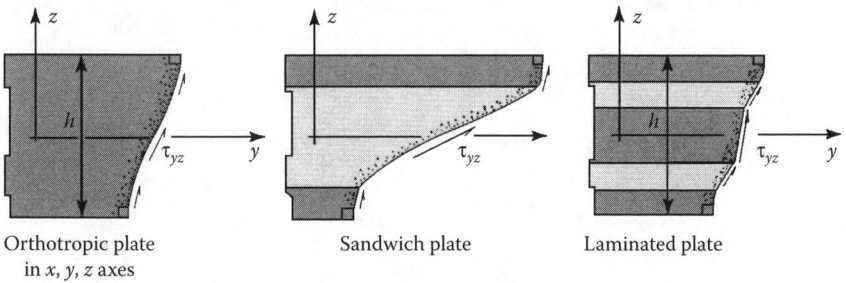

Orthotropic plate in x, y, z axes   Sandwich plate   Laminated plate

Figure 17.1   Distortion of Section due to Transverse Shear τ_{yz}

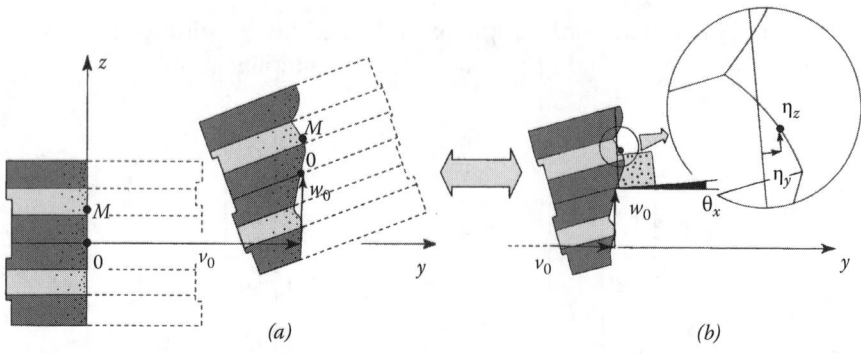

Figure 17.2 Flexural Displacements

In this figure, (a) represents a section before and after bending and (b) shows the evolution of the section as a rigid displacement (parameters v_0, w_0, and θ_x) to which are associated increments η_y and η_z in the plane y,z. Note: due to the existence of midplane symmetry, the antisymmetric manner with respect to z, these increments are small, but they cannot be neglected *a priori*. (At this stage, we do not have a definition for the equivalent rotation, noted as θ_x in b).

This justifies the interest in the definition of the displacement field relating these increments. A supplementary interest rests in the possibility, during the study, to look after the necessary approximations more closely to lead to a useful technical formulation.[2]

17.2 DISPLACEMENT FIELD

The elastic displacement at each point of the laminate has the components $u(x, y, z)$, $v(x, y, z)$ and $w(x, y, z)$. With simplified description of Paragraph 12.2.1, one sees in Figure 17.2 (b), average translations denoted as v_0 and w_0, and a rotation of the section denoted as θ_x, to which one superimposes the supplementary displacements η_y and η_z. We will define these averages in integral forms as follows:

■ **Translation along *x* direction:** By definition, this is u_0 such that:

$$u_0 = \frac{1}{b} \int_{-b/2}^{b/2} u(x, y, z) dz$$

■ **Rotation about the *y* axis:** By definition, this is $\theta_y(x,y)$ such that[3]:

$$\theta_y = \int_{-b/2}^{b/2} \left(\frac{\bar{E}_{11}}{\bar{EI}_{11}} + \frac{\bar{E}_{12}}{\bar{EI}_{12}} \right) u(x, y, z) \times z dz$$

---

[2] Approximations that do not appear neatly in the specialized literature.

[3] Such a definition of the "average rotation" θ_y will be fundamental in the following to ensure the energetic coherence of the formulation for the transverse shears (see Section 17.6.6).

where one has reused the notations of Section 12.1.6 for the terms $\frac{1}{EI_{ij}}$.[4] The longitudinal displacement $u(x, y, z)$ then takes the form:

$$u(x, y, z) = u_0(x, y) + z\theta_y(x, y) + \eta_x(x, y, z)$$

with:

$$\int_{-b/2}^{b/2} \left(\frac{\bar{E}_{11}}{\overline{EI}_{11}} + \frac{\bar{E}_{12}}{\overline{EI}_{12}} \right) \eta_x \, z \, dz = 0$$

In effect, one can obtain starting from this expression:

$$\int_{-b/2}^{b/2} u \, dz = b \times u_o + \theta_y \int_{-b/2}^{b/2} z \, dz + \int_{-b/2}^{b/2} \eta_x \, dz$$

the integrals disappearing due to antisymmetry in z:[5]

$$\int_{-b/2}^{b/2} \left(\frac{\bar{E}_{11}}{\overline{EI}_{11}} + \frac{\bar{E}_{12}}{\overline{EI}_{12}} \right) u \, z \, dz = u_o \int_{-b/2}^{b/2} \left(\frac{\bar{E}_{11}}{\overline{EI}_{11}} + \frac{\bar{E}_{12}}{\overline{EI}_{12}} \right) z \, dz \ldots$$

$$\ldots + \theta_y + \int_{-b/2}^{b/2} \left(\frac{\bar{E}_{11}}{\overline{EI}_{11}} + \frac{\bar{E}_{12}}{\overline{EI}_{12}} \right) \eta_x \, z \, dz$$

In the second member, the first integral disappears due to midplane symmetry. In addition, taking into account the definition of θ_y written above, the second integral is also nil.

■ **Translation along the y direction:** This is $v_0(x, y)$ such that:

$$v_0 = \frac{1}{b} \int_{-b/2}^{b/2} v(x, y, z) \, dz$$

■ **Rotation about the x axis:** This is θ_x such that:

$$\theta_x = -\int_{-b/2}^{b/2} \left(\frac{\bar{E}_{22}}{\overline{EI}_{22}} + \frac{\bar{E}_{12}}{\overline{EI}_{12}} \right) v(x, y, z) \times z \, dz$$

The longitudinal displacement $v(x, y, z)$ then takes the form:

$$v(x, y) = v_0(x, y) - z\theta_x(x, y) + \eta_y(x, y, z)$$

---

[4] Recall that (Section 12.1.6):

$$\left[\frac{1}{\overline{EI}} \right] = [C]^{-1}, \quad \text{where} \quad C_{ij} = \sum_{k=1^{st} \text{ply}}^{n^{th} \text{ply}} \bar{E}_{ij}^k \left(\frac{z_k^3 - z_{k-1}^3}{3} \right)$$

[5] The coefficient of θ_y is 1 because one can note that:

$$\int_{-b/2}^{b/2} \left(\frac{\bar{E}_{11}}{\overline{EI}_{11}} + \frac{\bar{E}_{12}}{\overline{EI}_{12}} \right) z^2 \, dz = \frac{C_{11}}{\overline{EI}_{11}} + \frac{C_{12}}{\overline{EI}_{12}} = \frac{C_{11}C_{22}}{C_{11}C_{22} - C_{12}^2} - \frac{C_{12}^2}{C_{11}C_{22} - C_{12}^2} = 1$$

with:

$$\int_{-b/2}^{(b/2)} \left(\frac{\bar{E}_{22}}{\overline{EI}_{22}} + \frac{\bar{E}_{12}}{\overline{EI}_{12}} \right) \eta_y \, z \, dz = 0$$

■ **Translation along the z direction:** This is $w_0(x,y)$ such that:

$$w_0(x,y) = \frac{1}{b} \int_{-b/2}^{(b/2)} w(x,y,z) dz$$

The vertical displacement takes the form:

$$w(x,y,z) = w_0(x,y) + \eta_z(x,y,z)$$

In summary, one obtains for the elastic displacement field:

$$\boxed{\begin{aligned} u &= u_0 + z\theta_y + \eta_x(x,y,z) \\ v &= v_0 - z\theta_x + \eta_y(x,y,z) \\ w &= w_0 \qquad + \eta_z(x,y,z) \end{aligned}} \tag{17.3}$$

η_x, η_y, η_z antisymmetric in z. $\qquad\qquad$ (17.4)

$$\int_{-b/2}^{b/2} \left(\frac{\bar{E}_{11}}{\overline{EI}_{11}} + \frac{\bar{E}_{12}}{\overline{EI}_{12}} \right) \eta_x \, z \, dz = \int_{-b/2}^{b/2} \left(\frac{\bar{E}_{22}}{\overline{EI}_{22}} + \frac{\bar{E}_{12}}{\overline{EI}_{12}} \right) \eta_y \, z dz = 0$$

$$\tag{17.5}$$

17.3 STRAINS

One deduces from the previous displacements the strains:

$$\begin{aligned} \varepsilon_x &= \varepsilon_{0x} + z\frac{\partial\theta_y}{\partial x} + \frac{\partial\eta_x}{\partial x} \\[2mm] \varepsilon_y &= \varepsilon_{0y} - z\frac{\partial\theta_x}{\partial y} + \frac{\partial\eta_y}{\partial y} \\[2mm] \gamma_{xy} &= \gamma_{0xy} + z\left(\frac{\partial\theta_y}{\partial y} - \frac{\partial\theta_x}{\partial x} \right) + \frac{\partial\eta_x}{\partial y} + \frac{\partial\eta_y}{\partial x} \\[2mm] \gamma_{xz} &= \frac{\partial w_0}{\partial x} + \theta_y + \frac{\partial\eta_x}{\partial z} + \frac{\partial\eta_z}{\partial x} \\[2mm] \gamma_{yz} &= \frac{\partial w_0}{\partial y} - \theta_x + \frac{\partial\eta_y}{\partial z} + \frac{\partial\eta_z}{\partial y} \end{aligned} \tag{17.6}$$

17.4 CONSTITUTIVE RELATIONS

17.4.1 Membrane Equations

Recall the method that was already used in Section 12.1.1.

■ stress resultant $N_x = \int_{-b/2}^{b/2} \sigma_x dx$: from [17.2] and [17.6][6]:

$$N_x = \int_{-b/2}^{b/2} \bar{E}_{11}\left(\varepsilon_{0x} + z\frac{\partial\theta_y}{\partial x} + \frac{\partial\eta_x}{\partial x}\right)dz + \int_{-b/2}^{b/2} \bar{E}_{12}\left(\varepsilon_{0y} - z\frac{\partial\theta_x}{\partial y} + \frac{\partial\eta_y}{\partial y}\right)dz$$

$$N_x = A_{11}\varepsilon_{0x} + A_{12}\varepsilon_{0y} + \frac{\partial}{\partial x}\int_{-b/2}^{b/2} \bar{E}_{11}\eta_x dz + \frac{\partial}{\partial y}\int_{-b/2}^{b/2} \bar{E}_{12}\eta_y dz$$

■ stress resultant $N_y = \int_{-b/2}^{b/2} \sigma_y \, dz$:

$$N_y = A_{21}\varepsilon_{0x} + A_{22}\varepsilon_{0y}$$

■ stress resultant $T_{xy} = \int_{-b/2}^{b/2} \tau_{xy} \, dz$:

$$T_{xy} = \int_{-b/2}^{b/2} \bar{E}_{33}\left(\gamma_{oxy} + z\left(\frac{\partial\theta_y}{\partial y} - \frac{\partial\theta_x}{\partial x}\right) + \frac{\partial\eta_x}{\partial y} + \frac{\partial\eta_y}{\partial x}\right)dz$$

$$T_{xy} = A_{33}\gamma_{oxy}$$

In summary, one finds again the relations already established in Chapter 12 (Equations 12.5) as:

$$\begin{Bmatrix} N_x \\ N_y \\ T_{xy} \end{Bmatrix} = \begin{bmatrix} A_{11} & A_{12} & 0 \\ A_{21} & A_{22} & 0 \\ 0 & 0 & A_{33} \end{bmatrix} \begin{Bmatrix} \varepsilon_{ox} \\ \varepsilon_{oy} \\ \gamma_{oxy} \end{Bmatrix}$$

or, in inverse form, by using the notations in Equation 12.9:

$$\begin{Bmatrix} \varepsilon_{ox} \\ \varepsilon_{oy} \\ \gamma_{oxy} \end{Bmatrix} = h[A]^{-1} \times \frac{1}{h}\begin{Bmatrix} N_x \\ N_y \\ T_{xy} \end{Bmatrix} = \frac{1}{h}\begin{bmatrix} 1/\bar{E}_x & -\bar{v}_{yx}/\bar{E}_y & 0 \\ -\bar{v}_{xy}/\bar{E}_x & 1/\bar{E}_y & 0 \\ 0 & 0 & 1/\bar{G}_{xy} \end{bmatrix}\begin{Bmatrix} N_x \\ N_y \\ T_{xy} \end{Bmatrix} \quad (17.7)$$

17.4.2 Bending Behavior

One has again the already known moment resultants (see Section 12.2.1).

■ Moment resultant $M_y = \int_{-b/2}^{b/2} \sigma_x z dz$:
  with [17.2] and [17.5]:

$$M_y = \int_{-b/2}^{b/2} \bar{E}_{11}\left(z\varepsilon_{ox} + z^2\frac{\partial\theta_y}{\partial x} + z\frac{\partial\eta_x}{\partial x}\right)dz \cdots$$

$$\cdots + \int_{-b/2}^{b/2} \bar{E}_{12}\left(z\varepsilon_{oy} - z^2\frac{\partial\theta_x}{\partial y} + z\frac{\partial\eta_y}{\partial y}\right)dz$$

$$M_y = C_{11}\frac{\partial\theta_y}{\partial x} + C_{12} \times -\frac{\partial\theta_x}{\partial y} + \frac{\partial}{\partial x}\int_{-b/2}^{b/2} \bar{E}_{11}\eta_x z dz + \frac{\partial}{\partial y}\int_{-b/2}^{b/2} \bar{E}_{12}\eta_y z dz$$

---

[6] The simplifications are due to the antisymmetry of the integrated functions (midplane symmetry).

In the last two terms there appear the nonzero integrals of even functions. If one neglects the contribution of the rates of variation along the x and y direction, respectively, of these terms, the previous equation is reduced to[7]

$$M_y = C_{11} \frac{\partial \theta_y}{\partial x} + C_{12} \times -\frac{\partial \theta_x}{\partial y}$$

■ Moment resultant $M_x = -\int_{-b/2}^{b/2} \sigma_y z \, dz$:

$$-M_x = \int_{-b/2}^{b/2} \bar{E}_{12} \left(z\varepsilon_{ox} + z^2 \frac{\partial \theta_y}{\partial x} + z \frac{\partial \eta_x}{\partial x} \right) dz \dots$$

$$\dots + \int_{-b/2}^{b/2} \bar{E}_{22} \left(z\varepsilon_{oy} - z^2 \frac{\partial \theta_x}{\partial y} + z \frac{\partial \eta_y}{\partial y} \right) dz$$

which is reduced to:

$$-M_x = C_{12} \frac{\partial \theta_y}{\partial x} + C_{22} \times -\frac{\partial \theta_x}{\partial y} + \frac{\partial}{\partial x} \int_{-b/2}^{b/2} \bar{E}_{12} \eta_x z \, dz + \frac{\partial}{\partial y} \int_{-b/2}^{b/2} \bar{E}_{22} \eta_y z \, dz$$

and in neglecting the contribution of the last two terms[7]:

$$-M_x = C_{12} \frac{\partial \theta_y}{\partial x} + C_{22} \times -\frac{\partial \theta_x}{\partial y}$$

■ Moment resultant $M_{xy} = -\int_{-b/2}^{b/2} \tau_{xy} z \, dz$

$$-M_{xy} = \int_{-b/2}^{b/2} \bar{E}_{33} \left(z\gamma_{oxy} + z^2 \left(\frac{\partial \theta_y}{\partial y} - \frac{\partial \theta_x}{\partial x} \right) + z \frac{\partial \eta_x}{\partial y} + z \frac{\partial \eta_y}{\partial x} \right) dz$$

which is reduced to:

$$-M_{xy} = C_{33} \left(\frac{\partial \theta_y}{\partial y} - \frac{\partial \theta_x}{\partial x} \right) + \frac{\partial}{\partial y} \int_{-b/2}^{b/2} \bar{E}_{33} \eta_x z \, dz + \frac{\partial}{\partial x} \int_{-b/2}^{b/2} \bar{E}_{33} \eta_y z \, dz$$

and in neglecting the contribution of the variations of the differences η_x and η_y[7]:

$$-M_{xy} = C_{33} \left(\frac{\partial \theta_y}{\partial y} - \frac{\partial \theta_x}{\partial x} \right)$$

---

[7] The existence of such approximation does not appear if one neglects *a priori* the increments η_x, η_y, η_z in Equation 17.3.

In summary, one finds again a form analogous to Equation 12.16 (with $C_{13} = C_{23} = 0$ due to the orientation of the plies [see Hypotheses in Section 17.1.3]):

$$\left\{ \begin{array}{c} M_y \\ -M_x \\ -M_{xy} \end{array} \right\} = \left[\begin{array}{ccc} C_{11} & C_{12} & 0 \\ C_{21} & C_{22} & 0 \\ 0 & 0 & C_{33} \end{array} \right] \left\{ \begin{array}{c} \dfrac{\partial \theta_y}{\partial x} \\ -\dfrac{\partial \theta_x}{\partial y} \\ \left(\dfrac{\partial \theta_y}{\partial y} - \dfrac{\partial \theta_x}{\partial x} \right) \end{array} \right\} \tag{17.8}$$

Or in the inverse form, by reusing the notations of Section 12.1.6.

$$\left\{ \begin{array}{c} \dfrac{\partial \theta_y}{\partial x} \\ -\dfrac{\partial \theta_x}{\partial y} \\ \left(\dfrac{\partial \theta_y}{\partial y} - \dfrac{\partial \theta_x}{\partial x} \right) \end{array} \right\} = \left[\begin{array}{ccc} \dfrac{1}{\overline{EI}_{11}} & \dfrac{1}{\overline{EI}_{12}} & 0 \\ \dfrac{1}{\overline{EI}_{21}} & \dfrac{1}{\overline{EI}_{22}} & 0 \\ 0 & 0 & \dfrac{1}{C_{33}} \end{array} \right] \left\{ \begin{array}{c} M_y \\ -M_x \\ -M_{xy} \end{array} \right\} \tag{17.9}$$

17.4.3 Transverse Shear Equation

We define here new stress resultants starting from the transverse shear stresses, which are denoted as transverse shear stress resultants:

- Shear stress resultant $Q_x = \int_{-b/2}^{b/2} \tau_{xz} \, dz$
  Using Equations 17.2 and 17.6:

$$Q_x = \int_{-b/2}^{b/2} G_{xz} \left(\frac{\partial w_0}{\partial x} + \theta_y + \frac{\partial \eta_x}{\partial z} + \frac{\partial \eta_z}{\partial x} \right) dz$$

in setting

$$\langle b G_{xz} \rangle = \int_{-b/2}^{b/2} G_{xz} \, dz$$

yields

$$Q_x = \langle b G_{xz} \rangle \left(\frac{\partial w_0}{\partial x} + \theta_y \right) + \int_{-b/2}^{b/2} G_{xz} \frac{d \eta_x}{dz} \, dz \tag{17.10}$$

where one can note the presence of the integral of an even function:

- Shear stress resultant $Q_y = \int_{-b/2}^{b/2} \tau_{yz} \, dz$

$$Q_y = \int_{-b/2}^{b/2} G_{yz} \left(\frac{\partial w_0}{\partial y} - \theta_x + \frac{\partial \eta_y}{\partial z} + \frac{\partial \eta_z}{\partial y} \right) dz$$

in setting

$$\langle bG_{yz}\rangle = \int_{-b/2}^{b/2} G_{yz}\, dz$$

yields

$$Q_y = \langle bG_{yz}\rangle\left(\frac{\partial w_0}{\partial y} - \theta_x\right) + \int_{-b/2}^{b/2} G_{yz}\frac{\partial \eta_y}{\partial z}\, dz \qquad (17.11)$$

17.5 EQUILIBRIUM EQUATIONS

These are the same for the plates in general, no matter what are their compositions and are, therefore, classical.

One recalls here only the equilibrium equations for bending.

17.5.1 Transverse Equilibrium

■ local equilibrium relation $\frac{\partial \tau_{zx}}{\partial x} + \frac{\partial \tau_{zy}}{\partial y} + \frac{\partial \sigma_z}{\partial z} + f_z = 0$

In integrating across the thickness, one reveals the transverse shear stresses Q_x and Q_y:

$$\frac{\partial Q_x}{\partial x} + \frac{\partial Q_y}{\partial y} + [\sigma_z]_{-b/2}^{b/2} + \int_{-b/2}^{b/2} f_z\, dz = 0$$

Then denoted by p_z, the transverse stress density:

$$\frac{\partial Q_x}{\partial x} + \frac{\partial Q_y}{\partial y} + p_z = 0$$

17.5.2 Equilibrium in Bending

■ local equilibrium relation $\frac{\partial \sigma_x}{\partial x} + \frac{\partial \tau_{xy}}{\partial y} + \frac{\partial \tau_{xz}}{\partial z} + f_x = 0$

After multiplication with z, the integration over the thickness leads to

$$\frac{\partial M_y}{\partial x} - \frac{\partial M_{xy}}{\partial y} + \int_{-b/2}^{b/2}\left[\frac{\partial}{\partial z}(z\tau_{xz}) - \tau_{xz}\right]dz + \int_{-b/2}^{b/2} zf_x\, dz = 0$$

$$\frac{\partial M_y}{\partial x} - \frac{\partial M_{xy}}{\partial y} - Q_x + [z\tau_{xz}]_{-b/2}^{b/2} + \int_{-b/2}^{b/2} zf_x\, dz = 0$$

In neglecting the moment density:

$$\frac{\partial M_y}{\partial x} - \frac{\partial M_{xy}}{\partial y} - Q_x = 0 \qquad (17.12)$$

■ local equilibrium relation $\frac{\partial \tau_{yx}}{\partial x} + \frac{\partial \sigma_y}{\partial y} + \frac{\partial \tau_{yz}}{\partial z} + f_y = 0$

An analogous calculation leads to

$$\frac{\partial M_{xy}}{\partial x} + \frac{\partial M_x}{\partial y} + Q_y = 0 \qquad (17.13)$$

17.6 TECHNICAL FORMULATION FOR BENDING

■ One can note in the preceding that midplane symmetry always leads to the decoupling of the membrane behavior from the bending behavior. As a consequence, in what follows, one will consider uniquely the stresses due to bending (one will make $N_x = N_y = T_{xy} = 0$).

■ In addition to the hypotheses in Section 17.1.3, one will neglect, for the calculation of the stresses, the variations of increments $\{\eta_x, \eta_y, \eta_z\}$ as functions of x and y.[8]

17.6.1 Plane Stresses Due to Bending

One can write successively for a ply number k:

■ $\sigma_x = \bar{E}_{11}^k \varepsilon_x + \bar{E}_{12}^k \varepsilon_y$

Then with [17.6]:

$$\sigma_x = \bar{E}_{11}^k \left(\varepsilon_{ox} + z\frac{\partial \theta_y}{\partial x} + \frac{\partial \eta_x'}{\partial x} \right) + \bar{E}_{12}^k \left(\varepsilon_{oy} - z\frac{\partial \theta_x}{\partial y} + \frac{\partial \eta_y'}{\partial y} \right)$$

and with [17.7] and [17.9]:

$$\sigma_x = \bar{E}_{11}^k \left[\frac{N_x'}{hE_x} - \frac{v_{yx}}{hE_y}N_y' + z\left(\frac{M_y}{\overline{EI}_{11}} - \frac{M_x}{\overline{EI}_{12}} \right) \right] \dots$$

$$\dots + \bar{E}_{12}^k \left[-\frac{v_{xy}}{hE_x}N_x' + \frac{N_y'}{hE_y} + z\left(\frac{M_y}{\overline{EI}_{12}} - \frac{M_x}{\overline{EI}_{22}} \right) \right]$$

$$\sigma_x = z\left(\frac{\bar{E}_{11}^k}{\overline{EI}_{11}} + \frac{\bar{E}_{12}^k}{\overline{EI}_{12}} \right)M_y + z\left(\frac{\bar{E}_{11}^k}{\overline{EI}_{12}} + \frac{\bar{E}_{12}^k}{\overline{EI}_{22}} \right) \times -M_x \tag{17.14}$$

■ $\sigma_y = \bar{E}_{12}^k \varepsilon_x + \bar{E}_{22}^k \varepsilon_y$

An analogous calculation leads to

$$\sigma_y = z\left(\frac{\bar{E}_{12}^k}{\overline{EI}_{11}} + \frac{\bar{E}_{22}^k}{\overline{EI}_{12}} \right)M_y + z\left(\frac{\bar{E}_{12}^k}{\overline{EI}_{12}} + \frac{\bar{E}_{22}^k}{\overline{EI}_{22}} \right) \times -M_x \tag{17.15}$$

■ $\tau_{xy} = \bar{E}_{33}^k \gamma_{xy} = G_{xy}^k \gamma_{xy}$

Then with Equation 17.6:

$$\tau_{xy} = G_{xy}^k \left(\gamma_{oxy} + z\left(\frac{\partial \theta_y}{\partial y} - \frac{\partial \theta_x}{\partial x} \right) + \frac{\partial \eta_x'}{\partial y} + \frac{\partial \eta_y'}{\partial x} \right)$$

---

[8] This simplification constitutes here the extension to plates of the generalized Navier-Bernoulli principle for beams (see Section 15.1.5).

and with Equations 17.7 and 17.9 and $T_{xy} = 0$:

$$\tau_{xy} = -z \frac{G_{xy}^k}{C_{33}} M_{xy} \tag{17.16}$$

17.6.2 Transverse Shear Stresses in Bending

■ $\tau_{xz} = \bar{E}_{44}^k \gamma_{xz} = G_{xz}^k \gamma_{xz}$ from [17.2]

and with Equation 17.6 and neglecting the variation $\partial \eta_z / \partial x$:

$$\tau_{xz} = G_{xz}^k \left(\frac{\partial w_0}{\partial x} + \theta_y \right) + G_{xz}^k \frac{\partial \eta_x}{\partial z} \tag{17.17}$$

■ $\tau_{yz} = \bar{E}_{55}^k \gamma_{yz} = G_{yz}^k \gamma_{yz}$

which leads in an analogous manner to

$$\tau_{yz} = G_{yz}^k \left(\frac{\partial w_0}{\partial y} - \theta_x \right) + G_{yz}^k \frac{\partial \eta_y}{\partial z} \tag{17.18}$$

Knowledge of the transverse shears, then, requires the previous calculation of the warping increments η_x and η_y.

17.6.3 Characterization of the Bending, Warping Increments η_x and η_y

■ **Warping $\eta_x (x, y, z)$:**
Starting from the first equation of local equilibrium:

$$\frac{\partial \tau_{xz}}{\partial z} = -\frac{\partial \sigma_x}{\partial x} - \frac{\partial \tau_{xy}}{\partial y}$$

Then with Equations 17.14, 17.16, and 17.17:

$$G_{xz}^k \frac{\partial^2 \eta_x}{\partial z^2} = -z \left(\frac{\bar{E}_{11}^k}{\overline{EI}_{11}} + \frac{\bar{E}_{12}^k}{\overline{EI}_{12}} \right) \frac{\partial M_y}{\partial x} + z \left(\frac{\bar{E}_{11}^k}{\overline{EI}_{12}} + \frac{\bar{E}_{12}^k}{\overline{EI}_{22}} \right) \frac{\partial M_x}{\partial x} + z \frac{G_{xy}^k}{C_{33}} \frac{\partial M_{xy}}{\partial y}$$

Taking into account the equilibrium Equation 17.12, one can rewrite

$$G_{xz}^k \frac{\partial^2 \eta_x}{\partial z^2} = -z \left(\frac{\bar{E}_{11}^k}{\overline{EI}_{11}} + \frac{\bar{E}_{12}^k}{\overline{EI}_{12}} \right) Q_x + z \left(\frac{\bar{E}_{11}^k}{\overline{EI}_{12}} + \frac{\bar{E}_{12}^k}{\overline{EI}_{22}} \right) \frac{\partial M_x}{\partial x} \cdots$$

$$\cdots + z \left(\frac{G_{xy}^k}{C_{33}} - \frac{\bar{E}_{11}^k}{\overline{EI}_{11}} - \frac{\bar{E}_{12}^k}{\overline{EI}_{12}} \right) \frac{\partial M_{xy}}{\partial y} \tag{17.19}$$

■ **Warping $\eta_y (x, y, z)$:**
In an analogous manner, starting from the second equation of the local equilibrium:

$$\frac{\partial \tau_{yz}}{\partial z} = -\frac{\partial \sigma_y}{\partial y} - \frac{\partial \tau_{yx}}{\partial x}$$

Then with Equations 17.15, 17.16, and 17.18:

$$G_{yz}^k \frac{\partial^2 \eta_y}{\partial z^2} = -z\left(\frac{\bar{E}_{12}^k}{\overline{EI}_{11}} + \frac{\bar{E}_{22}^k}{\overline{EI}_{12}}\right)\frac{\partial M_y}{\partial y} + z\left(\frac{\bar{E}_{12}^k}{\overline{EI}_{12}} + \frac{\bar{E}_{22}^k}{\overline{EI}_{22}}\right)\frac{\partial M_x}{\partial y} \cdots$$

$$\cdots + z\frac{G_{xy}^k}{C_{33}}\frac{\partial M_{xy}}{\partial x}$$

Taking into account the equilibrium Equation 17.13, one can rewrite

$$G_{yz}^k \frac{\partial^2 \eta_x}{\partial z^2} = -z\left(\frac{\bar{E}_{12}^k}{\overline{EI}_{11}} + \frac{\bar{E}_{22}^k}{\overline{EI}_{12}}\right)\frac{\partial M_y}{\partial y} - z\left(\frac{\bar{E}_{12}^k}{\overline{EI}_{12}} + \frac{\bar{E}_{22}^k}{\overline{EI}_{22}}\right)Q_y \cdots$$

$$\cdots + z\left(\frac{G_{xy}^k}{C_{33}} - \frac{\bar{E}_{12}^k}{\overline{EI}_{12}} - \frac{\bar{E}_{22}^k}{\overline{EI}_{22}}\right)\frac{\partial M_{xy}}{\partial x} \tag{17.20}$$

17.6.3.1 Particular Cases

Equations 17.19 and 17.20 simplify in the following particular cases:

- **Homogeneous orthotropic plate:**
  Then from relations 17.2, 17.8, and 17.9:

$$\bar{E}_{11}^k = \bar{E}_{11}; \ \bar{E}_{12}^k = \bar{E}_{12}; \ \bar{E}_{22}^k = \bar{E}_{22}$$

$$\frac{1}{\overline{EI}_{11}} = \frac{C_{22}}{C_{11}C_{22} - C_{12}^2} = \frac{\bar{E}_{22}}{\bar{E}_{11}\bar{E}_{22} - \bar{E}_{12}^2} \times \frac{12}{h^3};$$

$$\frac{1}{\overline{EI}_{22}} = \frac{\bar{E}_{11}}{\bar{E}_{11}\bar{E}_{22} - \bar{E}_{12}^2} \times \frac{12}{h^3}$$

$$\frac{1}{\overline{EI}_{12}} = -\frac{C_{12}}{C_{11}C_{22} - C_{12}^2} = \frac{-\bar{E}_{12}}{\bar{E}_{11}\bar{E}_{22} - \bar{E}_{12}^2} \times \frac{12}{h^3}; \quad \frac{1}{C_{33}} = \frac{1}{G_{xy}} \times \frac{12}{h^3}$$

Then Equation 17.19 and Equation 17.20 reduce to

$$\boxed{\begin{aligned} G_{xz}\frac{\partial^2 \eta_x}{\partial z^2} &= -z \times \frac{12}{h^3} \times Q_x \\ G_{yz}\frac{\partial^2 \eta_y}{\partial z^2} &= -z \times \frac{12}{h^3} \times Q_y \end{aligned}} \tag{17.21}$$

- **Cylindrical bending about x or y axis** of a multilayered plate characterized by identical Poisson coefficient in the x,y plane of the plate as:

$$\forall k: \ v_{xy}^k = v_{xy}; \ v_{yx}^k = v_{yx}$$

Then for any two plies k and m, one has (see Equations 17.2)[9]:

$$\frac{\bar{E}_{11}^{k}}{\bar{E}_{11}^{m}} = \frac{\bar{E}_{12}^{k}}{\bar{E}_{12}^{m}} = \frac{\bar{E}_{22}^{k}}{\bar{E}_{22}^{m}} = \alpha_{km}$$

then:

$$C_{ij} = \int_{-b/2}^{b/2} \bar{E}_{ij}^{k} z^{2} dz = \sum_{k=1^{st}\,ply}^{n^{th}\,ply} \left\{ \bar{E}_{ij}^{k} \int_{z_{k-1}}^{z_{k}} z^{2} dz \right\}$$

$$= \bar{E}_{ij}^{1} \int_{z_{0}}^{z_{1}} z^{2} dz + \bar{E}_{ij}^{2} \int_{z_{1}}^{z_{2}} z^{2} dz \cdots + \bar{E}_{ij}^{n} \int_{z_{n-1}}^{z_{n}} z^{2} dz$$

$$C_{ij} = \bar{E}_{ij}^{1} \left\{ \int_{z_{0}}^{z_{1}} z^{2} dz + \alpha_{12} \int_{z_{1}}^{z_{2}} z^{2} dz \cdots + \alpha_{n-1,\,n} \int_{z_{n-1}}^{z_{n}} z^{2} dz \right\} = \bar{E}_{ij}^{1} \times \frac{\alpha b^{3}}{12}$$

where α is a nondimensional coefficient. One then has

$$\frac{1}{\overline{EI}_{11}} = \frac{C_{22}}{C_{11}C_{22} - C_{12}^{2}} = \frac{\bar{E}_{22}^{1}}{\bar{E}_{11}^{1}\bar{E}_{22}^{1} - \left(\bar{E}_{12}^{1}\right)^{2}} \times \frac{12}{\alpha b^{3}} \;;$$

$$\frac{1}{\overline{EI}_{22}} = \frac{\bar{E}_{11}^{1}}{\bar{E}_{11}^{1}\bar{E}_{22}^{1} - \left(\bar{E}_{12}^{1}\right)^{2}} \times \frac{12}{\alpha b^{3}}$$

$$\frac{1}{\overline{EI}_{12}} = \frac{\bar{E}_{12}^{1}}{\bar{E}_{11}^{1}\bar{E}_{22}^{1} - \left(\bar{E}_{12}^{1}\right)^{2}} \times \frac{12}{\alpha b^{3}}$$

In [17.19], we have the simplification:

$$\frac{\bar{E}_{11}^{k}}{\overline{EI}_{12}} + \frac{\bar{E}_{12}^{k}}{\overline{EI}_{22}} = \frac{-\bar{E}_{11}^{k}\bar{E}_{12}^{1} + \bar{E}_{12}^{k}\bar{E}_{11}^{1}}{\bar{E}_{11}^{1}\bar{E}_{22}^{1} - \left(\bar{E}_{12}^{1}\right)^{2}} \times \frac{12}{\alpha b^{3}}$$

$$= \frac{\alpha_{k1}\left(-\bar{E}_{11}^{1}\bar{E}_{12}^{1} + \bar{E}_{12}^{1}\bar{E}_{11}^{1}\right)}{\bar{E}_{11}^{1}\bar{E}_{22}^{1} - \left(\bar{E}_{12}^{1}\right)^{2}} \times \frac{12}{\alpha b^{3}} = 0$$

as well as an analogous simplification in Equation 17.20:

$$\frac{\bar{E}_{12}^{k}}{\overline{EI}_{11}} + \frac{\bar{E}_{22}^{k}}{\overline{EI}_{12}} = 0$$

Equations 17.19 and 17.20 simplify as follows[10]:

---

[9] Recall the relation $v_{yx} E_x = v_{xy} E_y$ (see Equation 9.4).

[10] In the first case in Equation 17.22, $M_{xy} = Q_y = 0$ and Equation 17.20 disappears. In the second case, $M_{xy} = Q_y = 0$ and Equation 17.19 disappears.

1. cylindrical bending about y axis

$$G_{xz}^k \frac{\partial^2 \eta_x}{\partial z^2} = -z \left(\frac{\bar{E}_{11}^k}{\overline{EI}_{11}} + \frac{\bar{E}_{12}^k}{\overline{EI}_{12}} \right) Q_x$$

2. cylindrical bending about x axis   (17.22)

$$G_{yz}^k \frac{\partial^2 \eta_y}{\partial z^2} = -z \left(\frac{\bar{E}_{22}^k}{\overline{EI}_{22}} + \frac{\bar{E}_{12}^k}{\overline{EI}_{12}} \right) Q_y$$

■ **The case of a multilayer plate** such that for any two plies k and m one has in the plane of the plate[11]:

$$\frac{\bar{E}_{ij}^k}{\bar{E}_{ij}^m} = \alpha_{km} \qquad \forall i \quad \text{and} \quad j = 1, 2, 3$$

Then Equations 17.19 and 17.20 reduce to

$$G_{xz}^k \frac{\partial^2 \eta_x}{\partial z^2} = -z \left(\frac{\bar{E}_{11}^k}{\overline{EI}_{11}} + \frac{\bar{E}_{12}^k}{\overline{EI}_{12}} \right) Q_x$$

(17.23)

$$G_{yz}^k \frac{\partial^2 \eta_y}{\partial z^2} = -z \left(\frac{\bar{E}_{22}^k}{\overline{EI}_{22}} + \frac{\bar{E}_{12}^k}{\overline{EI}_{12}} \right) Q_y$$

The preceding particular cases constitute a severe restriction among the variety of practical laminations. Nevertheless we will conserve in the following the simplified forms of Equations 17.21, 17.22, and 17.23 because they well show the direct connection between the warpings η_x and η_y and the transverse shear forces Q_x and Q_y, respectively.

17.6.3.2 Consequences

Setting η_x and η_y in the forms:

$$\left. \begin{aligned} \eta_x(x, y, z) &= \frac{Q_x}{\langle bG_{xz} \rangle} \times g(z) \\[2mm] \eta_y(x, y, z) &= \frac{Q_y}{\langle bG_{yz} \rangle} \times p(z) \end{aligned} \right\}$$

(17.24)

The constitutive Equations 17.10 and 17.11 are written as:

■ $Q_x = \langle bG_{xz} \rangle \left(\frac{\partial w_0}{\partial x} + \theta_y \right) + \frac{Q_x}{\langle bG_{xz} \rangle} \int_{-b/2}^{b/2} G_{xz} \frac{dg}{dz} dz$

---

[11] Such a limiting case is rare in practice, because it imposes in particular: $\frac{E_{33}^k}{E_{33}^m} = \frac{G_{xy}^k}{G_{xy}^m} = \alpha_{km}$.

then by setting :

$$k_x = \left(1 - \frac{1}{\langle bG_{xz}\rangle}\int_{-b/2}^{b/2} G_{xz}\frac{dg}{dz}dz\right)$$

$$Q_x = \frac{\langle bG_{xz}\rangle}{k_x}\left(\frac{\partial w_0}{\partial x} + \theta_y\right)$$

(17.25)

■ $Q_y = \langle bG_{yz}\rangle\left(\frac{\partial w_0}{\partial y} - \theta_x\right) + \frac{Q_y}{\langle bG_{yz}\rangle}\int_{-b/2}^{b/2} G_{yz}\frac{dp}{dz}dz$

then by setting :

$$k_y = \left(1 - \frac{1}{\langle bG_{yz}\rangle}\int_{-b/2}^{b/2} G_{yz}\frac{dp}{dz}dz\right)$$

$$Q_y = \frac{\langle bG_{yz}\rangle}{k_y}\left(\frac{\partial w_0}{\partial y} - \theta_x\right)$$

(17.26)

There appear two transverse shear coefficients k_x and k_y which require the knowledge of the functions $g(z)$ and $p(z)$ for their calculations.

17.6.4 Warping Functions

■ **Boundary conditions:** We have assumed that the upper and lower faces of the plate were free of any shear. Then the transverse shear in Equations 17.17 and 17.18 leads to

■ $\left(\frac{\partial w_0}{\partial x} + \theta_y\right) + \frac{Q_x}{\langle bG_{xz}\rangle}\frac{dg}{dz} = 0$ for $z = \pm b/2$

then with Equation 17.25:

$$k_x + \frac{dg}{dz} = 0 \quad \text{for} \quad z = \pm b/2$$

■ $\left(\frac{\partial w_0}{\partial y} - \theta_x\right) + \frac{Q_y}{\langle bG_{yz}\rangle}\frac{dp}{dz} = 0$ for $z = \pm b/2$

then with [17.26]:

$$k_y + \frac{dp}{dz} = 0 \quad \text{for} \quad z = \pm b/2$$

■ **Continuity at the interfaces:** The continuity of the transverse shear at the interfaces between layers results from the assumed perfect bonding between two plies (see Paragraph 15.1.2). One then has at the interface between two consecutive plies k and $k + 1$:

$$\tau_{xz}^k = \tau_{xz}^{k+1}; \ \tau_{yz}^k = \tau_{yz}^{k+1}$$

then with Equations 17.17, 17.18 and Equations 17.25, 17.26:

$$G_{xz}^k\left(k_x + \frac{dg_k}{dz}\right) = G_{xz}^{k+1}\left(k_x + \frac{dg_{k+1}}{dz}\right)$$

$$G_{yz}^k\left(k_y + \frac{dp_k}{dz}\right) = G_{yz}^{k+1}\left(k_y + \frac{dp_{k+1}}{dz}\right)$$

- **Formulation of the warping functions:** Let us substitute to $g(z)$ and $p(z)$ the functions $g_0(z)$ and $p_0(z)$ such that:

$$g_0(z) = g(z) + z \times k_x; \; p_0(z) = p(z) + z \times k_y$$

$g_0(z)$ and $p_0(z)$ are called the **warping functions**. Then the boundary conditions and the interface conditions simplify, and Equations 17.23 allow one to formulate the problems that permit a simple calculation of warping functions $g_0(z)$ and $p_0(z)$. One obtains

$$\begin{cases} \dfrac{d^2 g_0}{dz^2} = -z \times \dfrac{\langle hG_{xz}\rangle}{G_{xz}^k}\left(\dfrac{\overline{E}_{11}^k}{\overline{EI}_{11}} + \dfrac{\overline{E}_{12}^k}{\overline{EI}_{12}}\right) \\[2mm] \dfrac{dg_0}{dz} = 0 \quad \text{for} \quad z = \pm h/2 \\[2mm] G_{xz}^k \dfrac{dg_{0k}}{dz} = G_{xz}^{k+1}\dfrac{dg_{0k+1}}{dz} \quad \text{for} \quad z = z_k \end{cases} \tag{17.27}$$

$$\begin{cases} \dfrac{d^2 p_0}{dz^2} = -z \times \dfrac{\langle hG_{yz}\rangle}{G_{yz}^k}\left(\dfrac{\overline{E}_{22}^k}{\overline{EI}_{22}} + \dfrac{\overline{E}_{12}^k}{\overline{EI}_{12}}\right) \\[2mm] \dfrac{dp_0}{dz} = 0 \quad \text{for} \quad z = \pm h/2 \\[2mm] G_{yz}^k \dfrac{dp_{0k}}{dz} = G_{yz}^{k+1}\dfrac{dp_{0k+1}}{dz} \quad \text{for} \quad z = z_k \end{cases} \tag{17.28}$$

The antisymmetric functions g_0 and p_0 are then defined in a unique manner.

17.6.5 Consequences

- **Form of the transverse shear stresses:** Equations 17.17 and 17.18 then take the simple forms:

$$\tau_{xz} = Q_x \times \frac{G_{xz}^k}{\langle hG_{xz}\rangle}\frac{dg_0}{dz}; \; \tau_{yz} = Q_y \times \frac{G_{yz}^k}{\langle hG_{yz}\rangle}\frac{dp_0}{dz} \tag{17.29}$$

■ **Transverse shear coefficients:** One obtains these coefficients from the Equation 17.5:

■ $\int_{-b/2}^{b/2} \left(\dfrac{\bar{E}_{11}}{\overline{EI}_{11}} + \dfrac{\bar{E}_{12}}{\overline{EI}_{12}} \right) \eta_x z\, dz = 0$

using equation 17.24 and definition of g_0 gives:

$$\int_{-b/2}^{b/2} \left(\frac{\bar{E}_{11}}{\overline{EI}_{11}} + \frac{\bar{E}_{12}}{\overline{EI}_{12}} \right) \times \frac{Q_x}{\langle bG_{xz} \rangle} (g_0 - k_x z) z\, dz = 0$$

noting that:

$$\int_{-b/2}^{b/2} \left(\frac{\bar{E}_{11}}{\overline{EI}_{11}} + \frac{\bar{E}_{12}}{\overline{EI}_{12}} \right) z^2\, dz = \frac{C_{11}}{\overline{EI}_{11}} + \frac{C_{12}}{\overline{EI}_{12}} = \frac{C_{11}C_{12} - C_{12}^2}{C_{11}C_{12} - C_{12}^2} = 1.$$

One obtains:

$$k_x = \int_{-b/2}^{b/2} \left(\frac{\bar{E}_{11}}{\overline{EI}_{11}} + \frac{\bar{E}_{12}}{\overline{EI}_{12}} \right) g_0 z\, dz \qquad (17.30)$$

■ $\int_{-b/2}^{b/2} \left(\dfrac{\bar{E}_{22}}{\overline{EI}_{22}} + \dfrac{\bar{E}_{12}}{\overline{EI}_{12}} \right) \eta_y z\, dz = 0$

using equation 17.24 and definition of p_0 gives:

$$\int_{-b/2}^{b/2} \left(\frac{\bar{E}_{22}}{\overline{EI}_{22}} + \frac{\bar{E}_{12}}{\overline{EI}_{12}} \right) \times \frac{Q_y}{\langle bG_{yz} \rangle} (p_0 - k_y z) z\, dz = 0$$

leading to:

$$k_y = \int_{-b/2}^{b/2} \left(\frac{\bar{E}_{22}}{\overline{EI}_{22}} + \frac{\bar{E}_{12}}{\overline{EI}_{12}} \right) p_0 z\, dz \qquad (17.31)$$

In summary, in the absence of body forces (inertia forces, example), the bending behavior uncoupled from the membrane behavior of a thick laminated plate can be simplified in a few particular cases noted below. The characteristic relations are summarized in the following table.

Bending Behavior (no in-plane stress resultants)

	homogeneous orthotropic plate/orthotropic axes : x, y, z
or	Laminated plate/midplane symmetry/orthotropic axes of plies: x,y,z/same Poisson ratios v_{xy} and v_{yx} for all plies/cylindrical bending about x or y axis.
or	Laminated plate/midplane symmetry/orthotropic axes of plies: x,y,z/elastic constants are proportional from one ply to another
●	**Equilibrium relation:** $\dfrac{\partial Q_x}{\partial x} + \dfrac{\partial Q_y}{\partial y} + p_z = 0; \quad \dfrac{\partial M_y}{\partial x} - \dfrac{\partial M_{xy}}{\partial y} - Q_x = 0; \quad \dfrac{\partial M_{xy}}{\partial x} + \dfrac{\partial M_x}{\partial y} + Q_y = 0$

● **Constitutive relations:**

$$\begin{Bmatrix} M_y \\ -M_x \\ -M_{xy} \\ Q_x \\ Q_y \end{Bmatrix} = \begin{bmatrix} C_{11} & C_{12} & 0 & 0 & 0 \\ C_{21} & C_{22} & 0 & 0 & 0 \\ 0 & 0 & C_{33} & 0 & 0 \\ 0 & 0 & 0 & \dfrac{\langle hG_{xz}\rangle}{k_x} & 0 \\ 0 & 0 & 0 & 0 & \dfrac{\langle hG_{yz}\rangle}{k_y} \end{bmatrix} \begin{Bmatrix} \dfrac{\partial \theta_y}{\partial x} \\ -\dfrac{\partial \theta_x}{\partial y} \\ \dfrac{\partial \theta_y}{\partial y} - \dfrac{\partial \theta_x}{\partial x} \\ \dfrac{\partial w_0}{\partial x} + \theta_y \\ \dfrac{\partial w_0}{\partial y} - \theta_x \end{Bmatrix}$$

with

$$[C]^{-1} = \left[\overline{\overline{\dfrac{1}{EI}}}\right]$$

● **Stresses**
  ■ stresses within the ply :

  σ_x: *cf.* [17.14]; σ_y: *cf.* [17.15]; τ_{xy}: *cf.* [17.16]

  ■ transverse shear stresses

(17.32)

$$\tau_{xz} = Q_x \frac{G_{xz}^k}{\langle hG_{xz}\rangle}\frac{dg_0}{dz}; \qquad \tau_{yz} = Q_y \frac{G_{yz}^k}{\langle hG_{yz}\rangle}\frac{dh_0}{dz}$$

● **Warping functions**
  ■ $g_0(z)$ is the solution of the problem:

$$\begin{cases} \dfrac{d^2 g_0}{dz^2} = -z\dfrac{\langle hG_{xz}\rangle}{G_{xz}^k}\left(\dfrac{\bar{E}_{11}^k}{\overline{\overline{EI}}_{11}} + \dfrac{\bar{E}_{12}^k}{\overline{\overline{EI}}_{12}}\right) \\[1em] \dfrac{dg_0}{dz} = 0 \quad \text{for} \quad z = \pm h/2 \\[1em] G_{xz}^k\dfrac{dg_{0k}}{dz} = G_{xz}^{k+1}\dfrac{dg_{0k+1}}{dz} \quad \text{for} \quad z = z_k \end{cases}$$

  ■ $p_0(z)$ is the solution of the problem:

$$\begin{cases} \dfrac{d^2 p_0}{dz^2} = -z\dfrac{\langle hG_{yz}\rangle}{G_{yz}^k}\left(\dfrac{\bar{E}_{22}^k}{\overline{\overline{EI}}_{22}} + \dfrac{\bar{E}_{12}^k}{\overline{\overline{EI}}_{12}}\right) \\[1em] \dfrac{dp_0}{dz} = 0 \quad \text{for} \quad z = \pm h/2 \\[1em] G_{yz}^k\dfrac{dp_{0k}}{dz} = G_{yz}^{k+1}\dfrac{dp_{0k+1}}{dz} \quad \text{for} \quad z = z_k \end{cases}$$

● **Transverse shear coefficients k_x and k_y:**
  ■ They are given by the formula:

$$k_x = \int_{-h/2}^{h/2}\left(\frac{\bar{E}_{11}}{\overline{\overline{EI}}_{11}} + \frac{\bar{E}_{12}}{\overline{\overline{EI}}_{12}}\right)g_0 z\, dz; \qquad k_y = \int_{-h/2}^{h/2}\left(\frac{\bar{E}_{22}}{\overline{\overline{EI}}_{22}} + \frac{\bar{E}_{12}}{\overline{\overline{EI}}_{12}}\right)p_0 z\, dz$$

17.6.6 Interpretation in Terms of Energy

We will limit ourselves to the surface energy density due to transverse shear stresses as:

$$W_\tau = \frac{1}{2}\int_{-b/2}^{b/2} (\tau_{xz}\gamma_{xz} + \tau_{yz}\gamma_{yz})\,dz = \frac{1}{2}\int_{-b/2}^{b/2} \left(\frac{\tau_{xz}^2}{G_{xz}} + \frac{\tau_{yz}^2}{G_{yz}}\right)dz$$

Substituting Equation 17.29, one obtains

$$W_\tau = \frac{1}{2}\int_{-b/2}^{b/2} Q_x^2 \frac{G_{xz}}{\langle bG_{xz}\rangle^2}\left(\frac{dg_0}{dz}\right)^2 dz + \frac{1}{2}\int_{-b/2}^{b/2} Q_y^2 \frac{G_{yz}}{\langle bG_{yz}\rangle^2}\left(\frac{dp_0}{dz}\right)^2 dz$$

The first integral can be rewritten as:

$$\frac{1}{2}\frac{Q_x^2}{\langle bG_{xz}\rangle^2}\int_{-b/2}^{b/2} G_{xz}\left[\frac{d}{dz}\left(g_0\frac{dg_0}{dz}\right) - g_0\frac{d^2 g_0}{dz^2}\right]dz$$

or, taking into account Equation 17.27:

$$\frac{1}{2}\frac{Q_x^2}{\langle bG_{xz}\rangle^2}\left\{G_{xz}\left[g_0\frac{dg_0}{dz}\right]_{-b/2}^{b/2} + \langle bG_{xz}\rangle\int_{-b/2}^{b/2}\left(\frac{\bar{E}_{11}}{\overline{EI}_{11}} + \frac{\bar{E}_{12}}{\overline{EI}_{12}}\right)g_0 z\,dz\right\}$$

where one recognizes Equation 17.30 of the transverse shear coefficient k_x, the first integral is reduced to

$$\frac{1}{2}k_x\frac{Q_x^2}{\langle bG_{xz}\rangle}$$

Following a similar approach for the second integral and taking into account Equations 17.28 and 17.31 for the transverse shear coefficient k_y, the surface energy due to transverse shear takes the form:

$$W_\tau = \frac{1}{2}k_x\frac{Q_x^2}{\langle bG_{xz}\rangle} + \frac{1}{2}k_y\frac{Q_y^2}{\langle bG_{yz}\rangle}$$

17.7 EXAMPLES

Examples for plates in bending are shown in details in Part Four of this book, in Chapter 18, "Applications." We give here a few useful elements to treat these examples.

17.7.1 Homogeneous Orthotropic Plate

■ **Warping functions:** Equation 17.27 becomes

$$\bar{E}_{11}^k = \bar{E}_{11}; \quad \bar{E}_{12}^k = \bar{E}_{12}; \quad \bar{E}_{22}^k = \bar{E}_{22} \; ; \; G_{xz}^k = G_{xz}$$

$$\frac{d^2 g_0}{dz^2} = -zh\left(\frac{\bar{E}_{11}\bar{E}_{22}}{\bar{E}_{11}\bar{E}_{22} - \bar{E}_{12}^2}\frac{12}{h^3} - \frac{\bar{E}_{12}^2}{\bar{E}_{11}\bar{E}_{22} - \bar{E}_{12}^2}\frac{12}{h^3}\right) = -z \times \frac{12}{h^2}$$

$$\frac{dg_0}{dz} = 0 \text{ for } z = \pm h/2$$

then[12] $\quad \dfrac{dg_0}{dz} = \dfrac{3}{2}\left(1 - 4\dfrac{z^2}{h^2}\right); \quad g_0 = \dfrac{3}{2}z\left(1 - \dfrac{4z^2}{3h^2}\right)$

■ **Transverse shear stresses and shear coefficients:** One deduces from Equation 17.32:

$$\tau_{xz} = \frac{Q_x}{h} \times \frac{3}{2}\left(1 - 4\frac{z^2}{h^2}\right) \tag{17.33}$$

$$k_x = \frac{12}{h^3}\int_{-h/2}^{h/2}\frac{3}{2}\left(1 - \frac{4z^2}{3h^2}\right)z^2 dz$$

$$k_x = \frac{6}{5} \tag{17.34}$$

In an analogous manner starting from Equation 17.28:

$$p_0(z) = g_0(z)$$

then:

$$\tau_{yz} = \frac{Q_y}{h} \times \frac{3}{2}\left(1 - 4\frac{z^2}{h^2}\right) \tag{17.35}$$

$$k_y = \frac{6}{5} \tag{17.36}$$

Remark: In Application 18.3.7 (Chapter 18), one treats the case of a thick homogeneous orthotropic plate in cylindrical bending about the y axis. The plate supports a uniformly distributed load. One can consider there the strong influence of transverse shear in bending. Two characteristics of the plate then apply directly on the deflection:

---

[12] g_0 is, as g, antisymmetric in z (see Equation 17.4).

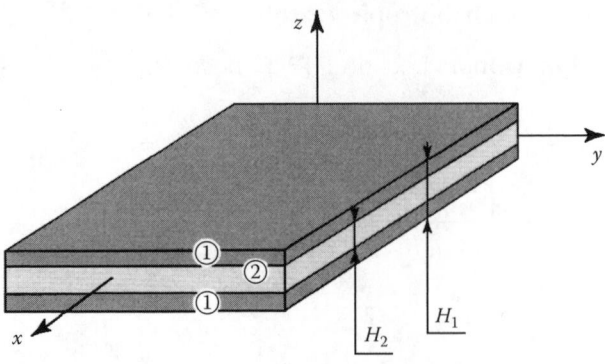

Figure 17.3 Sandwich Plate

- The relative thickness h/a, where a is the length of the bent side of the plate
- The ratio E_x/G_{xz} (for certain combinations of fiber/matrix, this ratio becomes large compared with unity; for example, for unidirectional)

17.7.2 Sandwich Plate

The plate consists of two orthotropic materials:

Material (1) for the skins
Material (2) for the core (see Figure 17.3)

Assuming the proportionality of elastic coefficients for the two materials leads to (see Section 17.6.3):

then

$$C_{ij} = \int_{-b/2}^{b/2} \bar{E}_{ij} z^2 dz = \bar{E}_{ij}^{\textcircled{1}} \int_{-H_1/2}^{-H_2/2} z^2 dz + \bar{E}_{ij}^{\textcircled{2}} \int_{-H_2/2}^{H_2/2} z^2 dz + \bar{E}_{ij}^{\textcircled{1}} \int_{H_2/2}^{H_1/2} z^2 dz$$

$$C_{ij} = \bar{E}_{ij}^{\textcircled{1}} \left(\frac{H_1^3 - H_2^3}{12} \right) + \bar{E}_{ij}^{\textcircled{2}} \frac{H_2^3}{12}$$

$$C_{ij} = \bar{E}_{ij}^{\textcircled{1}} \times \frac{\alpha H_1^3}{12} \text{ with } \frac{\alpha H_1^3}{12} = \frac{H_1^3 - H_2^3}{12} + \alpha_{12} \frac{H_2^3}{12}$$

one deduces from there:

$$\frac{1}{\overline{EI}_{11}} = \frac{C_{11}}{C_{11} C_{22} - C_{12}^2} = \frac{\bar{E}_{11}^{\textcircled{1}}}{\bar{E}_{11}^{\textcircled{1}} \bar{E}_{22}^{\textcircled{1}} - (\bar{E}_{12}^{\textcircled{1}})^2} \times \frac{12}{\alpha H_1^3}$$

$$\frac{1}{\overline{EI}_{12}} = \frac{-C_{12}}{C_{11} C_{22} - C_{12}^2} = \frac{-\bar{E}_{12}^{\textcircled{1}}}{\bar{E}_{11}^{\textcircled{1}} \bar{E}_{22}^{\textcircled{1}} - (\bar{E}_{12}^{\textcircled{1}})^2} \times \frac{12}{\alpha H_1^3}$$

17.7.2.1 Warping Functions

- From the above one can write in Equation 17.27^{13}:

$$\left(\frac{\bar{E}_{11}^{k}}{\overline{EI}_{11}} + \frac{\bar{E}_{12}^{k}}{\overline{EI}_{12}}\right) = \frac{E_{x}^{k}}{E_{x}^{①}} \times \frac{12}{\alpha H_{1}^{3}} = \frac{E_{x}^{k}}{E_{x}^{①}\frac{\left(H_{1}^{3} - H_{2}^{3}\right)}{12} + E_{x}^{②}\frac{H_{2}^{3}}{12}}$$

In addition :

$$\langle bG_{xz}\rangle = G_{xz}^{①}(H_{1} - H_{2}) + G_{xz}^{①}H_{2}$$

Equation 17.27 then can be written as:

$$\begin{cases} \dfrac{d^{2}g_{0}}{dz^{2}} = -z \times \dfrac{E_{x}^{k}}{G_{xz}^{k}} \times 12\dfrac{G_{xz}^{①}(H_{1} - H_{2}) + G_{xz}^{②}H_{2}}{E_{x}^{①}(H_{1}^{3} - H_{2}^{3}) + E_{x}^{②}\,H_{2}^{3}} \\[2mm] \dfrac{dg_{0}}{dz} = 0 \text{ for } z = \pm H_{1}/2 \\[2mm] G_{xz}\dfrac{dg_{0}}{dz} \text{ continuous for } z = \pm H_{2}/2 \end{cases}$$

- In Equation 17.28, one obtains an analogous formulation. In effect, one can write

$$\left(\frac{\bar{E}_{22}^{k}}{\overline{EI}_{22}} + \frac{\bar{E}_{12}^{k}}{\overline{EI}_{12}}\right) = \frac{E_{y}^{k}}{E_{y}^{①}} \times \frac{12}{\alpha b^{3}} = \frac{E_{y}^{k}}{E_{y}^{①}\frac{\left(H_{1}^{3} - H_{2}^{3}\right)}{12} + E_{y}^{②}\frac{H_{2}^{3}}{12}}$$

The problem [17.28] is then written as:

$$\begin{cases} \dfrac{d^{2}p_{0}}{dz^{2}} = -z \times \dfrac{E_{y}^{k}}{G_{yz}^{k}} \times 12\dfrac{G_{yz}^{①}(H_{1} - H_{2}) + G_{yz}^{②}H_{2}}{E_{y}^{①}(H_{1}^{3} - H_{2}^{3}) + E_{y}^{②}H_{2}^{3}} \\[2mm] \dfrac{dp_{0}}{dz} = 0 \qquad z = \pm H_{1}/2 \\[2mm] G_{yz}\dfrac{dp_{0}}{dz} \text{ continuous for } z = \pm H_{2}/2 \end{cases}$$

- **Remark:** These problems are identical to that which allows the calculation of the warping function for the bending of a sandwich beam, and one can consider it in Chapter 18, application 18.3.5. One can then carry out the same steps of calculation. The results obtained are shown below.

---

[13] See Equations 17.2.

17.7.2.2 Transverse Shear Stresses

■ Stress τ_{xz}:

$$-\frac{H_2}{2} \le z \le \frac{H_2}{2}; \quad \tau_{xz} = Q_x \times 6 \times \frac{E_x^{②}\left(\frac{H_2^2}{4} - z^2\right) + E_x^{①}\left(\frac{H_1^2}{4} - \frac{H_2^2}{4}\right)}{E_x^{①}(H_1^3 - H_2^3) + E_x^{②}H_2^3}$$

$$\frac{H_2}{2} \le z \le \frac{H_1}{2}: \quad \tau_{xz} = Q_x \times 6 \times \frac{E_x^{①}\left(\frac{H_1^2}{4} - z^2\right)}{E_x^{①}(H_1^3 - H_2^3) + E_x^{②}H_2^3}$$

(17.37)

■ Stress τ_{yz}:

$$-\frac{H_2}{2} \le z \le \frac{H_2}{2}; \quad \tau_{yz} = Q_y \times 6 \times \frac{E_y^{②}\left(\frac{H_2^2}{4} - z^2\right) + E_y^{①}\left(\frac{H_1^2}{4} - \frac{H_2^2}{4}\right)}{E_y^{①}(H_1^3 - H_2^3) + E_y^{②}H_2^3}$$

$$\frac{H_2}{2} \le z \le \frac{H_1}{2}; \quad \tau_{yz} = Q_y \times 6 \times \frac{E_y^{①}\left(\frac{H_1^2}{4} - z^2\right)}{E_y^{①}(H_1^3 - H_2^3) + E_y^{②}H_2^3}$$

(17.38)

17.7.2.3 Transverse Shear Coefficients

$$k_x = \frac{a_x}{8[E_x^{①}(H_1^3 - H_2^3) + E_x^{②}H_2^3]}\left\{\frac{E_x^{②}}{G_{xz}^{②}}H_2^3\left[E_x^{①}H_1^2 + \left(\frac{4}{5}E_x^{②} - E_x^{①}\right)H_2^2\right] \ldots\right.$$

$$\ldots + \frac{(E_x^{①})^2}{G_{xz}^{①}}\left(\frac{4}{5}H_1^5 + \frac{H_2^5}{5} - H_1^2H_2^3\right)\right\} + \frac{3b_x E_x^{①}(H_1^2 - H_2^2)}{E_x^{①}(H_1^3 - H_2^3) + E_x^{②}H_2^3}$$

$$\text{with} \quad a_x = 12 \times \frac{G_{xz}^{①}(H_1 - H_2) + G_{xz}^{②}H_2}{E_x^{①}(H_1^3 - H_2^3) + E_x^{②}H_2^3}$$

$$b_x = \frac{a_x}{16}H_2\frac{E_x^{①}}{G_{xz}^{①}}\left\{\frac{H_2^2}{3} + H_1^2\left(\frac{G_{xz}^{①}}{G_{xz}^{②}} - 1\right) - H_2^2\frac{G_{xz}^{①}}{G_{xz}^{②}}\left(1 - \frac{2}{3}\frac{E_x^{②}}{E_x^{①}}\right)\right\}$$

(17.39)

k_y is given by an expression formally identical to that in which the index x is replaced by y.

In Application 18.3.8 we treat the case of a rectangular sandwich plate in cylindrical bending, clamped on one side and subjected to a uniform linear force on another. The plate is free on the other sides. One shows the influence of transverse shear on the deflection. This influence increases when:

■ The mechanical characteristics (moduli) of the core are weaker than those of the skins.
■ The relative thickness of the core is important (thin skins).
■ The relative thickness of the plate is large (thick plate).

PART IV

APPLICATIONS

We have grouped in this last part of the book exercises and examples for applications. These have various objectives and different degrees of difficulties. Leaving aside (except for special cases) the cases that are too academic, we will concern ourselves with applications of concrete nature, with an emphasis on the numerical aspect of the results. A few of these applications should be used as validation tests for numerical models.

18

APPLICATIONS

18.1 LEVEL 1

18.1.1 Simply Supported Sandwich Beam

Problem Statement:

1. The following figure represents a beam made of duralumin that is supported at two points. It is subjected to a transverse load of $F = 50$ daN. Calculate the deflection—denoted as Δ—of the beam under the action of the force F.

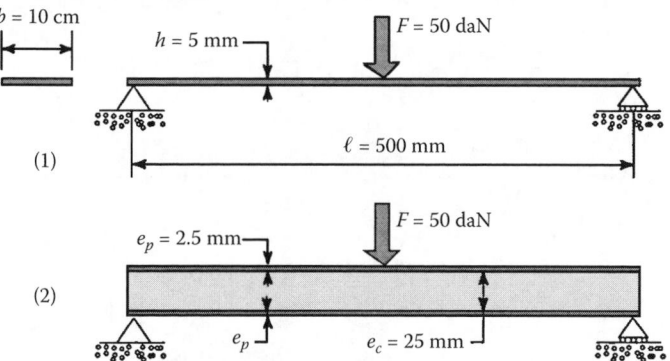

2. We separate the beam of duralumin into two parts with equal thickness $e_p = 2.5$ mm, by imaginarily cutting the beam at its midplane. Each half is bonded to a parallel pipe made of polyurethane foam, making the skins of a sandwich beam having essentially the same mass as the initial beam (in neglecting the mass of the foam and the glue). The beam is resting on the same supports and is subjected to the same load F. Calculate the deflection caused by F, denoted by Δ'. Compare with the value of Δ found in Part 1. (Take the shear modulus of the foam to be: $G_c = 20$ MPa.)

Solution:

1. We will use the classical formula that gives the deflection at the center of the beam on two supports:

$$\Delta = \frac{Fl^3}{48EI} \quad \text{with} \quad I = \frac{bh^3}{12}$$

For duralumin (see Section 1.6): $E = 75,000$ MPa. One finds

$$\boxed{\Delta = 16.7 \text{ mm}}$$

2. Denoting by W the elastic energy due to flexure, one has[1]

$$W = \int_{\text{beam}} \frac{1}{2} \frac{M^2}{\langle EI \rangle} dx + \int_{\text{beam}} \frac{1}{2} \frac{k}{\langle GS \rangle} T^2 dx$$

with[2]:

$$\frac{k}{\langle GS \rangle} \# \frac{1}{G_c(e_c + 2e_p) \times b}$$

Using Castigliano theorem, one has $\Delta' = \dfrac{\partial W}{\partial F}$
then:

$$\Delta' = \int_{\text{beam}} \frac{M}{\langle EI \rangle} \frac{dM}{dF} dx + \int_{\text{beam}} \frac{k}{\langle GS \rangle} T \frac{dT}{dF} dx$$

$$0 \le x \le \ell/2: \quad M = Fx/2; \quad T = -F/2$$

$$\ell/2 \le x \le \ell: \quad M = \frac{F}{2}(\ell - x); \quad T = F/2$$

$$\Delta' = \frac{1}{\langle EI \rangle} \left\{ \int_0^{\ell/2} \frac{Fx}{2} \times \frac{x}{2} dx + \int_{\ell/2}^{\ell} \frac{F}{2}(\ell - x) \frac{(\ell - x)}{2} dx \cdots \right\}$$

$$\cdots + \frac{k}{\langle GS \rangle} \left\{ \int_0^{\ell/2} -\frac{F}{2} \times -\frac{dx}{2} + \int_{\ell/2}^{\ell} \frac{F}{2} \times \frac{dx}{2} \right\}$$

$$\boxed{\Delta' = \frac{F\ell^3}{48 \langle EI \rangle} + \frac{F\ell}{4} \frac{k}{\langle GS \rangle}}$$

Approximate calculation:

$$\langle EI \rangle \# E_p \times e_p \times b \times \frac{(e_c + e_p)^2}{2} + E_c \times \frac{e_c^3 b}{12}$$

then:

$$\langle EI \rangle = 7090 \text{ MKS} + \underset{\text{negligible}}{7.8 \text{ MKS}} \quad \text{with} \quad E_c = 60 \text{ MPa } (cf. \text{ 1.6})$$

---

[1] To establish this relation, see Chapter 15, Equation 15.17.
[2] See calculation of this coefficient in 18.2.1, and more precise calculation in 18.3.5.

one obtains for Δ':

$$\Delta' = \underset{\substack{\text{bending} \\ \text{moment}}}{0.18 \text{ mm}} + \underset{\text{shear}}{1.04 \text{ mm}}$$

$$\boxed{\Delta' = 1.22 \text{ mm}}$$

Comparing with the deflection Δ found in Part 1 above:

$$\boxed{\frac{\Delta}{\Delta'} = \frac{14}{1}}$$

Remarks:

■ The sandwich configuration has allowed us to divide the deflection by 14 without significant augmentation of the mass: with adhesive film thickness 0.2 mm and a specific mass of 40 kg/m^3 for the foam, one obtains a total mass of the sandwich:

$$m = 700 \text{ g (duralumin)} + 50 \text{ g (foam)} + 48 \text{ g (adhesive)}$$

This corresponds to an increase of 14% with respect to the case of the full beam in Question 1.

■ The deflection due to the shear energy term is close to 6 times more important than that due to the bending moment only. In the case of the full beam in question 1, this term is negligible. In effect one has:

$k = 1.2$ for a homogeneous beam of rectangular section, then:

$$\frac{k}{GS} = 8.27 \times 10^{-8}$$

(with $G = 29,000$ MPa, Section 1.6). The contribution to the deflection Δ of the shear force is then:

$$\int \frac{k}{GS} T \frac{dT}{dF} dx = 0.02 \text{ mm} \ll \Delta$$

18.1.2 Poisson Coefficient of a Unidirectional Layer

Problem Statement:

Consider a unidirectional layer with thickness e as shown schematically in the figure below. The moduli of elasticity are denoted as E_ℓ (longitudinal direction) and E_t (transverse direction).

Show that two distinct Poisson coefficients $v_{\ell t}$ and $v_{t\ell}$ are necessary to characterize the elastic behavior of this unidirectional layer. Numerical application: a layer of glass/epoxy. $V_f = 60\%$ fiber volume fraction.

Solution:

Let the plate be subjected to two steps of loading as follows:

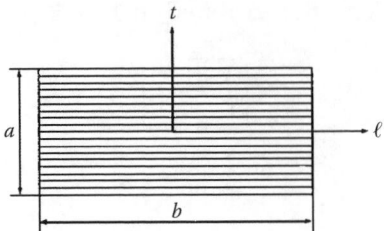

1. A uniform stress σ_ℓ along the ℓ direction: the changes in lengths of the sides can be written as:

$$\frac{\Delta b_1}{b} = \frac{\sigma_\ell}{E_\ell}; \quad \frac{\Delta a_1}{a} = -\frac{v_{\ell t}}{E_\ell}\sigma_\ell$$

2. A uniform stress σ_t along the t direction: for a relatively important elongation of the resin, one can only observe a weak shortening of the fibers along ℓ. Using then another notation for the Poisson coefficient, the change in length can be written as:

$$\frac{\Delta b_2}{b} = -\frac{v_{t\ell}}{E_t}\sigma_t; \quad \frac{\Delta a_2}{a} = \frac{\sigma_t}{E_t}$$

Now calculating the accumulated elastic energy under the two loadings above:

■ When σ_ℓ is applied first, and then σ_t is applied,

$$W = \frac{1}{2}\sigma_\ell \times a \times e \times \Delta b_1 + \frac{1}{2}\sigma_t \times b \times e \times \Delta a_2 + \sigma_\ell \times a \times e \times \Delta b_2$$

■ When σ_t is applied first, and then σ_ℓ is applied,

$$W' = \frac{1}{2}\sigma_t \times b \times e \times \Delta a_2 + \frac{1}{2}\sigma_\ell \times a \times e \times \Delta b_1 + \sigma_t \times b \times e \times \Delta a_1$$

The final energy is the same:

$$W = W'$$

then:

$$\sigma_\ell \times a \times e \times \Delta b_2 = \sigma_t \times b \times e \times \Delta a_1$$

with the values obtained above for Δb_2 and Δa_1:

$$\sigma_\ell \times a \times e \times -\frac{v_{t\ell}}{E_t} \sigma_t \times b = \sigma_t \times b \times e \times -\frac{v_{\ell t}}{E_\ell} \sigma_\ell \times a$$

$$\boxed{\frac{v_{t\ell}}{E_t} = \frac{v_{\ell t}}{E_\ell}}$$

Numerical application: $v_{\ell t} = 0.3$, $E_\ell = 45{,}000$ MPa, $E_t = 12{,}000$ MPa (see Section 3.3.3):

$$v_{t\ell} = 0.3 \times \frac{12{,}000}{45{,}000}$$

$$\boxed{v_{t\ell} = 0.08}$$

Remark: The same reasoning can be applied to all balanced laminates having midplane symmetry, by placing them in the symmetrical axes.[3] However, depending on the composition of the laminate, the Poisson coefficients in the two perpendicular directions vary in more important ranges:

- in absolute value and
- one with respect to the other.

One can find in Section 5.4.2 in Table 5.14 the domain of evolution of the global Poisson coefficient v_{xy} of the glass/epoxy laminate, from which one can deduce the Poisson coefficient v_{yx} using a formula analogous to the one above, as:

$$v_{yx}/E_y = v_{xy}/E_x$$

18.1.3 Helicopter Blade

This study has the objective of bringing out some important particularities related to the operating mode of the helicopter blade, notably the behavior due to normal load.

Problem Statement:

Consider a helicopter blade mounted on the rotor mast as shown schematically in the following figure.

---

[3] Or the orthotropic axes: see Chapter 12, Equation 12.9.

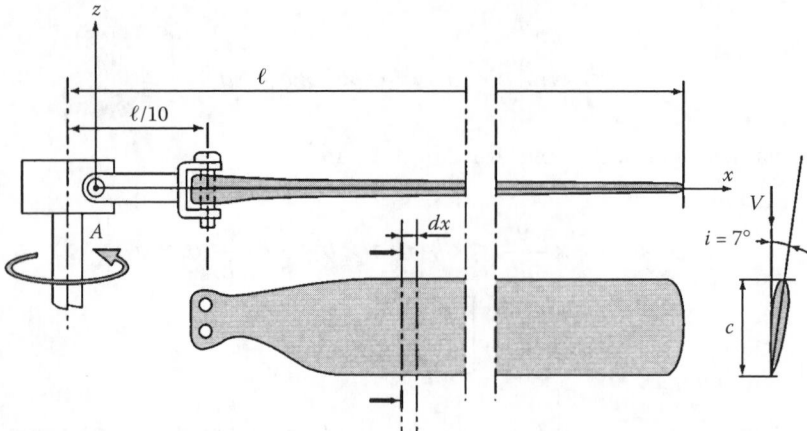

The characteristics of the rotor are as follows:

■ Rotor with three blades; rotational speed: 500 revolutions per minute.
■ The mass per unit length of a blade at first approximation is assumed to have a constant value of 3.5 kg/m.
■ $\ell = 5$ m; $c = 0.3$ m.
■ The elementary lift of a segment dx of the blade (see figure above) is written as:

$$dF_z = \frac{1}{2}\rho(cdx)C_zV^2$$

in which V is the relative velocity of air with respect to the profile of the blade. In addition, C_z (7°) = 0.35 (lift coefficient).

$$\rho = 1.3 \text{ kg/m}^3 \text{ (specific mass of air in normal conditions)}.$$

We will not concern ourselves with the drag and its consequences. One examines the helicopter as immobile with respect to the ground (stationary flight in immobile air). In neglecting the weight of the blade compared with the load application and in assuming infinite rigidity, the relative equilibrium configuration in uniform rotation is as follows:

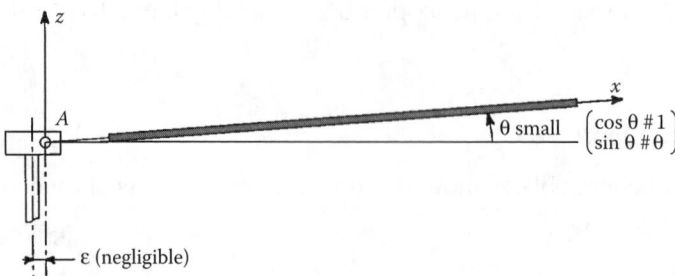

1. Justify the presence of the angle called "flapping angle" θ and calculate it.
2. Calculate the weight of the helicopter.
3. Calculate the normal force in the cross section of the blade and at the foot of the blade (attachment area).

The spar of the blade[4] is made of unidirectional glass/epoxy with 60% fiber volume fraction "R" glass ($\sigma_{\ell\ \text{rupture}} \# 1700$ MPa). The safety factor is 6. Calculate:

4. The longitudinal modulus of elasticity E_ℓ of the unidirectional.
5. The cross section area for any x value of the spar, and its area at the foot of the blade.
6. The total mass of the spar of the blade.
7. The elongation of the blade assuming that only the spar of the blade is subject to loads.
8. The dimensions of the two axes to clamp the blade onto the rotor mast. Represent the attachment of the blade in a sketch.

Solution:

1. The blade is subjected to two loads, in relative equilibrium:
   - Distributed loads due to inertia, or centrifugal action, radial (that means in the horizontal plane in the figure, with supports that cut the rotor axis.
   - Distributed loads due to lift, perpendicular to the direction of the blade (Ax in the figure).

From this there is an intermediate equilibrium position characterized by the angle θ.

Joint A does not transmit any couple. The moment of forces acting on the blade about the y axis is nil, then:

$$\int_{\ell/10}^{\ell} dF_z \times x = \int_{\ell/10}^{\ell} dF_c \times x \sin\theta \# \theta \times \int_{\ell/10}^{\ell} dF_c \times x$$

with:

$$dF_z = \frac{1}{2}\rho c\ dx\ C_z V^2 = \frac{1}{2}\rho c\ dx\ C_z(x\cos\theta \times \omega)^2 \# \frac{1}{2}\rho c\ dx\ C_z x^2 \omega^2$$

$$dF_c = dm\ \omega^2 x\cos\theta \# m\ dx\ \omega^2 x \quad \text{(centrifugal load)}$$

then after the calculation:

$$\frac{1}{2}\rho c C_z \omega^2 \frac{(\ell^4 - \ell^4/10^4)}{4} = \theta\ m\omega^2 \frac{(\ell^3 - \ell^3/10^3)}{3}$$

$$\boxed{\theta \# \frac{3}{8}\frac{\rho c C_z}{m} \times \ell}$$

---

[4] See Section 7.2.3.

or numerically:

$$\boxed{\theta = 0.073 \text{ rad} = 4°11'}$$

Remarks:

- One verifies that $\sin\theta = 0.073 \# \theta$ and $\cos\theta = 0.997 \# 1$.
- When the helicopter is not immobile, but has a horizontal velocity, for example v_0, the relative velocity of air with respect to the blade varies between $v_0 + \omega x$ for the blade that is forward, and $-v_0 + \omega x$ for a blade that is backward. If the incidence i does not vary, the lift varies in a cyclical manner, and there is vertical "flapping motion" of the blade. This is why a mechanism for cyclic variation of the incidence is necessary.
- We have not taken into account the drag to simplify the calculations. This can be considered similarly to the case of the lift. It then gives rise to an equilibrium position with a second small angle, called φ, with respect to the radial direction from top view, as in the following figure. This is why a supplementary joint, or a drag joint, is necessary.

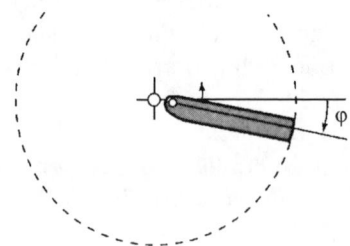

2. Weight of the helicopter: The lift and the weight balance themselves out. The lift of the blade is then:

$$F_z = \int_{\ell/10}^{\ell} dF_z \cos\theta \# \int_{\ell/10}^{\ell} dF_z = \frac{1}{2}\rho c C_z \omega^2 \frac{(\ell^3 - \ell^3/10^3)}{3}$$

then for the 3 rotor blades:

$$Mg = 3F_z$$

$$\boxed{Mg \# \frac{1}{2}\rho c C_z \omega^2 \ell^3}$$

numerically:

$$\boxed{Mg = 2340 \text{ daN}}$$

3. Normal load: It is denoted as $N(x)$:

$$N(x) = \int_x^\ell dF_c \cos\theta \neq \int_x^\ell dF_c = \int_x^\ell m\omega^2 x \ dx$$

$$\boxed{N(x) = \frac{m\omega^2}{2}(\ell^2 - x^2)}$$

at the foot of the blade ($x = l/10$):

$$\boxed{N(\ell/10) \neq 12{,}000 \ \text{daN}}$$

4. Longitudinal modulus of elasticity:
Using the relation of Section 3.3.1:

$$E_\ell = E_f V_f + E_m V_m$$

with (Section 1.6): $E_f = 86{,}000$ MPa; $E_m = 4{,}000$ MPa.

$$\boxed{E_\ell = 53{,}200 \ \text{MPa}}$$

5. Section of the spar of the blade made of glass/epoxy:
The longitudinal rupture tensile stress of the unidirectional is

$$\sigma_{\ell \ \text{rupture}} \neq 1700 \ \text{MPa}$$

With a factor of safety of 6, the admissible stress at a section $S(x)$ becomes

$$\sigma = \frac{N(x)}{S(x)} = \frac{1700}{6} = 283 \ \text{MPa}$$

then:

$$S(x) = \frac{N(x)}{\sigma}$$

$$\boxed{S(x) = \frac{m\omega^2}{2\sigma}(\ell^2 - x^2)}$$

at the foot of the blade:

$$\boxed{S(\ell/10) = 4.24 \ \text{cm}^2}$$

6. Mass of the spar (longeron) of the blade:

$$m_{spar} = \int_{\ell/10}^{\ell} \rho_{unidirect.} S(x) dx$$

$$m_{spar} = \rho_{unidirect.} \times \frac{m\omega^2}{\sigma} \times \frac{1.7}{6} \ell^3$$

Specific mass of the unidirectional layer (see Section 3.2.3):

$$\rho_{unidirect.} = V_f \rho_f + V_m \rho_m = 1980 \text{ kg/m}^3$$

Then:

$$\boxed{m_{spar} = 2.38 \text{ kg}}$$

7. Elongation of the spar of the blade: The longitudinal constitutive relation is written as (see Section 3.1):

$$\varepsilon_x = \frac{\sigma_x}{E_x} = \frac{N(x)}{E_\ell \times S(x)} = \frac{\sigma}{E_\ell}$$

Elongation of a segment dx : $\varepsilon_x (x) dx$.
   For the whole spar of blade:

$$\Delta\ell = \int_{\ell/10}^{\ell} \varepsilon_x \, dx$$

$$\boxed{\Delta\ell = 0.9 \frac{\ell\sigma}{E_\ell}}$$

then:

$$\boxed{\Delta\ell = 2.4 \text{ cm}}$$

One has to reinforce the spar of the blade to diminish the elongation to resist the centrifugal force.

8. Clamped axes: For 2 axes in 30 NCD16 steel (rupture shear strength $\tau_{rupt} = 500$ MPa; bearing strength $\sigma_{bearing} = 1600$ MPa); 4 sheared sections; factor of safety = 6:
   - diameter: $N(\ell/10)/\pi\phi^2 \leq \tau_{rupt}/6 \rightarrow \phi \geq 21.4$ mm
   - length: $N(\ell/10)/2b\phi \leq \sigma_{bearing}/6 \rightarrow b \geq 10.5$ mm

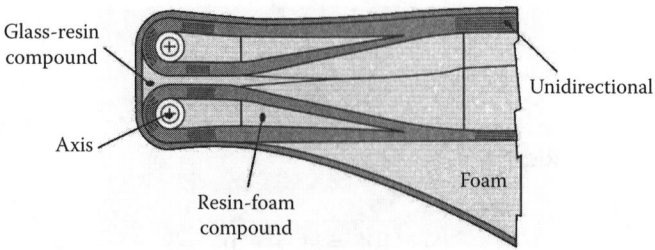

Glass-resin compound
Unidirectional
Axis
Foam
Resin-foam compound

18.1.4 Transmission Shaft for Trucks

Problem Statement:

One proposes to replace the classical transmission shaft made of universal cardan joint and intermediary thrust bearing as shown below:

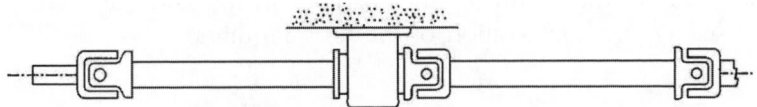

with a solution consisting of a long shaft made of carbon/epoxy, with the following dimensions:

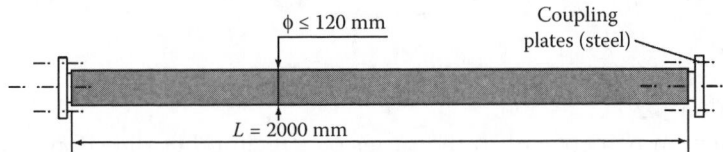

The characteristics of the transmission shaft are as follows:

- Maximum torsional couple: M_t = 300 m daN
- Maximum rotation speed: N = 4000 revolutions/minute
- The first resonant flexural frequency of a beam on two supports is given by:

$$f_1 = \frac{\pi}{2}\sqrt{\frac{EI}{mL^3}}$$

where m is the mass of the beam, and I is its flexure moment of inertia. It corresponds to a "critical speed" for a beam in rotation, which should not be reached during the operation.

- The carbon/epoxy unidirectional has V_f = 60% fiber volume fraction. The thickness of a cured ply is 0.125 mm.
  1. Give the characteristics of a suitable shaft of carbon/epoxy composite. One will make use of the tables in Section 5.4.2 and will use a factor of safety of 6.
  2. Study the adhesive fitting of the coupling plates to the shaft.
  3. Carry out an assessment on the saving in weight with respect to the "shaft in steel" solution (not including the coupling plates).

Solution:

1. Characteristics of the shaft: The shaft is assumed to be thin and hollow (thickness e is small compared with the average radius r as in the following figure).

The shear stress τ is as follows:

$$\tau = \frac{M_t}{2\pi r^2 e}$$

Taking into account the nature of the loading on the laminate making up the tube (pure shear), the composition of the tube requires

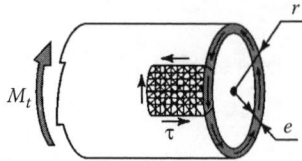

- ■ Important percentages of unidirectionals in the direction of ±45° (see Section 5.2.2).
- ■ Minimum percentages in the order of 10% in other directions (see Section 5.2.3.6).

This leads, for example, to the following distribution:

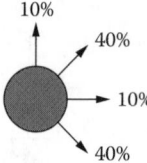

In Section 5.4, one finds Table 5.3, which gives the maximum shear stress that can be applied to a laminate subject to pure shear, as a function of the proportions of the plies at 0°, 90°, +45°, -45°. One reads for the proportions above:

$$\tau_{\text{max}} = 327 \text{ MPa}$$

from which the admissible stress taking into account a safety factor of 6:

$$\tau_{\text{admis.}} = 327/6 \text{ MPa}$$

One then has

$$\frac{M_t}{2\pi r^2 e} \leq \tau_{\text{admis.}}$$

or numerically:

$$r^2 e \geq 8\ 760 \text{ mm}^3$$

For a minimum specified radius $r = 60$ mm, one obtains

$$e \geq 2.43 \text{ mm}$$

The corresponding number of plies of carbon/epoxy is

$$\frac{2.43}{0.125} \text{ \# } 20 \text{ plies}$$

giving a thickness of:

$$e = 2.5 \text{ mm}$$

One can verify that a number of 20 plies allows one to satisfy the following:
(a) The required proportions

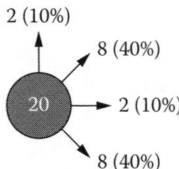

(b) The midplane symmetry, with the sequence:

$$\lfloor 90°/0°/\pm 45°_4 \rfloor_s$$

Critical speed of such a shaft:

$$f_1 = \frac{\pi}{2} \sqrt{\frac{EI}{mL^2}}$$

■ The longitudinal modulus E of the laminate (in the direction of the shaft) is (see Table 5.4 [longitudinal modulus] in Section 5.4.2)

$$E = 31,979 \text{ MPa}$$

■ The specific density of the laminate is (see Section 3.2.3)

$$\rho_{\text{lam}} = V_f \rho_f + V_m \rho_m$$

with (Section 1.6): $\rho_f = 1750$ kg/m^3; $\rho_m = 1,200$ kg/m^3. Then: $\rho_{\text{lam}} = 1,530$ kg/m^3 (see also Table 3.4 in Section 3.3.3).

- The second moment of inertia in flexure is $I = \pi r^3 e$ from which the first frequency is: $f_1 = 76$ Hz

It corresponds to a critical speed of 4,562 rev/minute, superior to the maximum speed of rotation of the shaft.[5]

2. Bonded fittings:

We will use the relation of Paragraph 6.2.3 (Figure 6.26) for simplification. This implies identical thicknesses for the tube making up the shaft and the coupling plate made of steel.[6] The maximum shear stress has an order of magnitude of:

$$\tau_{max} = \frac{a}{\text{th } a} \times \tau_{average} = \frac{a}{\text{th } a} \times \frac{M_t}{2\pi r^2 \ell}$$

where ℓ is the bond length, and

$$a = \ell \sqrt{\frac{G_c}{2 G e e_c}}$$

with G_c as the shear modulus of araldite, then $G_c = 1,700$ MPa (Section 1.6).

$$G_{\text{laminate}} = 28430 \text{ MPa (Section 5.4.2; Table 5.5)}$$

$$e_c = \text{bond thickness (Section 6.2.3): } e_c = \# 0.2 \text{ mm.}$$

- Thickness at bond location:

If one conserves the thickness found for the tube, as $e = 2.5$ mm, one obtains

$$a = l(m) \times 244.5$$

The resistance condition can then be written as:

$$\tau_{max} \leq \tau_{rupture} \text{ (15 MPa for araldite; see Section 6.2.3).}$$

Then:

$$\frac{a}{\text{th } a} \times \frac{M_t}{2\pi r^2 \ell} \leq \tau_{rupture}$$

$$\frac{244.5}{\text{th } a} \times \frac{M_t}{2\pi r^2} \leq \tau_{rupture}$$

numerically: th $a \geq 2.16 \rightarrow$ impossible (th $x \in] - 1, + 1 [$).

---

[5] One also has to verify the absence of buckling due to torsion of the shaft, see annex 2 for this subject.

[6] For different thicknesses for the tube made of carbon/epoxy and for the coupling plate part, one can use the more general relation established in application 18.3.1. This also allows different shear moduli for each material.

It is then necessary to augment the thickness of the tube at the bond location. One starts from the relation:

$$\frac{a}{\text{th } a} \times \frac{M_t}{2\pi r^2 \ell} \le \tau_{\text{rupture}}$$

placed in the form:

$$\frac{\sqrt{\dfrac{G_c}{2Gee_c}}}{(1-\varepsilon)} \times \frac{M_t}{2\pi r^2} \le \tau_{\text{rupture}} \text{ with } \varepsilon \ll 1$$

then:

$$\sqrt{\frac{G_c}{2Gee_C}} \le \tau_{\text{rupture}} \times \frac{2\pi r^2}{M_t} \times (1-\varepsilon)$$

One finds numerically:

$$e > 11.7 \text{ mm;}$$

we retain

$$e = 12 \text{ mm (one then has th } a = 1 - \varepsilon = 0.987)$$

■ Bond length
In accordance with Section 6.2.3, the resistance condition is written as:

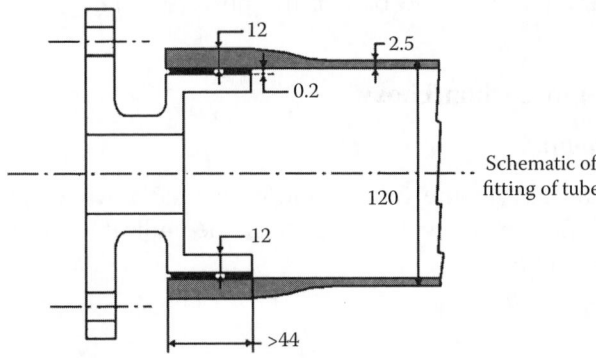

Schematic of fitting of tube

$$\tau_{\text{average}} = \frac{M_t}{2\pi r^2 \ell} \le 0.2 \times \tau_{\text{rupture}}$$

then:

$$\boxed{\ell \ge 44 \text{ mm}}$$

3. Mass assessment:
  ■ Mass of the shaft in carbon/epoxy

$$m_{\text{laminate}} = \rho \times 2\pi re \times L$$

with numerical values already cited:

$$m_{\text{laminate}} = 2.8 \text{ kg.}$$

  ■ If one takes a tubular shaft made of steel ($\tau_{\text{rupture}} = 300$ MPa) with a factor of safety that is 2 times less, say 3, and a minimum thickness of 2.5 mm, the resistance condition:

$$\frac{M_t}{2\pi r^2 e} \leq \frac{300}{3} \text{ MPa}$$

leads to a radius of the tube of

$$r \geq 43 \text{ mm.}$$

From this we find a mass of ($\rho_{\text{steel}} = 7,800$ kg/m^3):

$$m_{\text{steel}} = 10.5 \text{ kg}$$

The saving in mass of the composite solution over the steel solution is 73%. The real saving is higher because it takes into account the disappearance of the intermediate bearing and of one part of the universal joint.

18.1.5 Flywheel in Carbon/Epoxy

Problem Statement:

We show schematically in the figure below an inertia wheel made of carbon/epoxy with 60% fiber volume fraction, with the indicated proportions for the orientation of the fibers.

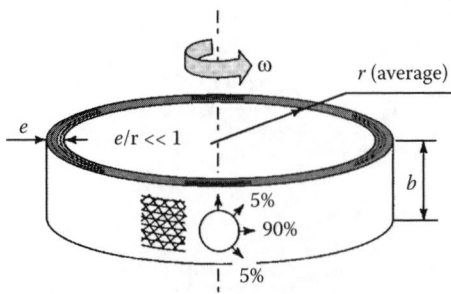

1. Calculate the maximum kinetic energy that one can obtain with such a flywheel with a mass of 1 kg.
2. Compare the maximum kinetic energy that one can obtain with a steel flywheel with a mass of 1 kg. One will take: $\sigma_{\text{rupture steel}} = 1,000$ MPa.

Solution:

1. The equilibrium of an element of the wheel (shown below) reveals inertia forces and cohesive forces.

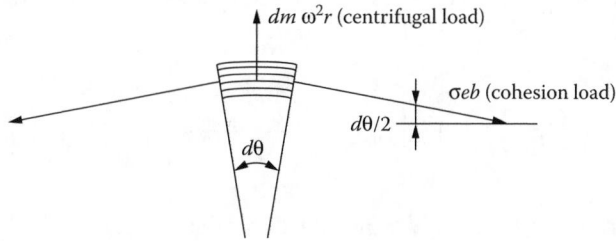

One deduces from there the equilibrium equation along the radial direction:

$$dm \times \omega^2 r = 2\sigma e b \frac{d\theta}{2}$$

Denoted by ρ the specific mass:

$$\rho r\, d\theta\, e b \omega^2 r = \sigma e b d\theta$$
$$\rho(r\omega)^2 = \sigma$$

Denoted by $V = r\omega$ the circumferential speed, one obtains the maximum for the rupture strength of carbon/epoxy, as:

$$\boxed{V_{\text{max}} = \sqrt{\frac{\sigma_{\text{rupture}}}{\rho}}}$$

Numerical application: For the composition of the carbon/epoxy laminate indicated above, one reads in Section 5.4.2, Table 5.1:

$$\sigma_{\text{rupture}} = 1,059 \text{ MPa}$$

and with $\rho = 1,530$ kg/m^3 (Table 3.4 of Section 3.3.3, or the calculation in Section 3.2.3):

$$V_{\text{max}} = 832 \text{ m/s}$$

from this the maximum kinetic energy obtained with 1kg of composite[7]:

$$W_{\text{Kinetic}} = \frac{1}{2} \times 1 \text{ kg} \times V_{\max}^2$$

then:

$$\boxed{W_{\text{Kinetic}} = 346 \text{ kjoules}}$$

2. The maximum circumferential speed that one can obtain with a steel flywheel can be written as:

$$V_{\text{max. steel}} = \sqrt{\frac{\sigma_{\text{rupture steel}}}{\rho_{\text{steel}}}}$$

Therefore, the ratio of kinetic energies of composite/steel is

$$\frac{W_{\text{Kinetic carbon}}}{W_{\text{Kinetic steel}}} = \frac{V_{\text{max carbon}}^2}{V_{\text{max steel}}^2} = \frac{\sigma_{\text{rupt. carbon}} \times \rho_{\text{steel}}}{\sigma_{\text{rupt. steel}} \times \rho_{\text{carbon}}}$$

with $\rho_{\text{steel}} = 7800$ kg/m^3 and $\sigma_{\text{rupt. steel}} = 1000$ MPa, one obtains

$$\boxed{\frac{W_{\text{Kinetic carbon}}}{W_{\text{Kinetic steel}}} = 5.4}$$

With respect to the same mass, it appears then possible to accumulate 5 times more kinetic energy with a flywheel in carbon/epoxy composite.

18.1.6 Wing Tip Made of Carbon/Epoxy

Problem Statement:

Wing tip refers to a part of airplane wing as shown in Figure 18.1. It is made of a sandwich structure with carbon/epoxy skins (Figure 18.2) fixed to the rest of the wing by titanium borders as shown. Under the action of the aerodynamic forces (Figure 18.3), the wing tip is subjected to bending moments, torsional moments, and shear forces as shown in Figure 18.4(a).

One can assume that the core of the sandwich structure transmits only shear forces, and the skins support the flexural moments. This is represented in Figure 18.4(b); the skins resist in their respective planes the in-plane stress resultants: N_x, N_y, and T_{xy}. Figure 18.5 shows the values of these stress resultants

---

[7] Recall the expression for the rotational kinetic energy of a mass m placed at a radius r and rotating at a speed of ω: $W_{\text{Kinetic}} = \frac{1}{2}I\omega^2 = \frac{1}{2}mr^2\omega^2 = \frac{1}{2}mV_{\text{circonfer}}^2$.

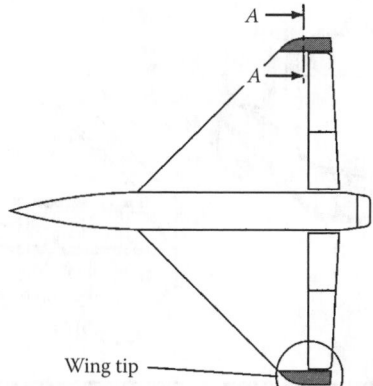

Figure 18.1

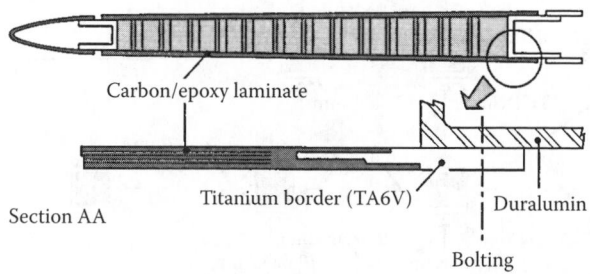

Carbon/epoxy laminate

Section AA

Titanium border (TA6V)

Duralumin

Bolting

Figure 18.2

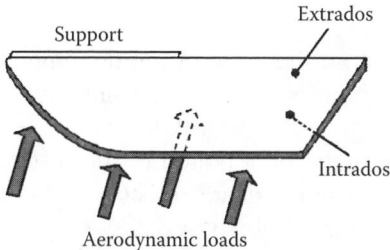

Support

Extrados

Intrados

Aerodynamic loads

Figure 18.3

at a few points of the upper skin (or extrados).

1. According to Figures 18.4(a) and 18.4(b), deduce the elements of the stress resultants N_x, N_y, and T_{xy} from the knowledge of the moment resultants M_x, M_y, and M_{xy}.
2. Using a factor of safety of 2, define the carbon/epoxy skin that is suitable at the surrounding of the support made of titanium alloy (proportions, thickness, number of plies). One will use unidirectional plies with $V_f = 60\%$ fiber volume fraction.

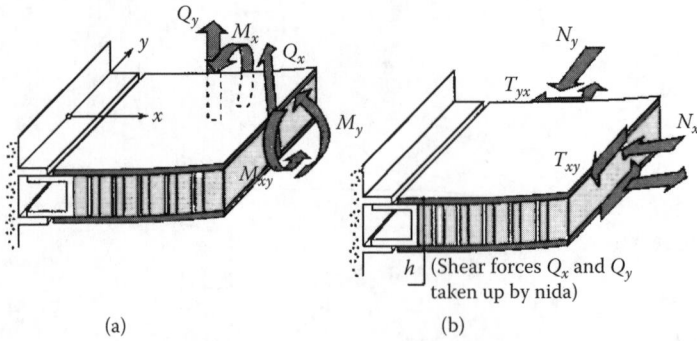

(a)   (b)

Figure 18.4

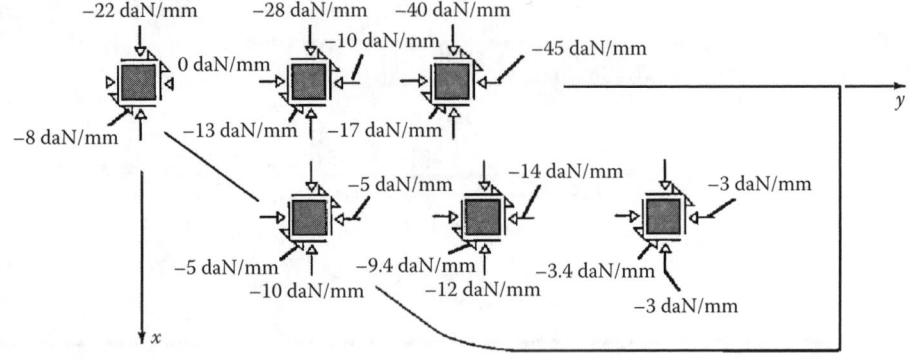

Figure 18.5

3. The skin is bonded on the edge of the titanium (Figure 18.2). Provide the dimensions of the bonded surface by using an average shear stress in the adhesive (araldite: $\tau_{\text{rupture}} = 30$ MPa).

4. The border of the titanium is bolted to the rest of the wing (Figure 18.2). Determine the dimensional characteristics of the joint: "pitch" of the bolts, thickness, foot, with the following data:

   ■ Bolts: 30 NCD 16 steel: $\varnothing = 6.35$ mm, adjusted, negligible tensile loading, $\sigma_{\text{rupture}} = 1{,}100$ MPa; $\tau_{\text{rupture}} = 660$ MPa; $\sigma_{\text{bearing}} = 1600$ MPa

   ■ TA6V titanium alloy: $\sigma_{\text{rupture}} = 900$ MPa; $\tau_{\text{rupture}} = 450$ MPa; $\sigma_{\text{bearing}} = 1100$ MPa

   ■ Duralumin: $\sigma_{\text{rupture}} = 420$ MPa; $\sigma_{\text{bearing}} = 550$ MPa

Solution:

1. The moment resultants M_x, M_y, M_{xy} (and M_{yx}, not shown in Figure 18.4a) are taken up by the laminated skins. One then has in the upper skin (Figure 18.4b), h being the mean distance separating the two skins:

$$N_x = \frac{M_y}{h}; \ N_y = \frac{-M_x}{h}; \ T_{xy} = -\frac{M_{xy}}{h}$$

Remark: The moment resultants, that means the moments per unit width of the skin – 1mm in practice – have units of daN × mm/mm. The stress resultants N_x, N_y, T_{xy} have units of daN/mm.

2. Looking at the most loaded region of the skin, we can represent the principal directions and stresses by constructing Mohr's circle (shown in the following figure). Then we note that there must be a nonnegligible proportion of the fibers at ±45°. However, the laminate has to be able to resist compressions along the axes x and y. The estimation of the proportions can be done following the method presented in Section 5.4.3. One then obtains the following composition[8]:

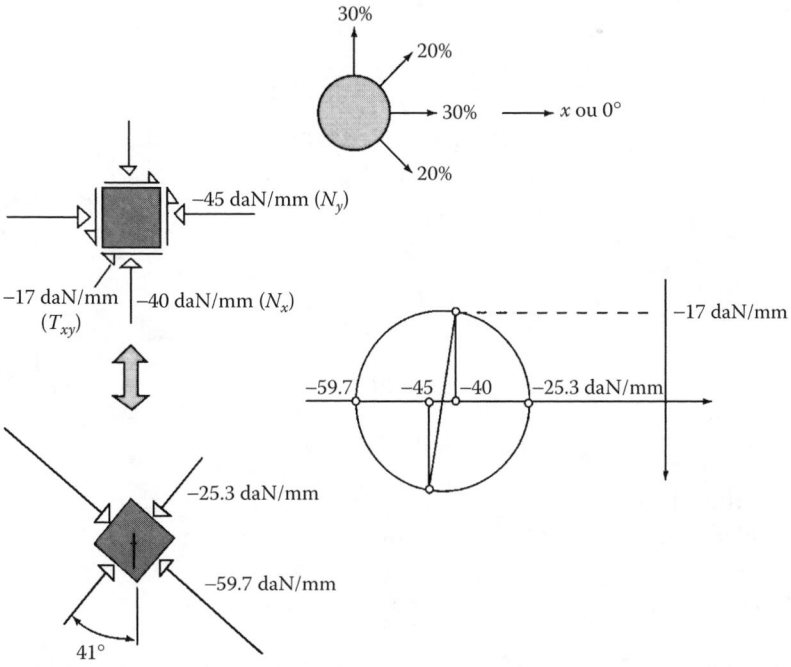

Let σ_ℓ, σ_t, $\tau_{\ell t}$ be the stresses along the principal axes l, t of one of the plies for the state of loading above, the thickness e of the laminate (unknown *a priori*) such that one finds the limit of the Hill–Tsai criterion of failure.[9] One then has

$$\frac{\sigma_\ell^2}{\sigma_{\ell\,\text{rupture}}^2} + \frac{\sigma_t^2}{\sigma_{t\,\text{rupture}}^2} - \frac{\sigma_\ell \sigma_t}{\sigma_{\ell\,\text{rupture}}^2} + \frac{\tau_{\ell t}^2}{\tau_{\ell t\,\text{rupture}}^2} = 1$$

---

[8] The calculation to estimate these proportions is shown in detail in the example of Section 5.4.3, where one has used the same values of the resultants with a factor of safety of 2, as: $N_x = -800$ N/mm; $N_y = -900$ N/mm; $T_{xy} = -340$ N/mm.

[9] See Section 5.3.2 and also Chapter 14.

If one multiplies the two sides by the square of the thickness e:

$$\boxed{\frac{(\sigma_\ell e)^2}{\sigma_{\ell \text{ rupture}}^2} + \frac{(\sigma_t e)^2}{\sigma_{t \text{ rupture}}^2} - \frac{(\sigma_\ell e)(\sigma_t e)}{\sigma_{\ell \text{ rupture}}^2} + \frac{(\tau_{\ell t} e)^2}{\tau_{\ell t \text{ rupture}}^2} = e^2} \qquad [1]$$

one will obtain the values $(\sigma_\ell\, e)$, $(\sigma_t\, e)$, $(\tau_{\ell t}\, e)$, by multiplying the global stresses σ_x, σ_y, τ_{xy} with the thickness e, as $(\sigma_x\, e)$, $(\sigma_y\, e)$, $(\tau_{xy}\, e)$, which are just the stress resultants defined previously:

$$N_x = (\sigma_x e); \quad N_y = (\sigma_y e); \quad T_{xy} = (\tau_{xy} e)$$

Units: the rupture resistances are given in MPa (or N/mm^2) in Appendix 1. As a consequence:

$$N_x = -400 \text{ MPa} \times \text{mm}$$

$$N_y = -450 \text{ MPa} \times \text{mm}$$

$$T_{xy} = -170 \text{ MPa} \times \text{mm}$$

with a factor of safety of 2, one then has

$$N_x' = -800 \text{ MPa} \times \text{mm}$$

$$N_y' = -900 \text{ MPa} \times \text{mm}$$

$$T_{xy}' = -340 \text{ MPa} \times \text{mm}$$

We use the Plates in annex 1 which show the stresses σ_ℓ, σ_t, $\tau_{\ell t}$ in each ply for an applied stress resultant of unit value (1 MPa, for example):
(a) Plies at 0°:
■ Loading $N_x' = -800$ MPa × mm only:

For the proportions defined in the previous question, one reads on Plate 1:

$$\left.\begin{array}{l} \sigma_\ell = 2.4 \\ \sigma_t = 0.0 \\ \tau_{\ell t} = 0 \end{array}\right] \rightarrow \left\{\begin{array}{l} (\sigma_\ell e) = 2.4 \times -800 = -1920 \text{ MPa} \times \text{mm} \\ (\sigma_t e) = 0 \\ (\tau_{\ell t} e) = 0 \end{array}\right.$$

■ Loading $N_y' = -900$ MPa × mm only:

One reads from Plate 5:

$$\left.\begin{array}{l} \sigma_\ell = -0.54 \\ \sigma_t = 0.12 \\ \tau_{\ell t} = 0 \end{array}\right] \rightarrow \left\{\begin{array}{l} (\sigma_\ell e) = -0.54 \times -900 = 486 \text{ MPa} \times \text{mm} \\ (\sigma_t e) = 0.12 \times -900 = -108 \text{ MPa} \times \text{mm} \\ (\tau_{\ell t} e) = 0 \end{array}\right.$$

■ Loading $T'_{xy}\ 1 = -340$ MPa × mm only:

One reads from Plate 9:

$$\left.\begin{array}{l} \sigma_\ell = 0 \\ \sigma_t = 0 \\ \tau_{\ell t} = 0.26 \end{array}\right\} \rightarrow \left\{\begin{array}{l} (\sigma_\ell e) = 0 \\ (\sigma_t e) = 0 \\ (\tau_{\ell t} e) = 0.26 \times -340 = -89 \text{ MPa} \times \text{mm} \end{array}\right.$$

The superposition of the three loadings will then give the plies at 0° a total state of stress of

$$(\sigma_\ell e) = -1920 + 486 = -1434 \text{ MPa} \times \text{mm}$$

$$(\sigma_t e) = -108 \text{ MPa} \times \text{mm}$$

$$(\tau_{\ell t} e) = -89 \text{ MPa} \times \text{mm}$$

Then we can write the Hill–Tsai criterion in the modified form written above (relation denoted as [1]) in which one notes the denominator with values of the rupture strengths indicated at the beginning of annex 1:

$$e^2 = \frac{1434^2}{1130^2} + \frac{108^2}{141^2} - \frac{1434 \times 108}{1130^2} + \frac{89^2}{63^2} = 4.07$$

$$\underset{(0°)}{e} = 2.02 \text{ mm}$$

One resumes the previous calculation as follows:

plies at 0°	$(\sigma_\ell e)$	$(\sigma_t e)$	$(\tau_{\ell t} e)$	
N'_x	−1920	0	0	
N'_y	486	−108	0	$e = 2.02$ mm
T'_{xy}	0	0	−89	
total (MPa × mm)	−1434	−108	−89	

(b) Plies at 90°: One repeats the same calculation procedure by using the Plates 2, 6, and 10. This leads to the following analogous table, with a thickness e calculated as previously (this is the minimum thickness of the laminate below which there will be rupture of the 90° plies).

plies at 90°	$(\sigma_\ell e)$	$(\sigma_t e)$	$(\tau_{\ell t} e)$	
N'_x	432	−96	0	
N'_y	−2160	0	0	$e = 2.16$ mm
T'_{xy}	0	0	89	
total (MPa × mm)	−1728	−96	89	

(c) Plies at +45°: Using Plates 3, 7, and 11 one obtains:

plies at 45°	$(\sigma_\ell e)$	$(\sigma_t e)$	$(\tau_{\ell t} e)$	
N'_x	−752	−48	72	
N'_y	−846	−54	−81	
T'_{xy}	−1384	55	0	e = 2.64 mm
total (MPa × mm)	−2982	−47	−9	

(d) Plies at −45°: Using Plates 4, 8, 12 one obtains:

plies at −45°	$(\sigma_\ell e)$	$(\sigma_t e)$	$(\tau_{\ell t} e)$	
N'_x	−752	−48	−72	
N'_y	−846	−54	81	
T'_{xy}	1384	−55	0	e = 1.13 mm
total (MPa × mm)	−214	−157	9	

Then the theoretical thickness to keep here is the largest out of the four thicknesses found, as:

$$e = 2.64 \text{ mm (rupture of the plies at } +45°).$$

The thickness of each ply is 0.13 mm. It takes 2.64/0.13 = 20 plies minimum from which is obtained the following composition allowing for midplane symmetry:

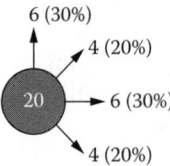

Remark: Optimal composition of the laminate: For the complex loading considered here, one can directly obtain the composition leading to the minimum thickness by using the tables in Section 5.4.4. One then uses the reduced stress resultants, deduced from the resultants taken into account above, to obtain

$$\bar{N}_x = -800/(|800| + |900| + |340|) = -39\%$$
$$\bar{N}_y = -900/(|800| + |900| + |340|) = -44\%$$
$$\bar{T}_{xy} = -17\%$$

Table 5.19 of Section 5.4.4 allows one to obtain an optimal composition close to

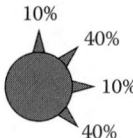

If one uses the previous exact stress resultants, the calculation by computer of the optimal composition leads to the following result, which can be interpreted as described in Section 5.4.4.

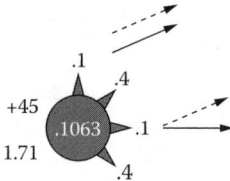

Then one has for the minimum thickness of the laminate:

$$\textit{thickness: } e = 0.1063 \times \frac{(|800| + |900| + |340|)}{100} = 2.17 \text{ mm}$$

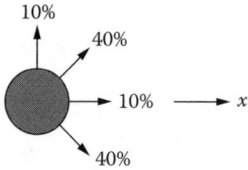

and for the two immediate neighboring laminates:

$$\textit{thickness: } e = 0.1068 \times \frac{(|800| + |900| + |340|)}{100} = 2.18 \text{ mm}$$

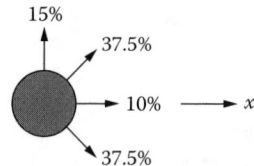

$$\textit{thickness: } e = 0.1096 \times \frac{(|800| + |900| + |340|)}{100} = 2.24 \text{ mm}$$

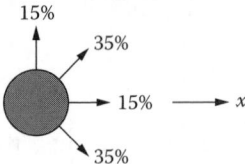

One notes a sensible difference between the initial composition estimated by the designer and the optimal composition. This difference in composition leads to a relative difference in thickness:

$$\frac{2.64 - 2.17}{2.17} = 21\%$$

which indicates a moderate sensibility concerning the effect of thickness and, thus, the mass. One foresees there a supplementary advantage: the possibility to reinforce the rigidity in given directions without penalizing very heavily the thickness. We can note this if we compare the moduli of elasticity obtained starting from the estimated design composition (Section 5.4.3) with the optimal composition, we obtain (Section 5.4.2, Tables 5.4 and 5.5) very different values noted below:

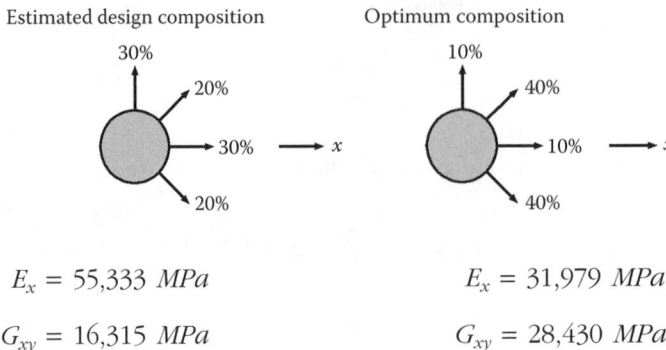

Estimated design composition

$E_x = 55{,}333 \text{ } MPa$

$G_{xy} = 16{,}315 \text{ } MPa$

Optimum composition

$E_x = 31{,}979 \text{ } MPa$

$G_{xy} = 28{,}430 \text{ } MPa$

3. Bonding of the laminate: We represent here after the principal loadings deduced from the values of the stress resultants in Figure 18.5, in the immediate neighborhood of the border of titanium:

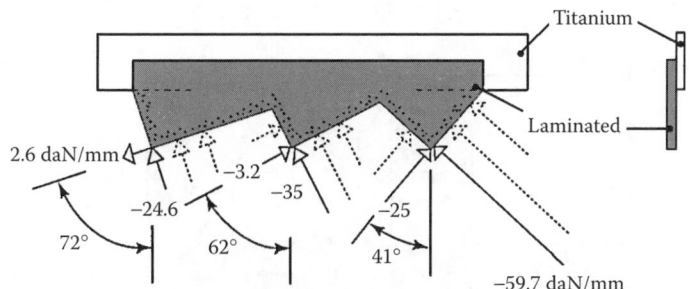

One can for example, overestimate these loadings by substituting them with a fictitious distribution based on the most important component among them. Taking −59.7 daN/mm, one obtains then the simplified schematic below:

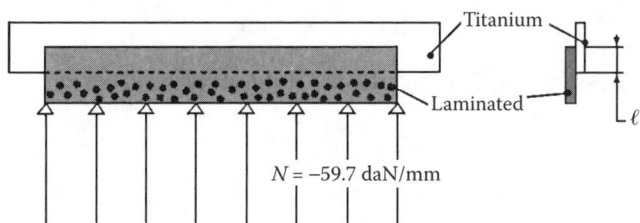

One must evaluate the width of bonding noted as ℓ. For a millimeter of the border, this corresponds to a bonding surface of $\ell \times 1$ mm. For an average rupture criterion of shear of the adhesive, one can write (see Section 6.2.3):

$$\frac{N}{\ell \times 1} \leq 0.2 \times \tau_{\text{adhesive rupture}}$$

then with $\tau_{\text{rupture}} = 30$ MPa:

$$\ell \geq \frac{597}{0.2 \times 30} \, \# \, 100 \text{ mm}$$

From this one obtains the following configuration such that $\ell_1 + \ell_2 + \ell_3 = 100$ mm.

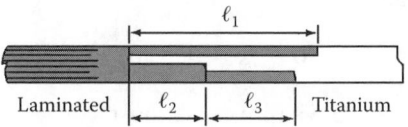

4. Bolting on the rest of the wing:
   ■ "Pitch of bolts": The tensile of bolting is assumed to be low, then bolting strength is calculated based on shear. The bolt load transmitted by a bolt is denoted as ΔF, and one has (*cf.* following figure):

$$\Delta F = N \times \text{pitch} \leq \frac{\pi \varnothing^2}{4} \times \tau_{\text{bolt rupture}}$$

where $\varnothing$ is the diameter of the bolt. One finds a pitch equal to 35 mm.

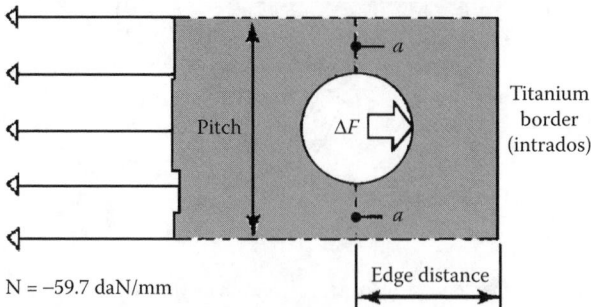

This value is a bit high. In practice, one takes pitch $\leq 5\,\varnothing$, for example, here:

$$\text{Pitch} = 30 \text{ mm.}$$

■ Thickness of the border: the bearing condition is written as:

$$\frac{N \times \text{pitch}}{\varnothing e_{\text{titanium}}} \leq \sigma_{\text{bearing}}$$

then:

$$\boxed{e_{\text{titanium}} \geq 2.55 \text{ mm}}$$

■ Verification of the resistance of the border in the two zones denoted a in the previous figure: the stress resultant in this zone, noted as N', is such that:

$$N \times \text{pitch} = N' \times (\text{pitch} - \varnothing)$$

then:

$$N' = N \frac{\text{pitch}}{\text{pitch} - \varnothing} = 75.4 \text{ daN/mm}$$

The rupture stress being:

$$\sigma_{\text{rupture}} = 900 \text{ MPa}$$

and the minimum thickness 2.55 mm, one must verify

$$\frac{N'(\text{daN/mm})}{e(\text{mm})} \leq \sigma_{\text{rupture}}(\text{daN/mm})$$

One effectively has

$$\frac{75.4}{2.55} \leq 90$$

■ Verification of the edge distance (see previous figure): One has to respect the following condition:

$$\frac{\Delta F}{2 \times \text{edge distance} \times e} \leq \tau_{\text{titanium rupture}}$$

then:

$$\text{edge distance} \geq 7.8 \text{ mm}$$

from which the configuration (partial) of the joint can be shown as in the following figure:

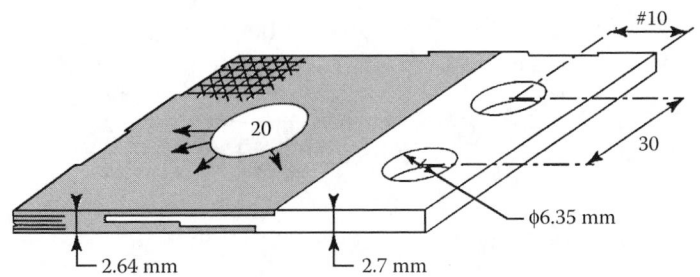

18.1.7 Carbon Fiber Coated with Nickel

Problem Statement:

With the objective of enhancing the electrical and thermal conductivity of a laminated panel in carbon/epoxy, one uses a thin coat of nickel with a thickness e for the external coating of the carbon fibers by electrolytic plating process (see following figure).

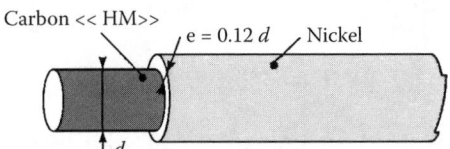

1. Calculate the longitudinal modulus of elasticity of a coated fiber.
2. Calculate the linear coefficient of thermal expansion in the direction of the coated fiber.

Solution:

1. Hooke's law applied to a fiber with length ℓ subject to a load F (following figure) can be written as:

$$F = E_f s \frac{\Delta \ell}{\ell}$$

where E_f is the modulus of the coated fiber that one wishes to determine, and

$$s = \pi \, (d/2 + e)^2$$

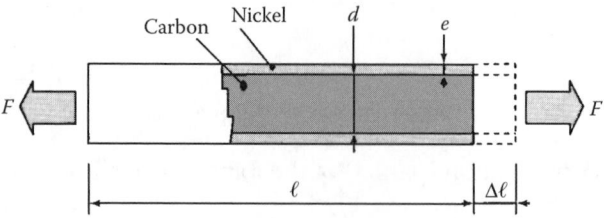

The load F is divided into F_C on the carbon fiber and F_N on the nickel coating. The equality of the elongations of the two components allow one to write

$$F_c = E_c \pi \frac{d^2}{4} \frac{\Delta \ell}{\ell}; \quad F_N = E_N \pi \left[\left(\frac{d}{2} + e \right)^2 - \frac{d^2}{4} \right] \frac{\Delta \ell}{\ell}$$

where, taking into account that $F = F_C + F_N$,

$$E_f \pi \left(\frac{d}{2} + e \right)^2 = E_c \pi \frac{d^4}{4} + E_N \pi \left[\left(\frac{d}{2} + e \right)^2 - \frac{d^2}{4} \right]$$

$$\boxed{E_f = E_c \frac{1}{\left(1 + \frac{2e}{d} \right)^2} + E_N \left[1 - \frac{1}{\left(1 + \frac{2e}{d} \right)^2} \right]}$$

Numerical application:

$$E_C = 390{,}000 \text{ MPa}; \; E_N = 220{,}000 \text{ MPa}; \; d = 6.5 \; \mu\text{m (Section 1.6)}$$

$$E_f = 330{,}500 \text{ MPa.}$$

2. Thermal expansion of an unloaded rod with length $\ell = 1$ m and corresponding to a temperature variation ΔT can be written as:

$$\Delta \ell_1 = \alpha \, \Delta T \times 1$$

where α is the thermal expansion coefficient of the material making up the rod. In addition, when this rod is subjected to a longitudinal stress σ, Hooke's law indicates a second expansion:

$$\Delta \ell_2 = \frac{\sigma}{E} \times 1$$

Superposition of the two cases simultaneously applied can be written as:

$$\Delta \ell = \Delta \ell_1 + \Delta \ell_2$$

or:

$$\Delta\ell = \left(\frac{\sigma}{E} + \alpha\Delta T\right) \times 1$$

When the coated fiber is subjected to a variation in temperature ΔT, each of the constituents will elongate an identical amount $\Delta\ell$. The whole coated fiber is not subjected to any external forces. The difference in the coefficients of thermal expansion of carbon and of nickel, which should lead to different free thermal expansions, then leads to the equilibrium of loads inside the coated fiber.

Let α_f be the thermal coefficient of expansion of the coated fiber. One has

$$\Delta\ell = \alpha_f \Delta T \times 1$$

Then for the carbon and for the nickel:

$$\Delta\ell = \frac{\sigma_c}{E_c} + \alpha_c\Delta T = \frac{\sigma_N}{E_N} + \alpha_N\Delta T \qquad [2]$$

The forces being in equilibrium

$$\pi\left[\left(\frac{d}{2} + e\right)^2 - \frac{d^2}{4}\right]\sigma_N + \pi\frac{d^2}{4}\sigma_c = 0 \qquad [3]$$

Equations [2] and [3] lead to

$$\sigma_c = (\alpha_N - \alpha_c)\Delta T \Big/ \left[\frac{1}{E_c} + \frac{1}{E_N} \times \frac{1}{\left(1 + \frac{2e}{d}\right)^2 - 1}\right]$$

and taking into account that

$$\alpha_f\Delta T = \Delta\ell = \frac{\sigma_c}{E_c} + \alpha_c\Delta T$$

one obtains

$$\alpha_f = \frac{\alpha_N + \alpha_c\dfrac{E_c}{E_N}\dfrac{1}{\left[\left(1 + \frac{2e}{d}\right)^2 - 1\right]}}{1 + \dfrac{E_c}{E_N}\dfrac{1}{\left[\left(1 + \frac{2e}{d}\right)^2 - 1\right]}}$$

18.1.8 Tube Made of Glass/Epoxy under Pressure

Problem Statement:

Consider a thin tube made by filament winding of glass/epoxy with a winding angle of $\pm45°$. The fiber volume fraction is $V_f = 0.6$. The tube is fixed at one end to a rigid undeformable mass and mounted to a sliding joint at the other end (see following figure).

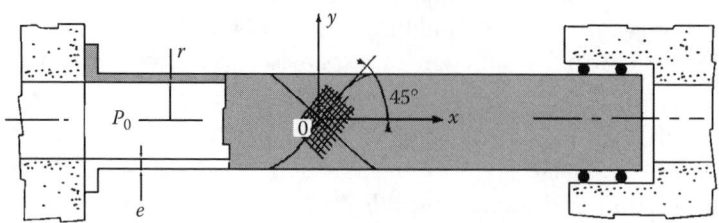

The thickness e is considered to be small as compared with the average radius ($e/r \ll 1$). One applies on the inside of the tube a unit pressure of $p_o = 1$ MPa (or 10 bars). Use a safety factor of 8 to take into account the aging effect.

1. Calculate the stresses (σ_x, σ_y) along the axes x and y in the tangent plane at point O of the tube.
2. What is the maximum stress allowable for the winding considered? From that deduce the minimum thickness of the tube for an average radius of $r = 100$ mm.
3. What are the moduli E_x, E_y, and G_{xy} of the laminate, and the Poisson coefficients v_{xy} and v_{yx}? Write the stress–strain behavior for the laminate in the coordinates $x - y$.
4. Calculate the strains ε_x and ε_y of the composite tube. From there deduce the strain in the direction that is perpendicular to the fiber direction at $+45°$, denoted as ε_t, which characterizes the strain in the resin.

This strain has to be less than 0.1% to avoid microfracture that can lead to the leakage of the fluid across the thickness of the tube (**weeping** phenomenon).

Solution:

1. The tube being free in the axial direction and neglecting the thickness e,
   $\sigma_x = 0$.

The equilibrium of a half-cylinder with unit length, represented in the figure below, allows one to write:

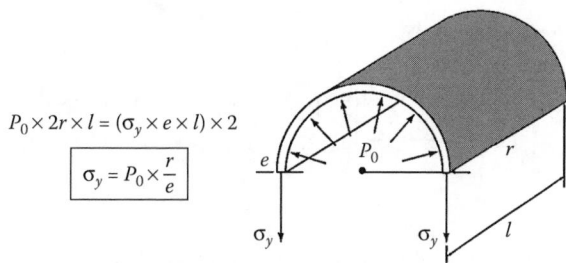

$$P_0 \times 2r \times l = (\sigma_y \times e \times l) \times 2$$

$$\boxed{\sigma_y = P_0 \times \frac{r}{e}}$$

2. Maximum admissible stress: One reads on Table 5.12, Section 5.4.2, for proportions of plies as 50% in the directions + and −45°:

$$\sigma_{y\ max\ (tension)} = 94\ \text{MPa}$$

then with $\sigma_{y\ max} = p_o\ (r/e)$, the theoretical minimum thickness is

$$e_{theoretical} = \frac{p_o \times r}{\sigma_{y\ max}} = \frac{1\ \text{MPa} \times 100\ \text{mm}}{94\ \text{MPa}} = 1.064\ \text{mm}$$

Taking into account the factor of safety of 8 for aging effect

$$e = 8.5\ \text{mm}$$

3. Moduli of the laminate: One reads on Table 5.14, Section 5.4.2:

$$E_x = 14{,}130\ \text{MPa} = E_y$$

$$\nu_{xy} = 0.57 = \nu_{yx}$$

and from Table 5.15:

$$G_{xy} = 12{,}760\ \text{MPa}.$$

Recalling the stress–strain relation for an anisotropic material described in Section 3.1, which is repeated here as:

$$\begin{Bmatrix} \varepsilon_x \\ \varepsilon_y \\ \gamma_{yx} \end{Bmatrix} = \begin{bmatrix} \dfrac{1}{E_x} & -\dfrac{\nu_{yx}}{E_y} & 0 \\ \dfrac{\nu_{xy}}{E_x} & \dfrac{1}{E_y} & 0 \\ 0 & 0 & \dfrac{1}{G_{xy}} \end{bmatrix} \begin{Bmatrix} \sigma_x \\ \sigma_y \\ \tau_{xy} \end{Bmatrix}$$

one has

$$\begin{Bmatrix} \varepsilon_x \\ \varepsilon_y \\ \gamma_{xy} \end{Bmatrix} = \frac{1}{14130} \begin{bmatrix} 1 & -0.57 & 0 \\ -0.57 & 1 & 0 \\ 0 & 0 & 1.107 \end{bmatrix} \begin{Bmatrix} \sigma_x \\ \sigma_y \\ \tau_{xy} \end{Bmatrix}$$

4. Strains: For $p_o = 1$ MPa and $e = 8.5$ mm, one has

$$\sigma_y = \frac{1 \text{ MPa} \times 100}{8.5} = 11.8 \text{ MPa}$$

then

$$\begin{Bmatrix} \varepsilon_x \\ \varepsilon_y \\ \gamma_{xy} \end{Bmatrix} = \frac{1}{14\,130} \begin{bmatrix} 1 & -0.57 & 0 \\ -0.57 & 1 & 0 \\ 0 & 0 & 1.107 \end{bmatrix} \begin{Bmatrix} 0 \\ 11.8 \\ 0 \end{Bmatrix}$$

from which

$$\varepsilon_x = -4.76 \times 10^{-4}$$

$$\varepsilon_y = 8.35 \times 10^{-4}$$

The Mohr's circle of strains, shown below, allows one to obtain the strain ε_t in the direction perpendicular to the fibers.

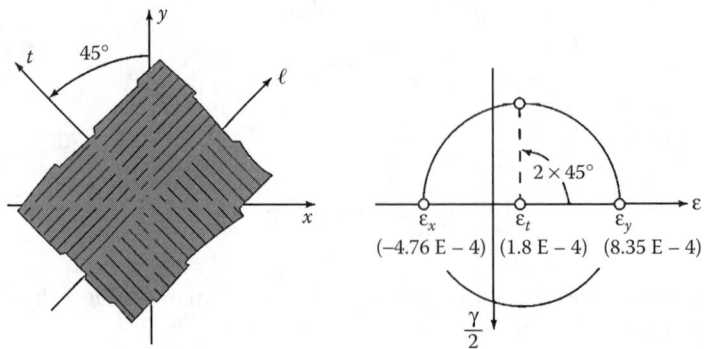

One obtains

$$\varepsilon_t = \frac{\varepsilon_x + \varepsilon_y}{2} = 1.8 \times 10^{-4}$$

$$\varepsilon_t = 0.018\%$$

One verifies that the strain in the matrix is less than 0.1%, the maximum limit.

18.1.9 Filament Wound Vessel, Winding Angle

Problem Statement:

One considers a vessel having the form of a thin shell of revolution, wound of "R" glass/epoxy rovings. In the cylindrical portion (see figure) the thickness is e_o which is small compared with the average radius R. This vessel is loaded by an internal pressure of p_o.

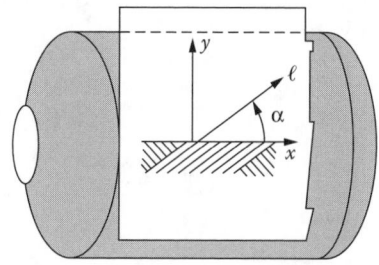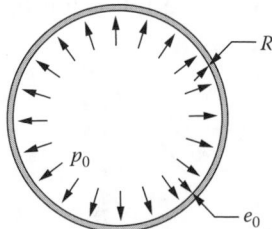

1. The resin epoxy is assumed to bear no load. Denoting by e the thickness of the reinforcement alone, calculate in the plane x,y (see figure) the stresses σ_{ox} and σ_{oy} in the wall, due to pressure p_o.
2. In the cylindrical part of the vessel, the winding consists of layers at alternating angles $\pm\alpha$ with the generator line (see figure). One wishes that the tension in each fiber along the direction ℓ could be of a uniform value σ_ℓ. (This uniform tension in all the fibers gives the situation of **isotensoid**.)
   (a) Evaluate the stresses σ_x and σ_y in the fibers as functions of σ_ℓ.
   (b) Deduce from the above the value of the helical angle α and the tension σ_ℓ in the fibers as functions of the pressure p_o.
   (c) What will be the thickness e_o for a reservoir of 80 cm in diameter that can support a pressure of 200 bars with 80% fiber volume fraction?

Solution:

1. Preliminary remark: The elementary load due to a pressure p_o acting on a surface dS has a projection on the x axis as (see figure):

$$p_o \, dS \cos \theta = p_o \, dS_o$$

where dS_o is the x axis projection of dS in a plane perpendicular to this axis.

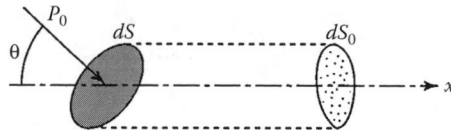

■ Equilibrium along the axial direction: The equilibrium represented in the following figure leads to

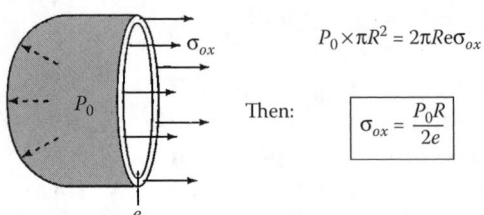

$$P_0 \times \pi R^2 = 2\pi Re\sigma_{ox}$$

Then:

$$\sigma_{ox} = \frac{P_0 R}{2e}$$

■ Equilibrium along the circumferential direction: The equilibrium represented in the following figure leads to

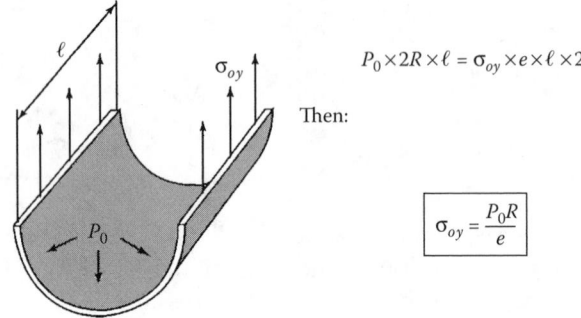

$$P_0 \times 2R \times \ell = \sigma_{oy} \times e \times \ell \times 2$$

Then:

$$\sigma_{oy} = \frac{P_0 R}{e}$$

2. (a) Stresses σ_x and σ_y in the fibers: one can represent as follows the Mohr's circle of stresses starting from the pure normal stress σ_ℓ on a face normal to axis ℓ (see following figure). From there, the construction leading to the stress σ_x (see figure below) is geometrically as[10]:

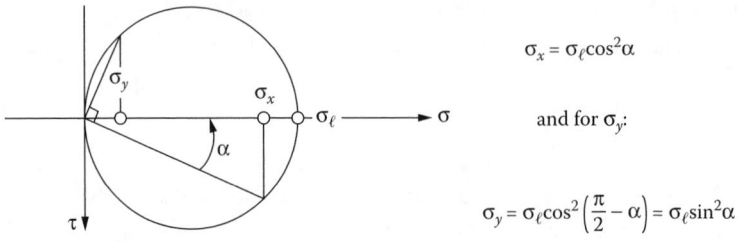

$$\sigma_x = \sigma_\ell \cos^2\alpha$$

and for σ_y:

$$\sigma_y = \sigma_\ell \cos^2\left(\frac{\pi}{2} - \alpha\right) = \sigma_\ell \sin^2\alpha$$

Value of the helical angle α: Identification of these stresses with the values σ_{ox} and σ_{oy} found above leads to

$$\sigma_\ell \cos^2\alpha = \frac{P_0 R}{2e}; \quad \sigma_\ell \sin^2\alpha = \frac{P_0 R}{e}$$

---

[10] One obtains this result immediately by using the Equation 11.4.

from which:

$$tg^2 \alpha = 2$$

then:

$$\boxed{\sin \alpha = \sqrt{\frac{2}{3}}; \ \alpha = 54.7°}$$

Tension in the fibers is then:

$$\sigma_\ell = \frac{3}{2} p_0 \frac{R}{e}$$

(c) Thickness e_0: One has for "R" glass[11]: σ_ℓ rupture = 3200 MPa. The thickness e of the reinforcement is such as:

$$e = \frac{3}{2} \frac{p_0 R}{\underset{\text{rupture}}{\sigma_\ell}} = 3.75 \text{ mm}$$

and the thickness of the glass/epoxy composite, V_f being the fiber volume fraction, is

$$e_0 = e/V_f = 4.7 \text{ mm}$$

18.1.10 Filament Wound Reservoir, Taking the Heads into Account

Problem Statement:

A reservoir having the form of a thin shell of revolution is wound with fibers and resin. It is subjected to an internal pressure p_o. The circular heads at the two ends of the reservoir have radius of r_o. We study the cylindrical part of the reservoir, with average radius R.

One part of the winding consists of filaments in helical windings making angles of $\pm\alpha_1$ with the generator (see figure). The other part consists of similar filaments in circumferential windings ($\alpha_2 = \pi/2$).

The resin is assumed to carry no load. The tension in the filaments of the helical layers is denoted by $\sigma_{\ell 1}$ and the tension in the filaments of the circumferential layers by $\sigma_{\ell 2}$.

---

[11] See Section 1.6.

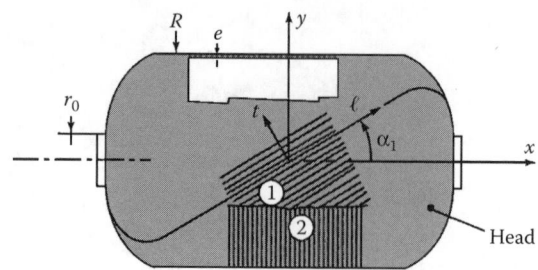

1. What has to be the value of α_1 so that the filaments can elongate on the heads along the lines of shortest distance?
2. Calculate the thickness e_1 of fibers of the helical layers and thickness e_2 of fibers of the circumferential layer as a function of p_o, R, α_1, $\sigma_{\ell 1}$, $\sigma_{\ell 2}$.
3. What is the minimum total thickness of fibers e_m that the envelope can have? What then are the ratios e_1/e_m and e_2/e_m? What is the real corresponding thickness of the envelope if the percentage of fiber volume, denoted as V_f, is identical for the two types of layers?

Note: It can be shown—and one admits—that on a surface of revolution the lines of shortest distance, called the **geodesic** lines, follow the relation (see following figure for the notations):

$$r \sin \alpha = \text{constant}$$

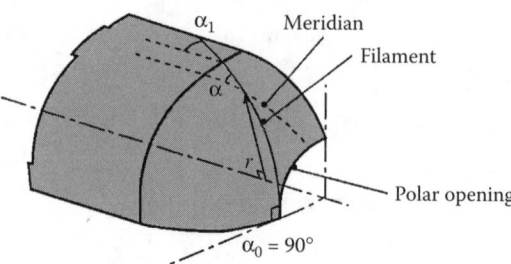

Solution:

1. The filaments wound helically (angle $\pm\alpha_1$) in the cylindrical part follow over the heads along the geodesic lines such that $r \sin \alpha = \text{constant}$. The circle making up the head is a geodesic for which $r = r_o$. Then $\alpha = \text{constant}$. Thus:

$$\alpha_o = \frac{\pi}{2}$$

One then has for the filaments joining the cylindrical part to the head:

$$r_o \sin \frac{\pi}{2} = R \sin \alpha_1$$

$$\sin \alpha_1 = \frac{r_o}{R}$$

2. Thicknesses of the layers: For an internal pressure p_o, the state of stresses in the cylindrical part of the thin envelope is defined in the tangent plane x, y (following figure) by[12]

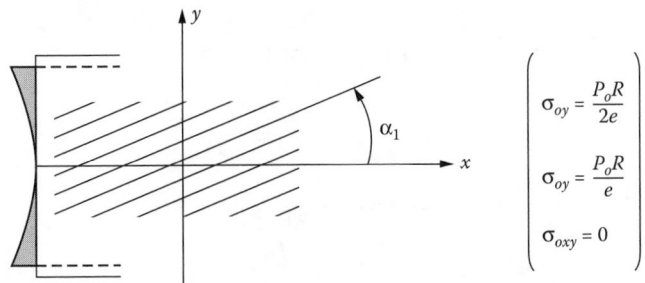

$$\left(\begin{array}{l} \sigma_{oy} = \dfrac{P_o R}{2e} \\[2mm] \sigma_{oy} = \dfrac{P_o R}{e} \\[2mm] \sigma_{oxy} = 0 \end{array}\right)$$

The resin being assumed to bear no load, e represents the thickness of the reinforcement alone. One can follow by direct calculation.[13] The state of stress in the helical layers reduce to

$$\sigma_\ell \; (\sigma_{t1} = \tau_{\ell t1} = 0).$$

One obtains for the state of stresses in plane x, y starting from the Mohr's circle (see following figure).[14]

$$\sigma_{x1} = \cos^2 \alpha_1 \times \sigma_{\ell 1}; \quad \sigma_{y1} = \sin^2 \alpha_1 \times \sigma_{\ell 1};$$

$$\tau_{xy1} = \cos \alpha_1 \sin \alpha_1 \sigma_{\ell 1}$$

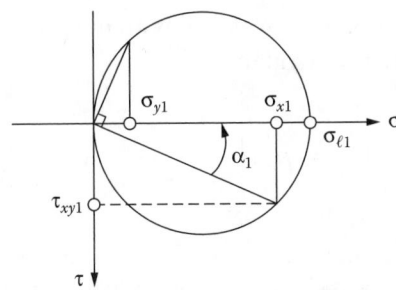

and for the circumferential layers ($\alpha_2 = \pi/2$)

$$\sigma_{x2} = 0; \quad \sigma_{y2} = \sigma_{\ell 2}; \quad \tau_{xy2} = 0$$

---

[12] See Section 18.1.9.

[13] One can also consider a balanced laminate with the ply angles of $+\alpha_1$, $-\alpha_1$, and $\pi/2$. The role of the matrix is neglected. The elastic coefficients of a ply (see Equation 11.11) reduce to only one nonzero E_ℓ. The calculation is done as shown in detail in Section 12.1.3. It is more laborious than the direct method shown above here.

[14] See also the Equation 11.4 inverted.

One then has the following equivalents, in calculating the resultant forces on sections of unit width and normal x and y respectively:

■ Along x:

$$\sigma_{x1} \times e_1 \times 1 + \sigma_{x2} \times e_2 \times 1 = \sigma_{ox} \times e \times 1$$

then:

$$e_1 \times \cos^2 \alpha_1 \times \sigma_{\ell 1} = e\sigma_{ox} = e \times p_o \frac{R}{2e}$$

from which:

$$\boxed{e_1 = \frac{p_o}{\sigma_{\ell 1}} \times \frac{R}{2\cos^2 \alpha_1}}$$

■ Along y:

$$\sigma_{y1} \times e_1 \times 1 + \sigma_{y2} \times e_2 \times 1 = \sigma_{oy} \times e \times 1$$

$$e_1 \times \sin^2 \alpha_1 \times \sigma_{\ell 1} + e_2 \times \sigma_{\ell 2} = e \times \sigma_{oy} = e \times \frac{p_o R}{e}$$

$$\boxed{e_2 = \frac{p_o}{\sigma_{\ell 2}} R\left(1 - \frac{\text{tg}^2 \alpha_1}{2}\right)}$$

3. Minimum thickness of the envelope: With the previous results, the thickness of the reinforcement is written as:

$$e = e_1 + e_2 = p_o R\left\{\frac{1}{2\sigma_{\ell 1}\cos^2 \alpha_1} + \frac{2 - \text{tg}^2 \alpha_1}{2\sigma_{\ell 2}}\right\}$$

The reinforcements for the helical layers and for the circumferential layers are of the same type. They can be subjected to identical maximum tension. Therefore, at fracture, one has

$$\sigma_{\ell 1} = \sigma_{\ell 2} = \sigma_{\ell \text{ rupture}}$$

Then:

$$e_{\min} = \frac{p_o R}{2\sigma_{\ell \text{ rupture}}}\left(\frac{1}{\cos^2 \alpha_1} + 2 - \text{tg}^2 \alpha_1\right)$$

$$\boxed{e_{\min} = \frac{3}{2} \times \frac{p_o R}{\sigma_{\ell \text{ rupture}}}}$$

Then for the ratios of thicknesses:

$$\frac{e_1}{e_{min}} = \frac{1}{3\cos^2\alpha_1}; \quad \frac{e_2}{e_{min}} = \frac{2 - tg^2\alpha_1}{3}$$

Real thickness of the envelope taking into account the percentage of fiber volume V_f:

$$\frac{dv_{reinforcement}}{dv_{real}} = V_f = \frac{2\pi Re_{min}\, dx}{2\pi Re_{real}\, dx}$$

$$e_{real} = \frac{3}{2} \times \frac{p_oR}{\sigma_\ell} \times \frac{1}{V_f}$$
$$\underset{rupture}{}$$

18.1.11 Composite Reservoir; Use of Standards

Problem Statement:

Composite reservoirs are usually designed to bleach pulp and paper. Chlorinated water solutions are used to bleach the pulp and some pressure and vacuum are required for the operation. The construction of these reservoirs consists usually of a liner made of flexible non-structural thermoplastics such as PolyVinyl Chloride (PVC), and a laminate made up of a layer of Hand Lay Up (HLU) which consists of alternate layers of mat and woven roving, and a layer of Filament Winding (FW) with fiber angles at about +/-54.7°with the generator line of the cylindrical shell.[15] The composite material used is usually glass/vinyl ester composites. The design usually takes into account the consideration of internal pressure, external pressure (vacuum), wind load (if the reservoir is located outside) and earthquake load (depending on the earthquake zone).

This application consists in a simplified verification of a typical reservoir as shown in Figure 18.6. The help of Standards formulations will be required for vacuum loading case. Some of the relevant equations provided by the British Standard are shown below[16]

Defining:

L: Length of the cylindrical shell between stiffeners.
L_c: Critical length
d_i: Inside structural diameter of the cylinder (the value of which can be considered close to the mean diameter's one)
E_{LAM}: Modulus of the laminate
P: External pressure

---

[15] See Application 18.1.9.

[16] For more elaborate treatment of the problem, see "Analysis for Design of Fiber Reinforced Plastic Vessels and Piping," by Hoa, S.V., Technomic, 1991.

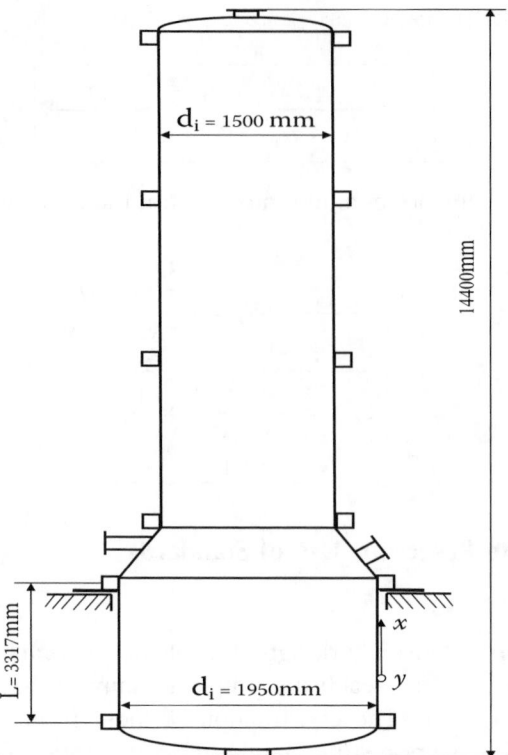

Figure 18.6 Configuration of the Tower

Equations to calculate minimum thickness t_m at a particular pressure P or critical pressure corresponding to a particular thickness depend on the length of the cylinder.

If: $\dfrac{L}{d_i} \geq \dfrac{L_c}{d_i}$ then

$$t_m = d_i \left[\frac{P \times F}{2E_{LAM}} \right]^{0.33} \quad \text{or} \quad P = \frac{2E_{LAM}}{F} \left[\frac{t_m}{d_i} \right]^3$$

and if $\dfrac{L}{d_i} \prec \dfrac{L_c}{d_i}$ then

$$t_m = d_i \left[\frac{0.4P \times F}{E_{LAM}} \times \frac{L}{d_i} \right]^{0.4} \quad \text{or} \quad P = 2.5 \frac{E_{LAM}}{F} \times \frac{d_i}{L} \times \left[\frac{t_m}{d_i} \right]^{2.5} \quad [17]$$

---

[17] See in Appendix 2, "Buckling of Orthotropic Tubes," the more accurate analogous formula which takes into account the different values of E_{LAM} in longitudinal x and circumferential y direction.

where t_m is the minimum thickness and F is the design factor (normally taken to be 5).

Numerical Application:

Consider the reservoir as shown in Figure 18.6. The empty weight of the reservoir is 64,960N, and the full weight (with liquid inside) is W= 435,456 N. The operating internal pressure is 750 mm Hg and the secure design value is 600 mmHg (absolute) and/or the situation of full of water. It is located in the earthquake zone 2 according to the National Building Code. The support for the reservoir is located as shown in Figure 18.6.

The thickness consists of a PVC liner of 3.18 mm thick (non structural), a 3.18 mm Hand Lay Up (HLU) and 19.05 mm Filament Winding. The material is made of glass/vinyl ester. The modulus of elasticity for Hand Lay Up and Filament Winding are given as:

$$E_{x_{HLU}} = E_{y_{HLU}} = 5442\,MPa$$

$$E_{x_{FW}} = 5442\,MPa \quad ; \quad E_{y_{FW}} = 10544\,MPa$$

The maximum allowable strengths for Hand Lay Up (Mat and Woven Roving layers) are

$$\sigma_{x_{max\,HLU}} = \sigma_{y_{max\,HLU}} = 6.8\,MPa$$

and for Filament Winding layer:

$$\sigma_{x_{max\,FW}} = 6.8\,MPa \quad ; \quad \sigma_{y_{max\,FW}} = 10.2\,MPa$$

1. Assuming that only Filament Winding resists to circumferential stress, verify the thickness of the shell in its lower portion when it is full of water.
2. Assuming that the whole laminate resists, verify the thickness of the lower portion of the shell for external pressure (vacuum) with a safety factor of 5.
3. Verify the thickness of the upper part of the shell to resist to an earthquake load as shown on Figure 18.7.

Solution:

1. Shell full of water:

The part that is subjected to the largest stress would be the lower portion of the reservoir, with the diameter of 1950 mm. The thickness of the shell at this section will be verified.

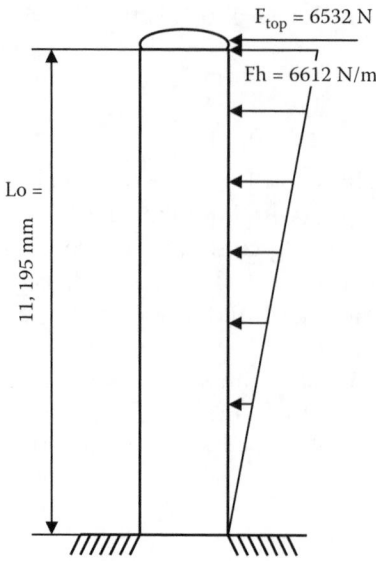

Figure 18.7   Configuration of Support and Resultant Due to Earthquake Load

■ *Circumferential stress due to internal pressure:*
Internal pressure is due to the height of the liquid column, with the maximum value:

$$p_0 = \rho \times g \times h = 1000 \times 9.81 \times 14.4 = 141264\,Pa = 0.141\,MPa$$

Using thin shell theory,[18] and assuming that the circumferential stress is supported by Filament Winding only leads to the circumferential stress in FW as:

$$\sigma_y = \frac{p_0 \times r}{t} = \frac{0.141 \times 1.95}{2 \times 19.05 \times 10^{-3}} = 7.21\,MPa$$

■ *Longitudinal stress due to weight load:*
Noting that the weight of the empty reservoir is only 15% of the full weight, we can assume, in view of a conservative estimation of the stress, that the full weight works on this area of the shell. Moreover, the whole laminate supports this weight. Then along vertical direction:

$$\sigma_x = \frac{W}{\pi \times d_i \times t} = \frac{435456}{\pi \times 1.95 \times 0.0222} = 3.2 \times 10^6\,Pa = 3.2\,MPa$$

Then:

---
[18] See Application 18.1.9.

For the Hand Lay Up: $\sigma_{x_{HLU}} = 3.2\,MPa \prec 6.8\,MPa$ (maximum allowed)
For the Filament Winding:
In this area, the Filament Winding is subjected to a biaxial state of stress:

$$\sigma_{x_{FW}} = 3.2\,MPa \quad ; \quad \sigma_{y_{FW}} = 7.21\,MPa$$

One can remark that each of these stresses is clearly less than the maximum allowed stress:

$$\sigma_{x_{FW}} \prec 6.8\,MPa \quad ; \quad \sigma_{y_{FW}} \prec 10.2\,MPa$$

Nevertheless, it is not sufficient to ensure that the Filament Winding will resist. A complete study would lead to evaluate σ_ℓ and σ_t in order to verify the Hill–Tsai criterion. To avoid too heavy calculus, one can examine, in a rough simplified manner, the application of a quadratic resistance criterion for the orthotropic Filament Winding in x,y axis. A Hill–Tsai analogy can be written with the conservative form:

$$\left(\frac{\sigma_x}{\sigma_{x_{\max FW}}} \right)^2 + \left(\frac{\sigma_y}{\sigma_{y_{\max FW}}} \right)^2 - \frac{\sigma_x \sigma_y}{\sigma^2_{y_{\max FW}}}$$

or with values above:

$$\left(\frac{3.2}{6.8} \right)^2 + \left(\frac{7.21}{10.2} \right)^2 - \frac{3.2 \times 7.21}{10.2^2} = 0.5 \prec 1$$

Since this value is decidedly less than the unity, we can conclude that the shell thickness is OK for internal pressure.[19]

2. External pressure:
The reservoir should be able to contain an internal absolute pressure of 600 mmHg. This means that there is vacuum in the operation and the level of the vacuum is:

---

[19] One can remark that if we suppress the Hand Lay Up and neglect the empty weight of the reservoir, the only Filament Winding resists to internal pressure in the same manner as in Application 18.1.9. This explains the fiber angles values of +/- 54°7. Then (see Application 18.1.9) *in this lower portion only*, the state of stress in the composite is reduced to the stress along the fibers:

$$\sigma_\ell = \frac{3}{2} \times \frac{p_0 \times r}{t} = 1.5 \times 7.21\,MPa = 10.8\,MPa\,,$$

which is a very little value compared to rupture value in traction of glass/vinylester (see for example an order of magnitude in Table 3.4).

$$760 \text{ mmHg-}600 \text{ mHg} = 160 \text{ mmHg. or } 0.021 \text{ MPa.}$$

For design to withstand vacuum, normally a factor of safety F= 5 is used. The pressure with the factor of safety is:

$$P \times F = 0.021\,MPa \times 5 = 0.105\,MPa$$

Length between two stiffeners:

$$L = 3.317m$$

Using the British Standard BS 4994, first one has to check the range of the length over diameter ratio:

$$\frac{L}{d_i} = \frac{3.317}{1.95} = 1.7$$

Then, using a conservative form with the minimum value of the modulus of elasticity:

$$\frac{L_c}{d_i} = 1.35 \times \left[\frac{E_{LAM}}{P \times F} \right]^{0.17} = 1.35 \times \left[\frac{5442\,MPa}{0.105\,MPa} \right] = 8.54$$

Since

$$\frac{L}{d_i} \prec \frac{L_c}{d_i}$$

the following equation is used to determine the minimum thickness required (with an analogous conservative form -minimum value of elasticity modulus-):[20]

$$t_m = 1.2 \times d_i \times \left[\frac{0.4 \times P \times F}{E_{LAM}} \times \frac{L}{d_i} \right]^{0.4} = 1.2 \times 1.95 \times \left[\frac{0.4 \times 0.105\,MPa}{5442\,MPa} \times \frac{3.317}{1.95} \right]^{0.4}$$

$$= 0.026m = 26mm$$

The structural thickness proposed is 22.23 mm. It is not sufficient and should be increased by 3.8mm.

---

[20] Using the more accurate formula for "Buckling due to External Pressure" given in Appendix 2 for orthotropic tubes gives a thickness of 18.4 mm.

3. Earthquake loading:

Refering to the National Building Code of Canada section 4.1.9., the earthquake load is given as:

$$V = A \times (SKI) \times F \times W$$

where:

      V is the total load acting on the tower

      A is a constant depending on the earthquake zone. For zone 2, A = 0.04.

      (SKI) = 2.5

      F is a foundation factor = 1.

      W is the full weight of the tower.

Then:

$$V = 0.04 \times 2.5 \times 1 \times 435456 = 43545.6 N$$

According to common practice, 15% of the load is assumed to act at the top:

$$F_{top} = 0.15 \times 43545.6 = 6532 N$$

The remaining load is assumed to be distributed over the height of the reservoir as shown in Figure 18.7. The maximum magnitude of the distributed load is calculated to be:

$$\frac{1}{2} \times F_b \times L_0 = 43545 N - 6532 N$$

From which:

$$F_b = \frac{2 \times \left(43545 - 6532\right)}{11.195} = 6612 N / m$$

Bending moment M_B at the lower part of the column:

$$M_B = F_{top} \times L_0 + \frac{1}{2} F_b \times L_0 \times \frac{2}{3} L_0$$

$$M_B = 6532 N \times 11.195 m + \left(43545 N - 6532 N\right) \times \frac{2}{3} \times 11.195 = 349370 N.m$$

Outside diameter of reservoir:

$$d_e = 1500 + 2 \times \left(3.18 + 19.05\right) = 1544.5 mm$$

Longitudinal bending stress (see Figure 5.31 and associated notations):

$$\sigma_{x_{max}} = \frac{M_B}{I} \times r_e$$

where

$$\frac{I}{r_e} = \frac{\pi}{64} \times \frac{\left(d_e^{\,4} - d_i^{\,4}\right)}{\dfrac{d_e}{2}}$$

$$\frac{I}{r_e} = \frac{\pi}{32} \times \frac{\left(d_e^{\,4} - d_i^{\,4}\right)}{d_e} = \frac{\pi}{32} \times \frac{\left(1.5445^4 - 1.5^4\right)}{1.5445} = 0.04\,m^3$$

From which

$$\sigma_{x_{max}} = \frac{349371\,Nm}{0.04\,m^3} = 8.73 \times 10^6\,Pa = 8.73\,MPa$$

This value is more than the allowed one (6.8Mpa), and the total structural thickness t should be increased in such a way that:

$$\frac{I}{r_e} = \frac{349371}{6.8 \times 10^6} = 0.051\,m^3$$

Noting as a first approximation that:

$$d_e^{\,4} = \left(d_i + 2t\right)^4 \approx d_i^{\,4}\left(1 + 4 \times \frac{2t}{d_i}\right)$$

We have:

$$\frac{I}{r_e} = \frac{\pi}{32} \cdot \frac{d_i^{\,4}\left(1 + 8 \times \dfrac{t}{d_i}\right) - d_i^{\,4}}{d_i + 2t} = 0.051$$

Which leads to:

$$t = 30 \text{ mm}$$

In the lower area of the column, the thickness needs to be increased to 7.8mm to support the specified earthquake. This thickness may be reduced along the height of the reservoir for upper elevations, under the control of analogous calculus.

Remark: There is a possibility that the earthquake load may be combined with the internal pressure due to the weight of the column of liquid. This gives rise in this area of the column of Figure 18.7 to circumferential stresses but not axial stresses. Then, the calculus above is not modified. Nevertheless, the resultant bi-axial state of stress needs to be verified as already seen in question 1.

18.1.12 Determination of the Volume Fraction of Fibers by Pyrolysis

Problem Statement:

One removes a sample from a carbon/epoxy laminate made up of identical layers of balanced fabric. The measured specific mass of the laminate is denoted as ρ. The specific mass of carbon is denoted as ρ_f, that of the matrix is denoted as ρ_m.

One burns completely the epoxy matrix in an oven. The mass of the residual fibers is compared with the initial mass of the sample. One then obtains the fiber mass denoted as M_f (see Section 3.2.1).

1. Express as a function of ρ, ρ_f, ρ_m, M_f:
   (a) The fiber volume fraction V_f
   (b) The matrix volume fraction, V_m
   (c) The volume fraction of porosities or voids, V_p
2. Numerical application:

$$\rho = 1{,}500 \text{ kg/m}^3; \quad \rho_f = 1{,}750 \text{ kg/m}^3; \quad \rho_m = 1{,}200 \text{ kg/m}^3; \quad M_f = 0.7$$

Solution:

1. (a) By definition (Section 3.2.2) one has

$$V_f = \frac{v_{\text{fibers}}}{v_{\text{total}}} = \frac{m_{\text{fibers}}}{\rho_f} \times \frac{\rho}{m_{\text{total}}} = M_f \times \frac{\rho}{\rho_f}$$

$$\boxed{V_f = M_f \times \frac{\rho}{\rho_f}}$$

(b) In an analogous manner:

$$V_m = M_m \times \frac{\rho}{\rho_m}$$

and with $M_f + M_m = 1$:

$$V_m = (1 - M_f) \times \frac{\rho}{\rho_m}$$

(c) Noting (Section 3.2.2) that:

$$V_f + V_m + V_p = 1$$

one deduces

$$V_p = 1 - \rho \times \left(\frac{M_f}{\rho_f} + \frac{(1 - M_f)}{\rho_m} \right)$$

2. Numerical application:

$$V_f = 60\%; \quad V_m = 37.5\%; \quad V_p = 2.5\%$$

Remark: In practice, a small amount of carbon fibers is also pyrolyzed. About 0.125% of its mass is pyrolyzed per hour.

18.1.13 Lever Arm Made of Carbon/PEEK Unidirectional and Short Fibers

Problem Statement:

The following sketch shows a lever arm pinned at A, B, C. It is subjected to the loads indicated. The external skin is obtained from a plate of thermoformed unidirectional carbon/PEEK,[21] 2.8 mm in thickness, and placed in a mold into which one injects short fibers of carbon/PEEK at high temperature.

		DENSITY (kg/m³)	$\sigma_{rupture}$	MODULUS OF ELASTICITY (mPa)
carbon/PEEK unidirectional	$V_f = 65\%$	1600	2100	$E_\ell = 125{,}000$ $G_{\ell t} = 4000$
short fibers carbon/PEEK	$V_f = 18\%$	1400	127	$E = 21{,}000$ $G = 8000$

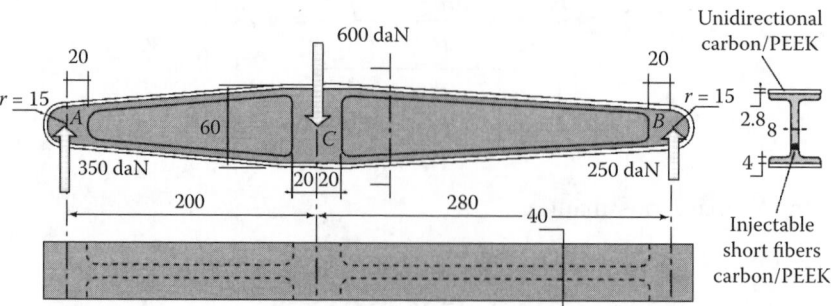

---

1. Verify the resistance of this piece by a simplified calculation.
2. Estimate the order of magnitude of the displacements due to loads at A and B.
3. Make an assessment of the mass of the piece.

Solution:

1. Verification of the resistance of the piece:

   ■ Unidirectional: Assume (simplified calculation) that the applied moment is resisted essentially by the unidirectional skins.[22] Then, in the section where the moment is maximum, one has (see following figure):

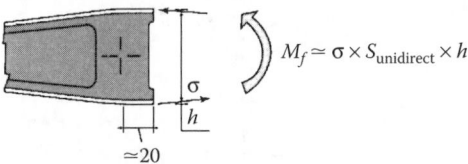

with

$$S_{\text{unidirectional}} = 2.8 \times 40 \text{ mm}^2; \quad h = 60 - 2 - 2.8 \text{ \# } 55 \text{ mm};$$

$$M_f = 650 \times 10^3 \text{ N} \times \text{mm}.$$

$$\boxed{\sigma = 106 \text{ MPa}}$$

Factor of safety: $\sigma_{\text{rupture}}/\sigma - 1 = 1880\%$.

Remark: In the injected layer, under the unidirectional skin, the order of magnitude of the normal stress is six times smaller.[22]

   ■ Injected core: We assume that the shear stress due to the shear force is taken up essentially by the web, with an order of magnitude (see following figure):

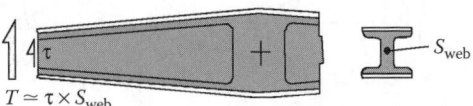

with

$$S_{\text{web}} = (33 - 5.6 - 8) \times 8 \text{ mm;}^2 \quad T = 3500 \text{ N}$$

$$\tau \text{ \# } 23 \text{ MPa}$$

---

[22] This is because the longitudinal modulus of elasticity E_ℓ of the unidirectional is six times higher than that of the injected resin. For more "exact" calculation of the stresses, see Equation 15.16.

Remark: In fact, the distribution of the shear stresses is expanded in the flanges (injected part and unidirectional part in the figure below). The bonding being assumed to be perfect, the distortion is the same in the injected part and in the unidirectional part, as:

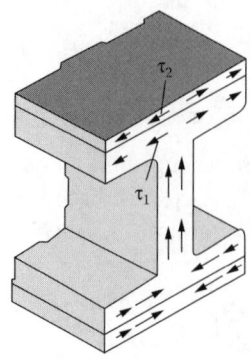

$$\gamma = \frac{\tau_2}{G_{\ell t}} = \frac{\tau}{G}$$

$$\tau_2 = 23 \times \frac{G_{\ell t}}{G} = 12 \text{ MPa}$$

2. Displacements under load: Keeping the central part C fixed in translation and in rotation, the deformation energy of each arm (right or left) is written as:

$$W = \frac{1}{2} \int\limits_{\text{arm}} \sigma \varepsilon \, dV + \frac{1}{2} \int\limits_{\text{arm}} \tau \gamma \, dV$$

then (with the previous approximations):

$$W = \frac{1}{2} \int\limits_{\text{unidirect}} \frac{\sigma^2}{E} dS \times dx + \frac{1}{2} \int\limits_{\text{web}} \frac{\tau^2}{G} \, dS \times dx$$

$$W = \frac{1}{2} \int \frac{M_f^2}{E \underset{\text{unidirect.}}{} (S_{\text{unidirect.}} \times b)^2} \times 2 S_{\text{unidirect.}} dx \dots$$

$$\dots + \frac{1}{2} \int \frac{T^2}{G \times S_{\text{web}}^2} \times S_{\text{web}} \times dx$$

with $M_f = F(l - x)$; $T = F$; $b = b_{\text{average}}$; $S_{\text{web}} = S_{\text{average web}}$ at midlength of the arm in view of an estimation:

$$W = \frac{1}{2} \int \frac{F^2 \ell^3 / 3}{E \underset{\text{unidirect.}}{} \left(S \underset{\text{unidirect.}}{} \times \frac{b_{\text{average}}^2}{2} \right)} + \frac{1}{2} \frac{F^2 \ell}{G S_{\text{average web}}}$$

One obtains for the displacement at the point of application of the load F (Castigliano theorem):

$$\Delta = \frac{\partial W}{\partial F} = \left\{ \frac{\ell^3/3}{E_{\underset{\text{unidirect.}}{}} \times S_{\underset{\text{unidirect.}}{}} \times \frac{b^2_{\text{average}}}{2}} + \frac{\ell}{G \times S_{\text{web ave.}}} \right\} \times F$$

- Right arm: $\ell = 280$ mm; $F = 2{,}500$ N; $b_{\text{average}} = 45$ mm $-$ 2.8 mm

$$\boxed{\Delta_B = 1.8 \text{ mm}}$$

- Left arm: $\ell = 200$ mm; $F = 3{,}500$ N; $b_{\text{average}} = 45$ mm -2.8 mm; $S_{\text{web average}} = 31.4 \times 8$ (mm^2)

$$\boxed{\Delta_A = 1.1 \text{ mm}}$$

3. Mass assessment: Unidirectional: 189 g; short fibers: 525 g; total mass before drilling

$$\boxed{m = 714 \text{ g}}$$

Remarks:

- Taking into account the low levels of stress in the unidirectional, the piece may be lightened in decreasing—uniformly and progressively—its thickness (here 40 mm). A reduction from 40 to 30 mm leads to a reduction of mass of 18% and an increase in displacements from 22 to 26% at A and B.
- To obtain a comparable mass in light alloys, one has to use folded and welded sheet. The price of the piece is higher. The composite piece is obtained by one single operation of injection after preforming of the unidirectional reinforcements.

18.1.14 Telegraphic Mast in Glass/Resin

Problem Statement:

A telegraphic mast 8 m long (of which 80 cm is buried in the ground) of glass/epoxy with 60% fiber volume fraction has the characteristics shown below.

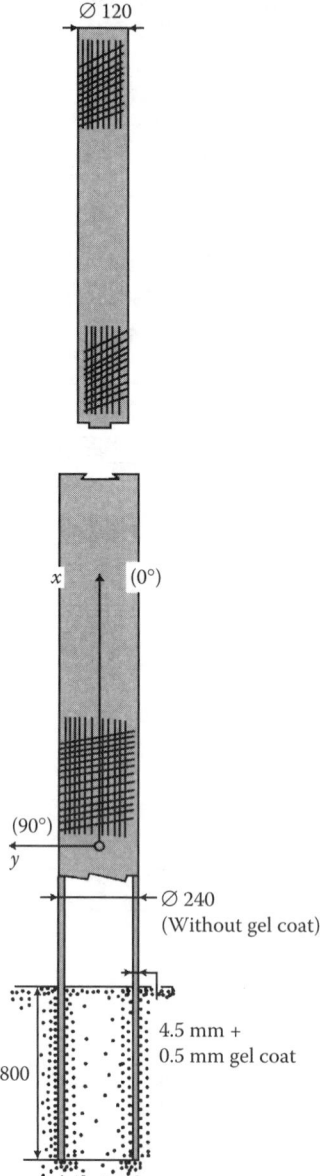

One has in the lower part of the mast:

- ■ 27 layers at 0°—or along x direction
- ■ three layers oriented in helix with an angle that will be taken practically equal to 90°

1. Give the elastic constants of the laminate in this zone.
2. What maximum horizontal load at the top is admissible for this lower zone?
3. Estimate the displacement of the top subject to this load.

Solution:

1. The composition of the laminate in the lower part is as:

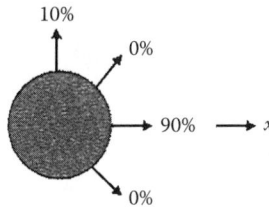

Tables 5.14 and 5.15 of Section 5.4.2 give for this composition:

$$\boxed{\begin{array}{l} E_x = 41{,}860 \text{ MPa}; \quad E_y = 15{,}360 \text{ MPa} \\ v_{xy} = 0.23; \quad v_{yx} = 0.09 \\ G_{xy} = 4500 \text{ MPa} \end{array}}$$

2. For maximum load at the top, three risks need to be taken into account:

- ▪ Risk of rupture due to classical flexure in this zone where the bending moment is maximum
- ▪ Risk of rupture due to shear load
- ▪ Risk of buckling by ovalization and then flattening of the tube

(a) Flexure moment: One has (see figure below)[23]:

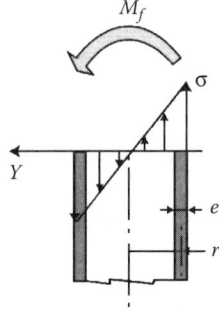

$$\sigma = -\frac{M_f}{I} \times Y \quad \text{with} \quad I = \pi r^3 e$$

---

[23] See at the end of Section 5.4.5, Figure 5.31 the distribution of stresses in a composite beam. See also Equations 15.16 in Chapter 15.

The maximum is obtained when $Y = -r$:

$$\sigma_{max} = \frac{M_f}{\pi r^2 e}$$

Note that for the laminate considered (Table 5.11, Section 5.4.2), the first ply fractures at a stress of

$$\sigma_{\text{tensile rupture}} = 128 \text{ MPa}$$

Then

$$M_f \leq 26 \times 10^6 \text{ N} \times \text{mm}$$

corresponding to a horizontal load at the top of the mast:

$$F_{\substack{max \\ (M_f)}} = \frac{26 \times 10^6}{7200} = 3600 \text{ N.}$$

(b) Shear load: On an average diameter of the tubular section (neutral plane), one can write

$$\tau = \alpha \ T/S$$

where T is the shear load, S is the area of the cross section, and α is the amplification factor ($\alpha > 1$).[24] Note that for the laminate considered (Table 5.13, Section 5.4.2), the first ply rupture occurs at $\tau_{\text{rupture}} = 63$ MPa, from which by taking $T = F_{max(M_f)} = 3{,}600$ N,

$$\alpha < \frac{63 \times 3329}{3600} = 58$$

This condition is well certified (recall that for a thin tube of isotropic material, one has $\alpha = 2$).

(c) Ovalization of the mast: One has (Appendix 2, b)

$$M_{\text{critical}} = \frac{2\sqrt{2}}{9} \pi re^2 \left[\frac{E_x E_y}{1 - \nu_{xy}\nu_{yx}}\right]^{1/2}$$

Here we have

$$M_{\text{critical}} = 6 \times 10^7 \text{ N mm}$$

---

[24] The exact value of α should be obtained from the complete study of the shear stresses in a composite beam (Equation 15.16).

which corresponds to a horizontal load at the top:

$$F_{\text{critical ovalization}} = 8360 \text{ N}$$

One can then retain the maximum load as:

$$F_{\text{max}} = 3600 \text{ N}$$

4. Deflection at the top: If the characteristics of the mast (dimension of the section, composition) were constant along the midline, in taking the average diameter of 180 mm, one would obtain for the previous maximum load the following deflection at the top:

$$\Delta = \frac{F_{\text{max}} \times L^3}{3E_x I_z} \# 1 \text{ m}$$

To obtain a more precise value, it is necessary to discretize the mast into finite elements of shorter beams (four or five) with corresponding sections and moduli (helical angle increasing due to the decreasing diameter, the moduli E_x and E_y vary a little).

18.1.15 Unidirectional Ply of HR Carbon

Problem Statement:

Consider a unidirectional ply made of HR (high strength) carbon/epoxy. What is the fiber volume fraction one can predict to obtain a modulus of elasticity in the longitudinal direction that is comparable to duralumin (AU4G – 2024)?

Solution:

In the fiber direction, the modulus of elasticity E_ℓ is given by the relation (see Section 3.3.1):

$$E_\ell = E_f V_f + E_m (1 - V_f)$$

One reads in the tables in Section 1.6:

HR carbon: $E_f = 230,000$ MPa

Epoxy resin: $E_m = 4500$ MPa

Duralumin: $E_{2024} = 75,000$ MPa.

The fiber volume fraction V_f has to be such that:

$$E_{2024} = E_f V_f + E_m(1 - V_f)$$

where:

$$V_f = \frac{E_{2024} - E_m}{E_f - E_m}$$

$$\boxed{V_f = 31\%}$$

18.1.16 Manipulator Arm of Space Shuttle

Problem Statement:

A manipulator arm is made of two identical tubular columns in carbon/epoxy ($V_f = 60\%$; thin cylindrical tubes of revolution) with pins as shown in Figure 18.8.

Among the different geometric configurations found when the arm is deployed, one can consider the geometries noted as (a), (b), and (c) in Figure 18.9. F represents the concentrated inertial force.

Note the following:

E_x = Longitudinal modulus of elasticity of the tube in x direction (Figure 18.8)

G_{xy} = Shear modulus in the tangent plane x, y (Figure 18.8)

I = Quadratic moment of flexure of a cross section (annular) of the tube with respect to its diameter

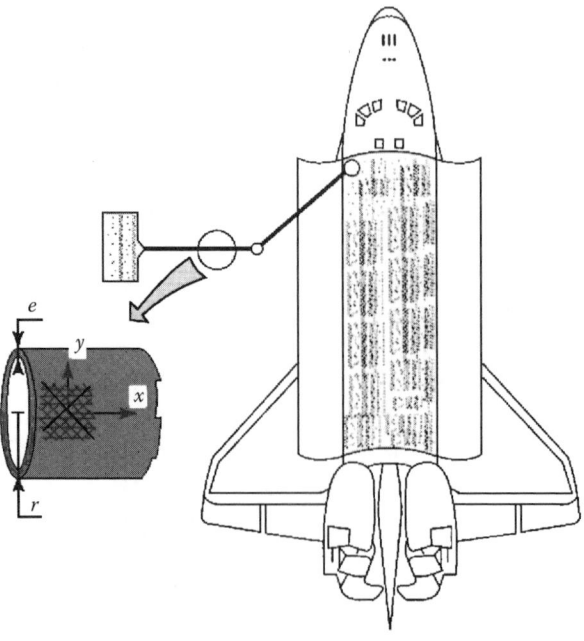

Figure 18.8

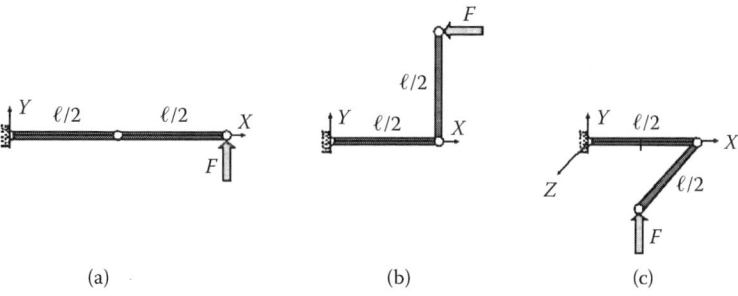

Figure 18.9

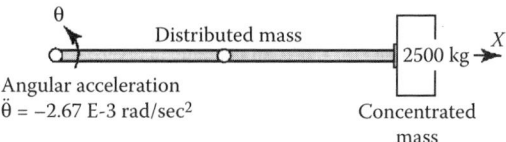

Figure 18.10

1. Calculate the deflection components along the directions X, Y, Z (Figure 18.9) at the point of application of the force F for each of the configurations (a), (b), (c) as function of F, ℓ, I, E_x, and G_{xy} (neglect the strains due to shear force and normal force). Comment on the relative values of these displacements.

2. What should be the ratio between E_x and G_{xy} to obtain identical deflections in the configurations (a) and (c)?

3. The tube is laminated starting from unidirectional layers. By means of the tables giving the moduli E_x and G_{xy} (Section 5.4.2), indicate by simple reading and without interpolation the composition of the laminate that verifies the ratio found in the previous question within a few percentages (choose G_{xy} as large as possible), as well as the values of the elastic characteristics.

4. Verify that this composition is preferable, for the mass assessment, to that of another tube of the same diameter but with different thickness, which has a modulus of elasticity E_x' as large as possible and with the same deflection as previously for configuration (c).

5. Keep the properties determined for the laminate in question 3. The arm has an average diameter of 0.3 m. Each of the two columns has a length of 7.5 m. One imposes a minimum stiffness for the arm $(F/\Delta)_{minimum} = 10^4$ N/m, where Δ is the deflection under the load F. Calculate the thickness of the tube, indicate the number of total unidirectional layers and the number of layers in each of the four orientations.

6. With the data given in Figure 18.10, verify that the distributed mass of the arm does not significantly influence the previous results during the adjustment in position of the apparatus.

Solution:

1. Starting from the relations of flexure and torsion of composite tubes (see Section 5.4.5, Figure 5.31):

$$E_x I \frac{d^2 v}{dX^2} = M_f; \quad G_{xy} I_o \frac{d\theta_X}{dX} = M_t$$

one obtains for the components of displacement at the end:

■ Configuration (a):

$$\Delta_Y = \frac{F\ell^3}{3E_x I}$$

■ Configuration (b):

$$\Delta_x = \frac{F(\ell/2)}{E_x I} \times \frac{\ell}{2} \times \frac{\ell}{2} - \frac{F(\ell/2)^3}{3E_x I} = -\frac{F\ell^3}{6E_x I}$$

$$\Delta_y = \frac{F(\ell/2)}{2E_x I} \times \left(\frac{\ell}{2}\right)^2 = \frac{F\ell^3}{16E_x I}$$

■ Configuration (c):

$$\Delta_y = \frac{F(\ell/2)^3}{3E_x I} \times 2 + \frac{F(\ell/2)}{G_{xy} I_o} \times \frac{\ell}{2} \times \frac{\ell}{2} = \cdots$$

$$\cdots \frac{F\ell^3}{8E_x I} \left(\frac{2}{3} + \frac{E_x}{2G_{xy}}\right)$$

Remark: For configurations (a) and (b), one obtains a displacement that is as small as the modulus E_x is large. Then (see Section 5.4.2, Tables 5.4 and 5.5), G_{xy} is relatively small, which means that $E_x/G_{xy} \gg 1$. The displacement of configuration (c) is much larger than the others. This will create problems when operating the arm.

2. The deflections are identical for configurations (a) and (c) if

$$\frac{1}{3} = \frac{1}{8}\left(\frac{2}{3} + \frac{E_x}{2G_{xy}}\right)$$

then:

$$\boxed{\frac{E_x}{G_{xy}} = 4}$$

3. In looking for the modulus G_{xy} to be as high as possible, one reads on Tables 5.4 and 5.5 (Section 5.4.2) a ratio $E_x/G_{xy} = 3.9\ (\simeq 4)$ for the composition:

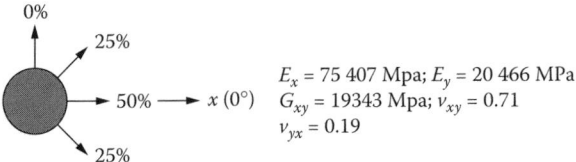

$E_x = 75\ 407$ Mpa; $E_y = 20\ 466$ MPa
$G_{xy} = 19343$ Mpa; $\nu_{xy} = 0.71$
$\nu_{yx} = 0.19$

4. The maximum value of the longitudinal modulus of elasticity observed on Table 5.4 is

$$E'_x\ =\ 134{,}000\ \text{MPa}$$

This corresponds to a shear modulus (Table 5.5):

$$G'_{xy}\ =\ 4{,}200\ \text{MPa}$$

The same deflection as the previous one for the configuration (c) can be obtained by

$$\frac{F\ell^3}{8E'_x\ I'}\left(\frac{2}{3} + \frac{E'_x}{2G'_{xy}}\right) = \frac{F\ell^3}{3E_x I}$$

then:

$$\frac{I'}{I} = \frac{\pi r^3 e'}{\pi r^3 e} = \frac{3E_x}{8E'_x}\left(\frac{2}{3} + \frac{E'_x}{2G'_{xy}}\right) = 3.5$$

$$\frac{e'}{e} = 3.5$$

The tube with thickness e' and modulus E'_x will be more stiff for configuration (a) but will have a mass multiplied by 3.5 to keep the stiffness of configuration (c).

5. Configurations (a) and (c) are the more deformable. One then has to write

$$\frac{F}{\Delta_y} = \frac{3E_x I}{\ell^3} \geq \left(\frac{F}{\Delta}\right)_{min}$$

with $\ell = 15$ m; $I = \pi r^3 e$; $r = 0.15$ m; $(F/\Delta)_{min} = 10^4$ N/m; $E_x = 75{,}407$ MPa

$$e \geq 14\ \text{mm}$$

The ply thickness being 0.13 mm, one obtains 108 layers oriented as follows:

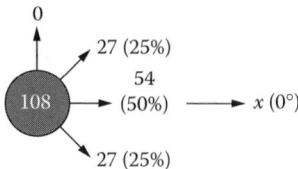

6. The specific mass of the laminate is indicated in Section 3.3.3, as $\rho = 1530$ kg/m.^{3} The distributed mass of the arm is then:

$$\frac{m}{\ell} = 2\pi re \times \rho = 20.2 \text{ kg/m}$$

with the angular acceleration indicated in Figure 18.8, one obtains the following inertial load:

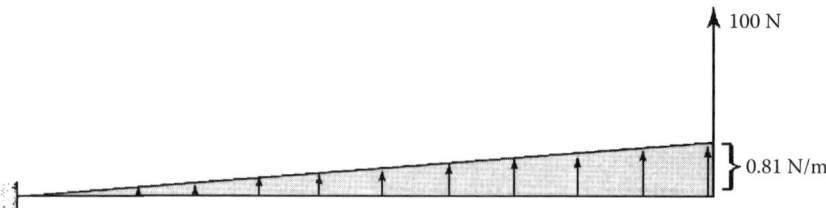

We then deduce from there:

■ The deflection at the end due to the concentrated mass:

$$\Delta_{\text{concent.}} = \frac{100\ell^3}{3E_x I}$$

■ The deflection at the end due to distributed load[25]:

$$\Delta_{\text{distributed}} = \frac{11}{120} \times \frac{0.81\ell^4}{E_x I}$$

from which we can obtain a total deflection:

$$\Delta_{\text{total}} = \frac{100\ell^3}{3E_x I}(1 + 0.033) \# \frac{100\ell^3}{3E_x I}$$

The rigidity (F/Δ_{total}) appears to be well related essentially to the concentrated inertial load at the extremity of the arm.

---

[25] Result obtained from the differential equation: $EI_x d^2 v/dX^2 = -\frac{0.81}{6}\ell^2[2 - 3(X/\ell) + (X/\ell)^3]$

18.2 LEVEL 2

18.2.1 Sandwich Beam: Simplified Calculation of the Shear Coefficient

Problem Statement:

Represented below is the cross section of a sandwich beam. The thickness of the skins is small compared with that of the core. Under the action of a shear load T, the shear stresses in the section are assumed to vary in a piecewise-linear fashion[26] along the y direction. The constitutive materials, denoted as 1 and 2, are assumed to be isotropic, or transversely isotropic. The shear moduli are denoted as G_1 for material 1 (skin) and G_2 for material 2 (core). The beam has a width of unity.

1. Calculate the shear coefficient k for flexure in the plane x,y.

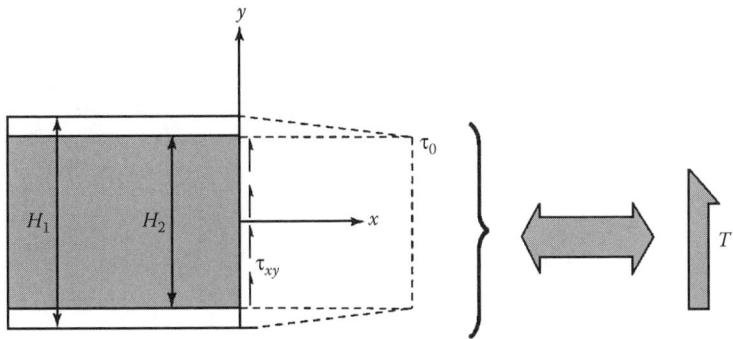

2. Give a simplified expression for the case—current in the applications— where $G_1 \gg G_2$ with the notations for the thicknesses:

$$e_p = \frac{H_1 - H_2}{2}; \quad e_c = H_2$$

Solution:

1. Let W be the strain energy due to shear stresses. One has (Equation 15.17):

$$\frac{dW}{dx} = \frac{1}{2} \frac{kT^2}{\langle GS \rangle} = \frac{1}{2} \int_{\text{section}} \frac{\tau_{xy}^2}{G_i} dy$$

In the upper skin, one has

$$\tau_{xy} = \frac{H_1 - 2y}{H_1 - H_2} \times \tau_o$$

---

[26] This representation of the shear stresses is only approximate. One will find in Application 18.3.5 the results concerning a more precise distribution of these stresses. In fact, the approximate representation of the shear proposed here is better approximated than the skins of the sandwich structure will have a small thickness as compared to that of the core.

On the other hand in the core: $\tau_{xy} = \tau_0$

Then with:

$$T = \int_{\text{section}} \tau_{xy}(dy \times 1)$$

One deduces from there the maximum shear stress τ_o:

$$\tau_o = T \times \frac{2}{H_1 + H_2}$$

Strain energy:

$$\frac{dW}{dx} = \frac{1}{2}\int \frac{\tau_{xy}^2}{Gi}dy = \int_o^{H_2/2} \frac{\tau_o^2}{G_2}dy + \int_{H_2/2}^{H_1/2} \frac{\tau_o^2(H_1 - 2y)^2}{G_1(H_1 - H_2)^2}dy$$

After calculation:

$$\frac{1}{2}\int \frac{\tau_{xy}^2}{G_i}dy = \frac{\tau_o^2}{2}\left(\frac{H_2}{G_2} + \frac{H_1 - H_2}{3G_1}\right) = \frac{2T^2}{(H_1 + H_2)^2}\left(\frac{H_2}{G_2} + \frac{H_1 - H_2}{3G_1}\right)$$

one then has

$$\frac{1}{2}\frac{kT^2}{\langle GS \rangle} = \frac{2T^2}{(H_1 + H_2)^2}\left(\frac{H_2}{G_2} + \frac{H_1 - H_2}{3G_1}\right)$$

Then:

$$k = \frac{4\langle GS \rangle}{(H_1 + H_2)^2}\left(\frac{H_2}{G_2} + \frac{H_1 - H_2}{3G_1}\right)$$

with (Equation 15.16): $\langle GS \rangle = G_1(H_1 - H_2) + G_2 H_2$:

$$\boxed{k = \frac{4[G_1(H_1 - H_2) + G_2 H_2]}{(H_1 + H_2)^2}\left(\frac{H_2}{G_2} + \frac{H_1 - H_2}{3G_1}\right)}$$

2. Case where $G_2 \ll G_1$: One can rewrite

$$k = \frac{4\langle GS \rangle}{(e_c + 2e_p + e_c)^2} \times \frac{e_c}{G_c}\left[1 + \underbrace{\frac{2}{3}\frac{e_p G_c}{e_c G_p}}_{\ll 1}\right]$$

then:

$$k \# \frac{\langle GS \rangle}{e_c^2\left(1 + \dfrac{e_p}{e_c}\right)^2} \times \frac{e_c}{G_c}$$

One obtains the following simplified form, valid if $e_p \ll e_c$ and $G_c \ll G_p$:

$$\boxed{\frac{k}{\langle GS \rangle} = \frac{1}{G_c(e_c + 2e_p)}}$$

18.2.2 Procedure for Calculation of a Laminate

Problem Statement:

Consider a balanced carbon/epoxy laminate with respect to the 0° direction (or x), having midplane symmetry. The plies have the orientations 0°, 90°, +45°, −45° with certain proportions (recall that there are as many plies of +45° as there are −45°). This laminate is subjected to a unit uniaxial stress $\sigma_{ox} = 1$ MPa (see following figure).

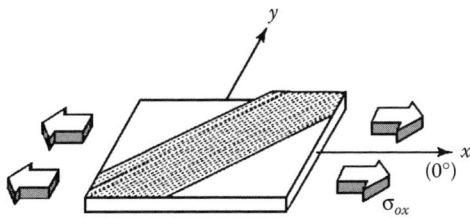

Propose a procedure to establish a simple program allowing one to obtain the following:

1. The modulus of elasticity E_x of the laminate and the Poisson coefficient v_{xy}[27]
2. The stresses in each ply and in the orthotropic axes of this ply[28]
3. The Hill–Tsai[29] expression for each ply
4. The largest stress $\sigma_{ox\ max}$ admissible without failure of any ply

One gives in the following the characteristics of the unidirectional plies (identical) making up the laminate:

Carbon/epoxy ply with $V_f = 60\%$ fiber volume fraction

$$E_\ell = 134{,}000 \text{ MPa}[30]; \ E_t = 7000 \text{ MPa}; \ G_{lt} = 4200 \text{ MPa}; \ v_{lt} = 0.25$$

Fracture strengths:
$\sigma_{\ell\ \text{tension}} = 1{,}270$ MPa; $\sigma_{\ell\ \text{compression}} = 1130$ MPa
$\sigma_{t\ \text{tension}} = 42$ MPa; $\sigma_{t\ \text{compression}} = 141$ MPa
$\tau_{\ell t} = 63$ MPa

---

[27] See Equation 12.8.
[28] These are the stresses σ_ℓ, σ_t, $\tau_{\ell t}$ (see, for example, Equation 11.1).
[29] See Chapter 14.
[30] See Section 3.3.3, Table 3.4.

Solution:

Recall first the procedure for calculation (see also Section 12.1.3.):

1. Modulus E_x and Poisson coefficient v_{xy}: The behavior of a laminate having midplane symmetry and working in its plane can be written as (Equation 12.7):

$$\left\{ \begin{array}{c} \sigma_{ox} \\ \sigma_{oy} \\ \tau_{oxy} \end{array} \right\} = \frac{1}{h} \begin{bmatrix} A_{11} & A_{12} & A_{13} \\ A_{21} & A_{22} & A_{23} \\ A_{31} & A_{32} & A_{33} \end{bmatrix} \left\{ \begin{array}{c} \varepsilon_{ox} \\ \varepsilon_{oy} \\ \gamma_{oxy} \end{array} \right\} \qquad [a]$$

with:

$$\frac{1}{h} A_{ij} = \sum_{k=1^{st} \text{ply}}^{n^{th} \text{ply}} \bar{E}_{ij}^k \frac{e_k}{h}$$

where e_k is thickness of ply k, and h is the total thickness of the laminate. $[\bar{E}_{ij}]_k$ is the stiffness matrix for the ply k in the x,y axes (see Equation 11.8), as:

$$\left\{ \begin{array}{c} \sigma_x \\ \sigma_y \\ \tau_{xy} \end{array} \right\} = \begin{bmatrix} \bar{E}_{11} & \bar{E}_{12} & \bar{E}_{13} \\ \bar{E}_{21} & \bar{E}_{22} & \bar{E}_{23} \\ \bar{E}_{31} & \bar{E}_{32} & \bar{E}_{33} \end{bmatrix} \left\{ \begin{array}{c} \varepsilon_{ox} \\ \varepsilon_{oy} \\ \gamma_{oxy} \end{array} \right\} \qquad [b]$$

$$\text{ply } k \qquad\qquad \text{ply } k \qquad \text{ply } k$$

Note that $p^{0°}$ (%), $p^{90°}$ (%), $p^{45°}$ (%), $p^{-45°}$ (%) are the respective proportions of the plies in the directions 0°, 90°, +45°, −45°. The previous terms $(1/h) A_{ij}$ can be written as:

$$\frac{1}{h} A_{ij} = \bar{E}_{ij}^{0°} p^{0°} + \bar{E}_{ij}^{90°} p^{90°} + \bar{E}_{ij}^{45°} p^{45°} + \bar{E}_{ij}^{-45°} p^{-45°} \qquad [c]$$

Here the terms $\frac{1}{h} A_{13}$, $\frac{1}{h} A_{23}$, and their symmetrical counterparts are zero because the laminate is balanced (see Equation 11.8).

The relation denoted as [a] above is then inverted and can be written as:

$$\left\{ \begin{array}{c} \varepsilon_{ox} \\ \varepsilon_{oy} \\ \gamma_{oxy} \end{array} \right\} = \begin{bmatrix} \dfrac{1}{E_x} & -\dfrac{\bar{v}_{yx}}{E_y} & 0 \\ -\dfrac{\bar{v}_{xy}}{E_x} & \dfrac{1}{E_y} & 0 \\ 0 & 0 & \dfrac{1}{G_{xy}} \end{bmatrix} \left\{ \begin{array}{c} \sigma_{ox} \\ \sigma_{oy} \\ \tau_{oxy} \end{array} \right\} \qquad [d]$$

where $\bar{E}_x$, $\bar{E}_y$, $\bar{G}_{xy}$, $\bar{v}_{xy}$, $\bar{v}_{yx}$ are the global moduli and Poisson coefficients of the laminate.

Here this laminate is subjected to a uniaxial stress $\sigma_{ox} = 1$ MPa, then:

$$\varepsilon_{ox} = \frac{\sigma_{ox}}{\bar{E}_x} = \frac{1 \text{ MPa}}{\bar{E}_x \text{ (MPa)}}; \quad \varepsilon_{oy} = -\frac{\bar{v}_{xy}}{\bar{E}_x}\sigma_{ox} = -\frac{\bar{v}_{xy}}{\bar{E}_x \text{(MPa)}} \times 1 \text{ MPa}$$

One obtains as well the modulus and the Poisson coefficient required:

$$\bar{E}_x \text{(MPa)} = \frac{1}{\varepsilon_{ox}}$$

$$\bar{v}_{xy} = -\varepsilon_{oy} \times \frac{\bar{E}_x \text{(MPa)}}{1 \text{ MPa}}$$

2. Stresses in the ply: The previous result gives us the global strains of the laminate, strains that each ply should follow as:

$$\varepsilon_{ox} = \frac{1}{\bar{E}_x}\sigma_{ox}; \quad \varepsilon_{oy} = \frac{\bar{v}_{xy}}{\bar{E}_x}\sigma_{ox}; \quad \gamma_{oxy} = 0$$

For a ply k, the relation mentioned above in [b] is then written as:

$$\left\{\begin{array}{c} \sigma_x \\ \sigma_y \\ \tau_{xy} \end{array}\right\}_{\text{ply } k} = \begin{bmatrix} \bar{E}_{11} & \bar{E}_{12} & \bar{E}_{13} \\ \bar{E}_{21} & \bar{E}_{22} & \bar{E}_{23} \\ \bar{E}_{31} & \bar{E}_{32} & \bar{E}_{33} \end{bmatrix}_{\text{ply } k} \left\{\begin{array}{c} \varepsilon_{ox} \\ \varepsilon_{oy} \\ 0 \end{array}\right\} \quad \text{[e]}$$

This gives the stresses in ply k, expressed in the coordinates x, y. One can express them in the orthotropic axes of the ply (axes ℓ, t of the following figure, and Equation 11.4 recalled below):

$$\left\{\begin{array}{c} \sigma_\ell \\ \sigma_t \\ \tau_{\ell t} \end{array}\right\}_{\text{ply } k} = \begin{bmatrix} c^2 & s^2 & -2cs \\ s^2 & c^2 & 2cs \\ sc & -sc & (c^2 - s^2) \end{bmatrix} \left\{\begin{array}{c} \sigma_x \\ \sigma_y \\ \tau_{xy} \end{array}\right\}_{\text{ply } k} \quad \begin{array}{l} c = \cos\theta \\ s = \sin\theta \end{array} \quad \text{[f]}$$

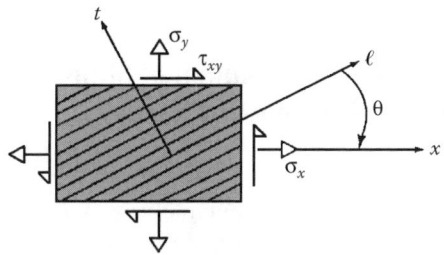

3. Hill–Tsai expression: It is written as (Equation 14.7):

$$\alpha^2 = \frac{\sigma_\ell^2}{\sigma_\ell^2_{\text{rupture}}} + \frac{\sigma_t^2}{\sigma_t^2_{\text{rupture}}} - \frac{\sigma_\ell \sigma_t}{\sigma_\ell^2_{\text{rupture}}} + \frac{\tau_{\ell t}^2}{\tau_{\ell t}^2_{\text{rupture}}}$$

One can then calculate the values $(\alpha^2)_k$ required for each ply k.

4. The largest stress $\sigma_{ox\,max}$ without fracture:
The stresses σ_ℓ, σ_t, and $\tau_{\ell t}$ are calculated for a uniaxial stress: $\sigma_{ox} = 1$ MPa. Now apply the maximum stress found $\sigma_{ox\,max}$ (MPa). The stresses σ_ℓ, σ_t, and $\tau_{\ell t}$ in the ply k are multiplied by the ratio:

$$\frac{\sigma_{ox\,max}}{1\,\text{MPa}}$$

and the critical value of the Hill–Tsai expression is obtained as:

$$\frac{\sigma_{ox\,max}^2}{(1\,\text{MPa})^2} \left\{ \frac{\sigma_\ell^2}{\sigma_\ell^2_{\text{rupture}}} + \frac{\sigma_t^2}{\sigma_t^2_{\text{rupture}}} - \frac{\sigma_\ell \sigma_t}{\sigma_\ell^2_{\text{rupture}}} + \frac{\tau_{\ell t}^2}{\tau_{\ell t}^2_{\text{rupture}}} \right\}_k = 1$$

With the values $(\alpha^2)_k$ found in the previous question for the Hill–Tsai expression between brackets, one obtains

$$\sigma_{ox\,max}^2 \alpha_k^2 = (1\,\text{MPa})^2$$

Then:

$$\sigma_{ox\,max} = \frac{1\,\text{MPa}}{\alpha_k}$$

Examination of each ply will lead to a different value for $\sigma_{ox\,max}$. One has to keep the minimum value as the critical stress that should initialize damage (failure of a ply) as:

$$\sigma_{ox\,max} = \min \frac{1}{\alpha_k}$$

18.2.3 Kevlar/Epoxy Laminates: Evolution of Stiffness Depending on the Direction of the Load

Problem Statement:

Consider the balanced laminates of Kevlar/epoxy with $V_f = 60\%$ fiber volume fraction, working in their planes, with the following compositions:

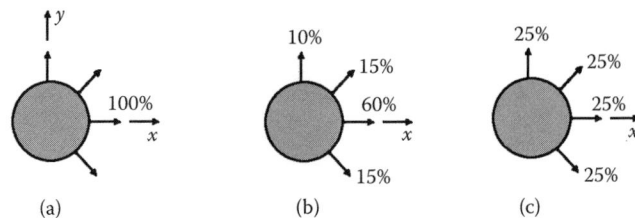

(a)   (b)   (c)

1. Give the expression of the longitudinal modulus of elasticity for these laminates denoted as $E(\theta)$ for a direction i in the plane xy making an angle θ with the direction x.
2. Give for each of the laminates the expression for the "specific modulus" $E(\theta)/\rho$, ρ being the mass density. Use the tables in Section 5.4.2.
3. Represent in polar coordinates the variations of the specific modulus with θ for each of the laminates.
4. Compare with the specific moduli of conventional materials, steel, aluminum alloys Duralumin-2024, and titanium alloy TA6V.

Solution:

Each of the balanced laminates constitutes a thin plate of orthotropic material, with orthotropic axes x, y, z (see figures above and below). The constitutive relation corresponds with Equation 12.9.

For a balanced laminate, this law is reduced to the following expression:

$$
\left\{
\begin{array}{c}
\varepsilon_{ox} \\
\varepsilon_{oy} \\
\gamma_{oxy}
\end{array}
\right\}
=
\left[
\begin{array}{ccc}
\dfrac{1}{E_x} & -\dfrac{v_{yx}}{E_y} & 0 \\
-\dfrac{v_{xy}}{E_x} & \dfrac{1}{E_y} & 0 \\
0 & 0 & \dfrac{1}{G_{xy}}
\end{array}
\right]
\left\{
\begin{array}{c}
\sigma_{ox} \\
\sigma_{oy} \\
\tau_{oxy}
\end{array}
\right\}
$$

1. E_x, E_y, G_{xy} are the moduli in the orthotropic axes x, y, which means the moduli of the laminate. In the axes i, j (see following figure) making an angle θ with the axes x, y, these coefficients are transformed according to the Equation 13.8. The modulus in the direction i is[31]

$$
\boxed{E(\theta) = \cfrac{1}{\cfrac{\cos^4\theta}{E_x} + \cfrac{\sin^4\theta}{E_y} + \cos^2\theta\sin^2\theta\left(\cfrac{1}{G_{xy}} - \cfrac{2v_{xy}}{E_x}\right)}}
$$

---

[31] Recalling (Section 9.3 and Application 18.1.2), the relation: $v_{xy}/E_x = v_{yx}/E_y$.

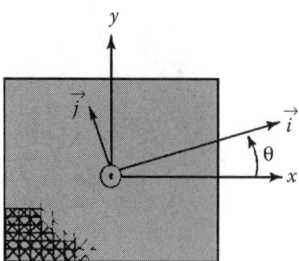

2. Specific modulus: One finds in the tables of Section 5.4.2 the coefficients E_x, ν_{xy}, G_{xy} of the Kevlar/epoxy laminates (Tables 5.9 and 5.10). Table 5.9 also allows one to obtain the value E_y. For this it is sufficient to permute the 0° percentage and 90° percentage.

The specific mass ρ is shown in Table 3.4 in Section 3.3.3. It can also be calculated using the relation in Section 3.2.3.

One has $\rho = 1350$ kg/m³. One then has for the expressions of the specific modulus:

■ Laminate (a):

$E_x = 85{,}000$ MPa
$E_y = 5600$ MPa
$G_{xy} = 2100$ MPa
$\nu_{xy} = 0.34$

$$\frac{E(\theta)}{\rho}(m/s)^2 = \frac{10^6/1350}{\dfrac{\cos^4\theta}{85{,}000} + \dfrac{\sin^4\theta}{5600} + \cos^2\theta\sin^2\theta\left(\dfrac{1}{2100} - 2\times\dfrac{0.34}{85{,}000}\right)}$$

■ Laminate (b):

$E_x = 56{,}600$ MPa
$E_y = 18{,}680$ MPa
$G_{xy} = 8030$ MPa
$\nu_{xy} = 0.4$

$$\frac{E(\theta)}{\rho}(m/s)^2 = \frac{10^6/1350}{\dfrac{\cos^4\theta}{56{,}600} + \dfrac{\sin^4\theta}{18{,}680} + \cos^2\theta\sin^2\theta\left(\dfrac{1}{8030} - 2\times\dfrac{0.4}{56{,}600}\right)}$$

■ Laminate (c): The proportions of 25% along the directions 0° and 90° can be obtained from Table 5.9. In this view one has to evaluate by extrapolation, starting from the values corresponding to the percentages of 20% and 30%, as[32]:

$$E_x = (1/2)\,(28{,}260 + 35{,}400) = 31{,}830 \text{ MPa}$$

---

[32] See also Application 18.2.14.

$$E_y = E_x$$
$$G_{xy} = 11,980 \text{ MPa}$$
$$v_{xy} = 0.335$$

$$\frac{E(\theta)}{\rho}(\text{m/s})^2 = \frac{10^6/1350}{\dfrac{\cos^4\theta + \sin^4\theta}{31,830} + \cos^2\theta\sin^2\theta\left(\dfrac{1}{11,980} - 2 \times \dfrac{0.335}{31,830}\right)}$$

3. One obtains the evolutions in the following figure for the specific modulus. One verifies well that it is possible to "control" the anisotropy of the laminate by modifying the percentages of the plies at 0°, 90°, +45°, −45°.

4. For the other materials, one obtains immediately (Section 1.6):
$$E/\rho \text{ (steel)} = 26.3 \times 10^6 \text{ (m/s)}^2$$
$$E/\rho \text{ (Duralumin-2024)} = 26.8 \times 10^6 \text{ (m/s)}^2$$
$$E/\rho \text{ (Titanium-TA6V)} = 23.9 \times 10^6 \text{ (m/s)}^2$$

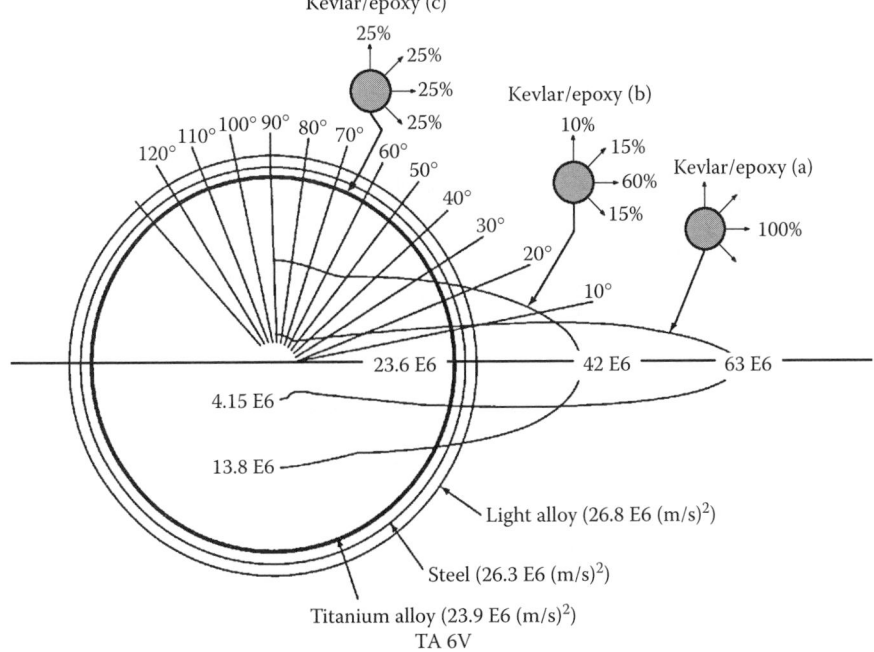

Remark: The notion of specific modulus is particularly important for aeronautical construction. When one compares on the above diagram the performances of Kevlar/epoxy with those of steel, Duralumin, and titanium, one sees clearly the advantage for the laminate inside angular borders for the directions of application of the loads.

18.2.4 Residual Thermal Stresses due to Curing of the Laminate

Problem Statement:

Consider a laminated panel in carbon/epoxy with $V_f = 60\%$ fiber volume fraction, with midplane symmetry, and a composition shown in the following figure:

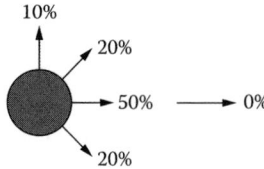

It is cured in an autoclave at 180°C and demolded at 20°C.

1. Calculate the thermal deformations due to demolding.
2. Calculate the thermal residual stresses in the 90° plies.

Solution:

1. Thermal deformations:

The thermomechanical behavior of the laminate can be written as (see Equation 12.19):

$$
\left\{
\begin{array}{c}
\varepsilon_{ox} \\
\varepsilon_{oy} \\
\gamma_{oxy}
\end{array}
\right\}
=
\begin{bmatrix}
\dfrac{1}{\bar{E}_x} & -\dfrac{\bar{v}_{yx}}{\bar{E}_y} & \dfrac{\bar{\eta}_{xy}}{\bar{G}_{xy}} \\
-\dfrac{\bar{v}_{xy}}{\bar{E}_x} & \dfrac{1}{\bar{E}_y} & \dfrac{\bar{\mu}_{xy}}{\bar{G}_{xy}} \\
\dfrac{\bar{\eta}_x}{\bar{E}_x} & \dfrac{\bar{\mu}_y}{\bar{E}_y} & \dfrac{1}{\bar{G}_{xy}}
\end{bmatrix}
\left\{
\begin{array}{c}
\sigma_{ox} \\
\sigma_{oy} \\
\tau_{oxy}
\end{array}
\right\}
+ \Delta T
\left\{
\begin{array}{c}
\alpha_{ox} \\
\alpha_{oy} \\
\alpha_{oxy}
\end{array}
\right\}
$$

The panel is not subjected to any external mechanical loading. This law can then be written as:

$$
\left\{
\begin{array}{c}
\varepsilon_{ox} \\
\varepsilon_{oy} \\
\gamma_{oxy}
\end{array}
\right\}
= \Delta T
\left\{
\begin{array}{c}
\alpha_{ox} \\
\alpha_{oy} \\
\alpha_{oxy}
\end{array}
\right\}
$$

The laminate being balanced, Equations 12.18, 12.17, and 11.10 lead to

$$\alpha_{oxy} = 0$$

Then Table 5.4 of Section 5.4.2 indicates for the laminate with the corresponding percentages of the composition above:

$$\alpha_{ox} = -0.072 \times 10^{-5}$$

One also deduces from Table 5.4, by permutation between 0° and 90°:

$$\alpha_{oy} = 0.44 \times 10^{-5}$$

Therefore, the thermal strains corresponding to $\Delta T = -160$ °C are

$$\varepsilon_{ox} = -160 \times (-0.0072 \times 10^{-5}); \; \varepsilon_{oy} = -160 \times (0.44 \times 10^{-5})$$

or:

$$\varepsilon_{ox} = 115 \times 10^{-6}$$

$$\varepsilon_{oy} = -704 \times 10^{-6}$$

$$\gamma_{oxy} = 0$$

2. Thermal residual stresses in the 90° plies:
The Equation 11.10 allows one to write

$$\sigma_x = \bar{E}_{11}^{90°}\varepsilon_{ox} + \bar{E}_{12}^{90°}\varepsilon_{oy} - \Delta T \overline{\alpha E}_1^{90°}$$

where:

$$\overline{\alpha E}_1^{90°} = \bar{E}_t(v_{\ell t}\alpha_\ell + \alpha_t)$$

with (Equation 11.8):

$$\bar{E}_{11}^{90°} = \bar{E}_t \quad \text{and} \quad \bar{E}_{12}^{90°} = v_{t\ell}\bar{E}_\ell$$

The moduli of elasticity and coefficients of expansion are given in Section 3.3.3, Table 3.4.[33] Then:

$$\overline{\alpha E}_1^{90°} = 0.237$$

With the known values ε_{ox} and ε_{oy}, one has

$$\sigma_x = 7021 \times 115 \times 10^{-6} + 1717 \times (-704 \times 10^{-6}) - (-160)(0.237) = 37.5 \text{ MPa}$$

In an analogous manner:

$$\sigma_y = \bar{E}_{21}^{90°}\varepsilon_{ox} + \bar{E}_{22}^{90°}\varepsilon_{oy} - \Delta T \overline{\alpha E}_2^{90°}$$

---

[33] Recall also the property $v_{t\ell}/E_t = v_{\ell t}/E_\ell$ (see Sections 3.1 and 3.2 and Application 18.1.2).

with (Equation 11.8):

$$\bar{E}_{22}^{90°} = \bar{E}_\ell \quad \text{and} \quad \overline{\alpha E_2}^{90°} = \bar{E}_\ell(\alpha_\ell + \nu_{t\ell}\alpha_t)$$

One obtains

$$\sigma_y = -110.2 \text{ MPa}$$

and

$$\tau_{xy} = 0$$

One then has in the axes ℓ, t of the 90° plies (see Equation 11.4):

$$\sigma_\ell = -110.2 \text{ MPa}$$

$$\sigma_t = 37.5 \text{ MPa}$$

$$\tau_{\ell t} = 0$$

Remark: If one writes the Hill–Tsai expression (Section 5.3.2) for the 90° plies, one obtains with the rupture strengths of Section 3.3.3, Table 3.4:

$$\left(\frac{110.2}{1130}\right)^2 + \left(\frac{37.5}{42}\right)^2 - \left(\frac{(-110.2)(37.5)}{1130^2}\right) = 0.81$$

The factor of safety[34] is only:

$$\frac{1}{\sqrt{0.81}} - 1 = 11\%$$

This is due to high value of σ_t close to the rupture strength and explains the phenomenon of **microfracture** of the resin that happens during cooling. Subsequently, the microcracks favor the absorption of moisture by the resin and the fibers, which provoke expansions analogous to those induced by heating, with coefficients of expansion of hygrometric nature. Then, the residual stresses in the plies will be generally weaker.

18.2.5 Thermoelastic Behavior of a Tube Made of Filament-Wound Glass/Polyester

Problem Statement:

Obtain the thermoelastic behavior of a cylindrical tube made by filament winding E glass/polyester, with ±45° balanced composition, with a fiber volume fraction of $V_f = 25\%$.

---

[34] See Section 14.2.3.

Solution:

In the x,y axes (following figure), the stress–strain law takes the form (see Equation 12.19):

$$\left\{\begin{array}{c} \varepsilon_{ox} \\ \\ \varepsilon_{oy} \\ \\ \gamma_{oxy} \end{array}\right\} = \left[\begin{array}{ccc} \dfrac{1}{\bar{E}_x} & -\dfrac{\bar{v}_{yx}}{\bar{E}_y} & \dfrac{\bar{\eta}_{xy}}{\bar{G}_{xy}} \\ -\dfrac{\bar{v}_{xy}}{\bar{E}_x} & \dfrac{1}{\bar{E}_y} & \dfrac{\bar{\mu}_{xy}}{\bar{G}_{xy}} \\ \dfrac{\bar{\eta}_x}{\bar{E}_x} & \dfrac{\bar{\mu}_y}{\bar{E}_y} & \dfrac{1}{\bar{G}_{xy}} \end{array}\right] \left\{\begin{array}{c} \sigma_{ox} \\ \\ \sigma_{oy} \\ \\ \tau_{oxy} \end{array}\right\} + \Delta T \left\{\begin{array}{c} \alpha_{ox} \\ \\ \alpha_{oy} \\ \\ \alpha_{oxy} \end{array}\right\}$$

■ Calculation of moduli:

First we have to evaluate the terms of the matrix $h^{-1}[A_{ij}]$ (see Equation 12.7). This calculation requires the knowledge of the stiffness coefficients for each ply $\bar{E}_{ij}$ (see Equation 11.18).

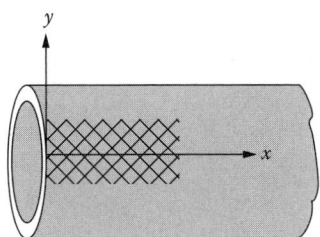

In this view, first calculate the elastic moduli of a ply in its principal axes (ℓ,t); one has (Equation 10.2 and those that follow and numerical values in Tables 1.3 and 1.4 in Section 1.6)

$$E_\ell = 74{,}000 \times 0.25 + 4{,}000 \times 0.75 = 21{,}500 \text{ MPa}$$

$$v_{\ell t} = 0.25 \times 0.25 + 0.4 \times 0.75 = 0.36$$

$$E_t = 4000\frac{1}{0.75 + \frac{4000}{74{,}000} \times 0.25} = 5240 \text{ MPa}$$

$$G_{\ell t} = 1400\frac{1}{0.75 + \frac{1400}{30{,}000} \times 0.25} = 1840 \text{ MPa}$$

$$v_{t\ell} = (5240/21{,}500) \times 0.36 = 0.088$$

$$\bar{E}_\ell = 22{,}200 \text{ MPa}; \quad \bar{E}_t = 5410 \text{ MPa}$$

then (Equation 11.8):

$$\bar{E}_{11}^{+45°} = \bar{E}_{11}^{-45°} = \bar{E}_{22}^{+45°} = \bar{E}_{22}^{-45°} = 9720 \text{ MPa}$$

$$\bar{E}_{33}^{+45°} = \bar{E}_{33}^{-45°} = 5928 \text{ MPa}; \quad \bar{E}_{12}^{+45°} = \bar{E}_{12}^{-45°} = 6040 \text{ MPa}$$

$$\bar{E}_{13}^{+45°} = -\bar{E}_{13}^{-45°}; \quad \bar{E}_{23}^{+45°} = -\bar{E}_{23}^{-45°}$$

from which one can write (Equation 12.8)

$$\frac{1}{b}[A_{ij}] = \begin{bmatrix} 9720 & 6040 & 0 \\ 6040 & 9720 & 0 \\ 0 & 0 & 5928 \end{bmatrix} (\text{MPa})$$

The inversion of this matrix leads to (see Equation 12.9)

$$\begin{bmatrix} \dfrac{1}{\bar{E}_x} & -\dfrac{\bar{v}_{yx}}{\bar{E}_y} & 0 \\ -\dfrac{\bar{v}_{xy}}{\bar{E}_x} & \dfrac{1}{\bar{E}_y} & 0 \\ 0 & 0 & \dfrac{1}{\bar{G}_{xy}} \end{bmatrix} = \begin{bmatrix} 1.676 \times 10^{-4} & -1.041 \times 10^{-4} & 0 \\ -1.041 \times 10^{-4} & 1.676 \times 10^{-4} & 0 \\ 0 & 0 & \dfrac{1}{5928} \end{bmatrix} (\text{MPa}^{-1})$$

from which by identification:

$$\boxed{\begin{aligned} \bar{E}_x &= \bar{E}_y = 5966 \text{ MPa} \\ \bar{v}_{yx} &= \bar{v}_{xy} = 0.62 \\ \bar{G}_{xy} &= 5928 \text{ MPa} \end{aligned}}$$

■ Calculation of coefficient of thermal expansion:
One has to calculate first $b^{-1}(\alpha Eb)_x$, $b^{-1}(\alpha Eb)_y$, and $b^{-1}(\alpha Eb)_{xy}$ from the Equation 12.18. This calculation requires knowledge of the terms $\overline{\alpha E_1}$, $\overline{\alpha E_2}$, and $\overline{\alpha E_3}$ of each ply (Equations 12.17 and 11.10 and numerical values in Tables 1.3 and 1.4 of Section 1.6). For that, one has to know the coefficients of expansion α_ℓ and α_t of a ply in its principal axes (ℓ, t). It can be written (Equations 10.7 and 10.8 and numerical values in Tables 1.3 and 1.4 of Section 1.6):

$$\alpha_\ell = 1.55 \times 10^{-5}; \quad \alpha_t = 7.86 \times 10^{-5}$$

$$\overline{\alpha E_1}^{+45°} = \overline{\alpha E_1}^{-45°} = \overline{\alpha E_2}^{+45°} = \overline{\alpha E_2}^{-45°} = 0.476 \text{ MPa/°C}$$

$$\overline{\alpha E_3}^{+45°} = -\overline{\alpha E_3}^{-45°}$$

From which (Equation 12.17):

$$\begin{Bmatrix} \dfrac{1}{b}(\alpha Eb)_x \\ \dfrac{1}{b}(\alpha Eb)_y \\ \dfrac{1}{b}(\alpha Eb)_{xy} \end{Bmatrix} = \begin{Bmatrix} 0.476 \\ 0.476 \\ 0 \end{Bmatrix} (\text{MPa/°C})$$

then (Equation 12.18):

$$
\left\{
\begin{array}{c}
\alpha_{ox} \\
\alpha_{oy} \\
\alpha_{oxy}
\end{array}
\right\}
=
\begin{bmatrix}
1.676 \times 10^{-4} & -1.041 \times 10^{-4} & 0 \\
-1.041 \times 10^{-4} & 1.676 \times 10^{-4} & 0 \\
0 & 0 & \dfrac{1}{5928}
\end{bmatrix}
\left\{
\begin{array}{c}
0.476 \\
0.476 \\
0
\end{array}
\right\}
=
\left\{
\begin{array}{c}
3.02 \times 10^{-5} \\
3.02 \times 10^{-5} \\
0
\end{array}
\right\}
$$

In summary, the thermoelastic behavior of the filament-wound tube in glass/polyester can be written as:

$$
\left\{
\begin{array}{c}
\varepsilon_{ox} \\
\varepsilon_{oy} \\
\gamma_{oxy}
\end{array}
\right\}
=
\begin{bmatrix}
\dfrac{1}{5966} & -\dfrac{0.62}{5966} & 0 \\
-\dfrac{0.62}{5966} & \dfrac{1}{5966} & 0 \\
0 & 0 & \dfrac{1}{5928}
\end{bmatrix}
\left\{
\begin{array}{c}
\sigma_{ox} \\
\sigma_{oy} \\
\tau_{oxy}
\end{array}
\right\}
+ \Delta T
\left\{
\begin{array}{c}
3.02 \times 10^{-5} \\
3.02 \times 10^{-5} \\
0
\end{array}
\right\}
$$

$$
\left(\text{MPa}^{-1} \right) \qquad\qquad \left({}^{\circ}\text{C}^{-1} \right)
$$

18.2.6 Polymeric Tube Under Thermal Load and Creep

Consider a cylindrical tube of revolution made of polyvinylidene fluoride (PVDF) reinforced externally by filament winding of glass/polyester at ±45° (see figure below).

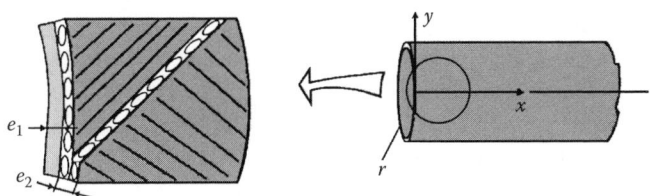

The characteristics of the constituents are as follows:

■ Polymer tube: thickness $e_1 = 10$ mm; isotropic material; modulus of elasticity $E_1 = 260$ MPa; Poisson coefficient ν_1; thermal expansion coefficient $\alpha_1 = 15 \times 10^{-5}$ (°C^{-1}).

■ Glass/polyester reinforcement: thickness $e_2 = 3$ mm; modulus of elasticity E_2; Poisson coefficient ν_2; coefficient of thermal expansion $\alpha_2 = 0.7 \times 10^{-5}$ (°C^{-1}). These coefficients are valid for the behavior in the coordinate axes x, y (see figure). Fiber volume fraction $V_f = 60\%$.

Problem Statement:

The thicknesses e_1 and e_2 are small relative to the average radius of the tube, denoted as r.

1. Give the numerical values of E_2 and v_2 (noting that the moduli of elasticity of epoxy resins and polyester resins are equivalent).
2. When taking into account the temperature variation, denoted as ΔT, the mechanical behavior of the polymer and of the reinforcement, respectively, can be written in the x, y axes as:

$$
\left\{
\begin{array}{c}
\varepsilon_{1x} \\
\varepsilon_{1y} \\
\gamma_{1xy}
\end{array}
\right\}
=
\left[
\begin{array}{ccc}
\dfrac{1}{E_1} & -\dfrac{v_1}{E_1} & 0 \\
-\dfrac{v_1}{E_1} & \dfrac{1}{E_1} & 0 \\
0 & 0 & \dfrac{1}{G_1}
\end{array}
\right]
\left\{
\begin{array}{c}
\sigma_{1x} \\
\sigma_{1y} \\
\tau_{1xy}
\end{array}
\right\}
+ \alpha_1 \Delta T
\left\{
\begin{array}{c}
1 \\
1 \\
0
\end{array}
\right\};
$$

$$
\left\{
\begin{array}{c}
\varepsilon_{2x} \\
\varepsilon_{2y} \\
\gamma_{2xy}
\end{array}
\right\}
=
\left[
\begin{array}{ccc}
\dfrac{1}{E_2} & -\dfrac{v_2}{E_2} & 0 \\
-\dfrac{v_2}{E_2} & \dfrac{1}{E_2} & 0 \\
0 & 0 & \dfrac{1}{G_2}
\end{array}
\right]
\left\{
\begin{array}{c}
\sigma_{2x} \\
\sigma_{2y} \\
\tau_{2xy}
\end{array}
\right\}
+ \alpha_2 \Delta T
\left\{
\begin{array}{c}
1 \\
1 \\
0
\end{array}
\right\}
$$

where one can recognize the strains and stresses in each of the materials. Starting with an assembly (polymer + reinforcement) not stressed nor strained at ambient temperature (20°C), which is heated up to 140°C.

(a) Write the equations for the external equilibrium of the assemblage.
(b) Write the equality of the strains. Deduce a system of equations that allows the calculation of stresses σ_{1x}, σ_{1y}, σ_{2x}, σ_{2y}.
(c) Numerical application: Calculate the stresses in each of the two components (polymer and glass/polyester reinforcement) as well as their strains.

3. Being subjected to high temperature, the internal tube in polymer obeys creep law. The stresses calculated previously do not remain constant in time. They evolve and stabilize at a certain final state. When this state is achieved, if one separates the internal polymer envelope (by imagination) from its reinforcement and cools it quickly from 140°C to 20°C, one will observe residual strains denoted as $\Delta \varepsilon_{1x} = \Delta \varepsilon_{1y} = \Delta \varepsilon$. Note that in the absence of creep, there are no residual strains.

(a) Write the four relations allowing the calculation of the stresses in the assembly at 140°C **after** creep in the polymer, denoted as σ'_{1x}, σ'_{1y}, σ'_{2x}, σ'_{2y}.

(b) Numerical application: From experiments one finds: $\Delta \varepsilon = -0.6 \times \alpha_1 \Delta T$. Calculate the stresses after creep.

4. Considering the assembly at 140°C already crept, one cools the whole reinforced tube quickly, from 140°C to 20°C. Calculate the final stresses in the assembly, denoted as σ''_{1x}, σ''_{1y}, σ''_{2x}, σ''_{2y} at the end of the cooling. Remark.

Solution:

1. We will use for the elastic characteristics of a unidirectional ply of glass/ polyester at $V_f = 0.6$ those of a glass/epoxy ply from Table 3.4. For a laminate at ±45°, Table 5.14 (Section 5.4.2) shows:

$$E_2 = 14{,}130 \text{ MPa}$$

$$v_2 = 0.57$$

2. (a) Equilibrium of the assembly: Sections cut from the tube do not show any external resultant force, in spite of the existence of stresses of thermal origin (see following figure).

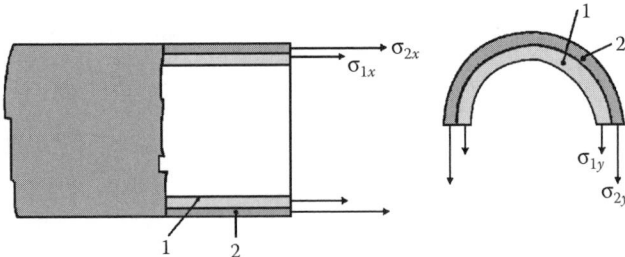

In addition, because the thicknesses are assumed to be small compared with the radius, the stresses will be taken to be uniform over the thicknesses. From there we have the relations:

$$2\pi r(\sigma_{1x} e_1 + \sigma_{2x} e_2) = 0; \quad 1 \times 2(\sigma_{1y} e_1 + \sigma_{2y} e_2) = 0$$

then:

$$\boxed{\begin{aligned} \sigma_{1x} e_1 + \sigma_{2x} e_2 &= 0 \\ \sigma_{1y} e_1 + \sigma_{2y} e_2 &= 0 \end{aligned}} \qquad \text{[1], [2]}$$

Due to the symmetry of revolution for the stress distribution, there are no shear stresses: $\tau_{1xy} = \tau_{2xy} = 0$.

(b) Equality of strains: This is assured by the assumed perfect bonding between the components 1 and 2 as:

$$\varepsilon_{1x} = \varepsilon_{2x}; \qquad \varepsilon_{1y} = \varepsilon_{2y}; \qquad \gamma_{1xy} = \gamma_{2xy}.$$

With the behavior as mentioned in the problem statement, the above equalities become:

$$\frac{\sigma_{1x}}{E_1} - \frac{\nu_1}{E_1}\sigma_{1y} + \alpha_1\Delta T = \frac{\sigma_{2x}}{E_2} - \frac{\nu_2}{E_2}\sigma_{2y} + \alpha_2\Delta T$$

$$-\frac{\nu_1}{E_1}\sigma_{1x} + \frac{\sigma_{1y}}{E_1} + \alpha_1\Delta T = -\frac{\nu_2}{E_2}\sigma_{2x} + \frac{\sigma_{2y}}{E_2} + \alpha_2\Delta T$$

[3], [4]

The above relations [1], [2], [3], [4] constitute a system of four equations for four unknowns σ_{1x}, σ_{1y}, σ_{2x}, σ_{2y}.

(c) In performing successively [3] − [4], [3] + [4], then substituting σ_{2x}, σ_{2y} obtained from [1] and [2], one obtains

$$\begin{cases} \sigma_{1x} - \sigma_{1y} = 0 \\ \sigma_{1x} + \sigma_{1y} = 2\Delta T\dfrac{(\alpha_2 - \alpha_1)}{\left(\dfrac{1 - \nu_1}{E_1}\right) + \dfrac{e_1}{e_2}\left(\dfrac{1 - \nu_2}{E_2}\right)} \end{cases}$$

from which:

$$\sigma_{1x} = \sigma_{1y} = \Delta T\frac{(\alpha_2 - \alpha_1)}{\left(\dfrac{1 - \nu_1}{E_1}\right) + \dfrac{e_1}{e_2}\left(\dfrac{1 - \nu_2}{E_2}\right)}$$

One deduces from there, with $\Delta T = 140 - 20 = 120°C$:

$$\sigma_{1x} = \sigma_{1y} = -6.14 \text{ MPa}$$

$$\sigma_{2x} = \sigma_{2y} = 20.4 \text{ MPa}$$

The internal envelope in polymer is in a state of biaxial compression. The external envelope in glass/polyester is in a state of biaxial tension. The mechanical behavior (in the Problem Statement) then indicates:

$$\varepsilon_{1x} = \varepsilon_{2x} = \varepsilon_{1y} = \varepsilon_{2y} = 1.47 \times 10^{-3}$$

3. Creep

(a) The equilibrium relations are formally unchanged as:

$$\sigma'_{1x}\, e_1 + \sigma'_{2x}\, e_2 = 0 \tag{5}$$

$$\sigma'_{1y}\, e_1 + \sigma'_{2y}\, e_2 = 0 \tag{6}$$

The property of the perfect bond is now written in conformity with the following figure:

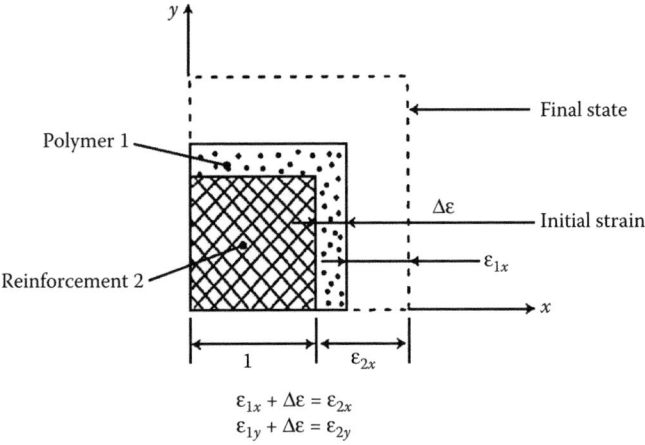

$$\varepsilon_{1x} + \Delta\varepsilon = \varepsilon_{2x}$$
$$\varepsilon_{1y} + \Delta\varepsilon = \varepsilon_{2y}$$

with the constitutive relations recalled in the Problem Statement, these equalities become

$$\frac{\sigma'_{1x}}{E_1} - \frac{v_1}{E_1}\sigma'_{1y} + \alpha_1\Delta T + \Delta\varepsilon = \frac{\sigma'_{2x}}{E_2} - \frac{v_2}{E_2}\sigma'_{2y} + \alpha_2\Delta T$$

$$-\frac{v_1}{E_1}\sigma'_{1x} + \frac{\sigma'_{1y}}{E_1} + \alpha_1\Delta T + \Delta\varepsilon = -\frac{v_2}{E_2}\sigma'_{2x} + \frac{\sigma'_{2y}}{E_2} + \alpha_2\Delta T \tag{7}, [8]$$

(b) Numerical application: In performing successively [7] − [8], [7] + [8], then substituting σ'_{2x} and σ'_{2y} obtained from [5] and [6], one obtains

$$\boxed{\sigma'_{1x} = \sigma'_{1y} = \Delta T \frac{(\alpha_2 - 0.4\alpha_1)}{\left(\dfrac{1-v_1}{E_1}\right) + \dfrac{e_1}{e_2}\left(\dfrac{1-v_2}{E_2}\right)}}$$

then:

$$\boxed{\begin{aligned}\sigma'_{1x} &= \sigma'_{1y} = -2.28\,\text{MPa}\\ \sigma'_{2x} &= \sigma'_{2y} = 7.6\,\text{MPa}\end{aligned}}$$

4. Cooling: It is sufficient to suppress the increase in temperature ΔT in the previous equations [7] and [8]. The system of equations becomes

$$
\left\{
\begin{array}{l}
\sigma''_{1x} e_1 + \sigma''_{2x} e_2 = 0 \\[2mm]
\sigma''_{1y} e_1 + \sigma''_{2y} e_2 = 0 \\[2mm]
\dfrac{\sigma''_{1x}}{E_1} - \dfrac{v_1}{E_1}\sigma''_{1y} + \Delta\varepsilon = \dfrac{\sigma''_{2x}}{E_2} - \dfrac{v_2}{E_2}\sigma''_{2y} \\[2mm]
-\dfrac{v_1}{E_1}\sigma''_{1x} + \dfrac{\sigma''_{1y}}{E_1} + \Delta\varepsilon = -\dfrac{v_2}{E_2}\sigma''_{2x} + \dfrac{\sigma''_{2y}}{E_2}
\end{array}
\right.
$$

In adopting a method of resolution analogous to that used in the previous problems, one obtains

$$
\sigma''_{1x} = \sigma''_{1y} = \Delta T \frac{0.6\alpha_1}{\left(\dfrac{1-v_1}{E_1}\right) + \dfrac{e_1}{e_2}\left(\dfrac{1-v_2}{E_2}\right)}
$$

then:

$$
\boxed{
\begin{array}{l}
\sigma''_{1x} = \sigma''_{1y} = 3.9\,\text{MPa} \\[2mm]
\sigma''_{2x} = \sigma''_{2y} = -12.9\,\text{MPa}
\end{array}
}
$$

It is worth noting that the polymer envelope is loaded now in biaxial tension. Subsequently, for one cycle of operation, the polymer envelope is successively compressed, released by creep, then extended, as shown in the following figure.

These cycles repeat themselves during the life of the tube, and this gives rise to fatigue. Therefore, an overdimension of the tube is necessary to lead to low admissible stresses in the polymer, to prevent the risk of buckling of the tube subject to compression at the location of bond defects or of failure in tension while cooling.

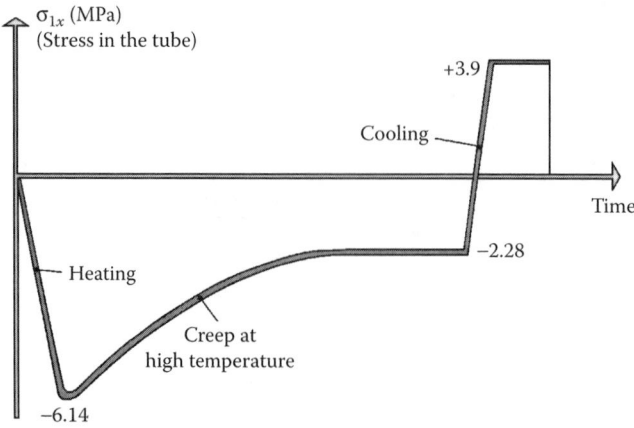

18.2.7 First Ply Failure of a Laminate—Ultimate Rupture

Problem Statement:

Consider a carbon/epoxy laminate with 60% fiber volume fraction and the following composition:

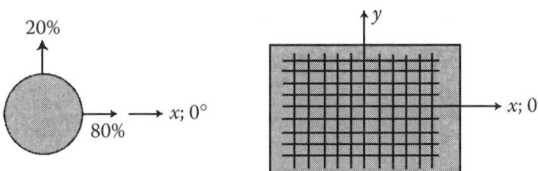

1. (a) Give the values of the moduli of elasticity and Poisson coefficients of this laminate.
   (b) What is the maximum tensile stress, denoted as $\sigma_{x\ maximum}$, that can be applied without deterioration?
2. When the value $\sigma_{x\ maximum}$ is reached, the 90° plies are deteriorated by microcracks of the epoxy resin ("first ply" failure). The characteristics of the 90° plies that are cracked are then decreased with respect to their values for intact plies. One admits the following damage factors:

$$\underset{\text{fractured}}{E'_\ell} \# \underset{\text{intact}}{E_\ell}; \quad \underset{\text{fractured}}{E'_t} \# 0.1 \times \underset{\text{intact}}{E_t}$$

$$\underset{\text{fractured}}{G'_{\ell t}} \# 0.1 \times \underset{\text{intact}}{G_{\ell t}}; \quad \underset{\text{fractured}}{v'_{\ell t}} \# 0.1 \ \underset{\text{intact}}{v_{\ell t}}$$

   (a) Calculate the new terms of the matrix $\frac{1}{p}[A]$ of the elastic behavior.[35] Deduce from there the new elastic moduli of the deteriorated laminate. Remark.
   (b) Calculate the maximum stress σ_{xM} leading to complete rupture of this laminate (rupture of 0° plies, or "last ply rupture").
3. What will be the rupture strength denoted as σ'_{xM} that one can obtain by eliminating all the elastic characteristics of the deteriorated 90° plies?

Remark: How one can obtain rapidly the value σ'_{xM}?

4. The design of an aeronautical piece is carried out using this laminate with the following considerations:
   ■ When this piece is subjected to a stress along the x direction called "limit load," the piece stays in a reversible elastic domain and is not altered in its structure.

---

[35] See Equation 12.7.

■ When this piece is subjected to a stress along the x direction called "extreme loading," one obtains total rupture.

Moreover, one has from common practice:

$$\text{Extreme loading} = 1.5 \times \text{limit loading}$$

Indicate the values of σ_x that should be kept here for extreme load and for limit load, respectively.

Solution:

1. (a) According to Tables 5.4 and 5.5 in Section 5.4.2, one notes for the indicated composition:

$$E_x = 108{,}860 \text{ MPa}; \quad E_y = 32{,}477 \text{ MPa}$$

$$v_{xy} = 0.054; \quad v_{yx} = 0.016$$

$$G_{xy} = 4200 \text{ MPa}$$

(b) Table 5.1, Section 5.4.2, indicates for the rupture limit of the first ply:

$$\sigma_x = 659 \text{ MPa}$$

2. (a) Terms of matrix $\frac{1}{b}[A]$ are written as (Equations 12.7 and 12.8):

$$\frac{1}{b}A_{ij} = \bar{E}_{ij}^{0°} \times p^{0°} + \bar{E}_{ij}^{90°} \times p^{90°}$$

Coefficients $\bar{E}_{ij}$ are given by Equation 11.8 which lead to[36]

$$\bar{E}_{11}^{0°} = 134\,440 \text{ MPa}; \quad \bar{E}_{22}^{0°} = 7\,023 \text{ MPa}; \quad \bar{E}_{12}^{0°} = 1748 \text{ MPa};$$
$$\bar{E}_{33}^{0°} = 4200 \text{ MPa}$$

The 90° plies are deteriorated. One then has[37]

$$\bar{E}_{11}^{90°} = \bar{E}_t' = 700 \text{ MPa}; \quad \bar{E}_{22}^{90°} = \bar{E}_\ell' = 134{,}000 \text{ MPa}$$
$$\bar{E}_{12}^{90°} = v_\ell' \, \bar{E}_\ell' = 17.5 \text{ MPa}; \quad \bar{E}_{33}^{90°} = 420 \text{ MPa}$$

---

[36] See Section 3.3.3 for the characteristics of a unidirectional ply of carbon/epoxy.
[37] $v_{t\ell}' = v_{\ell t}' \times E_t'/E_\ell'$ (see Application 18.1.2).

or after calculation:

$$\frac{1}{b}[A] = \begin{bmatrix} 107,692 & 1402 & 0 \\ 1402 & 32,418 & 0 \\ 0 & 0 & 3444 \end{bmatrix} (\text{MPa})$$

The new moduli of the deteriorated laminate are obtained by inversion of the above matrix. One has (Equation 12.9)

$$b[A]^{-1} = \begin{bmatrix} 1/E'_x & -v'_{yx}/E'_y & 0 \\ -v'_{xy}/E'_x & 1/E'_y & 0 \\ 0 & 0 & 1/G'_{xy} \end{bmatrix}$$

which leads to

$$\boxed{\begin{aligned} E'_x &= 107,630 \text{ MPa} \\ E'_y &= 32,400 \text{ MPa} \\ v'_{xy} &= 0.043; \quad v'_{yx} = 0.013 \\ G'_{xy} &= 3444 \text{ MPa} \end{aligned}}$$

Note that only the shear moduli G_{xy} has its value modified with respect to the value of the intact laminate.

(b) The 90° plies being deteriorated, total rupture of the laminate corresponds to rupture of the 0° plies. Let σ_{xM} be the corresponding ultimate rupture strength. The mechanical behavior of the deteriorated laminate is written, following what happens previously:

$$\begin{Bmatrix} \varepsilon_{ox} \\ \varepsilon_{oy} \\ \gamma_{oxy} \end{Bmatrix} = \begin{bmatrix} 9.29 \times 10^{-6} & -4.02 \times 10^{-7} & 0 \\ -4.02 \times 10^{-7} & 3.086 \times 10^{-5} & 0 \\ 0 & 0 & 2.9 \times 10^{-4} \end{bmatrix} \begin{Bmatrix} \sigma_{xM} \\ 0 \\ 0 \end{Bmatrix} = \begin{bmatrix} 9.29 \times 10^{-6} \times \sigma_{xM} \\ -4.02 \times 10^{-7} \times \sigma_{xM} \\ 0 \end{bmatrix}$$

from which the state of stress in the 0° plies (Equation 11.8):

$$\sigma_x = \bar{E}_{11}^{0°} \varepsilon_{ox} + \bar{E}_{12}^{0°} \varepsilon_{oy} = 1.248 \times \sigma_{xM} = \sigma_\ell$$

$$\sigma_y = \bar{E}_{12}^{0°} \varepsilon_{ox} + \bar{E}_{22}^{0°} \varepsilon_{oy} = 0.0134 \times \sigma_{xM} = \sigma_t$$

$$\tau_{xy} = 0 = \tau_{\ell t}$$

Writing that for σ_{xM} the Hill–Tsai criterion is saturated (Section 5.3.2) and using the rupture strength values of Section 3.3.3:

$$\left(\frac{1.248\sigma_{xM}}{1270}\right)^2 + \left(\frac{0.0134\sigma_{xM}}{42}\right)^2 - \frac{1.248 \times 0.0134\sigma_{xM}^2}{1270^2} = 1$$

one obtains:

$$\boxed{\sigma_{xM} = 973 \text{ MPa}}$$

3. If one cancels all elastic characteristics of the deteriorated plies at 90°, the matrix $\frac{1}{b}[A]$ becomes

$$\frac{1}{b}[A] = 0.8 \begin{bmatrix} \bar{E}_\ell & \nu_{t\ell}\bar{E}_\ell & 0 \\ \nu_{\ell t}\bar{E}_t & \bar{E}_t & 0 \\ 0 & 0 & G_{\ell t} \end{bmatrix} \text{ then } b[A]^{-1} = \frac{1}{0.8} \begin{bmatrix} \dfrac{1}{E_\ell} & -\dfrac{\nu_{t\ell}}{E_t} & 0 \\ -\dfrac{\nu_{\ell t}}{E_\ell} & \dfrac{1}{E_t} & 0 \\ 0 & 0 & \dfrac{1}{G_{\ell t}} \end{bmatrix}$$

under the loading of an ultimate stress denoted as σ'_{xM}, one will have for the strains:

$$\varepsilon_{ox} = \frac{1}{0.8}\frac{\sigma'_{xM}}{E_\ell}; \quad \varepsilon_{oy} = \frac{1}{0.8} \times -\frac{\nu_{\ell t}}{E_\ell}\sigma'_{xM}; \quad \gamma_{oxy} = 0$$

then in the 0° plies:

$$\sigma_x = \sigma_\ell = \bar{E}_{11}^{0°}\varepsilon_{ox} + \bar{E}_{12}^{0°}\varepsilon_{oy} = \frac{\sigma'_{xM}}{0.8}$$

$$\sigma_y = \sigma_t = \bar{E}_{12}^{0°}\varepsilon_{ox} + \bar{E}_{22}^{0°}\varepsilon_{oy} = 0$$

The saturated Hill–Tsai criterion then takes the form:

$$\left(\frac{\sigma'_{xM}}{0.8 \times 1270}\right)^2 = 1$$

then:

$$\boxed{\sigma'_{xM} = 1016 \text{ MPa}}$$

One immediately obtains this value when noting a stress resultant N_x written as:

$$N_x = \sigma_x \times b = \sigma_x^{0°} \times 0.8b + \sigma_x^{90°} \times 0.2b$$

then

$$\sigma'_{xM} = \sigma_{xM}^{0°} \times 0.8 = 1270 \times 0.8 = 1016 \text{ MPa}$$

Note that the rupture stress of the last ply calculated in the previous problem (σ_{xM}) is less than σ'_{xM}. It then appears to be dangerous to reason as if the $0°$ plies were the only ones to resist by occupying 80% of the thickness of the laminate.

(c) If one considers that the limit load corresponds to the rupture of the first ply, denoted as $\sigma_{x\,limit} = 659$ MPa, then the extreme load will be

$$\sigma_{x\,extreme} = 1.5 \times 659 = 988 \text{ MPa}$$

This is an excessive value because it is higher than the rupture strength of the last ply $\sigma_{xM} = 973$ MPa. One then is led to keep

■ For the extreme load: $\sigma_{x\,extreme} = \sigma_{xM} = 973$ MPa
■ For the limit load: $\sigma_{x\,limit} = \sigma_{xM}/1.5 = 649$ MPa (value less than the fracture strength of the first ply)

18.2.8 Optimum Laminate for Isotropic Stress State

Problem Statement:

Consider a laminate subjected to a state of plane uniform stresses:

$$\sigma_x = \sigma_y = \sigma_o; \ \tau_{xy} = 0 \text{ (state of isotropic stress)}$$

This laminate presents the following composition:

1. By means of a literal calculation show that the strain of the laminate is the same for any value of $p < 0.5$. Verify this property by means of Table 5.4 in Section 5.4.2 for $p = 0\%$, 30%, 50%.
2. Show that the Hill–Tsai criterion has the same value in each ply, no matter what the proportion p. Comment.
3. Verify the previous property for a carbon/epoxy laminate by means of the tables in annex 1 for $p = 0\%$, 30%, 50%.

Solution:

1. Determination of the apparent moduli of the carbon/epoxy laminate: We begin by calculating the terms of the matrix $\frac{1}{b}[A]$ (Equations 12.7 and 12.8):

$$\frac{1}{b}A_{11} = \bar{E}_{11}^{0°} \times p + \bar{E}_{11}^{90°} \times p + \bar{E}_{11}^{45°} \times \left(\frac{1}{2}-p\right) + \bar{E}_{11}^{-45°} \times \left(\frac{1}{2}-p\right)$$

then with the Equation 11.8:

$$\bar{E}_{11}^{0°} = \bar{E}_\ell; \quad \bar{E}_{11}^{90°} = E_t; \quad \bar{E}_{11}^{45°} = \bar{E}_{11}^{-45°} = \frac{\bar{E}_\ell + \bar{E}_t}{4} + \frac{1}{2}(v_{t\ell}\bar{E}_\ell + 2G_{\ell t})$$

$$\frac{1}{b}A_{11} = p(\bar{E}_\ell + \bar{E}_t) + 2\left(\frac{1}{2} - p\right)\left[\frac{\bar{E}_\ell + \bar{E}_t}{4} + \frac{1}{2}(v_{t\ell}\bar{E}_\ell + 2G_{\ell t})\right]$$

$$\frac{1}{b}A_{11} = p\left[\frac{\bar{E}_\ell + \bar{E}_t}{2} - v_{t\ell}\bar{E}_\ell - 2G_{\ell t}\right] + \frac{1}{2}\left[\frac{\bar{E}_\ell + \bar{E}_t}{2} + v_{t\ell}\bar{E}_\ell + 2G_{\ell t}\right]$$

$$\frac{1}{b}A_{22} = \frac{1}{b}A_{11}$$

$$\frac{1}{b}A_{12} = 2pv_{t\ell}\bar{E}_\ell + 2\left(\frac{1}{2} - p\right)\left[\frac{1}{4}(\bar{E}_\ell + \bar{E}_t - 4G_{\ell t}) + \frac{1}{2}v_{t\ell}\bar{E}_\ell\right]$$

$$\frac{1}{b}A_{12} = -p\left[\frac{\bar{E}_\ell + \bar{E}_t}{2} - v_{t\ell}\bar{E}_\ell - 2G_{\ell t}\right] + \frac{1}{2}\left[\frac{\bar{E}_\ell + \bar{E}_t}{2} + v_{t\ell}\bar{E}_\ell - 2G_{\ell t}\right]$$

$$\frac{1}{b}A_{13} = \frac{1}{b}A_{23} = 0$$

The constitutive law in Equation 12.7 here takes the form:

$$\left\{\begin{array}{c} \sigma_o \\ \sigma_o \\ 0 \end{array}\right\} = \frac{1}{b}\begin{bmatrix} A_{11} & A_{12} & 0 \\ A_{21} & A_{22} & 0 \\ 0 & 0 & A_{33} \end{bmatrix}\left\{\begin{array}{c} \varepsilon_{ox} \\ \varepsilon_{oy} \\ \gamma_{oxy} \end{array}\right\}$$

Its inverse is written as:

$$\left\{\begin{array}{c} \varepsilon_{ox} \\ \varepsilon_{oy} \\ \gamma_{oxy} \end{array}\right\} = \begin{bmatrix} 1/\bar{E}_x & -\bar{v}_{yx}/\bar{E}_y & 0 \\ -\bar{v}_{xy}/\bar{E}_x & 1/\bar{E}_y & 0 \\ 0 & 0 & 1/\bar{G}_{xy} \end{bmatrix}\left\{\begin{array}{c} \sigma_0 \\ \sigma_0 \\ 0 \end{array}\right\}$$

with:

$$\frac{1}{\bar{E}_x} = \frac{\frac{1}{b}A_{22}}{\frac{1}{b^2}(A_{11}A_{22} - A_{12}^2)} = \frac{1}{\bar{E}_y}; \quad \frac{\bar{v}_{yx}}{\bar{E}_y} = \frac{\frac{1}{b}A_{12}}{\frac{1}{b^2}(A_{11}A_{22} - A_{12}^2)}$$

and here:

$$\frac{1}{b^2}(A_{11}A_{22} - A_{12}^2) = 2\left[p\left(\frac{\bar{E}_\ell + \bar{E}_t}{2} - v_{t\ell}\bar{E}_\ell - 2G_{\ell t}\right) + G_{\ell t}\right]\left[\frac{\bar{E}_\ell + \bar{E}_t}{2} + v_{t\ell}\bar{E}_\ell\right]$$

then we can obtain the strains:

$$\varepsilon_{ox} = \sigma_o\left(\frac{1}{\bar{E}_x} - \frac{\bar{v}_{yx}}{\bar{E}_y}\right) = \frac{\sigma_o}{\left(\frac{\bar{E}_\ell + \bar{E}_t}{2} + v_{t\ell}\bar{E}_\ell,\right)} = \varepsilon_{oy}; \quad \gamma_{oxy} = 0$$

In summary[38]:

$$\boxed{\varepsilon_{ox} = \varepsilon_{oy} = \varepsilon_o = \frac{\sigma_o}{\left[\frac{\bar{E}_\ell + \bar{E}_t}{2} + v_{t\ell}\bar{E}_\ell\right]}; \quad \gamma_{oxy} = 0}$$

The strain ε_o is independent of the proportion p and of the shear modulus $G_{\ell t}$. Each elastic characteristic that appears has the same weight: $\bar{E}_\ell$, $\bar{E}_t$, $v_{t\ell}\bar{E}_\ell = v_{\ell t}\bar{E}_t$

- Verification: Table 5.4, Section 5.4.2:

$$\boldsymbol{p} = \mathbf{0\%} : \bar{E}_x = \bar{E}_y = 15\ 055\ \text{Mpa} ; \bar{v}_{xy} = 0.79 = \bar{v}_{yx}$$
$$\varepsilon_{ox} = \varepsilon_{oy} = \varepsilon_o = 1.39\ \text{E-5} \times \sigma_o(\text{MPa})$$
$$\boldsymbol{p} = \mathbf{30\%} : \bar{E}_x = \bar{E}_y = 55\ 333\ \text{Mpa} ; \bar{v}_{xy} = 0.23 = \bar{v}_{yx}$$
$$\varepsilon_{ox} = \varepsilon_{oy} = \varepsilon_o = 1.39\ \text{E-5} \times \sigma_o(\text{MPa})$$
$$\boldsymbol{p} = \mathbf{50\%} : \bar{E}_x = \bar{E}_y = 70\ 687\ \text{Mpa} ; \bar{v}_{xy} = 0.025 = \bar{v}_{yx}$$
$$\varepsilon_{ox} = \varepsilon_{oy} = \varepsilon_o = 1.38\ \text{E-5} \times \sigma_o(\text{MPa})$$

2. Hill–Tsai criterion:
   - 0° plies: Following the Equation 11.8:

$$\sigma_x^{0°} = \bar{E}_\ell\varepsilon_{ox} + v_{t\ell}\bar{E}_\ell\varepsilon_{oy} = \varepsilon_o\bar{E}_\ell(1 + v_{t\ell})$$
$$\sigma_y^{0°} = v_{t\ell}\bar{E}_\ell\varepsilon_{ox} + \bar{E}_t\varepsilon_{oy} = \varepsilon_o\bar{E}_t(1 + v_{\ell t})$$
$$\tau_{xy}^{0°} = 0$$

and following the Equation 11.4:

$$\sigma_\ell^{0°} = \sigma_x^o = \varepsilon_o\bar{E}_\ell(1 + v_{t\ell})$$
$$\sigma_t^{0°} = \sigma_y^o = \varepsilon_o\bar{E}_t(1 + v_{\ell t})$$
$$\tau_{\ell t}^{0°} = 0$$

---

[38] Recall (see Equation 11.8) that: $\bar{E}_\ell = E_\ell(1 - v_{\ell t}v_{t\ell})$; $\bar{E}_t = \bar{E}_t(1 - v_{\ell t}v_{t\ell})$.

■ 90° plies: Following Equations 11.8 and 11.4:

$$\sigma_\ell^{90°} = \sigma_y^{90°} = \varepsilon_o \bar{E}_\ell (1 + v_{t\ell})$$

$$\sigma_t^{90°} = \sigma_x^{90°} = \varepsilon_o \bar{E}_t (1 + v_{\ell t})$$

$$\tau_{\ell t}^{90°} = 0$$

■ 45° plies: Following Equations 11.8 and 11.4[39]:

$$\sigma_\ell^{45°} = \frac{1}{2}(\sigma_x^{45°} + \sigma_y^{45°}) + \tau_{xy}^{45°} = \varepsilon_o \bar{E}_\ell (1 + v_{t\ell})$$

$$\sigma_t^{45°} = \frac{1}{2}(\sigma_x^{45°} + \sigma_y^{45°}) - \tau_{xy}^{45°} = \varepsilon_o \bar{E}_t (1 + v_{\ell t})$$

$$\tau_{\ell t}^{45°} = 0$$

■ −45° plies: In an analogous manner:

$$\sigma_\ell^{-45°} = \varepsilon_o \bar{E}_\ell (1 + v_{t\ell})$$

$$\sigma_t^{-45°} = \varepsilon_o \bar{E}_t (1 + v_{\ell t})$$

$$\tau_{\ell t}^{-45°} = 0$$

■ The Hill–Tsai criterion (see Section 5.3.2 or Equation 14.7) then has the same value in each of the plies, no matter what the proportion p and the value of the shear modulus $G_{\ell t}$. Rupture occurs simultaneously in all plies.
■ One can also note that the minimum thickness h of the laminate that is capable of supporting the isotropic membrane load:

$$N_x = N_y; \quad T_{xy} = 0$$

will be independent of the proportion p (see Equation 12.10). One then can, for this particular case of loading, vary the modulus of elasticity $\bar{E}_x = \bar{E}_y$[40] **without varying the thickness.**

---

[39] Or still from Equation 11.7:

$$\varepsilon_\ell^{45} = \frac{1}{2}(\varepsilon_{ox} + \varepsilon_{oy}) = \varepsilon_o; \quad \varepsilon_t^{45} = \frac{1}{2}(\varepsilon_{ox} + \varepsilon_{oy}) = \varepsilon_o; \quad \gamma_{\ell t}^{45} = 0;$$

then following [11.6]:

$$\sigma_\ell^{45} = \varepsilon_o \bar{E}_\ell (1 + v_{t\ell}); \quad \sigma_t^{45} = \varepsilon_o \bar{E}_t (1 + v_{\ell t}); \quad \tau_{\ell t}^{45} = 0$$

[40] See Equation 12.9, or Tables 5.4, 5.9, 5.14 in Section 5.4.2.

- One automatically obtains such a laminate by using layers of balanced fabric at 0° and 45°. It is then convenient to calculate the thickness in considering the proper rupture resistances of the layer of fabric.[41]

3. Verification:

(cf. plates
annex 1)

p = 0 %

plate 3 and 7

plies at +45°	σ_ℓ	σ_t	$\tau_{\ell t}$
$\sigma_x = 1$ MPa	.94	.06	−.5
$\sigma_y = 1$ MPa	.94	.06	.5
total (MPa)	1.88	.12	.0
Hill–Tsai criterion: 1.02×10^{-5}			

plate 4 and 8

plies at −45°	σ_ℓ	σ_t	$\tau_{\ell t}$
$\sigma_x = 1$ MPa	.94	.06	.5
$\sigma_y = 1$ MPa	.94	.06	−.5
total (MPa)	1.88	.12	0
Hill–Tsai criterion: 1.02×10^{-5}			

p = 30 %

plate 1 and 5

plies at 0°	σ_ℓ	σ_t	$\tau_{\ell t}$
$\sigma_x = 1$ MPa	2.4	.0	.0
$\sigma_y = 1$ MPa	−.54	.12	.0
total (MPa)	1.86	.12	.0
Hill–Tsai criterion: 1.017×10^{-5}			

plate 2 and 6

plies at 90°	σ_ℓ	σ_t	$\tau_{\ell t}$
$\sigma_x = 1$ MPa	−.54	.12	.0
$\sigma_y = 1$ MPa	2.4	.0	.0
total (MPa)	1.86	.12	.0
Hill–Tsai criterion: 1.017×10^{-5}			

plate 3 and 7

plies at +45°	σ_ℓ	σ_t	$\tau_{\ell t}$
$\sigma_x = 1$ MPa	.94	.06	−.09
$\sigma_y = 1$ MPa	.94	.06	.09
total (MPa)	1.88	.12	.0
Hill–Tsai criterion: 1.02×10^{-5}			

plate 4 and 8

plies at −45°	σ_ℓ	σ_t	$\tau_{\ell t}$
$\sigma_x = 1$ MPa	.94	.06	.09
$\sigma_y = 1$ MPa	.94	.06	−.09
total (MPa)	1.88	.12	0
Hill–Tsai criterion: 1.02×10^{-5}			

p = 50 %

plate 1 and 5

plies at 0°	σ_ℓ	σ_t	$\tau_{\ell t}$
$\sigma_x = 1$ MPa	1.9	.02	.0
$\sigma_y = 1$ MPa	−.02	.1	.0
total (MPa)	1.88	.12	.0
Hill–Tsai criterion: 1.02×10^{-5}			

plate 2 and 6

plies at 90°	σ_ℓ	σ_t	$\tau_{\ell t}$
$\sigma_x = 1$ MPa	−.02	.1	.0
$\sigma_y = 1$ MPa	1.9	.02	.0
total (MPa)	1.88	.12	.0
Hill–Tsai criterion: 1.02×10^{-5}			

18.2.9 Laminate Made of Identical Layers of Balanced Fabric

Problem Statement:

A carbon/epoxy laminate consists of a stacking of identical layers of balanced fabric with the composition illustrated below. The fiber volume fraction is $V_f = 60\%$.

---

[41] See Applications 18.2.9 and 18.2.10.

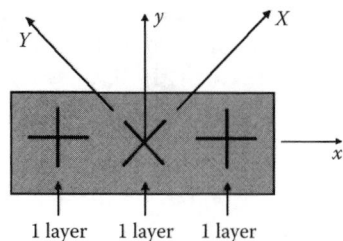

Give the elastic behavior law in the axes (x, y) and then in the axes (X, Y).

Solution:

- Axes x, y: The fabric being balanced, each layer can be replaced by two identical unidirectional plies crossed at 90°, with the resulting thicknesses (see Section 3.4.2):

$$e_{\text{warp}} = e_{\text{fill}} = e/2$$

The laminate is balanced and its composition is as follows (see figure):

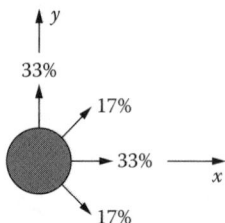

One then notes (Table 5.4, Section 5.4.2):

$$E_x = 55,333 + \Delta E_x (\text{MPa})$$

One can evaluate ΔE_x starting from the expression:

$$dE_x = \frac{\partial E}{\partial p^{0°}} \times dp^{0°} + \frac{\partial E}{\partial p^{90°}} \times dp^{90°}$$

as:

$$\Delta E_x = (65,888 - 55,333) \times \frac{3}{10} + (53,545 - 55,333) \times \frac{3}{10} = 2630 \text{ MPa}$$

then:

$$E_x = 57,960 \text{ MPa} = E_y$$

Poisson coefficient: $v_{xy} = 0.23 + \Delta v_{xy}$.
From an analogous calculation:

$$v_{xy} = 0.20 = v_{yx}$$

Shear modulus: One notes (Table 5.5, Section 5.4.2)

$$G_{xy} = 16,315 + \Delta G_{xy} \text{ (MPa)}$$

Then from an analogous calculation:

$$G_{xy} = 14,500 \text{ MPa}.$$

From which the elastic behavior relation in axes (x, y) can be written as (Equation 12.9):

$$\left\{ \begin{array}{c} \varepsilon_{ox} \\ \varepsilon_{oy} \\ \gamma_{oxy} \end{array} \right\} = \begin{bmatrix} \dfrac{1}{57,960} & -\dfrac{0.2}{57,960} & 0 \\ -\dfrac{0.2}{57,960} & \dfrac{1}{57,960} & 0 \\ 0 & 0 & \dfrac{1}{14,500} \end{bmatrix} \left\{ \begin{array}{c} \sigma_{ox} \\ \sigma_{oy} \\ \tau_{oxy} \end{array} \right\} \text{(MPa)}$$

■ Axes X, Y: The laminate is balanced and then has the composition:

In using the same tables as before, one obtains

$$E_X = E_Y = 31,979 + \Delta E_X = 41,400 \text{ MPa}$$
$$v_{XY} = v_{YX} = 0.56 + \Delta v_{XY} = 0.43$$
$$G_{XY} = 28,430 + \Delta G_{XY} = 24,190 \text{ MPa}$$

from which the law for the behavior in the axes X, Y can be written as (Equation 12.9):

$$
\begin{Bmatrix} \varepsilon_{oX} \\ \varepsilon_{oY} \\ \gamma_{oXY} \end{Bmatrix} = \begin{bmatrix} \dfrac{1}{41,400} & -\dfrac{0.43}{41,400} & 0 \\ -\dfrac{0.43}{41,400} & \dfrac{1}{41,400} & 0 \\ 0 & 0 & \dfrac{1}{24,190} \end{bmatrix} \begin{Bmatrix} \sigma_{oX} \\ \sigma_{oY} \\ \tau_{oXY} \end{Bmatrix} (\text{MPa})
$$

Remarks:

■ One can note that a laminate constituted of layers of balanced fabric with four orientations 0°, 90°, +45°, −45° admits two systems of orthotropic axes x,y and X,Y.

■ The elastic properties are suitably estimated when one uses Tables 5.4 and 5.5 in Section 5.4.2. **This is not the same** for the maximum admissible stresses indicated in Tables 5.1, 5.2, and 5.3 that are valid only for laminates made of unidirectional layers. In effect, the resistance to rupture for a layer of balanced fabric is clearly higher in tension than the first ply failure limit for an equivalent fabric, made up of layers at 0° (50%) and 90° (50%). For a calculation of first-ply failure or for the failure criterion of the laminate proposed in this application, it is convenient to consider a layer of fabric as an anisotropic ply with thickness e (see Section 3.4.2) with the values of rupture stresses $\sigma_{\ell\ \text{rupture}}$, $\sigma_{t\ \text{rupture}}$, and $\tau_{\ell t\ \text{rupture}}$ of the balanced fabric itself (see examples in Section 3.4.3).[42] One will then have the following equivalence[43]:

18.2.10 Wing Spar in Carbon/Epoxy

Problem Statement:

Consider an airplane flap with the internal structure (excluding facings) shown schematically in the following figure. It consists of a spar and several ribs. The spar is a laminate of carbon/epoxy fabric with $V_f = 45\%$ fiber volume fraction, the composition of which varies with the longitudinal coordinate axis x, in the flange and in the web. A preliminary calculation of the flap in isostatic equilibrium

---

[42] See also Application 18.2.10.

[43] See Section 5.2.3.

reveals the maximum stress resultants in the two zones of the spar indicated in the figure.

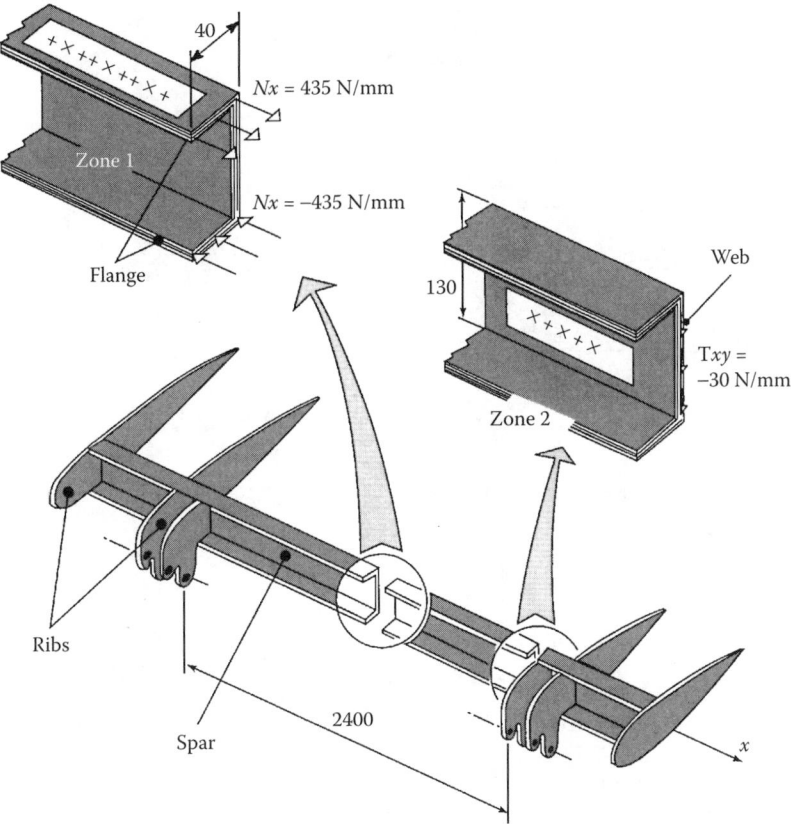

One proposes for each of these zones the compositions indicated in the figure.

1. Evaluate the elastic properties of the laminate in these two zones.
2. Verify the two corresponding laminates:
    a. At rupture.
    b. At buckling.

  - ■ Thickness of a layer of fabric: 0.24 mm.
  - ■ Properties of carbon/epoxy fabric: See Section 3.4.3.

Solution:

1. Elastic properties:
    (a) Zone 1: Composition of the laminate[44]: See sketch below.

----

[44] See Section 5.2.3 and remark at the end of previous Exercise 18.2.9.

Calculation of elastic moduli (Equations 12.7, 12.8, 12.9, and 11.8):

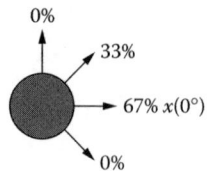

$$\bar{E}_{11}^{0°} = \bar{E}_\ell; \quad \bar{E}_{12}^{0°} = \nu_{t\ell}\bar{E}_\ell; \quad \bar{E}_{33}^{0°} = G_{\ell t}$$

$$\bar{E}_{11}^{45°} = \frac{\bar{E}_\ell + \bar{E}_t}{4} + \frac{1}{2}(\nu_{t\ell}\bar{E}_\ell + 2G_{\ell t}); \quad \bar{E}_{12}^{45°} = \frac{\bar{E}_\ell + \bar{E}_t}{4} - G_{\ell t} + \frac{1}{2}\nu_{t\ell}\bar{E}_\ell$$

$$\bar{E}_{33}^{45°} = \frac{\bar{E}_\ell + \bar{E}_t}{4} - \frac{1}{2}\nu_{t\ell}\bar{E}_\ell$$

with (Section 3.4.3):

$$\bar{E}_\ell = \bar{E}_t = E_x(1 - \nu_{xy} \times \nu_{yx}); \quad E_x = 54\,000 \text{ MPa}; \quad \nu_{xy} = \nu_{yx} = 0.045$$
$$G_{\ell t} = G_{xy} = 4000$$

Then:

$$\bar{E}_{11}^{0°} = 54,100 \text{ MPa}; \quad \bar{E}_{12}^{0°} = 2435 \text{ MPa}; \quad \bar{E}_{33}^{0°} = 4000 \text{ MPa}$$
$$\bar{E}_{11}^{45°} = 32,270 \text{ MPa}; \quad \bar{E}_{12}^{45°} = 24,270 \text{ MPa}; \quad \bar{E}_{33}^{45°} = 25,840 \text{ MPa}$$

One deduces from there:

$$\frac{1}{h}A_{11} = \bar{E}_{11}^{0°} \times 0.67 + \bar{E}_{11}^{45°} \times 0.33 = 46,900 \text{ MPa} = \frac{1}{h}A_{22}$$

$$\frac{1}{h}A_{12} = \bar{E}_{12}^{0°} \times 0.67 + \bar{E}_{12}^{45°} \times 0.33 = 9640 \text{ MPa}$$

$$\frac{1}{h}A_{33} = \bar{E}_{33}^{0°} \times 0.67 + \bar{E}_{33}^{45°} \times 0.33 = 11,210 \text{ MPa}$$

After the calculation of $h[A]^{-1}$, one obtains the law for the behavior in zone 1:

$$\left\{ \begin{array}{c} \varepsilon_{ox} \\ \varepsilon_{oy} \\ \gamma_{oxy} \end{array} \right\} = \left[\begin{array}{ccc} \dfrac{1}{44,920} & -\dfrac{0.2}{44,920} & 0 \\ -\dfrac{0.2}{44,920} & \dfrac{1}{44,920} & 0 \\ 0 & 0 & \dfrac{1}{11,210} \end{array} \right] \left\{ \begin{array}{c} \sigma_{ox} \\ \sigma_{oy} \\ \tau_{oxy} \end{array} \right\} \qquad [1]$$

(b) Zone 2: Composition of the laminate:

Following the same method as above:

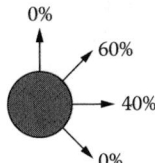

$$\frac{1}{h}A_{11} = \frac{1}{h}A_{22} = \bar{E}_{11}^{0°} \times 0.4 + \bar{E}_{11}^{45°} \times 0.6 = 41,010 \text{ MPa}$$

$$\frac{1}{h}A_{12} = 15,540 \text{ MPa}$$

$$\frac{1}{h}A_{33} = 17,100 \text{ MPa}$$

then, after inversion of the behavior law in zone 2:

$$\begin{Bmatrix} \varepsilon_{ox} \\ \varepsilon_{oy} \\ \gamma_{oxy} \end{Bmatrix} = \begin{bmatrix} \dfrac{1}{35,120} & -\dfrac{0.38}{35,120} & 0 \\ -\dfrac{0.38}{35,120} & \dfrac{1}{35,120} & 0 \\ 0 & 0 & \dfrac{1}{17,100} \end{bmatrix} \begin{Bmatrix} \sigma_{ox} \\ \sigma_{oy} \\ \tau_{oxy} \end{Bmatrix} \qquad [2]$$

2. (a) Verification of non rupture:

■ Zone 1: Compression in the lower skin: $N_x = -435$ N/mm, then with 9 layers of fabric of thickness 0.24 mm:

$$\sigma_{ox} = -202 \text{ MPa}$$

from which the strains are (Equation [1] above):

$$\varepsilon_{ox} = -4.497 \times 10^{-3}; \quad \varepsilon_{oy} = 9 \times 10^{-4}; \quad \gamma_{oxy} = 0$$

• Layers at 0°/90°: (Equation 11.8):

$$\sigma_x^{0°} = \bar{E}_{11}^{0°} \times \varepsilon_{ox} + \bar{E}_{12}^{0°} \times \varepsilon_{oy} = -241 \text{ MPa} = \sigma_\ell^{0°}$$

$$\sigma_y^{0°} = \bar{E}_{21}^{0°} \times \varepsilon_{ox} + \bar{E}_{22}^{0°} \times \varepsilon_{oy} = 38 \text{ MPa} = \sigma_t^{0°}$$

$$\tau_{xy}^{0°} = 0 = \tau_{\ell t}^{0°}$$

The Hill–Tsai expression: (Section 5.3.2 and Chapter 14)[45]:

$$\frac{-241^2}{360^2} + \frac{38^2}{420^2} - \frac{-241 \times 38}{360^2} = (0.72)^2 < 1$$

Factor of safety (Section 14.2.3.): $\frac{1}{0.72} - 1 = 38\%$.

  • Layers at 45°/–45°: One finds by an analogous calculation a much weaker value for the Hill–Tsai expression: $(0.49)^2$. The layers 0°/90° fail first.

  ■ Zone 2: With a shear stress resultant $T_{xy} = -30$ N/mm and 5 layers of fabric with 0.24 mm thickness, one has:

$$\tau_{oxy} = -25 \text{ MPa}.$$

From which the strains are (Equation [2] above):

$$\varepsilon_{ox} = 0; \ \varepsilon_{oy} = 0; \ \gamma_{oxy} = -1.46 \times 10^{-3}$$

  • Layers at 45°/–45° (Equation 11.8):

$$\sigma_x^{45°} = \sigma_y^{45°} = 0; \quad \tau_{xy}^{45°} = -38 \text{ MPa}$$

Equation 11.4:

$$\sigma_\ell^{45°} = -\tau_{xy}^{45°} = 38 \text{ MPa} = -\sigma_t^{45°}; \quad \tau_{\ell t}^{45°} = 0$$

Hill–Tsai expression:

$$\frac{38^2}{420^2} + \frac{-38^2}{360^2} - \frac{-38 \times 38}{420^2} = (0.17)^2$$

corresponding to a factor of safety of $\frac{1}{0.17} - 1 = 500\%$

  • Layers at 0°/90°: One finds a smaller value for the Hill–Tsai expression: $(0.1)^2$. It is the 45°/–45° layers that fail first.

2. (b) Verification for buckling: This is done starting from the graphs of Appendix 2. In this view, one has to evaluate the constants C_{11}, C_{22}, C_{12}, C_{33} that appear in the law of the bending behavior (Equation 12.16):

---

[45] As recalled from Section 14.2.2 (note 4 at bottom of the page), a balanced fabric is not transversely isotropic. The Hill–Tsai (Equation 14.6) can be rewritten in this case as:

$$\frac{\sigma_\ell^2}{\sigma_{\ell \text{ rupt}}^2} + \frac{\sigma_t^2}{\sigma_{t \text{ rupt}}^2} - \sigma_\ell \sigma_t \left(\frac{2}{\sigma_{\ell \text{ rupt}}^2} - \frac{1}{\sigma_{z \text{ rupt}}^2} \right) + \frac{\tau_{\ell t}^2}{\tau_{\ell t \text{ rupt}}^2} < 1$$

Without knowing $\sigma_{z \text{ rupture}}$ and taking into account the weak influence of the modified term, one uses the Equation 14.6.

■ Zone 1:

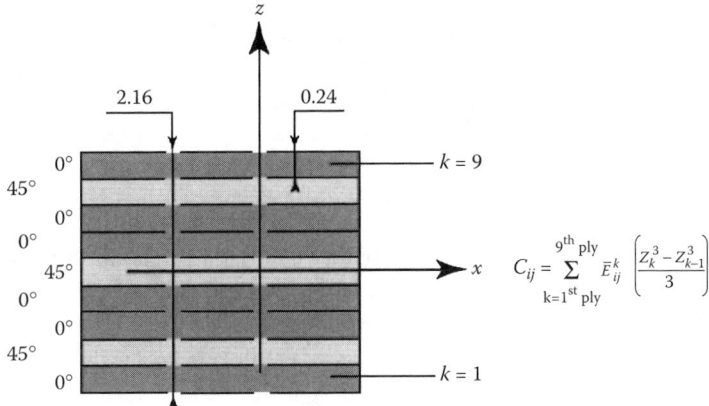

ply $n°k$	1	2	3	4	5	6	7	8	9
$\left(\dfrac{z_k^3 - z_{k-1}^3}{3}\right)$	0.2223	0.1256	0.0564	0.0150	1.152 E-3	0.0150	0.0564	0.1256	0.223

from which:

$$C_{11} = C_{22} = 39{,}930 \text{ N mm}$$

$$C_{12} = C_{21} = 7555 \text{ N mm}$$

$$C_{33} = 8870 \text{ N mm}$$

Then:

$$[C] = \begin{bmatrix} 39{,}930 & 7555 & 0 \\ 7555 & 39{,}930 & 0 \\ 0 & 0 & 8870 \end{bmatrix} (\text{N} \times \text{mm})$$

Consider the unfavorable case of a plate simply supported at two of its sides, clamped along the third side, and free on the fourth side (see figure in the Problem Statement). Using the Plate 16 in Appendix 2 with the values:

$$C = \frac{C_{21} + 2C_{33}}{\sqrt{C_{11} \times C_{22}}} = \frac{25{,}295}{39{,}930} = 0.63; \quad \frac{a}{b}\left(\frac{C_{22}}{C_{11}}\right)^{1/4} \gg 1$$

one obtains

$$k \# 1.15$$

from which the critical compressive stress resultant is

$$N_{x\ \text{critical}} = 1.15\ \pi^2\ (39930/40^2)$$

$$N_{x\ \text{critical}} = 283\ \text{N/mm} < 435\ \text{N/mm applied}$$

There is a risk of buckling, and one must reinforce the wing in the central part of the spar where the compressive stress resultant is maximum by means of exterior layers at 0°/90° in such a way to augment C_{11} and C_{22}. For example, with a supplementary external layer on either side:

$$C'_{22} = 77{,}475\ \text{N/mm}; \quad C'_{21} = 9245\ \text{N/mm}; \quad C'_{33} = 11{,}646\ \text{N/mm}.$$

from which $C = 0.42$, $k \# 1$, and:

$$N'_{x\ \text{critical}} = 477\ \text{N/mm} > 435\ \text{N/mm applied}.$$

■   Zone 2:

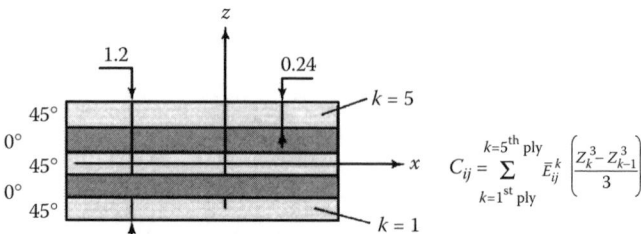

One obtains after calculation:

$$[C] = \begin{bmatrix} 5300 & 2840 & 0 \\ 2840 & 5300 & 0 \\ 0 & 0 & 3065 \end{bmatrix} (\text{N} \times \text{mm})$$

In the unfavorable case of a plate simply supported on four sides (see figure in the Problem Statement), one uses Plate 18 of Appendix 2 with the values:

$$C = 1.7; \quad \frac{a}{b}\left(\frac{C_{22}}{C_{11}}\right)^{1/4} \gg 1$$

One obtains

$$k \# 7$$

from which the critical shear stress resultant is

$$T_{xy \text{ critical}} = 7 \; \pi^2 \; (5300/130^2)$$

$$T_{xy \text{ critical}} = 21 \; \text{N/mm} < 30 \; \text{N/mm applied}$$

There is then a risk of buckling and one must reinforce the web in this part of the spar where the shear force is maximum. A supplementary external layer at 0°/90° on either side of this web gives

$$C'_{22} = 18{,}890 \; \text{N/mm}; \quad C'_{21} = 3450 \; \text{N/mm}; \quad C'_{33} = 4070 \; \text{N/mm}.$$

From which: $C = 0.6$, $k \; \# \; 4.3$, and:

$$T_{xy \text{ critical}} = 47 \; \text{N/mm} > 30 \; \text{N/mm applied.}$$

18.2.11 Determination of the Elastic Characteristics of a Carbon/Epoxy Unidirectional Layer from Tensile Test

Problem Statement:

Consider a unidirectional plate of carbon/epoxy, from which one cuts the two samples shown below. They are tested in a testing machine. One measures the strains using strain gages arranged as shown. The strains obtained under different loads are linearized. One shows their values corresponding to a uniform tensile stress σ_x equal to 20 MPa.

Calculate the elastic constants of the unidirectional layer subject to in-plane loading.

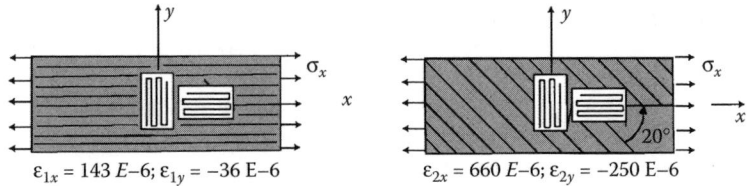

$$\varepsilon_{1x} = 143 \; E\text{-}6; \; \varepsilon_{1y} = -36 \; E\text{-}6 \qquad \varepsilon_{2x} = 660 \; E\text{-}6; \; \varepsilon_{2y} = -250 \; E\text{-}6$$

Solution:

One can use the Equation 11.5:

■ Sample No. 1: The axes x and y are ,coincident with the axes ℓ and t ($\theta = 0$). From which:

$$\varepsilon_{1x} = \frac{\sigma_x}{E_x} = \frac{\sigma_x}{E_\ell} \rightarrow E_\ell = \frac{20}{143 \; E\text{-}6} = 139{,}860 \; \text{MPa}$$

$$\varepsilon_{1y} = -\frac{\nu_{xy}}{E_x} \times \sigma_x = -\frac{\nu_{\ell t}}{E_\ell} \times \sigma_x \rightarrow \nu_{\ell t} = 0.25$$

■ Sample No. 2: The axes x and y make an angle of $\theta = 20°$ with the axes l and t, from which[46]:

$$\varepsilon_{2x} = \frac{\sigma_x}{E_x} = \left\{ \frac{c^4}{E_\ell} + \frac{s^4}{E_t} + c^2 s^2 \left(\frac{1}{G_{\ell t}} - 2\frac{\nu_{\ell t}}{E_\ell} \right) \right\} \times \sigma_x$$

$$\varepsilon_{2y} = -\frac{\nu_{xy}}{E_x} \times \sigma_x = -\left\{ \frac{\nu_{\ell t}}{E_\ell}(c^4 + s^4) - c^2 s^2 \left(\frac{1}{E_\ell} + \frac{1}{E_t} - \frac{1}{G_{\ell t}} \right) \right\} \times \sigma_x$$

leading to

$$\begin{cases} \dfrac{1}{G_{\ell t}} + \dfrac{0.1325}{E_t} = 2.69 \text{ E-4} \\[2mm] \dfrac{1}{G_{\ell t}} - \dfrac{1}{E_t} = 1.144 \text{ E-4} \end{cases}$$

from which: $E_t = 7320$ MPa; $G_{\ell t} = 3980$ MPa
In summary:

$$\boxed{\begin{aligned} &E_\ell = 139{,}860 \text{ MPa} \\ &E_t = 7320 \text{ MPa} \\ &\nu_{\ell t} = 0.25; \quad \nu_{t\ell} = 0.013 \\ &G_{\ell t} = 3980 \text{ MPa} \end{aligned}}$$

18.2.12 Sailboat Shell in Glass/Polyester

Consider a siding of a laminated shell for a sailboat made of glass/polyester. It is made up of a stack of layers of balanced fabric and glass mat. The reinforcements, in "E" glass, are in the following form:

Balanced fabric: $V_f = 20\%$, mass of the glass per square meter: $m_{of} = 500$ g.
Mat: $V_f = 15\%$, mass of glass per square meter: $m_{of} = 300$ g.

Problem Statement:

1. Calculate:
   (a) The thickness of one layer of fabric of glass/polyester.
   (b) The thickness of a layer of mat of glass/polyester.
2. Given the composition of the laminated siding as follows:

$$[M/F/M/F]_s \quad (M \leftrightarrow \text{Mat}; \ F \leftrightarrow \text{Fabric})$$

---

[46] One has $\dfrac{\nu_{t\ell}}{E_t} = \dfrac{\nu_{\ell t}}{E_\ell}$ See Exercise 18.1.2.

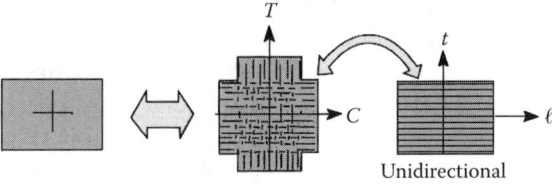

Figure 18.11

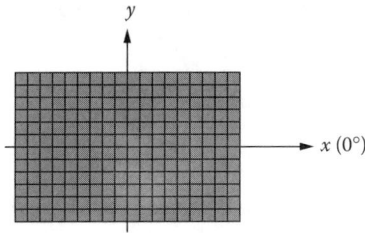

Figure 18.12

what is the total thickness, denoted as h, of the laminated siding?

3. Elastic characteristics of a fabric layer: One considers a layer of balanced fabric to be equivalent to two series of unidirectional plies crossed at 90°, each series possessing half of the total thickness of the fabric layer (see Figure 18.11).
The elastic characteristics of these unidirectional plies are as follows:

$$E_\ell = 18{,}000 \text{ MPa}; \; E_t = 4900 \text{ MPa}; \; G_{\ell t} = 1850 \text{ MPa}; \; \nu_{\ell t} = 0.3.$$

Calculate the elastic characteristics (moduli, Poisson coefficients) of a layer of fabric in axes C,T.

4. The layers of mat are considered as isotropic in their planes, with the elastic characteristics:

$$E_{\text{Mat}} = 8350 \text{ MPa}; \; \nu_{\text{Mat}} = 0.3.$$

Figure 18.12 represents one planar portion of the laminated siding. All the fabric plies are oriented at 0° – 90°. Calculate the global elastic characteristics (moduli, Poisson coefficients) of the siding working in its plane.

Remark: Tests done on samples made of this material indicate a modulus of elasticity along the x direction to be equal to 9200 MPa. What can be said about this?

5. Rupture: The rupture strengths, considered to be equal in tension and in compression, are as follows:

- Fabric layer: along C or T: $\sigma_{\text{rupture fabric}}$ = 139 MPa.
- Mat layer: $\sigma_{\text{rupture Mat}}$ = 113 MPa.
(a) Calculate the maximum stress σ_{ox} leading to first ply failure of the siding. What are the layers that fracture?
(b) Apply the maximum stress σ_{ox}. In the previous layers that fractured, the glass fibers are supposed all broken. What happens to the laminate?

Solution:

1. The thickness of a layer denoted as b is such that (see Section 3.2.4):

$$b = \frac{m_{of}}{V_f \rho_f}$$

The specific mass of "E" glass is (see Section 1.6): ρ = 2600 kg/m³, from which:

$$b_{\text{fabric}} = 0.96 \text{ mm}; \quad b_{\text{Mat}} = 0.77 \text{ mm}.$$

2. The siding is constituted of the following stacking sequence:

Mat   Fabric

The total thickness is

$$b = 0.77 \times 4 + 0.96 \times 3 = 5.96 \text{ mm}.$$

3. Elastic characteristics of a fabric layer: The moduli and Poisson coefficients can be evaluated starting from the simplified relations of Section 3.4.2. One obtains, with k = 0.5 (balanced fabric):

$$E_C = E_T = 11{,}450 \text{ MPa}$$

$$G_{CT} = 1850 \text{ MPa}; \quad \nu_{CT} = \nu_{TC} = 0.128$$

A more precise calculation of these characteristics requires to establish the matrix $b[A]^{-1}$ of Section 12.1.2. (Equation (12.9)). We calculate at first $\frac{1}{b}[A]$ (Equation 12.8):

$$\frac{1}{b}A_{ij} = (\bar{E}_{ij}^{0°})(0.5) + (\bar{E}_{ij}^{90°})(0.5)$$

The terms $\bar{E}_{ij}$ are given by the Equation 11.8. One will have, for example:

$$\frac{1}{b}A_{11} = (\bar{E}_\ell)(0.5) + (\bar{E}_t)(0.5) = \frac{1}{2}\frac{E_\ell + E_t}{1 - \nu_{\ell t}\nu_{t\ell}}$$

with

$$\nu_{t\ell} = \nu_{\ell t}\frac{E_t}{E_\ell};$$

One obtains

$$\frac{1}{b}[A] = \begin{bmatrix} 11{,}737 & 1507 & 0 \\ 1507 & 11{,}737 & 0 \\ 0 & 0 & 1850 \end{bmatrix} (\text{MPa})$$

$$b[A]^{-1} = \begin{bmatrix} \dfrac{1}{11{,}540} & -\dfrac{0.128}{11{,}540} & 0 \\ -\dfrac{0.128}{11{,}540} & \dfrac{1}{11{,}540} & 0 \\ 0 & 0 & \dfrac{1}{1850} \end{bmatrix}$$

from which:

$$\boxed{\begin{aligned} E_C &= E_T = 11{,}540 \text{ MPa} \\ G_{CT} &= 1850 \text{ MPa} \\ \nu_{CT} &= \nu_{TC} = 0.128 \end{aligned}}$$

(The difference between the values obtained above is small.)

4. Elastic characteristics of the siding: These are deduced from the matrix $b[A]^{-1}$ (Equation 12.9) calculated for all the laminate.

We calculate at first $\frac{1}{b}[A]$ (Equation 12.8):

$$\frac{1}{b}A_{ij} = (\bar{E}_{ij}^{\text{fabric}})(p^{\text{fabric}}) + (\bar{E}_{ij}^{\text{Mat}})(p^{\text{Mat}})$$

with

$$p^{\text{fabric}} = \frac{3 \times 0.96}{5.96} = 0.483; \quad p^{\text{Mat}} = 0.517.$$

$$\bar{E}_{11}^{\text{fabric}} = \bar{E}_{22}^{\text{fabric}} = \bar{E}_C = \frac{E_C}{1 - \nu_{CT}^2} \qquad (\text{see Equation 11.8})$$

$$\bar{E}_{12}^{\text{fabric}} = \nu_{CT}\bar{E}_C; \quad \bar{E}_{33}^{\text{fabric}} = G_{CT}$$

$$\bar{E}_{11}^{\text{Mat}} = \bar{E}_{22}^{\text{Mat}} = \frac{E_{\text{Mat}}}{1 - \nu_{\text{Mat}}^2}; \quad \bar{E}_{12}^{\text{Mat}} = \frac{\nu_{\text{Mat}}E_{\text{Mat}}}{1 - \nu_{\text{Mat}}^2}$$

$$\bar{E}_{33}^{\text{Mat}} = G_{\text{Mat}} = \frac{E_{\text{Mat}}}{2(1 + \nu_{\text{Mat}})}$$

We obtain

$$
\frac{1}{b}[A] = \begin{bmatrix} 10{,}410 & 2149 & 0 \\ 2149 & 10{,}410 & 0 \\ 0 & 0 & 2554 \end{bmatrix} [\text{MPa}]
$$

$$
b[A]^{-1} = \begin{bmatrix} \dfrac{1}{9966} & -\dfrac{0.206}{9966} & 0 \\ -\dfrac{0.206}{9966} & \dfrac{1}{9966} & 0 \\ 0 & 0 & \dfrac{1}{2554} \end{bmatrix}
$$

then:

$$
\boxed{\begin{aligned} E_x &= E_y = 9966 \text{ MPa} \\ G_{xy} &= 2554 \text{ MPa} \\ \nu_{xy} &= \nu_{yx} = 0.206 \end{aligned}}
$$

Remark: The real modulus (measured) 9200 MPa is a bit smaller than the one calculated. In effect, due to the curvature of fibers from weaving, a fabric layer is softer than the stacking of unidirectionals that are crossed at 90°. However, the approximation obtained by calculation is suitable (difference < 10%).

5. Fracture of the siding:
   (a) One subjects the siding to a stress σ_{ox}. The strains of this siding are given by the Equation 12.9 as:

$$
\left\{ \begin{array}{c} \varepsilon_{ox} \\ \varepsilon_{oy} \\ \gamma_{oxy} \end{array} \right\} = b[A]^{-1} \left\{ \begin{array}{c} \sigma_{ox} \\ 0 \\ 0 \end{array} \right\} = \left\{ \begin{array}{c} \sigma_{ox}/9966 \\ -0.206\,\sigma_{ox}/9966 \\ 0 \end{array} \right\}
$$

These strains give rise to the following stresses:

■ In the layers of fabric (see results from Question 3):

$$
\left\{ \begin{array}{c} \sigma_C \\ \sigma_T \\ \tau_{CT} \end{array} \right\} = \begin{bmatrix} 11{,}737 & 1507 & 0 \\ 1507 & 11{,}737 & 0 \\ 0 & 0 & 1850 \end{bmatrix} \left\{ \begin{array}{c} \sigma_{ox}/9966 \\ -0.206\,\sigma_{ox}/9966 \\ 0 \end{array} \right\} \cdots
$$

$$
\cdots = \left\{ \begin{array}{c} 1.15\,\sigma_{ox} \\ -0.09\,\sigma_{ox} \\ 0 \end{array} \right\}
$$

The Hill–Tsai criterion in these layers is satisfied for a stress σ_{ox} such that:

$$\left(\frac{1.15\sigma_{ox}}{139}\right)^2 + \left(\frac{-0.09\sigma_{ox}}{139}\right)^2 - \frac{-0.09 \times 1.15\sigma_{ox}^2}{139^2} = 1$$

as:

$$\sigma_{ox} = 116 \text{ MPa}$$

■ In the layers of Mat, with the values of the coefficients $\bar{E}_{ij}^{\text{Mat}}$ of Question 4, we have

$$\left\{ \begin{array}{c} \sigma_x \\ \sigma_y \\ \tau_{xy} \end{array} \right\} = [\bar{E}^{\text{Mat}}] \left\{ \begin{array}{c} \sigma_{ox}/9\ 966 \\ -0.206\ \sigma_{ox}/9\ 966 \\ 0 \end{array} \right\} = \left\{ \begin{array}{c} 0.86\sigma_{ox} \\ 0.087\sigma_{ox} \\ 0 \end{array} \right\}$$

The Hill–Tsai criterion in the mat layers is satisfied for a stress σ_{ox} such that[47]:

$$\left(\frac{0.86\sigma_{ox}}{113}\right)^2 + \left(\frac{0.087\sigma_{ox}}{113}\right)^2 - \frac{0.86 \times 0.087\sigma_{ox}^2}{113^2} = 1$$

then:

$$\boxed{\sigma_{ox} = 138 \text{ MPa}}$$

The fabric layers are the first to fail, for a stress of

$$\sigma_{ox\ max} = 116 \text{ MPa}$$

(b) This stress being applied, the rupture of the fabric layers translates into the rupture of the glass fibers. The stress resultant corresponding to this constraint as:

$$N_x = \sigma_{ox\ max} \times h = 116 \times 5.96 = 691 \text{ N/mm}$$

is then completely taken up by the layers of Mat. The stress in these layers is then:

$$\sigma_{ox\ \text{Mat}} = \frac{N_x}{4 \times h_{\text{Mat}}} = \frac{691}{4 \times 0.77} = 224 \text{ MPa}$$

---

[47] A mat layer does not have transverse isotropy in the axes y, z (or x, z). The Hill–Tsai expression is then modified. We use however the Equation 14.6 here (see remark 39 at the bottom of page in Application 18.2.10).

It exceeds the rupture strength of the Mat (113 MPa). Then this latter layer fractures. The siding is then completely broken under the stress:

$$\sigma_{ox\,max} = 116 \text{ MPa}$$

18.2.13 Determination of the In-Plane Shear Modulus of a Balanced Fabric Ply

Problem Statement:

Consider a sample cut from a laminated panel made of identical layers of balanced fabric, all oriented along the axes C (warp direction) and T (fill direction) in the following figure.

The sample is in a state of simple tension in its plane along the x axis as shown in the figure.

$$\sigma_{ox} \neq 0; \qquad \sigma_{oy} = \tau_{oxy} = 0$$

Two strain gages are bonded (see figure). These are denoted as 1 and 2:

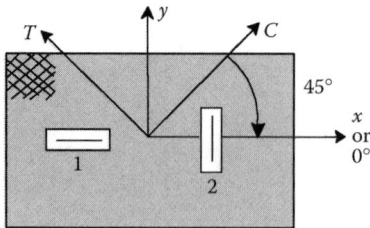

- From gage 1, one reads a strain ε_{ox}.
- From gage 2, one reads a strain ε_{oy}.

1. Noting that $\gamma_{oxy} = 0$, give the expression for the distortion γ_{CT} in the axes C and T as a function of ε_{ox} and ε_{oy}.
2. Give the expression for the stress τ_{CT} in the axes C and T as a function of σ_{ox}.
3. Deduce from the previous answer the shear modulus G_{CT} as a function of ε_{ox}, ε_{oy}, and σ_{ox}.

Solution:

1. Equation 11.7 allows one to write

$$\begin{Bmatrix} \varepsilon_C \\ \varepsilon_T \\ \gamma_{CT} \end{Bmatrix} = \begin{Bmatrix} c^2 & s^2 & -cs \\ s^2 & c^2 & cs \\ 2cs & -2cs & (c^2 - s^2) \end{Bmatrix} \begin{Bmatrix} \varepsilon_{ox} \\ \varepsilon_{oy} \\ \gamma_{oxy} \end{Bmatrix}$$

Here, we have a balanced laminate with midplane symmetry, loaded in its axes x,y. Then (Equation 12.9): $\gamma_{oxy} = 0$, from which:

$$\gamma_{CT} = 2cs\,\varepsilon_{ox} - 2cs\,\varepsilon_{oy} \text{ with } c = \frac{1}{\sqrt{2}}; \quad s = -\frac{1}{\sqrt{2}}$$

$$\gamma_{CT} = -\varepsilon_{ox} + \varepsilon_{oy}$$

2. According to Equation 11.4:

$$\left\{ \begin{array}{c} \sigma_C \\ \sigma_T \\ \tau_{CT} \end{array} \right\} = \left\{ \begin{array}{ccc} c^2 & s^2 & -2cs \\ s^2 & c^2 & 2cs \\ sc & -sc & (c^2 - s^2) \end{array} \right\} \left\{ \begin{array}{c} \sigma_{ox} \\ 0 \\ 0 \end{array} \right\}$$

then:

$$\tau_{CT} = sc\,\sigma_{ox} = -\frac{\sigma_{ox}}{2}$$

3. The constitutive behavior of the fabric in its axes can be written, starting from the Equation 11.5, as:

$$\left\{ \begin{array}{c} \varepsilon_C \\ \varepsilon_T \\ \gamma_{CT} \end{array} \right\} = \left\{ \begin{array}{ccc} \dfrac{1}{E_c} & -\dfrac{v_{CT}}{E_c} & 0 \\ -\dfrac{v_{CT}}{E_c} & \dfrac{1}{E_c} & 0 \\ 0 & 0 & \dfrac{1}{G_{CT}} \end{array} \right\} = \left\{ \begin{array}{c} \sigma_C \\ \sigma_T \\ \tau_{CT} \end{array} \right\}$$

$$\gamma_{CT} = \frac{\tau_{CT}}{G_{CT}}$$

From which:

$$\boxed{G_{CT} = \frac{\sigma_{ox}}{2(\varepsilon_{ox} - \varepsilon_{oy})}}$$

18.2.14 Quasi-Isotropic Laminate

Problem Statement:

Consider a laminate made up of a number of identical unidirectional plies, with midplane symmetry and the following composition:

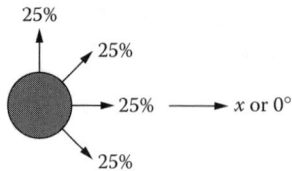

The elastic characteristics of a ply in its axes ℓ and t are denoted as:

$$E_\ell, \ E_t, \ G_{\ell t}, \ \nu_{\ell t}, \ \nu_{t\ell}$$

One examines the behavior of this laminate under in-plane loading, following the law (Equation 12.9):

$$\begin{Bmatrix} \varepsilon_{ox} \\ \varepsilon_{oy} \\ \gamma_{oxy} \end{Bmatrix} = h[A]^{-1} \begin{Bmatrix} \sigma_{ox} \\ \sigma_{oy} \\ \tau_{oxy} \end{Bmatrix}$$

1. Calculate the coefficients of the matrix $\frac{1}{h}[A]$.
2. By inversion, deduce from there the elastic moduli of the laminate.
3. What comment can one make? Deduce from there the law for the behavior of the laminate under in-plane loading in the axes (X,Y) derived from the (x,y) axes by a rotation angle θ.

Solution:

1. Coefficients $\frac{1}{h}A_{ij}$ are given by the Equation 12.8 as:

$$\frac{1}{h}A_{ij} = \frac{1}{4}\left[\bar{E}_{ij}^{0°} + \bar{E}_{ij}^{90°} + \bar{E}_{ij}^{+45°} + \bar{E}_{ij}^{-45°} \right]$$

The stiffness coefficients $\bar{E}_{ij}$ are obtained from the behavior of a ply (Equation 11.8). In using this relation for $\theta = 0°, 90°, +45°, -45°$, one obtains

$$\frac{1}{h}A_{11} = \frac{1}{h}A_{22} = \frac{1}{4}\left[\frac{3}{2}(\bar{E}_\ell + \bar{E}_t) + \nu_{t\ell}\,\bar{E}_\ell + 2G_{\ell t} \right]$$

$$\frac{1}{h}A_{12} = \frac{1}{4}\left[\frac{1}{2}(\bar{E}_\ell + \bar{E}_t) + 3\nu_{t\ell}\,\bar{E}_\ell - 2G_{\ell t} \right]$$

$$\frac{1}{h}A_{33} = \frac{1}{4}\left[\frac{1}{2}(\bar{E}_\ell + \bar{E}_t) - \nu_{t\ell}\,\bar{E}_\ell + 2G_{\ell t} \right]$$

$$\frac{1}{h}A_{13} = \frac{1}{h}A_{23} = 0$$

where one recalls that: $\bar{E}_\ell = \dfrac{E_\ell}{1 - \nu_{\ell t}\nu_{t\ell}}; \quad \bar{E}_t = \dfrac{E_t}{1 - \nu_{\ell t}\nu_{t\ell}}$

The matrix $\frac{1}{b}[A]$ reduces to

$$\frac{1}{b}\begin{bmatrix} A_{11} & A_{12} & 0 \\ A_{21} & A_{11} & 0 \\ 0 & 0 & A_{33} \end{bmatrix}$$

2. From the above, the modulus of elasticity of the laminate in directions x and y, and the associated Poisson coefficient are

$$\frac{1}{E} = \frac{\frac{1}{b}A_{11}}{\frac{1}{b^2}(A_{11}^2 - A_{12}^2)}; \qquad -\frac{v}{E} = -\frac{\frac{1}{b}A_{12}}{\frac{1}{b^2}(A_{11}^2 - A_{12}^2)}$$

One obtains after calculation:

$$E = \frac{[2(\bar{E}_\ell + \bar{E}_t) + 4v_{t\ell}\,\bar{E}_\ell][\bar{E}_\ell + \bar{E}_t - 2v_{t\ell}\,\bar{E}_\ell + 4G_{\ell t}]}{4\left[\frac{3}{2}(\bar{E}_\ell + \bar{E}_t) + v_{t\ell}\,\bar{E}_\ell + 2G_{\ell t}\right]}$$

$$v = \frac{\frac{1}{2}(\bar{E}_\ell + \bar{E}_t) + 3v_{t\ell}\,\bar{E}_\ell - 2G_{\ell t}}{\frac{3}{2}(\bar{E}_\ell + \bar{E}_t) + v_{t\ell}\,\bar{E}_\ell + 2G_{\ell t}}$$

The shear modulus is written as:

$$G = \frac{1}{4}\left[\frac{1}{2}(\bar{E}_\ell + \bar{E}_t) - v_{t\ell}\,\bar{E}_\ell + 2G_{\ell t}\right]$$

3. One can remark that:

$$G = \frac{E}{2(1 + v)}$$

This leads to an isotropic elastic behavior of the laminate in its plane. As a result, in all coordinate systems (X,Y) derived from (x,y) by any rotation angle, the constitutive behavior of the laminate is unchanged and is written as:

$$\left\{\begin{array}{c} \varepsilon_X \\ \varepsilon_Y \\ \gamma_{XY} \end{array}\right\} = \begin{bmatrix} \frac{1}{E} & -\frac{v}{E} & 0 \\ -\frac{v}{E} & \frac{1}{E} & 0 \\ 0 & 0 & \frac{1}{G} \end{bmatrix} = \left\{\begin{array}{c} \sigma_X \\ \sigma_Y \\ \tau_{XY} \end{array}\right\}$$

Remark: This result generalizes to other groups of orientations for plies such as:

$$\left[0, \frac{\pi}{3}, \frac{2\pi}{3}\right]; \quad \left[0, \frac{\pi}{5}, \frac{2\pi}{5}, \frac{3\pi}{5}, \frac{4\pi}{5}\right]$$

and so on.

More generally, a laminate made up of n orientations (a whole number $n > 2$), having the values of $\frac{\pi}{n}(p-1)$, with $p = 1,\ldots,$ n and with the same proportion of plies along each orientation as $(1/n)$ is elastically isotropic. One shows also that for all these laminates, E and ν are invariable.[48]

18.2.15 Orthotropic Plate in Pure Torsion

Problem Statement:

Consider a square plate ($a \times a$) of unidirectional glass/epoxy ($V_f = 60\%$), thickness h, welded at the center of its lower face on a support. It is subjected to a uniform and constant torsional moment density m_o (N mm/mm) along its perimeter.[49]

The directions ℓ, t of the unidirectional form an angle θ with the axes x,y of the plate (see figure).

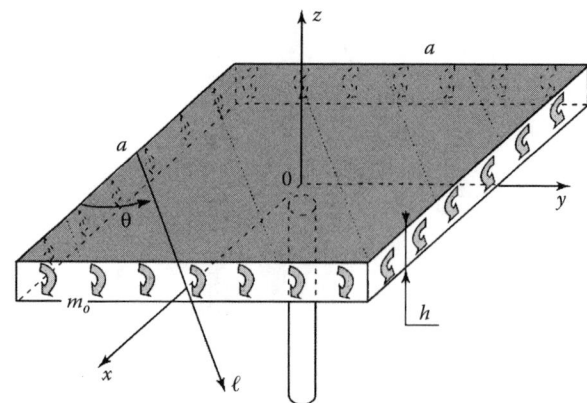

1. Assuming that all stress resultants in the plate are zero (except the torsional moment), determine the bending displacement at all points of the mid-plane.
2. Determine the state of stresses in the axes (x, y) then in the axes (ℓ, t) of the unidirectional layer.
3. Numerical application: $\theta = 45°$; $a = 1$ m; $h = 5$ mm; $m_o = -10$ N mm/mm.

---

[48] For more details, see Bibliography at the end of the book: "Stiffness isotropy and resistance quasi-isotropy of laminates with periodic orientations."

[49] The practical importance of such a load is very limited. It is better to consider this example as a means to validate a computer program using finite elements. It is one of the "patches" issued from the work "Computer programs for composite structures: Examples of reference for validation" (see Bibliography at the end of the book).

Solution:

1. In the constitutive Equation 12.16, one has

$$C_{ij} = \bar{E}_{ij} \int_{-b/2}^{b/2} z^2 dz = \bar{E}_{ij} \frac{b^3}{12}$$

then:

$$[C] = \frac{b^3}{12}[\bar{E}]$$

where $[\bar{E}]$ is the matrix shown in details in Equation 11.8.

By inverting the Equation 12.16 and noting that:

$$[C]^{-1} = \frac{12}{b^3}[\bar{E}]^{-1} = \frac{12}{b^3}\left[\frac{1}{E}\right]$$

where $\left[\frac{1}{E}\right]$ is the matrix shown in details in Equation 11.5, one has

$$\left\{ \begin{array}{c} -\dfrac{\partial^2 w_o}{\partial x^2} \\[2mm] -\dfrac{\partial^2 w_o}{\partial y^2} \\[2mm] -2\dfrac{\partial^2 w_o}{\partial x \partial y} \end{array} \right\} = \frac{12}{b^3}\left[\frac{1}{E}\right] \left\{ \begin{array}{c} M_y \\ -M_x \\ -M_{xy} \end{array} \right\} \qquad [1]$$

Assuming the unit stress resultants all to be zero except M_{xy},[50] one has

$$N_x = N_y = T_{xy} = M_x = M_y = 0; \quad M_{xy} = m_o$$

There remains (see Equation 11.5):

$$\frac{\partial^2 w_o}{\partial x^2} = \frac{12}{b^3}\frac{\eta_{xy}}{G_{xy}}m_o; \quad \frac{\partial^2 w_o}{\partial y^2} = \frac{12}{b^3}\frac{\mu_{xy}}{G_{xy}}m_o; \quad 2\frac{\partial^2 w_o}{\partial x \partial y} = \frac{12}{b^3}\frac{1}{G_{xy}}m_o$$

Therefore one can write $w_o(x,y)$ in the form:

$$w_o = \frac{12}{b^3}\frac{m_o}{G_{xy}}(Ax^2 + By^2 + Cxy + Dx + Ey + F)$$

---

[50] With this hypothesis, equations of equilibrium, constitutive equation, and boundary conditions are verified.

At the center of the plate: $w_o = 0;\ \dfrac{\partial w_o}{\partial x} = \dfrac{\partial w_o}{\partial y} = 0$

from which: $D = E = F = 0$. And by identification with the second derivatives:

$$2A = \eta_{xy};\quad 2B = \mu_{xy};\quad 2C = 1$$

The out-of-plane displacement takes the form:

$$w_o = \frac{6m_o}{b^3 G_{xy}}(\eta_{xy}x^2 + \mu_{xy}y^2 + xy) \qquad [2]$$

2. State of stresses: The strain field in the axes of the plate is written as (Equation 12.12, taking into account [1]):

$$\begin{Bmatrix} \varepsilon_x \\ \varepsilon_y \\ \gamma_{xy} \end{Bmatrix} = z \times \begin{Bmatrix} -\dfrac{\partial^2 w_o}{\partial x^2} \\ -\dfrac{\partial^2 w_o}{\partial y^2} \\ -2\dfrac{\partial^2 w_o}{\partial x \partial y} \end{Bmatrix} = z \times \frac{12}{b^3}\left[\frac{1}{E}\right] \begin{Bmatrix} 0 \\ 0 \\ -m_o \end{Bmatrix}$$

from which one can write the stresses in the axes (x, y) using Equation 11.8:

$$\begin{Bmatrix} \sigma_x \\ \sigma_y \\ \tau_{xy} \end{Bmatrix} = [\bar{E}]\begin{Bmatrix} \varepsilon_x \\ \varepsilon_y \\ \gamma_{xy} \end{Bmatrix} = z \times \frac{12}{b^3}[\bar{E}]\left[\frac{1}{E}\right]\begin{Bmatrix} 0 \\ 0 \\ -m_o \end{Bmatrix} = z \times \frac{12}{b^3}\begin{Bmatrix} 0 \\ 0 \\ -m_o \end{Bmatrix}$$

then:

$$\sigma_x = 0;\quad \sigma_y = 0;\quad \tau_{xy} = -z \times \frac{12}{b^3}m_o$$

■ Stresses in the axes of the unidirectional: These are obtained by using the Equation 11.4, which is[51]

$$\sigma_\ell = -2cs\,\tau_{xy} \quad = z \times cs \times \frac{24}{b^3}m_o$$

$$\sigma_t = 2cs\,\tau_{xy} \quad = -z \times cs \times \frac{24}{b^3}m_o$$

$$\tau_{\ell t} = (c^2 - s^2)\tau_{xy} = -z(c^2 - s^2) \times \frac{12}{b^3}m_o$$

---

[51] Note that here the angle $\theta = (\vec{\ell} \cdot \vec{x})$ since the Equation 11.4 is written with $\theta = (\vec{x} \cdot \vec{\ell})$.

3. Numerical application:

One has (Section 3.3.3) for the glass/epoxy:

$$E_\ell = 45,000 \text{ MPa}; \quad E_t = 12,000 \text{ MPa};$$

$$G_{\ell t} = 4500 \text{ MPa}; \quad v_{\ell t} = 0.3 \ (v_{t\ell} = 0.08)$$

from which (Equation 11.5 with $\theta = -45°$):

$$\frac{\eta_{xy}}{G_{xy}} = \frac{\mu_{xy}}{G_{xy}} = -\frac{0.1375}{4500}$$

and w_o takes the form:

$$w_o = -\frac{1}{9\ 375}[xy - 0.1375(x^2 + y^2)]$$

The deformed configuration is shown in the figure below:

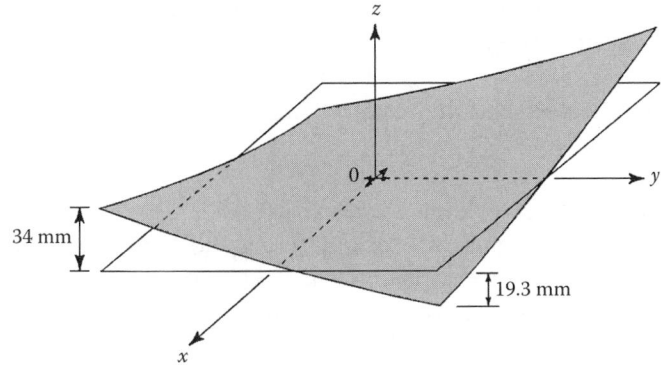

The stresses (in MPa) are written as:

$$\sigma_x = \sigma_y = 0; \quad \tau_{xy} = 0.96 \times z \text{ (mm)}.$$

$$\sigma_\ell = -\sigma_t = 0.96 \times z \text{ (mm)}; \quad \tau_{\ell t} = 0$$

18.2.16 Plate Made by Resin Transfer Molding (R.T.M.)

Problem Statement:

First part:
A roll of mat of carbon fibers has the following characteristics:

Areal mass density: $m_{of} = 30 \text{ g/m}^2$
Specific mass: $\rho_f = 1,750 \text{ kg/m}^3$

One deposits 21 layers of this mat over a plate in a rectangular mold. The mold is then closed and sealed, as shown in the figure below:

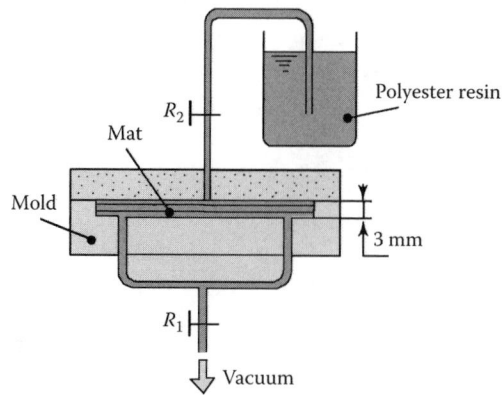

R_1 and R_2 represent two valves.

- R_2 is closed, R_1 is open. The mold is vacuumed.
- R_2 is open, R_1 is open. Polyester resin is filled into the cavity of the mold. Then resin begins to flow out through valve R_1.
- R_1 and R_2 are closed.

The mold is then heated, and the resin polymerizes. After demolding, one obtains a plate of mat/polyester.

1. Calculate the fiber volume fraction V_f (%).
2. Calculate the modulus of elasticity along the longitudinal and transverse directions, denoted respectively as E_ℓ and E_t, of a unidirectional of carbon/polyester, that would have the same amount of fiber volume fraction. The following is given

$$E_{f\ell} = 230,000 \text{ MPa}; \quad E_{ft} = 15,000 \text{ MPa (Section 3.3.1, Table 3.3)}$$

$$E_{resin} = 4000 \text{ MPa (Section 1.6)}$$

3. Starting from the relation in Section 3.5.1 giving the modulus of elasticity of mat (which is assumed to be isotropic in the plane of the plate), deduce from there the value of E_{mat}. Assume that $v_{mat} = 0.3$.

Second part:
One polymerizes on each face of the previous plate two plies of preimpregnated carbon/epoxy unidirectionals with $V_f = 60\%$ (see characteristics given in Section 3.3.3). Each ply has a thickness of 0.13 mm. The four plies (two above, two below) are oriented in the same direction denoted as x (or 0°). The midplane of the laminated plate coincides with axes x and y.

1. Write numerically for the unidirectional and for the mat the constitutive relation in the x, y axes in the form:

$$\left\{ \begin{array}{c} \sigma_x \\ \sigma_y \\ \tau_{xy} \end{array} \right\} = [\bar{E}] \left\{ \begin{array}{c} \varepsilon_x \\ \varepsilon_y \\ \gamma_{xy} \end{array} \right\}$$

2. Calculate in the axes x, y the coefficients of the in-plane constitutive relation of the laminated plate (matrix [A], Section 12.1.1). Deduce from there the moduli of elasticity and the Poisson coefficients of the plate.

3. Calculate in the axes x, y the coefficients for the bending behavior of the laminated plate (matrix [C], Section 12.1.4). Deduce from there the apparent bending moduli along the directions x and y.

4. This laminated plate is submitted to a tensile stress resultant along the x direction denoted as N_x (N/mm). The tensile rupture strength of mat is 100 MPa. Calculate the value of the stress resultant N_x that leads to first-ply failure of the laminate. In which component (unidirectional or mat) will this failure occur? This component is supposed to be completely broken (i.e., its mechanical characteristics are reduced to zero). What then is the state of stress in the other component? Make a conclusion.

Solution:

First part:

1. Fiber volume fraction of carbon:

$$V_f = \frac{\text{vol. fibers}}{\text{total volume}}$$

If s is the rectangular surface at the base of the mold, the volume of a layer of mat is

$$s \frac{m_{of}}{\rho_f}$$

from which, for 21 layers:

$$V_f = \frac{21 \times s \times m_{of}/\rho_f}{s \times 3 \times 10^{-3}} = 12\%$$

2. Moduli of elasticity (see Section 3.3.1): One has

$$E_\ell = E_f V_f + E_m V_m = 31{,}120 \text{ MPa}$$

$$E_t = E_m \left[\frac{1}{V_m + \dfrac{E_m}{E_{ft}} V_f} \right] = 4386 \text{ MPa}$$

3. One has (see Section 3.5.1)

$$E_{\text{mat}} = \frac{3}{8}E_\ell + \frac{5}{8}E_t = 14{,}410 \text{ MPa}$$

Second part:

1. Constitutive behavior:

■ Unidirectional:

$$\left\{ \begin{array}{c} \varepsilon_x \\ \varepsilon_y \\ \gamma_{xy} \end{array} \right\} = \left(\begin{array}{ccc} \dfrac{1}{134{,}000} & -\dfrac{0.25}{134{,}000} & 0 \\ -\dfrac{0.25}{134{,}000} & \dfrac{1}{7000} & 0 \\ 0 & 0 & \dfrac{1}{4200} \end{array} \right) \left\{ \begin{array}{c} \sigma_x \\ \sigma_y \\ \tau_{xy} \end{array} \right\}$$

After inversion:

$$\left\{ \begin{array}{c} \sigma_x \\ \sigma_y \\ \tau_{xy} \end{array} \right\} = \left[\begin{array}{ccc} 134{,}440 & 1756 & 0 \\ 1756 & 7023 & 0 \\ 0 & 0 & 4200 \end{array} \right] \left\{ \begin{array}{c} \varepsilon_x \\ \varepsilon_y \\ \gamma_{xy} \end{array} \right\}$$

■ Mat:

$$\left\{ \begin{array}{c} \varepsilon_x \\ \varepsilon_y \\ \gamma_{xy} \end{array} \right\} = \left[\begin{array}{ccc} \dfrac{1}{14{,}410} & -\dfrac{0.3}{14{,}410} & 0 \\ -\dfrac{0.3}{14{,}410} & \dfrac{1}{14{,}410} & 0 \\ 0 & 0 & \dfrac{2(1+0.3)}{14{,}410} \end{array} \right] \left\{ \begin{array}{c} \sigma_x \\ \sigma_y \\ \tau_{xy} \end{array} \right\}$$

After inversion:

$$\left\{ \begin{array}{c} \sigma_x \\ \sigma_y \\ \tau_{xy} \end{array} \right\} = \left[\begin{array}{ccc} 15{,}835 & 4750 & 0 \\ 4750 & 15{,}835 & 0 \\ 0 & 0 & 5542 \end{array} \right] \left\{ \begin{array}{c} \varepsilon_x \\ \varepsilon_y \\ \gamma_{xy} \end{array} \right\}$$

2. Membrane behavior of the laminated plate:

$$A_{ij} = \sum_{\text{ply } n°1}^{n} \bar{E}_{ij}^{(k)} e^{(k)}$$

$$A_{11} = 134,440 \times 4 \times 0.13 + 15,835 \times 3 = 117,408 \text{ (MPa mm)}$$

$$A_{22} = 7023 \times 4 \times 0.13 + 15,835 \times 3 = 51,151 \text{ (MPa mm)}$$

$$A_{12} = 1756 \times 4 \times 0.13 + 4,750 \times 3 = 15,163 \text{ (MPa mm)}$$

$$A_{13} = A_{23} = 0$$

$$A_{33} = 4200 \times 4 \times 0.13 + 5542 \times 3 = 18,810 \text{ (MPa mm)}.$$

From this, and with a total thickness of the plate of

$$h = 3 + 4 \times 0.13 = 3.52 \text{ mm}$$

we have

$$[A] = \begin{bmatrix} 117,408 & 15,163 & 0 \\ 15,163 & 51,151 & 0 \\ 0 & 0 & 18,810 \end{bmatrix};$$

$$h[A]^{-1} = \begin{bmatrix} \dfrac{1}{32,078} & -\dfrac{0.13}{13,975} & 0 \\ -\dfrac{0.3}{32,078} & \dfrac{1}{13,975} & 0 \\ 0 & 0 & \dfrac{1}{5344} \end{bmatrix}$$

then for the moduli of elasticity of the plate:

$$\bar{E}_x = 31,078 \text{ MPa}; \quad v_{xy} = 0.3; \quad \bar{G}_{xy} = 5344 \text{ MPa}$$

$$\bar{E}_y = 13,975 \text{ MPa}; \quad v_{yx} = 0.13$$

3. Bending behavior of the laminated plate:

$$C_{ij} = \sum_{\text{ply } n°1}^{n} \bar{E}_{ij}^{(k)} \left(\frac{z_k^3 - z_{k-1}^3}{3} \right)$$

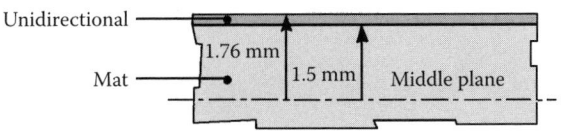

$$C_{11} = 134{,}440 \times \frac{1.76^3 - 1.5^3}{3} \times 2 + 15{,}835 \times \frac{1.5^3}{3} \times 2 = 221{,}763 \text{ MPa} \times \text{mm}^3$$

$$C_{22} = 7023 \times \frac{1.76^3 - 1.5^3}{3} \times 2 + 15{,}835 \times \frac{1.5^3}{3} \times 2 = 45{,}352 \text{ MPa} \times \text{mm}^3$$

$$C_{12} = 1756 \times \frac{1.76^3 - 1.5^3}{3} \times 2 + 4750 \times \frac{1.5^3}{3} \times 2 = 13{,}119 \text{ MPa} \times \text{mm}^3$$

$$C_{13} = C_{23} = 0$$

$$C_{33} = 4200 \times \frac{1.76^3 - 1.5^3}{3} \times 2 + 5542 \times \frac{1.5^3}{3} \times 2 = 18{,}284 \text{ MPa} \times \text{mm}^3$$

from which (see Section 12.1.6):

$$[C]^{-1} = \begin{bmatrix} \dfrac{1}{217{,}968} & -\dfrac{1}{753{,}509} & 0 \\ -\dfrac{1}{753{,}509} & \dfrac{1}{44{,}576} & 0 \\ 0 & 0 & \dfrac{1}{18{,}284} \end{bmatrix} = \begin{bmatrix} \dfrac{1}{\overline{EI}_{11}} & \dfrac{1}{\overline{EI}_{12}} & 0 \\ \dfrac{1}{\overline{EI}_{21}} & \dfrac{1}{\overline{EI}_{22}} & 0 \\ 0 & 0 & \dfrac{1}{\overline{EI}_{33}} \end{bmatrix}$$

Apparent bending modulus in the x direction:

$$\frac{1}{\overline{EI}_{11}} = \frac{1}{E_{fx} \times \dfrac{b^3}{12}} \rightarrow E_{fx} = 59{,}972 \text{ MPa}$$

The apparent bending modulus in the y direction:

$$\frac{1}{\overline{EI}_{22}} = \frac{1}{E_{fy} \times \dfrac{b^3}{12}} \rightarrow E_{fy} = 12{,}264 \text{ MPa}$$

4. Rupture: For a stress resultant N_x, the plate is deformed in its plane according to the relation:

$$\begin{Bmatrix} \varepsilon_x \\ \varepsilon_y \\ \gamma_{xy} \end{Bmatrix} = [A]^{-1} \begin{Bmatrix} N_x \\ 0 \\ 0 \end{Bmatrix}$$

then with the values found for $[A]^{-1}$:

$$\varepsilon_x = 8.856 \times 10^{-6} \times N_x; \quad \varepsilon_y = -2.66 \times 10^{-6} \times N_x; \quad \gamma_{xy} = 0$$

One then has for the stresses:

▪ In the unidirectional layer:

$$\sigma_\ell = \sigma_x = 134{,}440 \; \varepsilon_x + 1756 \; \varepsilon_y = 1.183 \; N_x$$

$$\sigma_t = \sigma_y = 1756 \; \varepsilon_x + 7023 \; \varepsilon_y = -0.003 \; N_x$$

$$\tau_{\ell t} = \tau_{xy} = 0.$$

▪ In the mat layer:

$$\sigma_x = 15{,}835 \; \varepsilon_x + 4750 \; \varepsilon_y = 0.128 \; N_x$$

$$\sigma_y = 4750 \; \varepsilon_x + 15{,}835 \; \varepsilon_y = 5.5 \times 10^{-5} \times N_x$$

$$\tau_{xy} = 0$$

From which the failure criteria are (see Section 14.2.3)

▪ In the unidirectional layer:

$$\frac{(1.183 N_x)^2}{1270^2} + \frac{(-0.003 N_x)^2}{141^2} - \frac{1.183 \times -0.003 N_x^2}{1270^2} < 1$$

Failure will not occur when: $N_x < 1072$ N/mm.

▪ In the mat layer:

$$\frac{(0.128 N_x)^2}{100^2} + \frac{(5.5 \; E\text{-}5 \times N_x)^2}{100^2} - \frac{0.128 \times 5.5 \; E\text{-}5 \times N_x^2}{100^2} < 1$$

Failure will not occur when $N_x < 781$ N/mm.
    Failure will first occur in the mat layer (first-ply rupture). The mat is supposed to be completely broken. The stress resultant $N_x = 781$ N/mm leads to a state of uniaxial stress in the laminate such that:

$$\sigma_\ell = \sigma_x = \frac{N_x}{4 \times 0.13} = \frac{781}{0.52} = 1502 \text{ MPa} > \sigma_{\ell \text{ rupture}}$$

The fibers in the unidirectional layer are broken.
Conclusion: The first-ply failure leads to ultimate rupture of the laminate.

18.2.17 Thermoelastic Behavior of a Balanced Fabric Ply

Problem Statement:

Consider a layer of balanced fabric made of carbon/epoxy ($V_f = 60\%$). The configuration of a unit cell ($a \times a$) is shown in Figure 18.13. One considers the layer of fabric as equivalent to two layers, each with a thickness e.

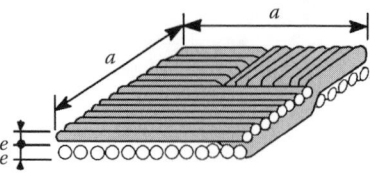

Figure 18.13

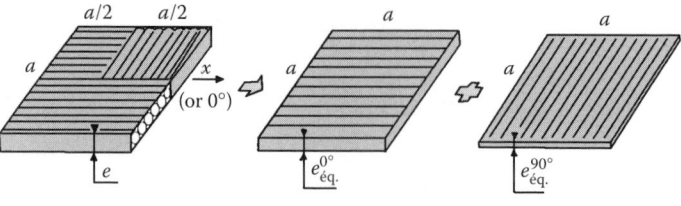

Figure 18.14

First part: Upper layer

We study the upper layer as shown schematically in Figure 18.14.

Assume that this upper layer behaves as if it consists of two equivalent unidirectional layers ($a \times a$) crossed at 0° and 90°. These layers have equivalent thicknesses denoted, respectively, as:

$$e_{\text{equi}}^{0°} \quad \text{and} \quad e_{\text{equi}}^{90°}$$

1. Show that $e_{\text{equi}}^{0°} = \frac{3}{4}e$; $\quad e_{\text{equi}}^{90°} = \frac{1}{4}e$
2. Deduce from the above the stiffness matrix $\frac{1}{b}[A]$ of this upper layer made up of the two previous unidirectionals, with the values of the moduli and Poisson coefficients for the unidirectional indicated in Section 3.3.3.
3. Deduce from the above the moduli of elasticity and Poisson coefficients of this upper layer, denoted as E_x, E_y, G_{xy}, v_{xy}.[52]
4. The coefficients of thermal expansion of the unidirectional are denoted as α_ℓ and α_t (see values in Section 3.3.3). What are the values of the coefficients of thermal expansion α_{ox}, α_{oy}, α_{oxy} of this layer? (One will at first calculate the terms denoted as $<\alpha E b>_i$ of Section 12.1.7.)

---

[52] Note here that:

$$\frac{1}{b}[A] \neq \begin{bmatrix} E_x & v_{yx}E_x & 0 \\ v_{xy}E_y & E_y & 0 \\ 0 & 0 & G_{xy} \end{bmatrix}$$

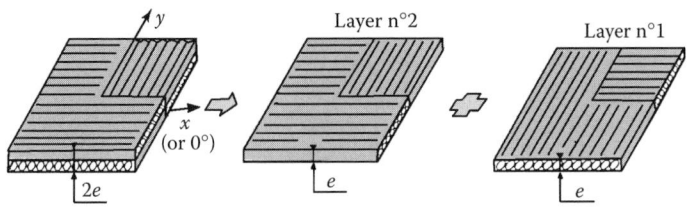

Figure 18.15

Second part: Complete fabric layer
Now we consider the complete fabric ply (thickness $2e$, see Figure 18.13) as the result of a simple superposition of two layers like the one that was studied in the previous part, these two layers being crossed at $0°$ (upper layer no. 2) and at $90°$ (lower layer no. 1).
   One retains in the following $e = 0.14$ mm.

1. Write numerically with the previous results the in-plane constitutive behavior for layer no. 2, then for layer no. 1 in Figure 18.15 in the form $\{\sigma\} = [\bar{E}]\{\varepsilon\}$.
2. Calculate the coefficients $\overline{\alpha E_i}$ (see Section 11.3.2) of layer no. 2, then of layer no. 1.
3. Calculate the matrix [A] characterizing the in-plane behavior of the double layer in Figure 18.15 (layer no. 1 + layer no. 2).

Third part: (Independent of the two previous parts until Question 9)
We consider a laminate which consists of two orthotropic plies noted as 2 and 1, each with a thickness e, crossed at $0°$ (or x) and at $90°$, respectively. We give below the respective thermomechanical behavior of these layers in axes x and y, which are written as:
Ply no. 1 (lower ply):

$$
\left\{
\begin{array}{c}
\sigma_x \\
\sigma_y \\
\tau_{xy}
\end{array}
\right\}_1
=
\begin{bmatrix}
a & c & 0 \\
c & b & 0 \\
0 & 0 & d
\end{bmatrix}
\left\{
\begin{array}{c}
\varepsilon_x \\
\varepsilon_y \\
\gamma_{xy}
\end{array}
\right\}
- \Delta T
\left\{
\begin{array}{c}
f \\
g \\
0
\end{array}
\right\}
$$

Ply no. 2 (upper ply):

$$
\left\{
\begin{array}{c}
\sigma_x \\
\sigma_y \\
\tau_{xy}
\end{array}
\right\}_2
=
\begin{bmatrix}
b & c & 0 \\
c & a & 0 \\
0 & 0 & d
\end{bmatrix}
\left\{
\begin{array}{c}
\varepsilon_x \\
\varepsilon_y \\
\gamma_{xy}
\end{array}
\right\}
- \Delta T
\left\{
\begin{array}{c}
g \\
f \\
0
\end{array}
\right\}
$$

Recalling that the thermomechanical behavior of a laminate is written as:

$$\begin{Bmatrix} N_x \\ N_y \\ T_{xy} \\ M_y \\ -M_x \\ -M_{xy} \end{Bmatrix} = \begin{bmatrix} A & B \\ B & C \end{bmatrix} \begin{Bmatrix} \varepsilon_{ox} \\ \varepsilon_{oy} \\ \gamma_{oxy} \\ -\dfrac{\partial^2 w_o}{\partial x^2} \\ -\dfrac{\partial^2 w_o}{\partial y^2} \\ -2\dfrac{\partial^2 w_o}{\partial x \partial y} \end{Bmatrix} - \Delta T \begin{bmatrix} \langle \alpha E h \rangle_x \\ \langle \alpha E h \rangle_y \\ \langle \alpha E h \rangle_{xy} \\ \langle \alpha E h^2 \rangle_x \\ \langle \alpha E h^2 \rangle_y \\ \langle \alpha E h^2 \rangle_{xy} \end{bmatrix}$$

1. Write the literal expression of matrix [A].
2. Write the literal expression of matrix [C].
3. Write the literal expression of matrix [B].
4. Calculate the terms $<\alpha E h>_x$, $<\alpha E h>_y$, $<\alpha E h>_{xy}$, $<\alpha E h^2>_x$, $<\alpha E h^2>_y$, $<\alpha E h^2>_{xy}$.
5. Write the thermomechanical behavior equation.
6. This plate is not externally loaded. It is subjected to a variation in temperature ΔT. Deduce from item 5 the corresponding system of equations.
7. Give the values of γ_{oxy} and $\dfrac{\partial^2 w_o}{\partial x \partial y}$.
8. Write the equations that allow the calculation of other strains.
9. Taking into account the results obtained in the second part, write numerically this system of equations with $\Delta T = -160°C$. Give the corresponding values of strains. Comment.

Solution:

1.1. Volume of fibers at 0°:

$$v^{0°} = \frac{3a^2}{4} \times e = a^2 \times e^{0°}_{\text{équiv.}}$$

Volume of fibers at 90°:

$$v^{90°} = \frac{a^2}{4} \times e = a^2 \times e^{90°}_{\text{équiv.}}$$

from which:

$$e^{0°}_{\text{équiv.}} = \frac{3e}{4}; \qquad e^{90°}_{\text{équiv.}} = \frac{e}{4}$$

1.2. Stiffness matrix $\frac{1}{h}[A]$: According to the Equation 11.8 and the values in Section 3.3.3:

$$\bar{E}_{11}^{0°} = \bar{E}_\ell = 134{,}439 \text{ MPa}; \quad \bar{E}_{12}^{0°} = v_{t\ell}\,\bar{E}_\ell = 1756 \text{ MPa}$$

$$\bar{E}_{22}^{0°} = \bar{E}_t = 7023 \text{ MPa}; \quad \bar{E}_{33}^{0°} = G_{\ell t} = 4200 \text{ MPa}$$

$$\bar{E}_{11}^{90°} = 7023 \text{ MPa}; \quad \bar{E}_{12}^{90°} = 1756 \text{ MPa}; \quad \bar{E}_{22}^{90°} = 134{,}439 \text{ MPa}; \quad \bar{E}_{33}^{90°} = 4200 \text{MPa}$$

$$A_{11} = \bar{E}_{11}^{0°} \times \frac{3e}{4} + \bar{E}_{11}^{90°} \times \frac{e}{4} = 102{,}585 \times e \text{ (MPa.mm)}$$

$$A_{22} = \bar{E}_{22}^{0°} \times \frac{3e}{4} + \bar{E}_{22}^{90°} \times \frac{e}{4} = 38{,}877 \times e \text{ (MPa.mm)}$$

$$A_{12} = 1756 \text{ MPa}; \quad A_{33} = 4200 \times e \text{ (MPa.mm)}$$

$$\frac{1}{h}[A] = \begin{bmatrix} 102{,}585 & 1756 & 0 \\ 1756 & 38{,}877 & 0 \\ 0 & & 4200 \end{bmatrix} \text{(MPa)}$$

1.3. One has, according to Equation 12.9:

$$h[A]^{-1} = \begin{bmatrix} \dfrac{1}{E_x} & -\dfrac{v_{yx}}{E_y} & 0 \\ -\dfrac{v_{xy}}{E_x} & \dfrac{1}{E_y} & 0 \\ 0 & 0 & \dfrac{1}{G_{xy}} \end{bmatrix}$$

from which:

$$E_x = 102{,}506 \text{ MPa}$$

$$E_y = 38{,}847 \text{ MPa}$$

$$v_{yx} = 0.017; \quad v_{xy} = 0.045$$

$$G_{xy} = 4200 \text{ MPa}$$

One then can verify that:

$$\frac{1}{h}[A]\# \begin{bmatrix} E_x & v_{yx}E_x & 0 \\ v_{xy}E_y & E_y & 0 \\ 0 & 0 & G_{xy} \end{bmatrix}$$

1.4. One has (Equation 12.18):

$$
\left\{
\begin{array}{c}
\alpha_{ox} \\
\alpha_{oy} \\
\alpha_{oxy}
\end{array}
\right\}
= b[A]^{-1}
\left\{
\begin{array}{c}
\frac{1}{b}\langle \alpha E b \rangle_x \\
\frac{1}{b}\langle \alpha E b \rangle_y \\
\frac{1}{b}\langle \alpha E b \rangle_{xy}
\end{array}
\right\}
$$

With (Equations 12.17 and 11.10):

$$
\langle \alpha E b \rangle_x = \overline{\alpha E_1}^{0°} \times \frac{3}{4}e + \overline{\alpha E_1}^{90°} \times \frac{e}{4} = \cdots
$$

$$
\cdots \bar{E}_\ell(\alpha_\ell + \nu_{t\ell}\alpha_t) \times \frac{3}{4}e + \bar{E}_t(\nu_{\ell t}\alpha_\ell + \alpha_t) \times \frac{e}{4}
$$

with (Section 3.3.3): $\alpha_\ell = -0.12 \times 10^{-5}$; $\alpha_t = 3.4 \times 10^{-5}$.

$$
\frac{1}{b}\langle \alpha E b \rangle_x = -1726 \times 10^{-5}. \text{ Then:}
$$

$$
\frac{1}{b}\langle \alpha E b \rangle_y = 15{,}203 \times 10^{-5}; \quad \frac{1}{b}\langle \alpha E b \rangle_{xy} = 0
$$

One then deduces:

$$
\alpha_{ox} = -2.3 \times 10^{-7}; \quad \alpha_{oy} = 39 \times 10^{-7}; \quad \alpha_{oxy} = 0
$$

2.1. Constitutive behavior: $\{\sigma\} = [\bar{E}]\{\varepsilon\}$: According to Equation 11.8 *Layer no. 2*:

$$
\bar{E}_{11}^{(2)} = \bar{E}_x = \frac{E_x}{1 - \nu_{yx}\,\nu_{xy}} = 102{,}584 \text{ MPa etc.}
$$

$$
[\bar{E}]^{(2)} =
\begin{bmatrix}
102{,}584 & 1744 & 0 \\
1744 & 38{,}877 & 0 \\
0 & 0 & 4200
\end{bmatrix}
$$

Layer no. 1:

$$
[\bar{E}]^{(1)} =
\begin{bmatrix}
38{,}877 & 1744 & 0 \\
1744 & 102{,}584 & 0 \\
0 & 0 & 4200
\end{bmatrix}
$$

2.2. Coefficients $\overline{\alpha E_i}$:
Layer no. 2:

$$\overline{\alpha E}_1^{(2)} = \overline{E}_x(\alpha_{ox} + v_{yx}\alpha_{oy}) = -0.0168$$

$$\overline{\alpha E}_2^{(2)} = 0.1512; \quad \overline{\alpha E}_3^{(2)} = 0$$

Layer no. 1 (rotation of 90°):

$$\overline{\alpha E}_1^{(1)} = 0.1512; \quad \overline{\alpha E}_2^{(1)} = -0.0168; \quad \overline{\alpha E}_3^{(1)} = 0$$

2.3. In-plane behavior of the double layer:

$$A_{11} = \overline{E}_{11}^{(1)} \times e + \overline{E}_{11}^{(2)} \times e = (102,584 + 38,877) \times 0.14, \text{etc.}$$

$$[A] = \begin{bmatrix} 19,804 & 488 & 0 \\ 488 & 19,804 & 0 \\ 0 & 0 & 1176 \end{bmatrix} (\text{MPa. mm})$$

3.1. Matrix [A]:

$$[A] = \begin{bmatrix} (a+b)e & 2ce & 0 \\ 2ce & (a+b)e & 0 \\ 0 & 0 & 2de \end{bmatrix}$$

3.2. Matrix [C]:

$$C_{11} = a\left(\frac{0 - (-e)^3}{3}\right) + b\left(\frac{e^3 - 0}{3}\right) = (a+b)\frac{e^3}{3}, \text{etc.}$$

$$[C] = \begin{bmatrix} (a+b)\frac{e^3}{3} & 2c\frac{e^3}{3} & 0 \\ 2c\frac{e^3}{3} & (a+b)\frac{e^3}{3} & 0 \\ 0 & 0 & 2d\frac{e^3}{3} \end{bmatrix}$$

3.3. Matrix [B]:

$$B_{11} = a\left(\frac{0-(-e)^2}{2}\right) + b\left(\frac{e^2-0}{2}\right) = (b-a)\frac{e^2}{2}, \text{etc.}$$

$$[B] = \begin{bmatrix} (b-a)\frac{e^2}{2} & 0 & 0 \\ 0 & (a-b)\frac{e^2}{2} & 0 \\ 0 & 0 & 0 \end{bmatrix}$$

3.4. Terms $\langle\alpha Eb\rangle_i$ and $\langle\alpha Eb^2\rangle_i$:

$$\langle\alpha Eb\rangle_x = fe+ge = (f+g)e$$

$$\langle\alpha Eb\rangle_y = (f+g)e; \quad (\alpha Eb)_{xy} = 0$$

$$\langle\alpha Eb^2\rangle_x = (g-f)\frac{e^2}{2}$$

$$\langle\alpha Eb^2\rangle_y = (f-g)\frac{e^2}{2}; \quad (\alpha Eb^2)_{xy} = 0$$

3.5. Thermomechanical behavior:

$$\begin{Bmatrix} N_x \\ N_y \\ T_{xy} \\ M_y \\ -M_x \\ -M_{xy} \end{Bmatrix} = \begin{bmatrix} (a+b)e & 2ce & 0 & (b-a)\frac{e^2}{2} & 0 & 0 \\ 2ce & (a+b)e & 0 & 0 & (a-b)\frac{e^2}{2} & 0 \\ 0 & 0 & 2de & 0 & 0 & 0 \\ (b-a)\frac{e^2}{2} & 0 & 0 & (a+b)\frac{e^3}{3} & 2c\frac{e^3}{3} & 0 \\ 0 & (a-b)\frac{e^2}{2} & 0 & 2c\frac{e^3}{3} & (a+b)\frac{e^3}{3} & 0 \\ 0 & 0 & 0 & 0 & 0 & 2d\frac{e^3}{3} \end{bmatrix} \begin{Bmatrix} \varepsilon_{ox} \\ \varepsilon_{oy} \\ \gamma_{oxy} \\ -\frac{\partial^2 w_o}{\partial x^2} \\ -\frac{\partial^2 w_o}{\partial y^2} \\ -2\frac{\partial^2 w_o}{\partial x\partial y} \end{Bmatrix} \cdots$$

$$\cdots - \Delta T \begin{Bmatrix} (f+g)e \\ (f+g)e \\ 0 \\ (g-f)\frac{e^2}{2} \\ (f-g)\frac{e^2}{2} \\ 0 \end{Bmatrix}$$

3.6.   Variation of temperature ΔT:

One has here:

$$N_x = N_y = T_{xy} = M_x = M_y = M_{xy} = 0$$

from which we have

$$
\begin{bmatrix}
(a+b)e & 2ce & 0 & (b-a)\frac{e^2}{2} & 0 & 0 \\
2ce & (a+b)e & 0 & 0 & (a-b)\frac{e^2}{2} & 0 \\
0 & 0 & 2de & 0 & 0 & 0 \\
(b-a)\frac{e^2}{2} & 0 & 0 & (a+b)\frac{e^3}{3} & 2c\frac{e^3}{3} & 0 \\
0 & (a-b)\frac{e^2}{2} & 0 & 2c\frac{e^3}{3} & (a+b)\frac{e^3}{3} & 0 \\
0 & 0 & 0 & 0 & 0 & 2d\frac{e^3}{3}
\end{bmatrix}
\begin{Bmatrix}
\varepsilon_{ox} \\
\varepsilon_{oy} \\
\gamma_{oxy} \\
-\dfrac{\partial^2 w_o}{\partial x^2} \\
-\dfrac{\partial^2 w_o}{\partial y^2} \\
-2\dfrac{\partial^2 w_o}{\partial x \partial y}
\end{Bmatrix}
= \Delta T
\begin{Bmatrix}
(f+g)e \\
(f+g)e \\
0 \\
(g-f)\frac{e^2}{2} \\
(f-g)\frac{e^2}{2} \\
0
\end{Bmatrix}
$$

3.7.   One can note that:

$$\gamma_{oxy} = 0; \quad \frac{\partial^2 w_o}{\partial x \partial y} = 0$$

3.8.   There remains

$$
\begin{bmatrix}
(a+b)e & 2ce & (b-a)\frac{e^2}{2} & 0 \\
2ce & (a+b)e & 0 & (a-b)\frac{e^2}{2} \\
(b-a)\frac{e^2}{2} & 0 & (a+b)\frac{e^3}{3} & 2c\frac{e^3}{3} \\
0 & (a-b)\frac{e^2}{2} & 2c\frac{e^3}{3} & (a+b)\frac{e^3}{3}
\end{bmatrix}
\begin{Bmatrix}
\varepsilon_{ox} \\
\varepsilon_{oy} \\
-\dfrac{\partial^2 w_o}{\partial x^2} \\
-\dfrac{\partial^2 w_o}{\partial^2 y^2}
\end{Bmatrix}
= \Delta T
\begin{Bmatrix}
(f+g)e \\
(f+g)e \\
(g-f)\frac{e^2}{2} \\
(f-g)\frac{e^2}{2}
\end{Bmatrix}
$$

Remark: According to the model studied, one truly must have

$$\varepsilon_{ox} = \varepsilon_{oy}; \quad \frac{\partial^2 w_o}{\partial x^2} = -\frac{\partial^2 w_o}{\partial y^2}$$

It is worthy to note that with this hypothesis one obtains two identical systems of equations which are written as:

$$
\begin{bmatrix}
(a+b+2c)e & (b-a)\dfrac{e^2}{2} \\[2ex]
(b-a)\dfrac{e^2}{2} & (a+b-2c)\dfrac{e^3}{3}
\end{bmatrix}
\begin{bmatrix}
\varepsilon_{ox} \\[2ex]
-\dfrac{\partial^2 w_o}{\partial x^2}
\end{bmatrix}
= \Delta T
\begin{bmatrix}
(f+g)e \\[2ex]
(g-f)\dfrac{e^2}{2}
\end{bmatrix}
$$

3.9. With the results of the second part, and $\Delta T = -160°C$ (corresponding to the cooling in the autoclave after the polymerization of the resin), one has (units: N and mm):

$$(a+b)e = 19{,}804; \quad 2ce = 488; \quad (a+b)\frac{e^3}{3} = 129; \quad 2c\frac{e^3}{3} = 3.2$$

$$(b-a)\frac{e^2}{2} = 624; \quad (f+g)e = 0.0188; \quad (g-f)\frac{e^2}{2} = -0.00164$$

from which we obtain the strains and curvatures:

$$\varepsilon_{ox} = \varepsilon_{oy} = -1.7 \times 10^{-4}$$

$$\frac{\partial^2 w_o}{\partial x^2} = -\frac{\partial^2 w_o}{\partial y^2} = -8.6 \times 10^{-4}$$

We can conclude that during the cooling the layer of balanced fabric not only contracts but also, due to its weave, takes the form of a double curvature surface along the warp and fill directions; that is, the form of a horse saddle.

18.3 LEVEL 3

18.3.1 Cylindrical Bonding

Problem Statement:

We propose to study, in a simplified approach, a bonded assembly of two cylindrical tubes (figure below). The shear moduli of the materials are denoted along with the figure:

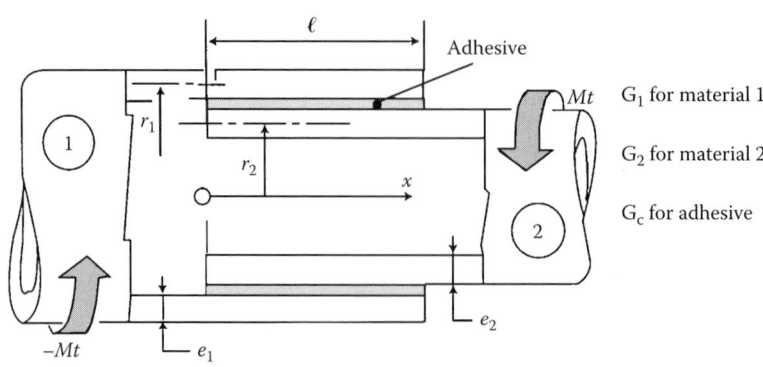

The deformed configuration of the generator of each of the tubes viewed from above is shown in the following figure, with the shear stresses τ_{10} and τ_{20} that are assumed to be uniform across the thickness of each tube. Also shown is the bonding element.

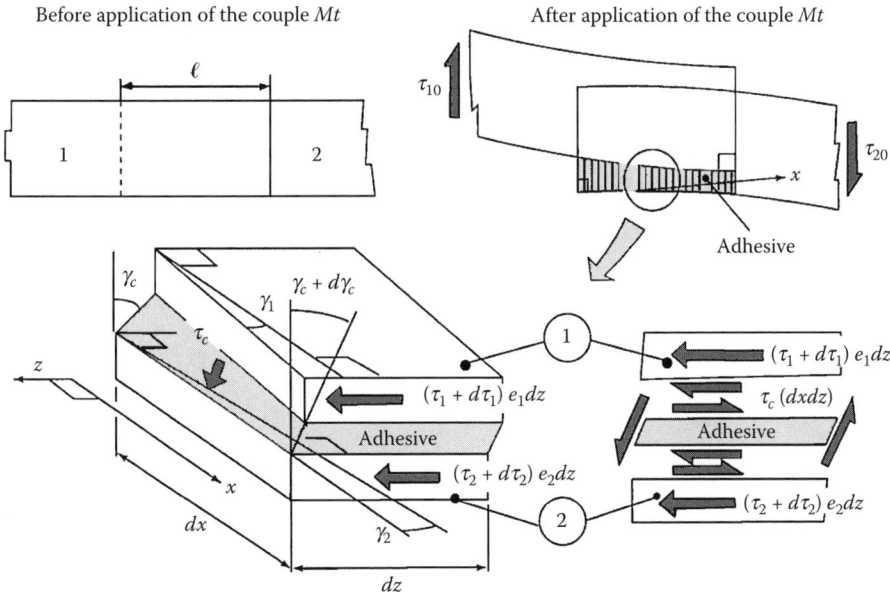

Before application of the couple Mt

After application of the couple Mt

1. Find the distribution of the shear stresses in the adhesive layer, denoted as τ_C in the previous figure.
2. Numerical application:

$G_1 = 28,430$ MPa; $\qquad$ $G_2 = 79,000$ MPa; $\qquad$ $G_C = 1700$ MPa;

$e_1 = e_2 = 12$ mm; $\qquad$ $e_C = 0.2$ mm; $\qquad$ $M_t = 300$ m.daN;

$r_1 = 63.5$ mm; $\qquad$ $r_2 = 51.5$ mm; $\qquad$ $\ell = 44$ mm.

3. Calculate the maximum shear stress in the particular case where the materials 1 and 2 are identical and have the same thickness, denoted as e, which is small compared with the radii.

Solution:

1. Shear stresses in the adhesive layer: In the previous figure that represents the bonding element, one reads the following equilibrium:
   ■ Equilibrium of material element 1:

$$d\tau_1 e_1\, dz + \tau_c\, dx\, dz = 0 \rightarrow \frac{d\tau_1}{dx} e_1 + \tau_c = 0 \qquad [a]$$

■ Equilibrium of material element 2:

$$d\tau_2 e_2\, dz - \tau_c\, dx\, dz = 0 \rightarrow \frac{d\tau_2}{dx} e_2 - \tau_c = 0 \qquad \text{[b]}$$

The shear stresses are proportional to the angular distortions, denoted here as γ_1 for material 1, γ_2 for material 2, and γ_c for the adhesive, from which:

$$\gamma_1 = \frac{\tau_1}{G_1}; \quad \gamma_2 = \frac{\tau_2}{G_2}; \quad \gamma_c = \frac{\tau_c}{G_c}$$

In addition one has the following geometric relation, by approximating the tangents and angles ($tg\ \theta \cong \theta$; see figure):

$$(\gamma_c + d\gamma_c) - \gamma_c \ \# \ \frac{-\gamma_1\, dx + \gamma_2\, dx}{e_c}$$

or:

$$\frac{d\gamma_c}{dx} = \frac{\gamma_2 - \gamma_1}{e_c}$$

In substituting the stresses:

$$\frac{d\tau_c}{dx}\frac{e_c}{G_c} = \frac{\tau_2}{G_2} - \frac{\tau_1}{G_1} \qquad \text{[c]}$$

One then obtains the 3 relations [a], [b], [c], from the unknowns τ_1, τ_2, τ_c. Eliminating τ_1 and τ_2 yields

$$\frac{d^2\tau_c}{dx^2}\frac{e_c}{G_c} = \frac{\tau_c}{e_2 G_2} + \frac{\tau_c}{e_1 G_1}$$

then:

$$\boxed{\frac{d^2\tau_c}{dx^2} - \lambda^2 \tau_c = 0 \quad \text{with} \quad \lambda^2 = \frac{G_c}{e_c}\left(\frac{1}{e_2 G_2} + \frac{1}{e_1 G_1}\right)}$$

The general solution for the above differential equation is:

$$\tau_C = A\ \text{ch}\ \lambda x + B\ \text{sh}\ \lambda x.$$

■ Boundary conditions:
   For $x = 0$: It is the free edge of material 2, where $\gamma_2 = 0$ and $\gamma_1 = \tau_{10}/G_1$.

from which:
$$\left.\frac{d\gamma_c}{dx}\right|_{x=0} = \frac{\gamma_2 - \gamma_1}{e_c} = -\frac{\tau_{10}}{e_c G_1} \qquad \text{[d]}$$

then:
$$\left.\frac{d\tau_c}{dx}\right|_{x=0} = -\frac{\tau_{10} G_c}{e_c G_1}$$

For $x = \ell$: It is the free edge of material 1, where $\gamma_1 = 0$ and $\gamma_2 = \tau_{20}/G_2$

from which:
$$\left.\frac{d\gamma_c}{dx}\right|_{x=\ell} = \frac{\gamma_2 - \gamma_1}{e_c} = \frac{\tau_{20}}{e_c G_2}$$

[e]

then:
$$\left.\frac{d\tau_c}{dx}\right|_{x=\ell} = \frac{\tau_{20} G_c}{e_c G_2}$$

The boundary conditions [d] and [e] allow the calculation of the constants A and B of the general solution. We obtain

$$\tau_c = \frac{G_c}{e_c \lambda}\left\{\left(\frac{\tau_{10}}{G_1}\frac{1}{\text{th }\lambda\ell} + \frac{\tau_{20}}{G_2}\frac{1}{\text{sh }\lambda\ell}\right)\text{ch }\lambda x - \frac{\tau_{10}}{G_1}\text{sh }\lambda x\right\}$$

2. Numerical application:

$$\tau_{10} = \frac{M_t}{2\pi r_1^2 e_1} = 9.86 \text{ MPa};$$

$$\tau_{20} = \frac{M_t}{2\pi r_2^2 e_2} = 15 \text{ MPa}$$

One obtains for the shear stress τ_C the following distribution, where the stress concentrations at the extremities of the assembly can be noted.

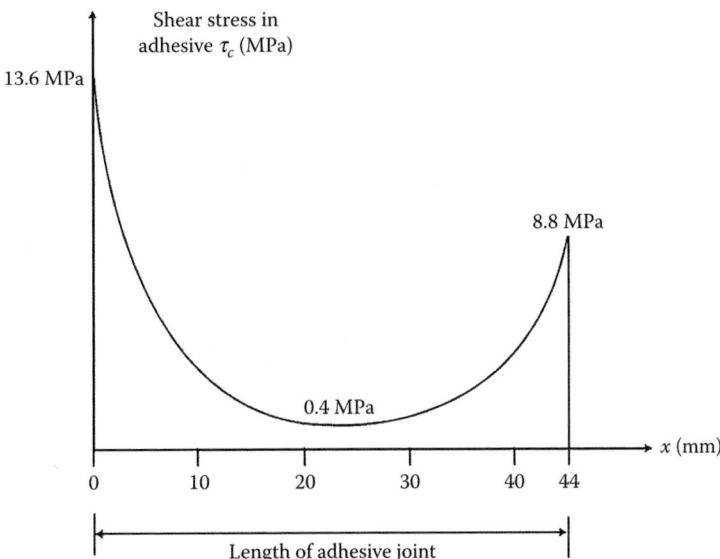

This explains that one should not design such a bonding assembly by basing on the average shear stress, which does not exist in reality.

Note: The proposed numerical values correspond here to those of Application 18.1.4 relative to the design of a transmission shaft in carbon/epoxy. One can note that the rupture strength of araldite, taken to be 15 MPa, is not effectively reached at the location of stress concentrations.

3. Particular case:

$$G_1 = G_2 = G; \quad e_1 = e_2 = e; \quad e/r_1 \# e/r_2.$$

The comparison:

$$\tau_{10} = \frac{M_t}{2\pi r_1^2 e} \quad \text{and} \quad \tau_{20} = \frac{M_t}{2\pi r_2^2 e}$$

allows one to write approximately:

$$\tau_{10} \# \tau_{20}$$

from which:

$$\tau_c = \frac{G_c}{\lambda e_c G} \tau_o \left\{ \left(\frac{1}{\text{th } \lambda\ell} + \frac{1}{\text{sh } \lambda\ell} \right) \text{ch } \lambda x - \text{sh } \lambda x \right\}$$

One notes the presence of peaks of identical stress at $x = 0$ and $x = \ell$ as:

$$\tau_{c \text{ max}} = \frac{G_c}{\lambda e_c G} \tau_o \frac{\text{ch } \lambda\ell + 1}{\text{sh } \lambda\ell} = \frac{G_c}{\lambda e_c G} \tau_o \frac{1}{\text{th } \frac{\lambda\ell}{2}}$$

Taking into account that:

$$\lambda^2 = \frac{2G_c}{e_c Ge}:$$

$$\tau_{c \text{ max}} = \tau_o \frac{\lambda^2 e}{2\lambda} \frac{1}{\text{th } \frac{\lambda\ell}{2}} = \tau_o e \frac{\lambda/2}{\text{th } \lambda\ell/2}$$

reveals the average stress in the adhesive (fictitious notion as mentioned above):

$$\tau_{\text{average}} = \frac{M_t}{2\pi r^2 \ell} = \frac{M_t}{2\pi r^2 e \ell} \frac{e}{} = \tau_o \frac{e}{\ell}$$

from which:

$$\tau_{c \text{ max}} = \tau_{\text{average}} \frac{\lambda\ell/2}{\text{th } \lambda\ell/2}$$

In setting $\lambda\ell/2 = a$, one finds again the relation of Section 6.2.3:

$$\tau_{max} = \frac{a}{th\,a} \times \tau_{average} ; \quad \text{with} \quad a = \sqrt{\frac{G_c\ell^2}{2Gee_c}}$$

18.3.2 Double Bonded Joint

Problem Statement:

Shown below is an assembly consisting of two identical plates of material 1 bonded to a central plate of material 2. This joint provides a plane of symmetry $(x - y)$. We will study approximately the shear stress in the adhesive. For that, assume that the stresses are just functions of x.

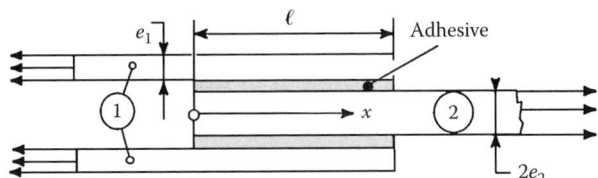

The configuration of a bonding element of length dx is shown below. The moduli of the materials are denoted as: E_1 for material 1, E_2 for material 2, G_C for the adhesive.

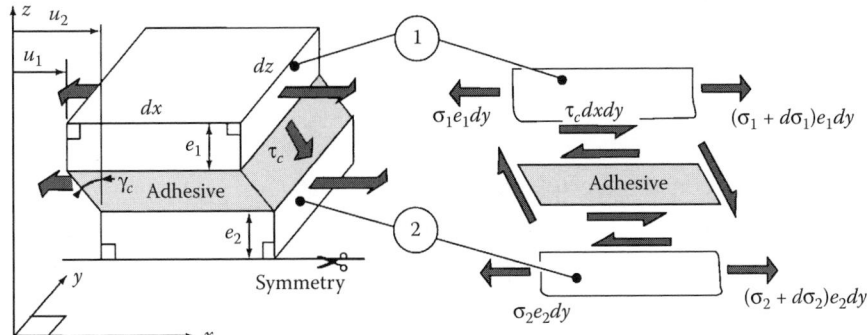

1. Determine the distribution of shear stresses in the adhesive, denoted as $\tau_C(x)$.
2. Numerical application: The two external plates are made of titanium alloy (TA 6V), with thickness 1.5 mm. The intermediate plate is a laminate of carbon/epoxy, with $V_f = 60\%$ fiber volume fraction and the following composition:

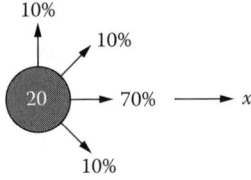

The thickness of one ply is 0.125 mm. The rupture strength of the adhesive (araldite) is taken to be 15 MPa. Its thickness is 0.2 mm. What length of bond ℓ will allow the bonding assembly to transmit a stress resultant of 20 daN/mm of width?

3. Calculate the maximum shear stress in the particular case where the materials 1 and 2 are identical and where $e_1 = e_2 = e$.

Solution:

1. Shear stress in the adhesive: In the previous figure showing an element of the bond, one reads the following equilibrium:
   ■ Equilibrium of element of material 1:

$$d\sigma_1 e_1 \, dy + \tau_c \, dx \, dy = 0 \rightarrow \frac{d\sigma_1}{dx} e_1 + \tau_c = 0 \qquad [a]$$

   ■ Equilibrium of element of material 2:

$$d\sigma_2 e_2 \, dy - \tau_c \, dx \, dy = 0 \rightarrow \frac{d\sigma_2}{dx} e_2 - \tau_c = 0 \qquad [b]$$

In addition, one also has the following geometric relation in approximating the tangents and angles:

$$\gamma_c \; \# \; \frac{u_2 - u_1}{e_c}$$

then with the constitutive relations:

$$\gamma_c = \frac{\tau_c}{G_c}; \quad \frac{du_1}{dx} = \frac{1}{E_1}\sigma_1; \quad \frac{du_2}{dx} = \frac{1}{E_2}\sigma_2$$

$$\frac{\tau_c}{G_c} \; \# \; \frac{u_2 - u_1}{e_c}$$

$$\frac{e_c \, d\tau_c}{G_c \, dx} = \frac{\sigma_2}{E_2} - \frac{\sigma_1}{E_1} \qquad [c]$$

One obtains three relations [a], [b], [c] for the three unknowns σ_1, σ_2, τ_c. One can write:

$$\frac{1}{E_1}\frac{d\sigma_1}{dx} = -\frac{\tau_c}{e_1 E_1}; \quad \frac{1}{E_2}\frac{d\sigma_2}{dx} = \frac{\tau_c}{e_2 E_2}$$

$$\frac{1}{E_1}\frac{d\sigma_1}{dx} - \frac{1}{E_2}\frac{d\sigma_2}{dx} = -\tau_c\left(\frac{1}{e_1 E_1} + \frac{1}{e_2 E_2}\right)$$

Taking into account the relation [c]:

$$\frac{d^2}{dx^2}\left(\frac{\sigma_1}{E_1} - \frac{\sigma_2}{E_2}\right) = \frac{G_c}{e_c}\left(\frac{1}{e_1 E_1} + \frac{1}{e_2 E_2}\right)\left(\frac{\sigma_1}{E_1} - \frac{\sigma_2}{E_2}\right)$$

$$\boxed{\frac{d^2}{dx^2}\left(\frac{\sigma_1}{E_1} - \frac{\sigma_2}{E_2}\right) - \lambda^2\left(\frac{\sigma_1}{E_1} - \frac{\sigma_2}{E_2}\right) = 0; \quad \text{with} \quad \lambda^2 = \frac{G_c}{e_c}\left(\frac{1}{e_1 E_1} + \frac{1}{e_2 E_2}\right)}$$

The solution of the differential equation can be written as:

$$\frac{\sigma_1}{E_1} - \frac{\sigma_2}{E_2} = A \text{ ch } \lambda x + B \text{ sh } \lambda x$$

■ Boundary conditions:

for $x = 0$; $\sigma_1 = \sigma_{10}$ and $\sigma_2 = 0$ then: $\left.\left(\frac{\sigma_1}{E_1} - \frac{\sigma_2}{E_2}\right)\right|_{x=0} = \frac{\sigma_{10}}{E_1}$

for $x = \ell$; $\sigma_1 = 0$ and $\sigma_2 = \sigma_{20}$ then: $\left.\left(\frac{\sigma_1}{E_1} - \frac{\sigma_2}{E_2}\right)\right|_{x=\ell} = -\frac{\sigma_{20}}{E_2}$

from which we can write the constant values:

$$A = \frac{\sigma_{10}}{E_1}; \quad B = -\left(\frac{\sigma_{20}}{E_2 \text{ sh } \lambda \ell} + \frac{\sigma_{10}}{E_1 \text{ th } \lambda \ell}\right)$$

In addition (relation [a] + [b]):

$$\frac{d\sigma_1}{dx}e_1 + \frac{d\sigma_2}{dx}e_2 = 0$$

where: $\quad \dfrac{d\sigma_1}{dx}\left[\dfrac{1}{E_1} + \dfrac{e_1}{e_2 E_2}\right] = A\lambda \text{ sh } \lambda x + B\lambda \text{ ch } \lambda x$

That is, according to [a]

$$-\tau_c\left[\frac{1}{e_1 E_1} + \frac{1}{e_2 E_2}\right] = A\lambda \text{ sh } \lambda x + B\lambda \text{ ch } \lambda x$$

$$\boxed{\tau_c = \frac{G_c}{e_c \lambda}\left\{\left(\frac{\sigma_{10}}{E_1}\frac{1}{\text{th } \lambda \ell} + \frac{\sigma_{20}}{E_2}\frac{1}{\text{sh } \lambda \ell}\right)\text{ch } \lambda x - \frac{\sigma_{10}}{E_1}\text{ sh } \lambda x\right\}} \qquad \text{[d]}$$

Remarks:

■ One obtains in this manner only an approximation for the shear stress τ_C. It should be possible to deduce directly from relations [a], [b], [c] a differential equation in τ_C. However, its integration will reveal at the limits $x = 0$ and $x = \ell$ the zero values of τ_C (free surface of the adhesive) making it impossible to obtain a nonzero solution. At the inverse, the expression found here for τ_C does not become zero for $x = 0$ and $x = \ell$. This contradicts with reality.

One can conclude from the above that the unidimensional approximation for the stresses σ_1, σ_2, τ_C is unwarranted. However, the form found here for τ_C gives an acceptable order of magnitude for this stress, except at the immediate vicinity of the free edge. Numerical modeling of the phenomenon (finite element method) shows in effect that the shear stress τ_C increases very rapidly from the free edge, up to a peak value very close to the value here. Apart from this particularity, there is a good correlation with the values given in relation [d].

■ It also appears in the adhesive normal peel stresses that are confined to a peak zone close to the free edge. They constitute another factor that is not taken into account in this study.

2. Numerical application:
Longitudinal modulus of titanium (see Section 1.6): $E_1 = 105,000$ MPa.
Shear modulus of the adhesive (araldite): $G_C = 1,700$ MPa.
Longitudinal modulus of the laminate: With the proportions of the previous plies along the directions 0°, 90°, ±45°, one finds (Table 5.4 in Section 5.4.2): $E_2 = 100,590$ MPa.
Thickness of the laminate: $2e_2 = 20$ plies × 0.125 mm = 2.5 mm from which $e_2 = 1.25$ mm.

A stress resultant of 20 daN/mm corresponds to the stresses:

■ In the titanium:

$$\sigma_{10} = \frac{200}{2 \times 1.5} = 66.66 \text{ MPa}$$

■ In the laminate:

$$\sigma_{20} = \frac{200}{2.5} = 80 \text{ MPa}$$

A numerical calculation of expression [d], for example, with a programmable calculator, allows one to verify easily the rupture criterion of the adhesive for a length of $\ell = 40$ mm, as shown in the following:

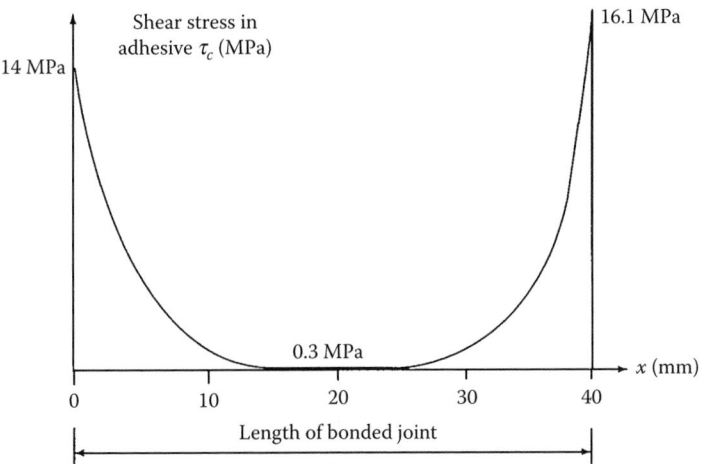

3. Particular case: The materials are identical: $e_1 = e_2 = e$. Then $\sigma_{10} = \sigma_{20} = \sigma_0$ and:

$$\tau_c = \frac{G_c}{\lambda e_c E}\sigma_0\left\{\left(\frac{1}{\text{th } \lambda\ell} + \frac{1}{\text{sh } \lambda\ell}\right)\text{ch } \lambda x - \text{sh } \lambda x\right\}$$

One notes identical peak values of stress for $x = 0$ or $x = \ell$ as:

$$\tau_{c\text{ max}} = \frac{G_c}{\lambda e_c E}\sigma_0\frac{\text{ch } \lambda\ell + 1}{\text{sh } \lambda\ell} = \frac{G_c}{\lambda e_c E}\sigma_0\frac{1}{\text{th } \frac{\lambda\ell}{2}}$$

Taking into account that:

$$\lambda^2 = \frac{2G_c}{e_c e E}$$

$$\tau_{c\text{ max}} = \sigma_0\frac{e\lambda^2}{2\lambda}\frac{1}{\text{th } \lambda\ell/2}$$

Introducing an average shear stress in the adhesive, which is a fictitious stress as one can consider in the previous figure:

$$\tau_{\text{average}} = \sigma_0\frac{e}{\ell}$$

then:

$$\tau_{c\text{ max}} = \frac{\lambda\ell/2}{\text{th } \lambda\ell/2}\tau_{\text{average}}$$

in posing:

$$\lambda \ell / 2 = a$$

$$\tau_c \atop \max = \frac{a}{\text{th } a} \times \tau_{average} \; ; \quad \text{with } a = \sqrt{\frac{G_c \ell^2}{2 E e e_c}}$$

18.3.3 Composite Beam with Two Layers

Problem Statement:

A composite beam is made up of two different materials, denoted as 1 and 2, that are bonded together. The cross section of the beam is shown in the figure below. The thickness of the adhesive is neglected. The materials are isotropic and elastic. The longitudinal and shear moduli of the two materials are denoted as E_1, G_1 and E_2, G_2.

The elements that allow the study of the bending behavior of this beam in its plane of symmetry are summarized in Table 15.16.

    1.   Determine the location of the elastic center denoted as O.

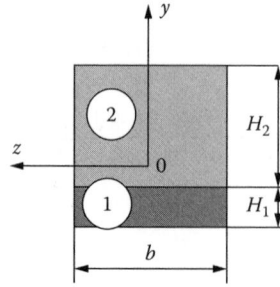

    2.   Write the expression for the equivalent stiffnesses (do not provide details for the shear coefficient k).

    3.   The shear force along the y direction for the considered section is denoted as T. Calculate the shear stress distribution τ_{xy}. Deduce from that the shear stress in the adhesive at the interface between the two materials.

Solution:

    1. Elastic Center: This is determined such that:

$$\int_{section} E_i y \, dS = 0$$

(see Equation 15.16). Let A be an arbitrary origin defining ordinate denoted as Y. The point O to be found is such that:

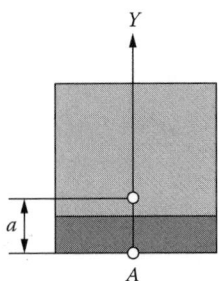

$$\int_{\text{section}} E_i(Y-a)\,dS = 0$$

then:

$$a = \frac{\displaystyle\int_{\text{section}} E_i Y\,dS}{\displaystyle\int_{\text{section}} E_i\,dS}$$

$$\int_{\text{section}} E_i Y\,dS = \int_0^{H_1} E_1 Yb\,dY + \int_{H_1}^{H_1+H_2} E_2 Yb\,dY$$

One finds after calculation:

$$a = \frac{1}{2}\left\{\frac{(E_1 - E_2)H_1^2 + E_2(H_1 + H_2)^2}{E_1 H_1 + E_2 H_2}\right\}$$

2. Equivalent stiffnesses:
   ■ Extensional stiffness:

$$\langle ES \rangle = \sum_i E_i S_i = b(E_1 H_1 + E_2 H_2)$$

   ■ Shear stiffness:

$$\frac{\langle GS \rangle}{k} = \sum_i \frac{G_i S_i}{k} = \frac{b}{k}(G_1 H_1 + G_2 H_2)$$

   ■ Bending stiffness:

$$\langle EI \rangle = \sum_i E_i I_i$$

$$\langle EI \rangle = bE_1 \int_{-a}^{H_1-a} y^2 \, dS + bE_2 \int_{H_1-a}^{H_1+H_2-a} y^2 \, dS$$

$$\langle EI \rangle = \frac{b}{3}\{E_1[(H_1-a)^3 + a^3] + E_2[(H_1+H_2-a)^3 - (H_1-a)^3]\}$$

3. Subject to a shear force T along the y direction, the shear stresses are assumed to be limited to the component τ_{xy}, given in the material "i" by the relation (see Equation 15.16):

$$\tau_{xy} = G_i \frac{T}{\langle GS \rangle} \frac{dg_{oi}}{dy}$$

in which $g_o(y)$ is the warping function due to shear and solution of the problem:

$$\begin{cases} \dfrac{d^2 g_o}{dy^2} = -\dfrac{E_i \langle GS \rangle}{G_i \langle EI \rangle} y \quad \text{throughout the cross section} \\[2mm] \dfrac{dg_o}{dy} = 0 \text{ for } y = -a \text{ and } y = H_1 + H_2 - a \text{ (free boundaries)} \end{cases}$$

The uniqueness of the function $g_o(y)$ is assured by the condition:

$$\int_{\text{section}} E_i g_o \, dS = 0.$$

One finds in material 1:

$$\frac{dg_{o1}}{dy} = \frac{1}{2}\frac{E_1}{G_1}\frac{\langle GS \rangle}{\langle EI \rangle}(a^2 - y^2)$$

and in material 2:

$$\frac{dg_{o2}}{dy} = \frac{1}{2}\frac{E_2}{G_2}\frac{\langle GS \rangle}{\langle EI \rangle}[(H_1 + H_2 - a)^2 - y^2]$$

from which one finds for shear the following parabolic distribution along the height of the beam

$$\boxed{\begin{aligned} -a \le y \le H_1 - a: \ \tau_{xy} &= \frac{T}{2}\frac{E_1}{\langle EI \rangle}(a^2 - y^2) \\[2mm] H_1 - a \le y \le H_1 + H_2 - a: \ \tau_{xy} &= \frac{T}{2}\frac{E_2}{\langle EI \rangle}[(H_1 + H_2 - a)^2 - y^2] \end{aligned}}$$

The corresponding variations are shown below. At the junction between the two materials ($y = H_1 - a$), one finds the shear in the adhesive:

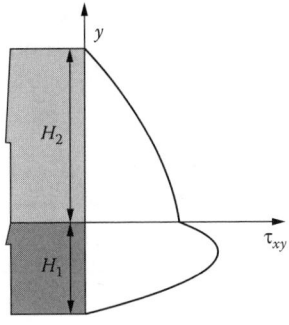

$$\boxed{\tau_{xy} = \frac{T}{2}\frac{E_1}{\langle EI \rangle}H_1(2a - H_1)}$$

adhesive

Remark: The integration of the function $g_o(y)$ allows the calculation of the shear coefficient k by Equation 15.16:

$$k = \frac{1}{\langle EI \rangle}\int_{\text{section}} E_i g_o\, y\, dS$$

The calculation is long but does not present any particular difficulty. The numerical values of k are shown in the following figure for different ratios of E_1/E_2 and H_2/H_1, for the particular case of identical Poisson coefficients.

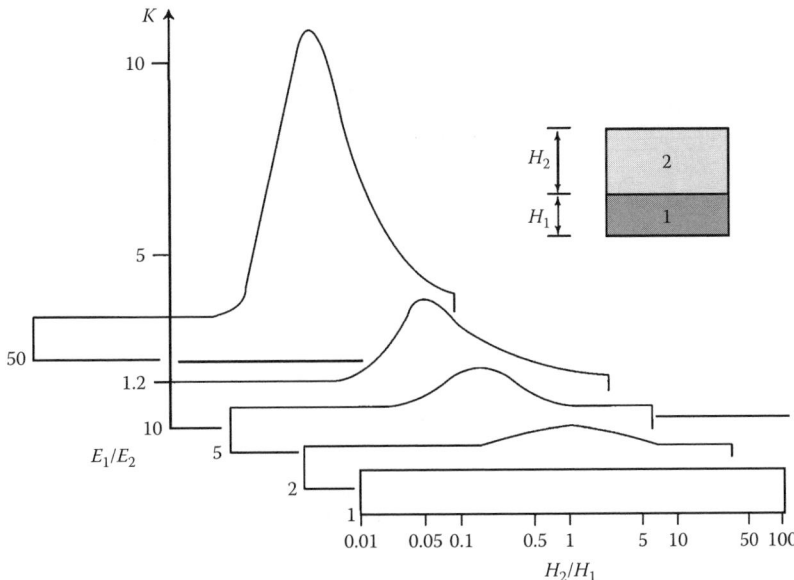

18.3.4 Buckling of a Sandwich Beam

Problem Statement:

A sandwich beam is compressed at its two ends by two opposite forces F. The two ends are constrained so that there is no rotation.

1. For what value of F, denoted as $F_{critical}$, can we obtain a deformed configuration for the beam other than the straight configuration, for example, the configuration shown in the figure below (adjacent-equilibrium)?

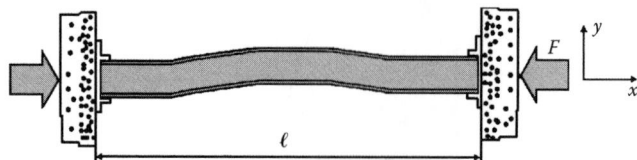

2. What error will be caused by neglecting shear deformation of the beam? (Assume that the dimensions and the material constitutive relations are known.)

Solution:

1. With the notation convention of Chapter 15 (bending of composite beams), recall the behavior equations for the beam Equation 15.16.

$$T_y = \frac{\langle GS \rangle}{k}\left(\frac{dv}{dx} - \theta_z\right); \quad M_z = \langle EI_z \rangle \frac{d\theta_z}{dx}$$

Referring to the figure below, one can write the following relations, in which C represents the moment due to the constraint on the right-hand side.

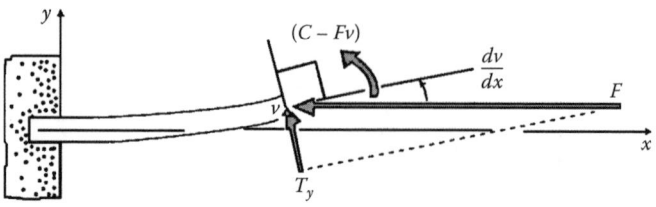

$$T_y \# F\frac{dv}{dx}; \quad M_z = C - Fv$$

from which, by substituting in the constitutive relations:

$$F\frac{dv}{dx} = \frac{\langle GS \rangle}{k}\left(\frac{dv}{dx} - \theta_z\right); \quad C - Fv = \langle EI_z \rangle \frac{d\theta_z}{dx}$$

Elimination of θ_z between these two relations leads to the following equation:

$$\boxed{\frac{d^2v}{dx^2} + \lambda^2 v = \lambda^2\frac{C}{F} \quad \text{with} \quad \lambda^2 = \frac{F}{\langle EI_z \rangle}\frac{1}{\left(1 - \frac{kF}{\langle GS \rangle}\right)}} \qquad [a]$$

from which the general solution can be written as: (with the condition that $F < \frac{\langle GS \rangle}{k}$)

$$v(x) = A\cos\lambda x + B\sin\lambda x + \frac{C}{F}$$

■ Boundary conditions:
For $x = 0$ one notes $v(0) = 0$ and $\theta_z(0) = 0$. This last condition leads to

$$\left.\frac{dv}{dx}\right|_{x=0} = 0$$

Due to:

$$\theta_z = \left(1 - F\frac{k}{\langle GS \rangle}\right)\frac{dv}{dx}$$

One then finds that:

$$B = 0; \quad A = -\frac{C}{F}$$

from which: $\quad v(x) = \frac{C}{F}(1 - \cos \lambda x)$

For $x = \ell$ one notes $v(\ell) = 0$ and $\theta_z(\ell) = 0$.

$$\cos \lambda \ell = 1$$

from which:

$$\lambda \ell = 2n\pi$$

one obtains for $v(x)$ the form:

$$\boxed{v(x) = \frac{C}{F}\left(1 - \cos 2n\pi\frac{x}{\ell}\right)} \qquad [b]$$

The critical value F_{critical} is given by:

$$\lambda^2 = \frac{4n^2\pi^2}{\ell^2}$$

where λ^2 has the form [a], leading to:

$$F_{\text{critical}} = \frac{4n^2\pi^2\langle EI_z\rangle}{\ell^2\left(1 + \dfrac{4n^2\pi^2\langle EI_z\rangle k}{\ell^2\langle GS\rangle}\right)}$$

The smallest value of F is obtained for $n = 1$ as:

$$F_{\text{critical}} = \frac{4\pi^2 \langle EI_z \rangle}{\ell^2 \left(1 + \frac{4\pi^2 \langle EI_z \rangle}{\ell^2} \frac{k}{\langle GS \rangle} \right)}$$

Remarks:

- One can verify that the value of F_{critical} is less than

$$\frac{\langle GS \rangle}{k}$$

the form of the general solution $v(x)$ written above is therefore legitimate.
- It is convenient to note that the deformed $v(x)$ written in [b] is only defined by a multiplication factor, because the constraining couple C is **indeterminate**. One can find this property by writing explicitly as a function of $v(x)$ the relation:

$$C = M_z(\ell) = \langle EI_z \rangle \frac{d\theta_z}{dx}\bigg|_{x=\ell}$$

2. Neglecting shear effect the assumed undeformability under shear leads to zero corresponding energy of deformation (Equation 15.16). In this case, one has: $k = 0$.

The critical force then takes the value:

$$F'_{\text{critical}} = \frac{4\pi^2 \langle EI_z \rangle}{\ell^2}$$

The error relative to its previous value is then:

$$\text{Error} = \frac{F'_{\text{critical}}}{F_{\text{critical}}} - 1$$

$$\text{Error} = \frac{4\pi^2 \langle EI_z \rangle k}{\ell^2 \langle GS \rangle}$$

For numerical value, we calculate this error for the beam in Section 4.2.2 (beam made of polyurethane foam and aluminum, 1 meter long). One has

$$\langle EI_z \rangle = 475 \times 10^2; \quad \frac{\langle GS \rangle}{k} = 650 \times 10^2$$

The error committed is spectacular:

$$Error = 28.84 = 2884\%!$$

18.3.5 Shear Due to Bending in a Sandwich Beam

Problem Statement:

One considers the cross section of a sandwich beam as shown in the following figure. The components, assumed to be isotropic (or transversely isotropic), are denoted as 1 and 2. They are perfectly bonded to each other with an adhesive with negligible thickness. The beam has a unit width. The moduli of elasticity are denoted as shown.

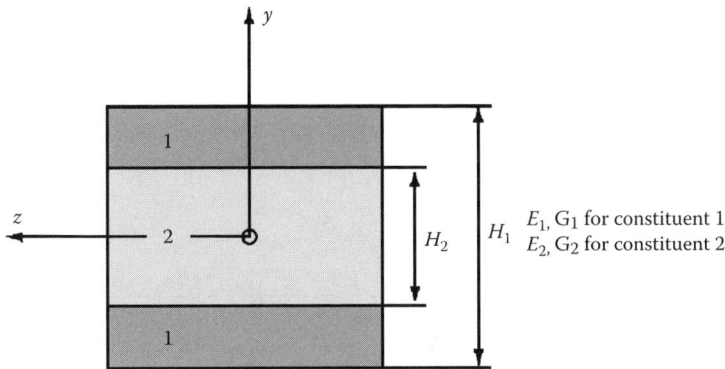

Using the formulation in Equation 15.16 for bending of composite beams:

1. Study the warping function g_o for this cross section.
2. Deduce from there the shear stress distribution.
3. Calculate the shear coefficient for bending in the plane xy, as well as the deformed configuration of a cross section under the effect of shear. Numerical application: Give the value of k for a beam made of polystyrene foam with thickness of 80.2 mm ($E_2 = 21.5$ MPa; $G_2 = 7.7$ MPa) and with aluminum skins with thickness of 2.15 mm ($E_1 = 65,200$ MPa; $G_1 = 24,890$ MPa).

Solution:

1. Warping due to bending:
   This is the solution of the problem described in Equation 15.16. Assuming here that g_o does not vary with the variable z:

$$\left|\begin{array}{l} \dfrac{d^2 g_o}{dy^2} = -\dfrac{E_i \langle GS \rangle}{G_i \langle EI_z \rangle} \times y \text{ in the domain of the section} \\[4mm] \dfrac{dg_o}{dy} = 0 \text{ for } y = \pm H_1/2. \end{array}\right.$$

in which both g_o and $G_i\, dg_o/dy$ are continuous as one crosses from material 1 to material 2.

Taking into account the antisymmetry of the function g_o with respect to variable y, one obtains

$$H_2/2 \leq y \leq H_1/2 \quad : \quad g_{o1} = -\frac{E_1}{G_1}\frac{a}{6}y^3 + A_1 y + B_1$$

$$-H_2/2 \leq y \leq H_2/2 \quad : \quad g_{o2} = -\frac{E_2}{G_2}\frac{a}{6}y^3 + A_2 y$$

$$-H_1/2 \leq y \leq -H_2/2 \quad : \quad g_{o3} = -\frac{E_1}{G_1}\frac{a}{6}y^3 + A_1 y - B_1$$

with:

$$a = \frac{\langle GS \rangle}{\langle EI_z \rangle} = 12\frac{G_2 H_2 + G_1(H_1 - H_2)}{E_2 H_2^3 + E_1(H_1^3 - H_2^3)}$$

$$A_1 = \frac{E_1}{G_1}\frac{a}{2}\frac{H_1^2}{4}$$

$$B_1 = a\frac{H_2}{16}\left\{ \left(\frac{1}{G_2} - \frac{1}{G_1}\right)E_1 H_1^2 - \left(\frac{E_1 - E_2}{G_2}\right)H_2^2 - \left(\frac{E_2}{G_2} - \frac{E_1}{G_1}\right)\frac{H_2^2}{3} \right\}$$

$$A_2 = A_1 + \frac{2B_1}{H_2} + \frac{aH_2^2}{3\,8}\left(\frac{E_2}{G_2} - \frac{E_1}{G_1}\right)$$

2. Shear stresses due to bending:

These are given by the relation (see Equation 15.16):

$$\vec{\tau} = G_i\frac{T_y}{\langle GS \rangle}\,\overrightarrow{\text{grad}}\,g_o$$

then:

$$\tau_{xy} = G_i\frac{T_y}{\langle GS \rangle}\frac{\partial g_o}{\partial y}; \quad \tau_{xz} = G_i\frac{T_y}{\langle GS \rangle}\frac{\partial g_o}{\partial z} = 0$$

One obtains

$$0 \leq y \leq H_2/2 \quad : \quad \tau_{xy} = \frac{1}{2}\frac{T_y}{\langle EI_z \rangle}\left\{ E_2\left(\frac{H_2^2}{4} - y^2\right) + E_1\left(\frac{H_1^2}{4} - \frac{H_2^2}{4}\right) \right\}$$

$$H_2/2 \leq y \leq H_1/2 \quad : \quad \tau_{xy} = \frac{1}{2}\frac{T_y}{\langle EI_z \rangle}E_1\left(\frac{H_1^2}{4} - y^2\right)$$

The corresponding distribution is illustrated below for two distinct designs of the components 1 and 2.[53]

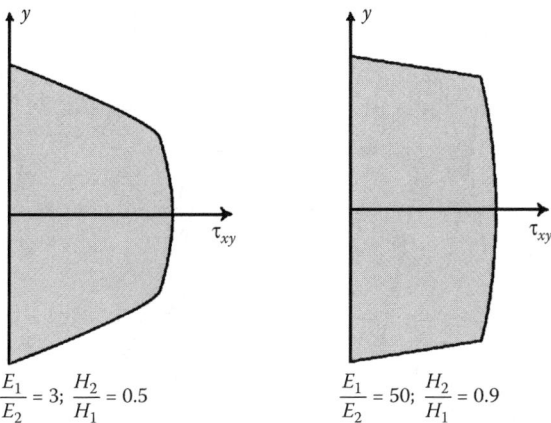

$$\frac{E_1}{E_2} = 3; \quad \frac{H_2}{H_1} = 0.5 \qquad\qquad \frac{E_1}{E_2} = 50; \quad \frac{H_2}{H_1} = 0.9$$

3. Shear coefficient:
   The calculation of k is done without difficulty starting from Equation 15.16:

$$k = \frac{1}{\langle EI_z \rangle} \int_D E_i g_\alpha y \ dS$$

One obtains

$$k = \frac{a}{8[E_2 H_2^3 + E_1(H_1^3 - H_2^3)]} \left\{ \frac{E_2}{G_2} H_2^3 \left[E_1 H_1^2 + \left(\frac{4}{5} E_2 - E_1 \right) H_2^2 \right] \cdots \right.$$

$$\left. \cdots + \frac{E_1^2}{G_1} \left(\frac{4}{5} H_1^5 + \frac{H_2^5}{5} - H_1^2 H_2^3 \right) \right\} + \frac{3 b E_1 (H_1^2 - H_2^2)}{E_2 H_2^3 + E_1 (H_1^3 - H_2^3)}$$

with: $\quad a = 12 \dfrac{G_2 H_2 + G_1 (H_1 - H_2)}{E_2 H_2^3 + E_1 (H_1^3 - H_2^3)}$

$$b = \frac{a}{16} H_2 \frac{E_1}{G_1} \left\{ \frac{H_2^2}{3} + H_1^2 \left(\frac{G_1}{G_2} - 1 \right) - H_2^2 \frac{G_1}{G_2} \left(1 - \frac{2 E_2}{3 E_1} \right) \right\}$$

The evolution of the shear coefficient k is represented in the following figure for different values of the ratios E_1/E_2 and with the same Poisson coefficient (0.3) when varying thickness of the skins.

---

[53] Observation of the evolution of τ_{xy} for the beam with thin skins justifies the simplification proposed in Application 18.2.1, "Sandwich Beam: Simplified Calculation of the Shear Coefficient."

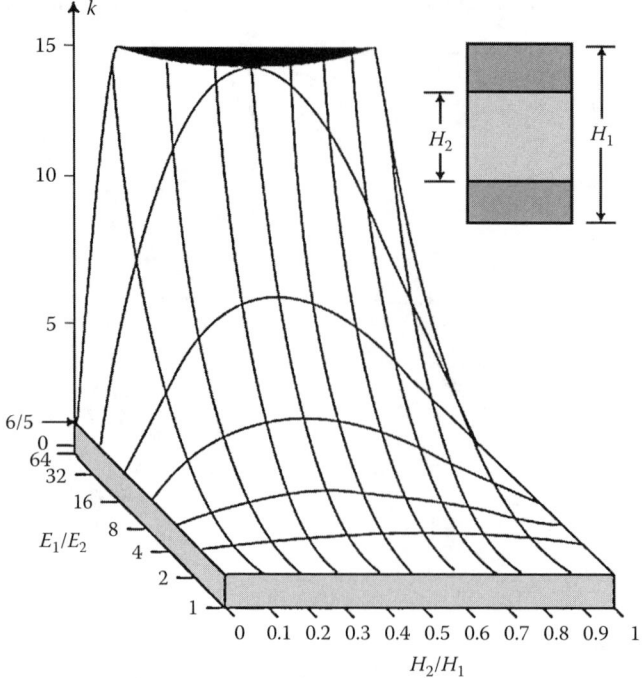

Remarks:

- The limiting cases $E_2 = E_1$; $H_2 = H_1$; $H_2 = 0$ correspond to a homogeneous beam with rectangular cross section for which one finds again the classical value $k = 6/5$ (or 1.2).
- The expression for the k coefficient written above is long. One can obtain a more simplified expression for easier manipulation if the skins are thin relative to the total thickness of the beam. One can refer to Application 18.2.1.
- Deformed configuration of a cross section: The displacement of each point of the cross section out of its initial plane is obtained starting from the function g_o by the relation (see Equations 15.12 and 15.15):

$$\eta_x = \frac{T_y}{\langle GS \rangle}(g_o - k \times y)$$

It is described graphically for two distinct sets of properties of components 1 and 2 in the following figure:

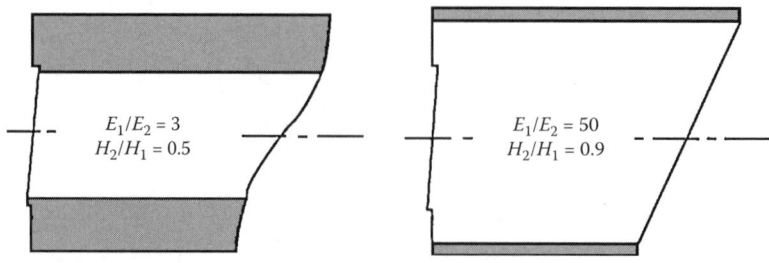

■ Numerical application:
One finds: $k = 165.7$. Note that for this type of beam, the shear coefficient can have very high values compared with those that characterize the homogeneous beams.

18.3.6 Shear Due to Bending in a Box-Beam and in a I-Beam

Box-Beam — Problem Statement:

One considers the cross section of a box-beam as shown in the figure below, bending in the plane (x,z), made of two distinct materials denoted as 1 (vertical parts) and 2 (horizontal parts), each of them being transversely isotropic in the plane (y,z).
Assuming that e_1 and e_2 have little values compared to h and b,

1. Calculate the shear stress distribution due to a shear force T_z.
2. Calculate the shear coefficient k_z for bending in the plane (x,z).

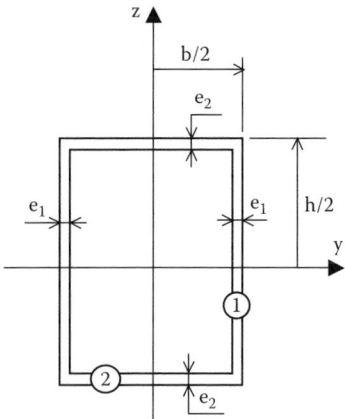

Solution:

1. Shear stresses due to bending:
These are given by Equation 15.19 as:

In area 1:

$$\tau_{xz} = \tau_1 = G_1 \frac{T_z}{\langle GS \rangle} \frac{db_{01}}{dz} \quad \text{[54]}$$

---

[54] Since $e_1 \ll b$, τ_1 is assumed to keep a constant value across the thickness e_1.

In area 2:

$$\tau_{xy} = \tau_2 = G_2 \frac{T_z}{\langle GS \rangle} \frac{db_{02}}{dy}$$

One must evaluate $\dfrac{db_{01}}{dz}$ and $\dfrac{db_{02}}{dy}$. From 15.19:

In area 1: $\dfrac{d^2 b_{01}}{dz^2} = -\dfrac{E_1}{G_1} \dfrac{\langle GS \rangle}{\langle EI_y \rangle} z$ which gives $\dfrac{db_{01}}{dz} = -\dfrac{E_1}{G_1} \dfrac{\langle GS \rangle}{\langle EI_y \rangle} \dfrac{z^2}{2} + a_1'$

In area 2: $\dfrac{d^2 b_{02}}{dy^2} = -\dfrac{E_2}{G_2} \dfrac{\langle GS \rangle}{\langle EI_y \rangle} z$ with $z = \dfrac{b}{2}$, which gives $\dfrac{db_{02}}{dy} = -\dfrac{E_2}{G_2} \dfrac{\langle GS \rangle}{\langle EI_y \rangle} \dfrac{b}{2} y$

(the integration constant is zero on axis of symetry z (y=0))

We can write an equilibrium condition for the corner of the box-beam ($y = -b/2$ and $z = b/2$) , as shown in the following figure. The longitudinal equilibrium along x-direction gives:

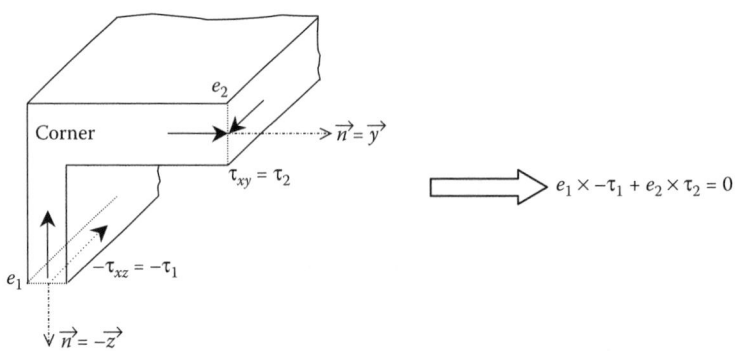

$$e_1 G_1 \frac{db_{01}}{dz} = e_2 G_2 \frac{db_{02}}{dy}$$

$$e_1 G_1 \frac{T_z}{\langle GS \rangle} \frac{db_{01}}{dz} = e_2 G_2 \frac{T_z}{\langle GS \rangle} \frac{db_{02}}{dy}$$

Then:

$$e_1 G_1 \left(-\frac{E_1}{G_1} \frac{\langle GS \rangle}{\langle EI_y \rangle} \frac{b^2}{8} + a_1' \right) = e_2 G_2 \left(\frac{E_2}{G_2} \frac{\langle GS \rangle}{\langle EI_y \rangle} \frac{bb}{4} \right)$$

From which:

$$a_1' = \frac{E_1}{G_1} \frac{\langle GS \rangle}{\langle EI_y \rangle} \frac{b^2}{8} + \frac{e_2 E_2}{e_1 G_1} \frac{\langle GS \rangle}{\langle EI_y \rangle} \frac{bb}{4}$$

One obtains for shear stress distribution:

$$Area\ 1:\ \tau_1 = G_1 \frac{T_z}{\langle GS \rangle} \frac{db_{01}}{dz} = \frac{E_1}{\langle EI_y \rangle} T_z \left[-\frac{z^2}{2} + \frac{b^2}{8} + \frac{e_2 E_2}{e_1 E_1} \frac{bb}{4} \right]$$

$$Area\ 2:\ \tau_2 = G_2 \frac{T_z}{\langle GS \rangle} \frac{db_{02}}{dy} = \frac{E_2}{\langle EI_y \rangle} T_z \times -\frac{b}{2} y$$

2. Shear coefficient k_z:
From 15.1.6.2:

$$\frac{dW_\tau}{dx} = \frac{1}{2} k_z \frac{T_z^2}{\langle GS \rangle} = \frac{1}{2} \left\{ \int\limits_{material.1} \frac{\tau_1^2}{G_1} e_1 dz \quad + \quad \int\limits_{material.2} \frac{\tau_2^2}{G_2} e_2 dy \right\}$$

$$\frac{dW_\tau}{dx} = \frac{1}{2} \left\{ 2 \int_{-\frac{b}{2}}^{\frac{b}{2}} \frac{\tau_1^2}{G_1} e_1 dz \quad + \quad 2 \int_{-\frac{b}{2}}^{\frac{b}{2}} \frac{\tau_2^2}{G_2} e_2 dy \right\} = \frac{e_1}{G_1} \int_{-\frac{b}{2}}^{\frac{b}{2}} \tau_1^2 dz + \frac{e_2}{G_2} \int_{-\frac{b}{2}}^{\frac{b}{2}} \tau_2^2 dy$$

With the expressions above for τ_1 and τ_2 one find after calculus:

$$\frac{dW_\tau}{dx} = \frac{1}{2} k_z \frac{T_z^2}{\langle GS \rangle} = \frac{e_1}{G_1} \frac{E_1^2}{\langle EI_y \rangle^2} T_z^2 \left[\frac{b^5}{120} + \left(\frac{e_2 E_2}{e_1 E_1} \right)^2 \frac{b^3 b^2}{16} + \frac{e_2 E_2}{e_1 E_1} \frac{b^4 b}{24} \right]$$

$$+ \frac{e_2}{G_2} \frac{E_2^2}{\langle EI_y \rangle^2} T_z^2 \frac{b^2 b^3}{48}$$

then

$$\boxed{\frac{k_z}{\langle GS \rangle} = 2 \frac{e_1}{G_1} \frac{E_1^2}{\langle EI_y \rangle^2} \left[\frac{b^5}{120} + \left(\frac{e_2 E_2}{e_1 E_1} \right)^2 \frac{b^3 b^2}{16} + \left(\frac{e_2 E_2}{e_1 E_1} \right) \frac{b^4 b}{24} \right] + 2 \frac{e_2}{G_2} \frac{E_2^2}{\langle EI_y \rangle^2} \frac{b^2 b^3}{48}}$$

I-Beam — Problem Statement:

One considers the cross section of a I-beam as shown in the figure below, bending in the plane (x,z), made of two distinct materials denoted as 1 (web) and 2 (flanges), each of them being transversely isotropic in the plane (y,z).

Assuming that e_1 and e_2 have little values compared to h and b,

1. Calculate the shear stress distribution due to a shear force T_z.
2. Calculate the shear coefficient k_z for bending in the plane (x,z).

Solution:

1. Shear stress distribution:
Equation 15.19: In area 1, the shear stress is:

$$\tau_{xz} = \tau_1 = G_1 \frac{T_z}{\langle GS \rangle} \frac{db_{01}}{dz} \quad {}^{55}$$

In area 2, the shear stress is:

$$\tau_{xy} = \tau_2 = G_2 \frac{T_z}{\langle GS \rangle} \frac{db_{02}}{dy}$$

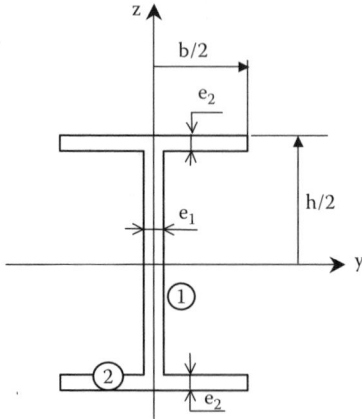

One needs to calculate $\dfrac{db_{01}}{dz}$ and $\dfrac{db_{02}}{dy}$. For this, one has from 15.19:

$$\text{In area 1: } \frac{d^2b_{01}}{dz^2} = -\frac{E_1}{G_1} \frac{\langle GS \rangle}{\langle EI_y \rangle} z \text{ from which } \frac{db_{01}}{dz} = -\frac{E_1}{G_1} \frac{\langle GS \rangle}{\langle EI_y \rangle} \frac{z^2}{2} + a_1'$$

---

[55] Since $e_1 \ll b$, τ_1 is assumed to keep a constant value across the thickness e_1.

In area 2 and for $0 < y \leq \dfrac{b}{2}$ [56]: $\quad \dfrac{d^2 h_{02}}{dy^2} = -\dfrac{E_2}{G_2} \dfrac{\langle GS \rangle}{\langle EI_y \rangle} z$, with $z = \dfrac{b}{2}$.

Then $\dfrac{dh_{02}}{dy} = -\dfrac{E_2}{G_2} \dfrac{\langle GS \rangle}{\langle EI_y \rangle} \dfrac{b}{2} y + a_2'$.

We remark that τ_2 is zero for $y = b/2$ (condition of shear stress reciprocity). Then:

$$\left. \frac{dh_{02}}{dy} \right|_{0 < y \leq \frac{b}{2}} = -\frac{E_2}{G_2} \frac{\langle GS \rangle}{\langle EI_y \rangle} \left[\frac{b}{2} y - \frac{hb}{4} \right]$$

for $-\dfrac{b}{2} \leq y < 0$, an analogous calculus gives:

$$\left. \frac{dh_{02}}{dy} \right|_{-\frac{b}{2} \leq y < 0} = -\frac{E_2}{G_2} \frac{\langle GS \rangle}{\langle EI_y \rangle} \left[\frac{b}{2} y + \frac{hb}{4} \right]$$

Equilibrium condition at the junction between web and flange

As shown in the following figure, the longitudinal equilibrium along x-direction for the junction ($y = 0$ and $z = h/2$) gives:

$$\left. e_1 \times -\tau_1 + e_2 \times -\tau_2 + e_2 \times \tau_2 = 0 \right|_{\frac{b}{2} \leq y < 0 \quad 0 < y \leq \frac{b}{2}}$$

$$\left. e_1 G_1 \frac{T_z}{\langle GS \rangle} \frac{dh_{01}}{dz} = 2 e_2 G_2 \frac{T_z}{\langle GS \rangle} \frac{dh_{02}}{dy} \right|_{0 < y \leq \frac{b}{2}}$$

$$\left. e_1 G_1 \frac{dh_{01}}{dz} = 2 e_2 G_2 \frac{dh_{02}}{dy} \right|_{0 < y \leq \frac{b}{2}}$$

---

[56] One can note that the hypothesis of a little value of the flange's thickness is not valid if $y = 0$.

The continuity condition can be re-written as:

$$e_1 G_1 \left(-\frac{E_1}{G_1} \frac{\langle GS \rangle}{\langle EI_y \rangle} \frac{b^2}{8} + a'_1 \right) = 2e_2 G_2 \left(-\frac{E_2}{G_2} \frac{\langle GS \rangle}{\langle EI_y \rangle} x - \frac{bb}{4} \right)$$

From which: $a'_1 = \dfrac{E_1}{G_1} \dfrac{\langle GS \rangle}{\langle EI_y \rangle} \dfrac{b^2}{8} + 2 \dfrac{e_2 E_2}{e_1 G_1} \dfrac{\langle GS \rangle}{\langle EI_y \rangle} \dfrac{bb}{4}$

Then, the shear stress distribution is obtained as:

In area 1: $\tau_1 = G_1 \dfrac{T_z}{\langle GS \rangle} \dfrac{db_{01}}{dz} = \dfrac{E_1}{\langle EI_y \rangle} T_z \left[-\dfrac{z^2}{2} + \dfrac{b^2}{8} + 2 \dfrac{e_2 E_2}{e_1 E_1} \dfrac{bb}{4} \right]$

In area 2: $0 < y \leq \dfrac{b}{2}$ $\tau_2 = G_2 \dfrac{T_z}{\langle GS \rangle} \dfrac{db_{02}}{dy} = \dfrac{E_2}{\langle EI_y \rangle} T_z \times \left[\dfrac{bb}{4} - \dfrac{b}{2} y \right]_{0 < y \leq \frac{b}{2}}$

$$-\frac{b}{2} \leq y < 0 \quad \tau_2 = G_2 \frac{T_z}{\langle GS \rangle} \frac{db_{02}}{dy} = -\frac{E_2}{\langle EI_y \rangle} T_z \times \left[\frac{bb}{4} + \frac{b}{2} y \right]_{-\frac{b}{2} \leq y < 0}$$

2. Shear coefficient k_z:
from 15.1.6.2:

$$\frac{dW_\tau}{dx} = \frac{1}{2} k_z \frac{T_z^2}{\langle GS \rangle} = \frac{1}{2} \left\{ \int_{material\,1} \frac{\tau_1^2}{G_1} e_1 dz \quad + \quad \int_{matérial\,2} \frac{\tau_2^2}{G_2} e_2 dy \right\}$$

$$\frac{dW_\tau}{dx} = \frac{1}{2} \left\{ \int_{-\frac{b}{2}}^{\frac{b}{2}} \frac{\tau_1^2}{G_1} e_1 dz \quad + \quad 2 \int_{-\frac{b}{2}}^{\frac{b}{2}} \frac{\tau_2^2}{G_2} e_2 dy \right\} = \frac{1}{2} \frac{e_1}{G_1} \int_{-\frac{b}{2}}^{\frac{b}{2}} \tau_1^2 dz + \frac{e_2}{G_2} \int_{-\frac{b}{2}}^{\frac{b}{2}} \tau_2^2 dy$$

With the expressions of τ_1 and τ_2 above, one can obtain after calculus the following expression for the shear energy density:

$$\frac{dW_\tau}{dx} = \frac{1}{2} k_z \frac{T_z^2}{\langle GS \rangle} =$$

$$\frac{1}{2} \left\{ \frac{e_1}{G_1} \frac{E_1^2}{\langle EI_y \rangle^2} T_z^2 \left[\frac{b^5}{120} + \left(\frac{2e_2 E_2}{e_1 E_1} \right)^2 \frac{b^3 b^2}{16} + \frac{2e_2 E_2}{e_1 E_1} \frac{b^4 b}{24} \right] + \frac{2e_2}{G_2} \frac{E_2^2}{\langle EI_y \rangle^2} T_z^2 \frac{b^2 b^3}{48} \right\}$$

From which:

$$\frac{k_z}{\langle GS \rangle} = \frac{e_1}{G_1} \frac{E_1^2}{\langle EI_y \rangle^2} \left[\frac{b^5}{120} + \left(\frac{2e_2E_2}{e_1E_1} \right)^2 \frac{b^3b^2}{16} + \left(\frac{2e_2E_2}{e_1E_1} \right) \frac{b^4b}{24} \right] + 2\frac{e_2}{G_2} \frac{E_2^2}{\langle EI_y \rangle^2} \frac{b^2b^3}{48}$$

Remarks:

- We note the close analogy of results for the two section-shapes above. The results turn identical if one gives the value $2e_1$ for the thickness of I-beam web. Then, the areas corresponding to material n°1 are identical for both sections-shapes.
- Moreover, one can note the role of the U-section shape of figure below:

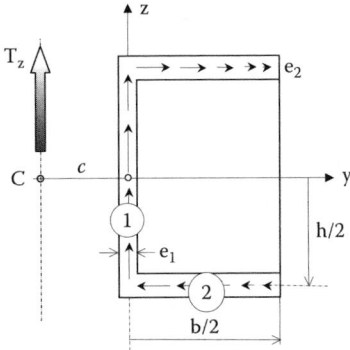

When this section is considered alone, the shear stress distribution due to T_z is from now on known as:

$$\tau_1 = \frac{E_1}{\langle EI_y \rangle} T_z \left[-\frac{z^2}{2} + \frac{b^2}{8} + \frac{e_2E_2}{e_1E_1} \frac{bb}{4} \right]$$

$$\tau_2 = \frac{E_2}{\langle EI_y \rangle} T_z \times \left[\frac{bb}{4} - \frac{b}{2}y \right]$$

Then the location of torsion center C can be readily obtained. At this point one must only write that the shear stress distribution (τ_1, τ_2) does not cause any torsional moment. This condition can be written as:

$$c \times \int_{-\frac{b}{2}}^{\frac{b}{2}} \tau_1 \times (e_1 dz) - b \times \int_0^{\frac{b}{2}} \tau_2 \times (e_2 dy) = 0$$

and leads after calculus to:

$$c = \frac{b}{4} \frac{1}{\left[1 + \dfrac{e_1 E_1}{e_2 E_2} \dfrac{b}{3b}\right]}$$

18.3.7 Column Made of Stretched Polymer

Problem Statement:

Consider a cylindrical column or revolution designed for use in the chemical industry (temperature can be high, and it may contain corrosive fluid under pressure) made of polyvinylidene fluoride (PVDF). It is reinforced on the outside by a filament-wound layer of "E" glass/polyester. The characteristics of the two layers of materials are as follows:

- Internal layer in PVDF: thickness e_1, isotropic material, modulus of elasticity E_1, Poisson coefficient ν_1.
- External layer in glass/polyester: To simplify the calculation, one will neglect the presence of the resin. As a consequence, E_t, $\nu_{t\ell}$, $G_{\ell t}$ (see Chapter 10) are neglected. The total thickness of the glass/polyester layer E_2 consists of a thickness denoted as b^{90} of windings along the 90° direction relative to the direction of the generator of the cylinder, and a thickness denoted as $b^{\pm45}$ of balanced windings along the +45° and −45° direction (as many fibers along the +45° as along the −45° direction). One then has $e_2 = b^{90} + b^{\pm45}$ (see figure below).

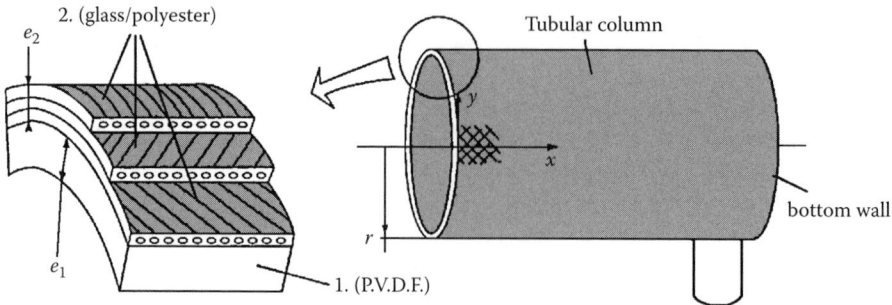

The longitudinal modulus of elasticity of the glass/polyester layer is denoted as E_ℓ. The thicknesses e_1 (internal) and e_2 (external) will be considered to be small relative to the average radius of the column.

1. The plane that is tangent to the midplane of the glass/polyester laminate is denoted as x,y (see figure). Calculate the equivalent moduli $\bar{E}_x$ and $\bar{E}_y$, the equivalent coefficients $\bar{\nu}_{yx}$ and $\bar{\nu}_{xy}$ of the reinforcement glass/polyester as function of E_ℓ, b^{90}, and $b^{\pm45}$.

2. One imposes a pressure of p_0 inside the column at room temperature (creep of the materials not considered). The resulting stresses are denoted in the axes x,y:

$$\sigma_{1x} \text{ and } \sigma_{1y} \text{ in the internal layer of PVDF}$$
$$\sigma_{2x} \text{ and } \sigma_{2y} \text{ in the external layer of glass/polyester}$$

(a) Write the equilibrium relation and the constitutive equation resulting from the assembly that is assumed to be perfectly bonded. Deduce from there the system that allows the calculation of σ_{1x} and σ_{2x}.

(b) Numerical applications: Internal pressure $p_0 = 3$ MPa (30 bars); $r = 100$ mm. PVDF: $E_1 = 260$ MPa; $v_1 = 0.3$; $e_1 = 10$ mm. Glass/polyester: $E_\ell = 74{,}000$ MPa; $e_2 = 0.75$ mm; $h^{90} = h^{\pm 45}/3$. Calculate σ_{1x}, σ_{1y}, σ_{2x}, σ_{2y}.

(c) Deduce from the previous results the stresses σ_ℓ^{90} in the glass fibers at 90°, and $\sigma_\ell^{\pm 45}$ in the fibers at ±45°. Comment.

3. We desire to modify the ratio $h^{90}/h^{\pm 45}$ such that the stresses are identical in the fibers at 90° and in the fibers at ±45° ("isotensoid" external layer).

(a) What are the relations that $h^{90}/h^{\pm 45}$, σ_{2x}, σ_{2y} have to verify?

(b) Indicate an iterative method that allows, starting from the results of Question 2b, the calculation of the suitable ratio $h^{90}/h^{\pm 45}$. Give the composition of the glass/polyester with the corresponding real thicknesses (use a mixture with $V_f = 25\%$ fiber volume fraction).

Solution:

1. Equivalent moduli:
The constitutive law of the laminate in the axes x,y is written as (see Equation 12.4):

$$\begin{Bmatrix} N_x \\ N_y \\ T_{xy} \end{Bmatrix} = [A] \begin{Bmatrix} \sigma_{ox} \\ \sigma_{oy} \\ \gamma_{oxy} \end{Bmatrix} \quad \text{with} \quad A_{ij} = \sum_{k=1^{st} \text{ply}}^{n^{th} \text{ply}} \bar{E}_{ij}^k e_k$$

The coefficients $\bar{E}_{ij}^k$ are given by Equation 11.8 as, neglecting E_t, $v_{t\ell}$, $G_{\ell t}$:

▪ For the plies at 90°:

$$\bar{E}_{11}^{90} = \bar{E}_{12}^{90} = \bar{E}_{33}^{90} = \bar{E}_{23}^{90} = \bar{E}_{13}^{90} = 0$$
$$\bar{E}_{22}^{90} = E_\ell$$

▪ For the plies at +45°:

$$\bar{E}_{11}^{+45} = \bar{E}_{22}^{+45} = \bar{E}_{33}^{+45} = \bar{E}_{12}^{+45} \cdots$$

$$\cdots = -\bar{E}_{13}^{+45} = -\bar{E}_{23}^{+45} = E_\ell/4$$

■ For the plies at $-45°$:

$$\bar{E}_{11}^{-45} = \bar{E}_{22}^{-45} = \bar{E}_{33}^{-45} = \bar{E}_{12}^{-45} \cdots$$

$$\cdots = \bar{E}_{13}^{-45} = \bar{E}_{23}^{-45} = E_\ell/4$$

from which one can find the coefficients A_{ij}. For example, one has

$$A_{11} = \bar{E}_{11}^{90} b^{90} + \bar{E}_{11}^{+45} b^{+45} + \bar{E}_{11}^{-45} b^{-45} = \frac{E_\ell}{4} b^{\pm45}$$

$$A_{12} = \bar{E}_{12}^{90} b^{90} + \bar{E}_{12}^{+45} b^{+45} + \bar{E}_{12}^{-45} b^{-45} = \frac{E_\ell}{4} b^{\pm45}$$

and so forth. One obtains

$$\begin{Bmatrix} N_x \\ N_y \\ T_{xy} \end{Bmatrix} = \frac{E_\ell}{4} b^{\pm45} \begin{bmatrix} 1 & 1 & 0 \\ 1 & \left(1 + 4\dfrac{b^{90}}{b^{\pm45}}\right) & 0 \\ 0 & 0 & 1 \end{bmatrix} \begin{Bmatrix} \varepsilon_{ox} \\ \varepsilon_{oy} \\ \gamma_{oxy} \end{Bmatrix}$$

In inverting and in denoting for the average stresses (fictitious) in the external laminated layer (index 2): $\sigma_{2x} = N_x/e_2$; $\sigma_{2y} = N_y/e_2$; $\tau_{2xy} = T_{xy}/e_2$

$$\begin{Bmatrix} \varepsilon_{ox} \\ \varepsilon_{oy} \\ \gamma_{oxy} \end{Bmatrix} = \frac{e_2}{E_\ell b^{90}} \begin{bmatrix} \left(1 + 4\dfrac{b^{90}}{b^{\pm45}}\right) & -1 & 0 \\ -1 & 1 & 0 \\ 0 & 0 & 1 \end{bmatrix} \begin{bmatrix} \sigma_{2x} \\ \sigma_{2y} \\ \tau_{2xy} \end{bmatrix}$$

The above relation can be also interpreted as follows (see Equation 12.9):

$$\begin{Bmatrix} \varepsilon_{ox} \\ \varepsilon_{oy} \\ \gamma_{oxy} \end{Bmatrix} = \begin{bmatrix} \dfrac{1}{\bar{E}_x} & -\dfrac{\bar{v}_{yx}}{\bar{E}_y} & 0 \\ -\dfrac{\bar{v}_{xy}}{\bar{E}_x} & \dfrac{1}{\bar{E}_y} & 0 \\ 0 & 0 & \dfrac{1}{\bar{G}_{xy}} \end{bmatrix} \begin{bmatrix} \sigma_{2x} \\ \sigma_{2y} \\ \tau_{2xy} \end{bmatrix}$$

where the equivalent moduli of the laminate appear. From this, by identification one has

$$\boxed{\bar{E}_x = \frac{E_\ell}{e_2\left(\frac{1}{b^{90}} + \frac{4}{b^{\pm45}}\right)}; \quad \bar{E}_y = E_\ell \frac{b^{90}}{e_2}}$$
$$\bar{v}_{xy} = \frac{1}{\left(1 + 4\frac{b^{90}}{b^{\pm45}}\right)}; \quad \bar{v}_{yx} = 1$$

[1]

Comment:

The obtained results are simple because:

- The polyester resin is not taken into account. The fibers work only in their direction.
- The voluntary decoupling between the external layer (glass/resin) and the internal layer (PVDF) is preferred to the consideration of a "global" laminate consisting of plies of glass/resin at 90°, +45°, −45° and a ply of PVDF, isotropic, with thickness e_1.

2. (a) Equilibrium relation:
    The isolation of the portions of the column shown below allows one to write

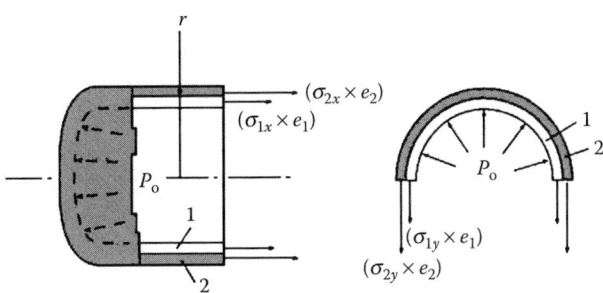

$$2\pi r(e_1\sigma_{1x} + e_2\sigma_{2x}) = \pi r^2 p_0$$
$$1 \times 2(e_1\sigma_{1y} + e_2\sigma_{2y}) = 1 \times 2r \times p_0$$

from which one has the equilibrium relations:

$$\boxed{\begin{aligned} e_1\sigma_{1x} + e_2\sigma_{2x} &= p_0\frac{r}{2} \\ e_1\sigma_{1y} + e_2\sigma_{2y} &= p_0 r \end{aligned}}$$

[2]

[3]

- Constitutive relations:

The elastic behavior of the internal layer of PVDF is described by the classical isotropic equation:

$$
\begin{Bmatrix} \varepsilon_{1x} \\ \varepsilon_{1y} \\ \gamma_{1xy} \end{Bmatrix} = \begin{bmatrix} \dfrac{1}{E_1} & -\dfrac{\nu_1}{E_1} & 0 \\ -\dfrac{\nu_1}{E_1} & \dfrac{1}{E_1} & 0 \\ 0 & 0 & \dfrac{1}{G_1} \end{bmatrix} \begin{Bmatrix} \sigma_{1x} \\ \sigma_{1y} \\ \tau_{1xy} \end{Bmatrix}
$$

The behavior of the external layer in composite is described by the relation obtained in the previous question as:

$$
\begin{Bmatrix} \varepsilon_{ox} \\ \varepsilon_{oy} \\ \gamma_{oxy} \end{Bmatrix} = \begin{bmatrix} \dfrac{1}{\bar{E}_x} & -\dfrac{\bar{\nu}_{yx}}{\bar{E}_y} & 0 \\ -\dfrac{\bar{\nu}_{xy}}{\bar{E}_x} & \dfrac{1}{\bar{E}_y} & 0 \\ 0 & 0 & \dfrac{1}{\bar{G}_{xy}} \end{bmatrix} \begin{Bmatrix} \sigma_{2x} \\ \sigma_{2y} \\ \tau_{2xy} \end{Bmatrix}
$$

Equality of strains under the action of stresses is written as:

$$
\varepsilon_{1x} = \varepsilon_{ox}; \quad \varepsilon_{1y} = \varepsilon_{oy}
$$

and leads to the relation:

$$
\frac{1}{E_1}\sigma_{1x} - \frac{\nu_1}{E_1}\sigma_{1y} = \frac{1}{\bar{E}_x}\sigma_{2x} - \frac{\bar{\nu}_{yx}}{\bar{E}_y}\sigma_{2y} \tag{4}
$$

$$
-\frac{\nu_1}{E_1}\sigma_{1x} + \frac{1}{E_1}\sigma_{1y} = -\frac{\bar{\nu}_{xy}}{\bar{E}_x}\sigma_{2x} + \frac{1}{\bar{E}_y}\sigma_{2y} \tag{5}
$$

Relations [2], [3], [4], [5] constitute a system of four equations for the four unknowns σ_{1x}, σ_{1y}, σ_{2x}, σ_{2y}. Performing the subtraction [4] − [5], one obtains

$$
\sigma_{1x}\left(\frac{1+\nu_1}{E_1}\right) - \sigma_{1y}\left(\frac{1+\nu_1}{E_1}\right) = \sigma_{2x}\left(\frac{1+\bar{\nu}_{xy}}{\bar{E}_x}\right) - \sigma_{2y}\left(\frac{1+\bar{\nu}_{yx}}{\bar{E}_y}\right)
$$

In performing the addition [4] + [5], one obtains

$$
\sigma_{1x}\left(\frac{1-\nu_1}{E_1}\right) - \sigma_{1y}\left(\frac{1-\nu_1}{E_1}\right) = \sigma_{2x}\left(\frac{1-\bar{\nu}_{xy}}{\bar{E}_x}\right) - \sigma_{2y}\left(\frac{1-\bar{\nu}_{yx}}{\bar{E}_y}\right)
$$

and with [2] and [3], by substitution, one obtains a system that allows the calculation of σ_{1x} and σ_{1y} as:

$$\sigma_{1x}\left[\left(\frac{1+v_1}{E_1}\right)+\frac{e_1}{e_2}\left(\frac{1+\bar{v}_{xy}}{\bar{E}_x}\right)\right]-\sigma_{1y}\left[\left(\frac{1+v_1}{E_1}\right)+\frac{e_1}{e_2}\left(\frac{1+\bar{v}_{yx}}{\bar{E}_y}\right)\right]\cdots$$

$$\cdots=\frac{p_0 r}{e_2}\left[\left(\frac{1+\bar{v}_{xy}}{2\bar{E}_x}\right)-\left(\frac{1+\bar{v}_{yx}}{\bar{E}_y}\right)\right]$$

$$\sigma_{1x}\left[\left(\frac{1-v_1}{E_1}\right)+\frac{e_1}{e_2}\left(\frac{1-\bar{v}_{xy}}{\bar{E}_x}\right)\right]+\sigma_{1y}\left[\left(\frac{1-v_1}{E_1}\right)+\frac{e_1}{e_2}\left(\frac{1-\bar{v}_{yx}}{\bar{E}_y}\right)\right]\cdots$$

$$\cdots=\frac{p_0 r}{e_2}\left[\left(\frac{1-\bar{v}_{xy}}{2\bar{E}_x}\right)+\left(\frac{1-\bar{v}_{yx}}{\bar{E}_y}\right)\right]$$

[6]

(b) Numerical application:
One has $b^{90}=b^{\pm45}/3$, from which we have

$$e_2=b^{90}+b^{\pm45}=0.75 \text{ mm}; \quad b^{\pm45}=0.56 \text{ mm}; \quad b^{90}=0.19 \text{ mm}.$$

Following the results [1], one finds

$$\bar{E}_x=7953 \text{ MPa}; \quad \bar{E}_y=18747 \text{ MPa}; \quad \bar{v}_{xy}=0.42$$

The system [6] has for solutions:

$$\boxed{\sigma_{1x}=1.71 \text{ MPa}; \quad \sigma_{1y}=3.07 \text{ MPa}}$$

Relations [4] and [5] allow the calculation of σ_{2x} and σ_{2y}. One finds

$$\boxed{\sigma_{2x}=188 \text{ MPa}; \quad \sigma_{2y}=386 \text{ MPa}}$$

(c) Stresses in the fibers:
Following Equation 11.8, one has for any ply k in the external layer:

$$\begin{Bmatrix}\sigma_x\\\sigma_y\\\tau_{xy}\end{Bmatrix}^k=E_\ell\begin{bmatrix}c^4 & c^2s^2 & -c^3s\\c^2s^2 & s^4 & -cs^3\\-c^3s & -cs^3 & c^2s^2\end{bmatrix}^k\begin{Bmatrix}\varepsilon_{ox}\\\varepsilon_{oy}\\\gamma_{oxy}\end{Bmatrix}$$

[7]

The strains ε_{ox} and ε_{oy} are obtained by means of the previous results (see Question 2a), for example:

$$\varepsilon_{ox}=\varepsilon_{1x}=\frac{\sigma_{1x}}{E_1}-\frac{v_1}{E_1}\sigma_{1y}=3.03\times10^{-3}$$

$$\varepsilon_{oy}=\varepsilon_{1y}=-\frac{v_1}{E_1}\sigma_{1x}+\frac{\sigma_{1y}}{E_1}=9.85\times10^{-3}$$

If one inverts Equation 11.4, taking into account that the only nonzero stress in the axes ℓ, t of the ply is σ_ℓ:

$$\left\{ \begin{array}{c} \sigma_x \\ \sigma_y \\ \tau_{xy} \end{array} \right\}^k = \left[\begin{array}{ccc} c^2 & s^2 & 2cs \\ s^2 & c^2 & -2cs \\ -sc & sc & (c^2 - s^2) \end{array} \right]^k \left\{ \begin{array}{c} \sigma_\ell \\ 0 \\ 0 \end{array} \right\} \qquad [8]$$

One then has

- In the fibers at 90°:
  Following [7], $\sigma_x^{90} = 0$; $\sigma_y^{90} = E_\ell\,\varepsilon_{oy}$
  Following [8], $\sigma_x^{90} = 0$; $\sigma_y^{90} = \sigma_\ell^{90}$
from which:

$$\boxed{\begin{array}{l} \sigma_\ell^{90} = E_\ell \varepsilon_{oy} \\[4pt] \sigma_\ell^{90} = 729 \text{ MPa} \end{array}}$$

- In the fibers at +45°:
  Following [7]: $\sigma_x^{+45} = \sigma_y^{+45} = \frac{E_\ell}{4}(\varepsilon_{ox} + \varepsilon_{oy})$
  Following [8]: $\sigma_x^{+45} = \sigma_y^{+45} = \frac{1}{2}\sigma_\ell^{+45}$
  from which one obtains

$$\boxed{\begin{array}{l} \sigma_\ell^{+45} = \dfrac{E_\ell}{2}(\varepsilon_{ox} + \varepsilon_{oy}) \\[8pt] \sigma_\ell^{+45} = 477 \text{ MPa} \end{array}}$$

One obtains an identical stress in the fibers at −45°. Note the disparity of the stresses in the fibers at 90° and at ±45°. In fact, the external layer is not suitably designed, because it is desirable to make all fibers work equally in order to obtain a uniform extension in the glass fibers.

3. (a) One desires that $\sigma_\ell^{90} = \sigma_\ell^{\pm 45}$:
    Referring to the results of the previous question, this equality is also written as:

$$E_\ell \varepsilon_{o\ell} = \frac{E_\ell}{2}(\varepsilon_{ox} + \varepsilon_{oy})$$

as:

$$\varepsilon_{oy} = \varepsilon_{ox}$$

The constitutive relation of the laminate (Question 1 and relation [1]) indicates then:

$$\frac{\sigma_{2x}}{\overline{E}_x} - \frac{\overline{v}_{yx}}{\overline{E}_y}\sigma_{2y} = -\frac{\overline{v}_{xy}}{\overline{E}_x}\sigma_{2x} + \frac{\sigma_{2y}}{\overline{E}_y}$$

Then after calculation:

$$\boxed{\frac{b^{90}}{b^{\pm 45}} = \frac{\sigma_{2y} - \sigma_{2x}}{\sigma_{2x}}}$$

[9]

(b) With the results of numerical application 2(b), relation [9] above indicates

$$\frac{\sigma_{2y} - \sigma_{2x}}{\sigma_{2x}} = 0.53$$

If one adopts this new value for the ratio $b^{90}/b^{\pm 45}$, one obtains for new results:

■ $b^{90}/b^{\pm 45} = 0.53$;

$$\bar{E}_x = 8216 \text{ MPa}; \quad \bar{E}_y = 25{,}653 \text{ MPa}; \quad \bar{v}_{xy} = 0.32; \quad \bar{v}_{yx} = 1$$
$$\sigma_{1x} = 2.42 \text{ MPa}; \quad \sigma_{1y} = 2.72 \text{ MPa};$$
$$\sigma_{2x} = 167 \text{ MPa}; \quad \sigma_{2y} = 364 \text{ MPa};$$

Relation [9] then indicates

$$\frac{\sigma_{2y} - \sigma_{2x}}{\sigma_{2x}} = 0.587$$

that one adopts for the new ratio $b^{90}/b^{\pm 45}$:

■ $b^{90}/b^{\pm 45} = 0.587$:

$$\bar{E}_x = 8166 \text{ MPa}; \quad \bar{E}_y = 27{,}627 \text{ MPa}; \quad \bar{v}_{xy} = 0.29; \quad \bar{v}_{yx} = 1$$
$$\sigma_{1x} = 2.63 \text{ MPa}; \quad \sigma_{1y} = 2.69 \text{ MPa};$$
$$\sigma_{2x} = 165 \text{ MPa}; \quad \sigma_{2y} = 364 \text{ MPa}.$$

Relation [9] then indicates

$$\frac{\sigma_{2y} - \sigma_{2x}}{\sigma_{2x}} = 0.6$$

that is, a relative variation of 2% with respect to the value of the ratio $(b^{90}/b^{\pm 45})$ taken to carry out the calculations. The iterative procedure then converges rapidly. One will obtain the external isotensoid layer and an internal layer of PVDF in biaxial tension for a ratio of

$$b^{90}/b^{\pm 45} \# 0.6$$

The composition of the glass/polyester reinforcement will be as follows:

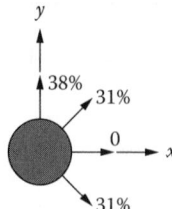

The real thickness of the windings in glass/polyester, taking into account the volume of the resin, will be (with $V_f = 0.25$):

$$e'_2 = e_2/0.25$$

$$e'_2 = 3 \text{ mm}$$

18.3.8 Cylindrical Bending of a Thick Orthotropic Plate under Uniform Loading

Problem Statement:

Consider a thick rectangular plate $b \times a$, with $b \gg a$ made of unidirectional glass/resin (see figure). It is supported at two opposite sides and is loaded by a constant transverse pressure of q_o.

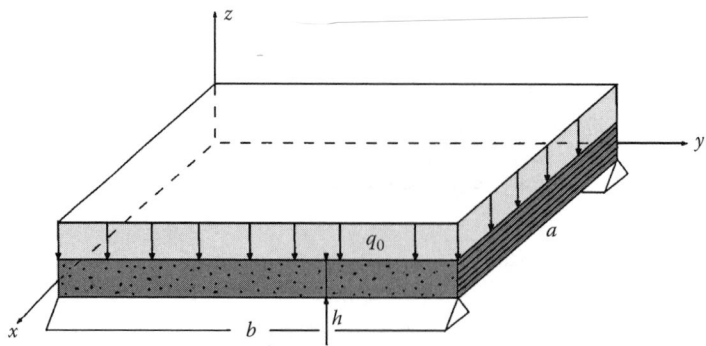

1. Calculate the deflection due to bending at the midline of the plate located at $x = a/2$ (maximum deflection).
2. Indicate the numerical values of the contributions from the bending moment and from transverse shear using the following: $E_x = 40,000$ MPa; $G_{xz} = 400$ MPa; $v_{xy} = 0.3$; $v_{yx} = 0.075$; $q_o = -1$ MPa; $a = 150$ mm; $b = 15$ mm. Comment.

Solution:

1. For the cylindrical bending considered, Equation 17.32 allows one to write

$$\frac{dQ_x}{dx} = -q_0; \quad \frac{dM_y}{dx} = Q_x; \quad M_y = C_{11}\frac{d\theta_y}{dx}; \quad Q_x = \frac{bG_{xz}}{k_x}\left(\frac{dw_o}{dx} + \theta_y\right)$$

Elimination of Q_x, M_y and θ_y leads to

$$\frac{d^4w_o}{dx^4} = \frac{q_o}{C_{11}}$$

then:

$$w_o = \frac{q_o}{C_{11}}\left(\frac{x^4}{24} + A\frac{x^3}{6} + B\frac{x^2}{2} + Cx + D\right)$$

The boundary conditions are written as:

$$\left.\begin{matrix} x = 0 \\ x = a \end{matrix}\right\} w_o = 0; \quad M_y = 0 \Rightarrow \frac{d\theta_y}{dx} = \frac{k_x}{bG_{xz}}q_0 - \frac{d^2w_o}{dx^2} = 0$$

After calculation of the constants A, B, C, D, one obtains for the deflection at $x = a/2$:

$$w_o\left(\frac{a}{2}\right) = q_0a^4\frac{12(1 - v_{xy}v_{yx})}{E_xb^3}\left\{\frac{5}{384} + k_x\left(\frac{b}{a}\right)^2\frac{E_x}{G_{xz}}\frac{1}{96(1 - v_{xy}v_{yx})}\right\}$$

The calculation of k_x was done in Section 17.7.1 for this type of plate. One has (see Equation 17.34)

$$k_x = 6/5 = 1.2$$

from which:

$$\boxed{w_o\left(\frac{a}{2}\right) = q_0a^4\frac{12(1 - v_{xy}v_{yx})}{E_xb^3}\left\{\frac{5}{384} + \left(\frac{b}{a}\right)^2\frac{E_x}{G_{xz}}\frac{1}{80(1 - v_{xy}v_{yx})}\right\}}$$

The first term in the brackets represents the contribution from the bending moment, and the second term represents that due to transverse shear.

2. Numerical values:

$$w_o(a/2) = -0.5727 \text{ mm} - 0.5625 \text{ mm}$$

$$\text{(moment)} \qquad \text{(transverse shear)}$$

$$w_o(a/2) = -1.13525 \text{ mm}$$

Note that 49.5% of this deflection is due to transverse shear. One can see from the above expression for $w_o(a/2)$ that the influence of transverse shear on the bending deflection increases with the value of the relative thickness b/a (here, $b/a = 1/10$ corresponds to a thick plate). One also notes the influence of the ratio E_x/G_{xz}.[57]

18.3.9 Bending of a Sandwich Plate

Problem Statement:

A rectangular sandwich plate ($a \times b$) is fixed on side b, and loaded along the opposite side by a constant distributed load f_o (N/mm). The two other sides are free (see figure).

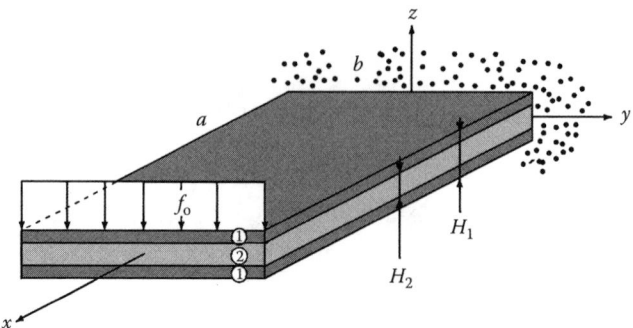

The plate consists of two identical orthotropic skins of material 1, and an orthotropic core made of material 2. The orthotropic axes are parallel to the axes x, y, z.

1. Calculate the deflection of the midplane at the extremity $x = a$ of the x axis, assuming cylindrical bending about y axis for the plate.
2. Numerical application:
   Given
   $f_o = -10$ N/mm
   $a = b = 1000$ mm

---

[57] This example of thick plate in bending constitutes a test case for the evaluation of computer programs using finite elements. For complementary information on this topic, see bibliography, "Computer programs for Composite Structures: Reference examples and Validation."

$H_1 = 2H_2 = 100$ mm

Material 1:
$E_x^{(1)} = 40,000$ MPa
$G_{xz}^{(1)} = 4000$ MPa

Material 2:
$E_x^{(2)} = 40$ MPa
$G_{xz}^{(2)} = 15$ MPa

For each of the materials
$v_{xy} = 0.3$
$v_{yx} = 0.075$

(a) Calculate the deflection at the extremity $x = a$ and show the contributions from the bending moment and from the transverse shear.
(b) Calculate the transverse shear stress τ_{xz}:

■ On the midplane of the plate.
■ At the interface between the core and the upper skin.
■ At the midthickness of the upper skin.

Solution:

1. In the case of cylindrical bending, Equation 17.32 allows one to write

$$\frac{dQ_x}{dx} = 0; \quad \frac{dM_y}{dx} = Q_x; \quad M_y = C_{11}\frac{d\theta_y}{dx}; \quad Q_x = \frac{\langle hG_{xz}\rangle}{k_x}\left(\frac{dw_0}{dx} + \theta_y\right)$$

Then $Q_x = f_o$, and elimination of Q_x, M_y, and θ_y leads to

$$\frac{d^3 w_o}{dx^3} = \frac{f_o}{C_{11}}$$

then:

$$w_o = -\frac{f_o}{C_{11}}\left(\frac{x^3}{6} + A\frac{x^2}{2} + Bx + C\right)$$

The boundary conditions are written as:

$$x = 0 : w_0 = 0 \text{ et } \theta_y = 0 \Rightarrow k_x\frac{f_o}{\langle hG_{xz}\rangle} - \frac{dw_o}{dx} = 0$$

$$x = a : M_y = 0 \Rightarrow \frac{d\theta_y}{dx} = -\frac{d^2 w_o}{dx^2} = 0$$

After calculation of the constants A, B, C, one obtains the deflection at $x = a$:

$$w_o(a) = \frac{f_o a^3}{3C_{11}} + k_x \frac{f_o a}{\langle bG_{xz}\rangle}$$

with (see Equation 12.16):

$$C_{11} = \bar{E}_{11}^{①}\left(\frac{H_1^3 - H_2^3}{12}\right) + \bar{E}_{11}^{②}\frac{H_2^3}{12}$$

and according to Equation 17.2:

$$C_{11} = \frac{E_x^{①}(H_1^3 - H_2^3) + E_x^{②}H_2^3}{12(1 - v_{xy}v_{yx})}$$

According to Equation 17.10:

$$\langle bG_{xz}\rangle = G_{xz}^{①}(H_1 - H_2) + G_{xz}^{②}H_2$$

from which one obtains

$$w_o(a) = \frac{4(1 - v_{xy}v_{yx})f_o a^3}{E_x^{①}(H_1^3 - H_2^3) + E_x^{②}H_2^3} + \frac{k_x f_o a}{G_{xz}^{①}(H_1 - H_2) + G_{xz}^{②}H_2}$$

The calculation of k_x was carried out in Section 17.7.2 for this type of plate. It was given by Equation 17.39.

2. Numerical application:
    (a) Deflection: Equation 17.39 gives $k_x = 110.8$
        from which:

$$w_o(a) = -1.177 \text{ mm} + (-5.519 \text{ mm})$$
$$\text{(moment)} \qquad \text{(transverse shear)}$$

$$w_o(a) = -6.636 \text{ mm}$$

Note that 83% of this deflection is due to transverse shear, and this happens in spite of very thick skins. This important influence is due to the very large value compared with unity (110.8) of the transverse shear coefficient and due to the fact that the plate is thick ($H_1/a = 1/10$).

   (b) Transverse shear stress τ_{xz} (see Section 17.7.2):

Midplane: ($z = 0$): $\tau_{xz} = 0.1286$ MPa

Interface ($z = H_2/2$): $\tau_{xz} = 0.12855$ MPa

Midthickness of the upper layer: $z = (H_1 + H_2)/4$: $\tau_{xz} = 0.075$ MPa

18.3.10 Bending Vibration of a Sandwich Beam[58]

Problem Statement:

Consider a sandwich beam of length ℓ and width d simply supported at its ends (see figure). It consists of two identical skins of material 1 (glass/resin) and a core of material 2 (foam). These materials are transversely isotropic in the plane y, z.

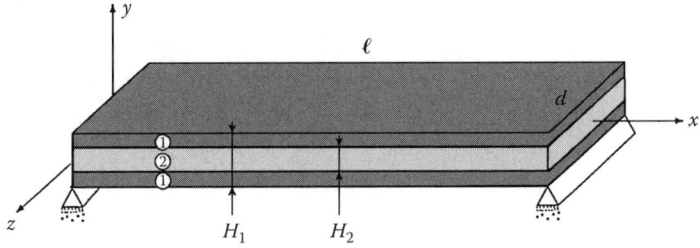

The elastic characteristics are denoted as:

$$E_x^{①}, G_{xy}^{①}, E_x^{②}, G_{xy}^{②}$$

Specific masses are ρ_1 and ρ_2.

1. Write the equation for the resonant frequencies for bending vibration in the plane of symmetry (x, y) of this beam.
2. Numerical application:

Given:

$$E_x^{①} = 40{,}000 \text{ MPa}; \quad G_{xy}^{①} = 4{,}000 \text{ MPa}; \quad \rho_1 = 2{,}000 \text{ kg/m}^3$$

$$E_x^{②} = 40 \text{ MPa}; \quad G_{xy}^{②} = 15 \text{ MPa}; \quad \rho_2 = 50 \text{ kg/m}^3$$

$$H_1 = 2H_2 = 100 \text{ mm}; \quad \ell = 1000 \text{ mm}; \quad d = 100 \text{ mm}$$

Calculate the first five frequencies in bending vibration.

Solution:

1. Equation for the vibration frequencies:
   At first one establishes the differential equation for the dynamic displacement $v(x, t)$ starting from the Equation 15.18, noting that for the example

---

[58] This application constitutes a test case for the validation of computer programs using finite elements, see in the bibiography, "Programs for the calculation of Composite Structures, Reference examples and Validation."

considered, the elastic center and the center of gravity of section are the same (decoupling between bending vibration and longitudinal vibrations).

$$\frac{\partial T_y}{\partial x} = \langle \rho S \rangle \frac{\partial^2 v}{\partial t^2}; \quad \frac{\partial M_z}{\partial x} + T_y = \langle \rho I_z \rangle \frac{\partial^2 \theta_z}{\partial t^2}$$

$$T_y = \frac{\langle GS \rangle}{k}\left(\frac{\partial v}{\partial x} - \theta_z\right); \quad M_z = \langle EI_z \rangle \frac{\partial \theta_z}{\partial x}$$

Elimination of T_y, M_z, θ_z between these four relations leads to the equation for $v(x, t)$:

$$\frac{\partial^4 v}{\partial x^4} - \frac{\langle \rho I_z \rangle}{\langle EI_z \rangle}(1+a)\frac{\partial^4 v}{\partial x^2 \partial t^2} + \frac{\langle \rho S \rangle}{\langle EI_z \rangle}\frac{\partial^2 v}{\partial t^2} + k\frac{\langle \rho I_z \rangle \langle \rho S \rangle}{\langle GS \rangle \langle EI_z \rangle}\frac{\partial^4 v}{\partial t^4} = 0$$

with

$$a = \frac{\langle \rho S \rangle \langle EI_z \rangle}{\langle GS \rangle \langle \rho I_z \rangle} \times k$$

In assuming that the solution takes the form $v(x, t) = v_o(x) \times \cos(\omega t + \varphi)$ one can rewrite the differential equation that defines the modal deformation $v_o(x)$ in the following nondimensional form:

$$\frac{d^4 \bar{v}_o}{d\bar{x}^4} + \bar{\omega}^2(1+a)\frac{d^2 \bar{v}_o}{d\bar{x}^2} + \bar{\omega}^2\left(a\bar{\omega}^2 - \frac{1}{\bar{r}^2}\right)\bar{v}_o = 0$$

in which

$$\bar{x} = \frac{x}{\ell}; \quad \bar{v}_o = \frac{v_o}{\ell}; \quad \bar{\omega}^2 = \frac{\langle \rho I_z \rangle}{\langle EI_z \rangle}\omega^2 \ell^2; \quad \bar{r}^2 = \frac{\langle \rho I_z \rangle}{\langle \rho S \rangle \ell^2}$$

After writing the characteristic equation, the reduced modal deformation takes the form:

$$\bar{v}_o = A\,cb\,X_1\bar{x} + B\,sb\,X_1\bar{x} + C\cos X_2\bar{x} + D\sin X_2\bar{x} \tag{1}$$

where:

$$\begin{matrix} X_1^2 \\ X_2^2 \end{matrix} = \mp\frac{\bar{\omega}^2(1+a)}{2} + \sqrt{\bar{\omega}^2\left[\bar{\omega}^2\left(\frac{1-a}{2}\right)^2 + \frac{1}{\bar{r}^2}\right]} \tag{2}$$

The boundary conditions corresponding to simply supported ends are written as:

$$x = 0 \quad \text{or} \quad \ell : v = 0; \quad M_z = \langle EI_z \rangle \frac{\partial \theta_z}{\partial x} = 0 \ \forall t$$

or:

$$\bar{x} = 0 \quad \text{or} \quad 1 : \bar{v}_o = 0; \quad \frac{d^2 \bar{v}_o}{d\bar{x}^2} + a\bar{\omega}^2 \bar{v}_o = 0$$

These four conditions allow one to obtain with [1] a linear and homogeneous system in A,B,C,D. By setting the determinant equal to zero, one obtains an equation for vibrations which reduces to

$$\sin X_2 = 0$$

Then the solution can be written as:

$$X_2 = n\pi, \ (n = 1, 2, 3,\ldots) \tag{3}$$

2. With the numerical values indicated in the Problem Statement, one can calculate the shear coefficient k, the literal expression for which has been established in Application 18.3.5. One finds $k = 110.8$. The frequencies can be written starting from the circular frequencies $\omega_1, \omega_2, \omega_3,\ldots$ extracted from equation [3], where X_2 takes the form [2].

$$f_i = \frac{\omega_i}{2\pi}(H_z)$$

one obtains:[59]

$$\boxed{\begin{array}{l} f_1 = 64.476 \text{ Hz}; \quad f_2 = 131.918 \text{ Hz}; \quad f_3 = 198.734 \text{ Hz} \\ f_4 = 265.383 \text{ Hz}; \quad f_5 = 331.963 \text{ Hz}. \end{array}}$$

---

[59] One keeps voluntarily the nonsignificant decimal, for the purpose of comparison with values obtained from numerical models.

APPENDICES, BIBLIOGRAPHY, AND INDEX

APPENDIX 1

STRESSES IN THE PLIES OF A LAMINATE OF CARBON/EPOXY LOADED IN ITS PLANE

The tables in this appendix give for each ply in the laminate the stresses along the principal orthotropic directions of the ply, denoted as ℓ and t. These stresses are denoted as σ_ℓ, σ_t, $\tau_{\ell t}$. The laminate is successively subjected to three cases of simple loading:

1. $\sigma_x = 1$ MPa: normal stress along the 0° direction.
2. $\sigma_y = 1$ MPa: normal stress along the 90° direction.
3. $\tau_{xy} = 1$ MPa: shear stress.

CHARACTERISTICS OF EACH PLY

- $V_f = 60\%$ fiber volume fraction.
- Thickness of each ply: 0.13 mm.
- Moduli:
    Modulus along the fiber direction: $E_\ell = 134,000$ MPa.
    Modulus along the transverse direction: $E_t = 7000$ MPa.
    Shear modulus: $G_{\ell t} = 4200$ MPa.
    Poisson coefficient: $\nu_{\ell t} = 0.25$.
- Fracture strength:
    Tension along the longitudinal direction: $\sigma_{\ell\,\text{rupture}} = 1270$ MPa.
    Compression along the longitudinal direction: $\sigma_{\ell\,\text{rupture}} = 1130$ MPa.
    Tension along the transverse direction: $\sigma_{t\,\text{rupture}} = 42$ MPa.
    Compression along the transverse direction: $\sigma_{t\,\text{rupture}} = 141$ MPa.
    Shear strength: $\tau_{\ell t\,\text{rupture}} = 63$ MPa.

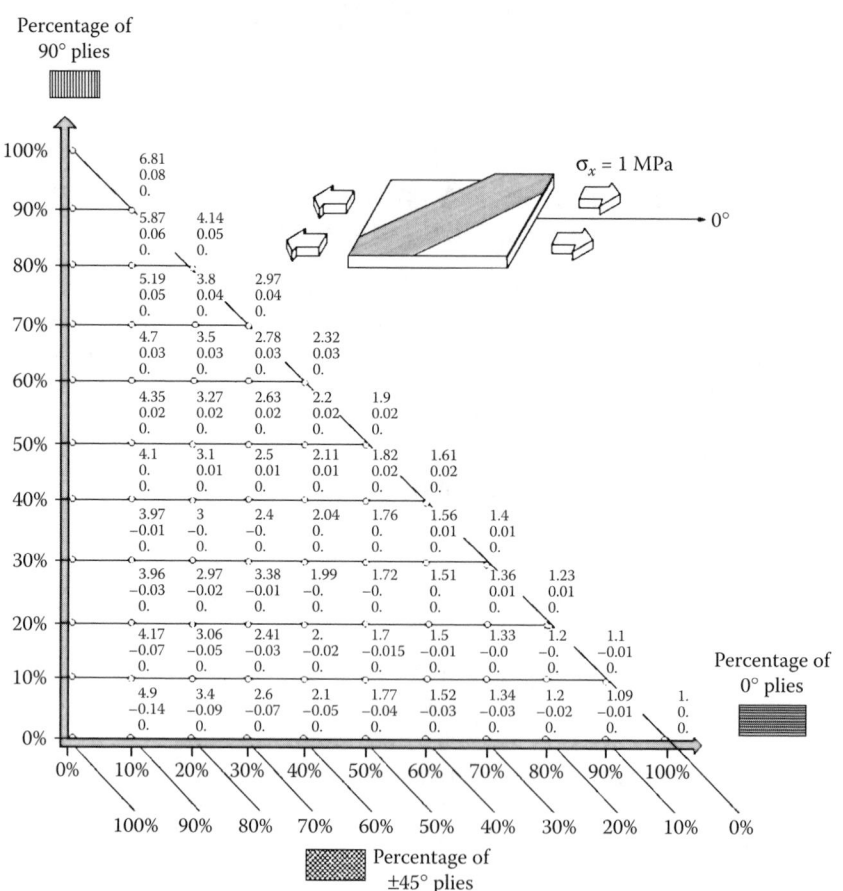

Stresses in 0° plies, respectively: as function of the percentage of plies in directions 0°, 90°, +45°, −45°, for an applied uniaxial stress $\sigma_x = 1$ MPa

$$\begin{cases} \sigma_\ell \\ \sigma_t \ (\text{MPa}) \\ \tau_{\ell t} \end{cases}$$

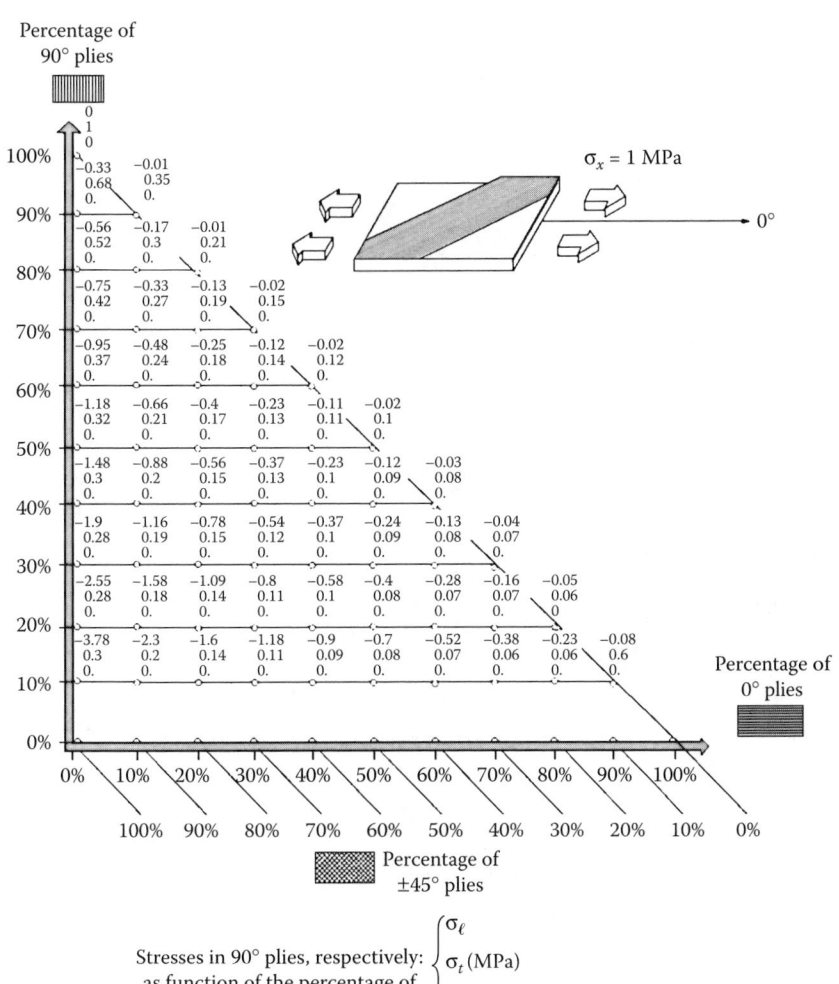

Percentage of
90° plies

Stresses in 90° plies, respectively:
as function of the percentage of
plies in directions 0°, 90°, +45°,
−45°, for an applied uniaxial
stress $\sigma_x = 1$ MPa
$$\begin{cases} \sigma_\ell \\ \sigma_t \text{ (MPa)} \\ \tau_{\ell t} \end{cases}$$

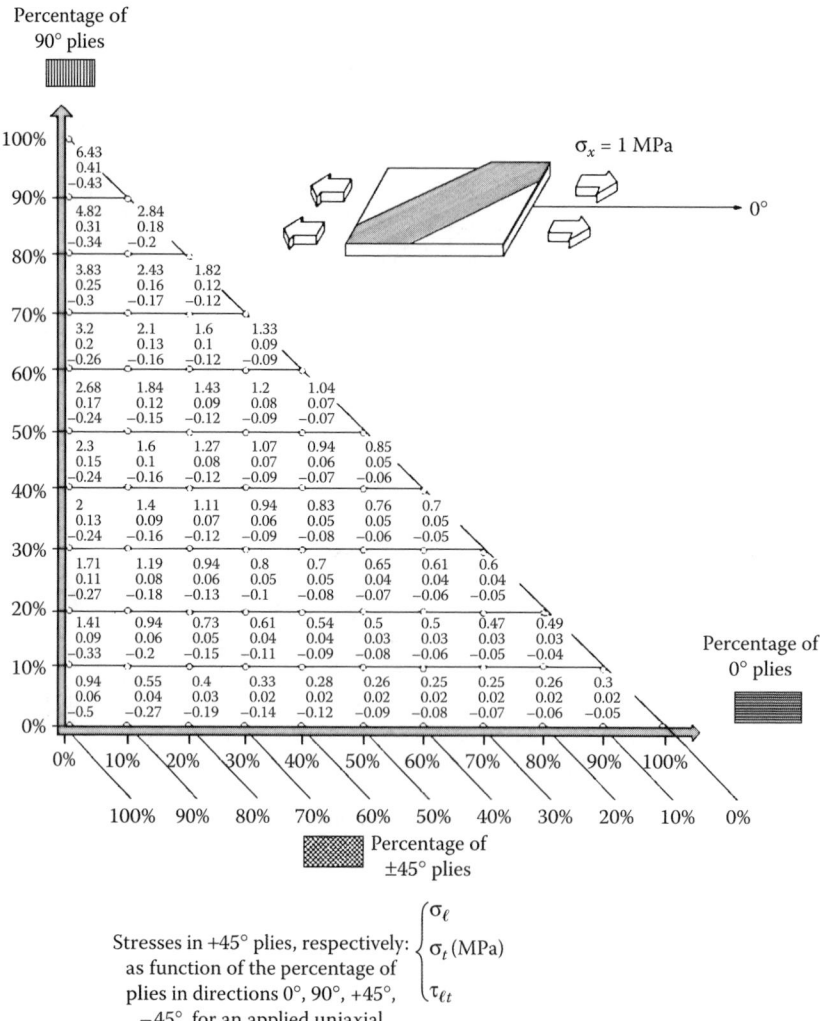

Stresses in +45° plies, respectively: $\begin{cases} \sigma_\ell \\ \sigma_t \text{ (MPa)} \\ \tau_{\ell t} \end{cases}$
as function of the percentage of plies in directions 0°, 90°, +45°, −45°, for an applied uniaxial stress $\sigma_x = 1$ MPa

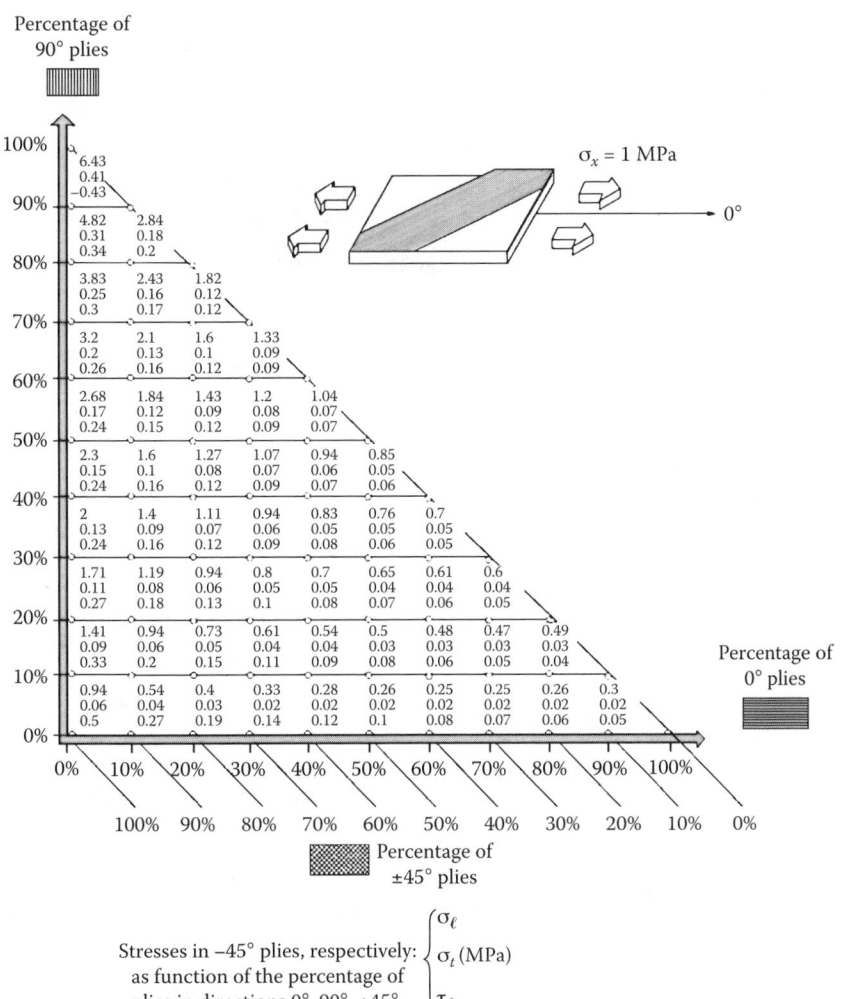

Percentage of
90° plies

$\sigma_x = 1$ MPa

0°

Percentage of
0° plies

Percentage of
±45° plies

Stresses in −45° plies, respectively:
as function of the percentage of
plies in directions 0°, 90°, +45°,
−45°, for an applied uniaxial
stress $\sigma_x = 1$ MPa

$\begin{cases} \sigma_\ell \\ \sigma_t \text{ (MPa)} \\ \tau_{\ell t} \end{cases}$

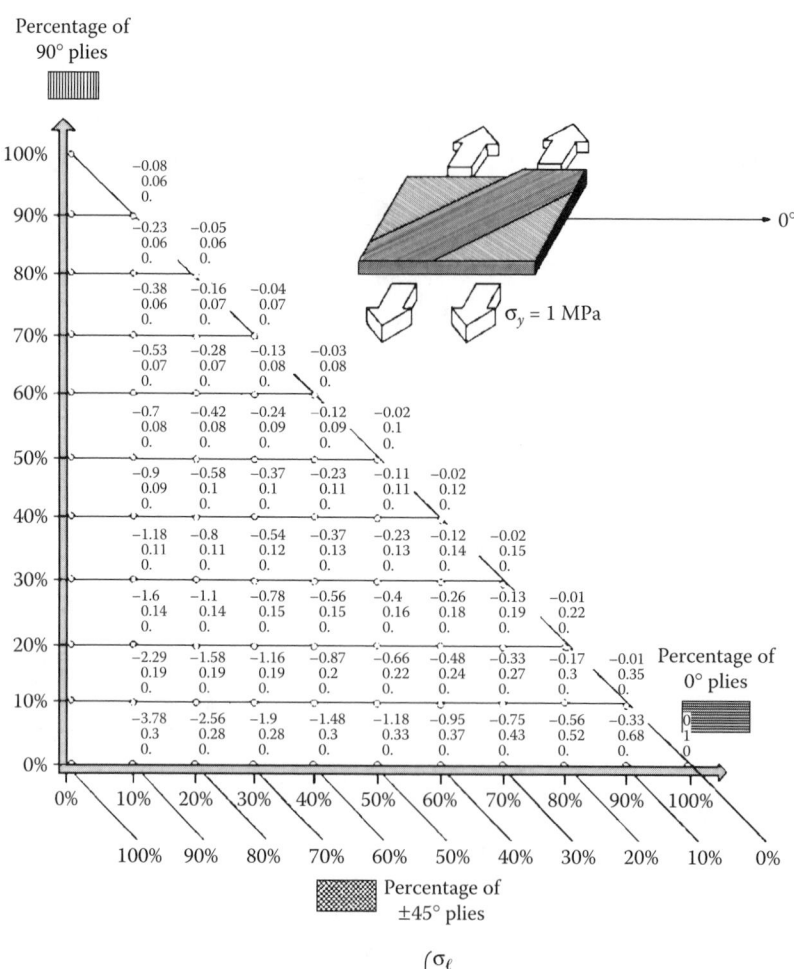

Stresses in 0° plies, respectively:
as function of the percentage of
plies in directions 0°, 90°, +45°,
−45°, for an applied uniaxial
stress $\sigma_y = 1$ MPa

$$\begin{cases} \sigma_\ell \\ \sigma_t \text{ (MPa)} \\ \tau_{\ell t} \end{cases}$$

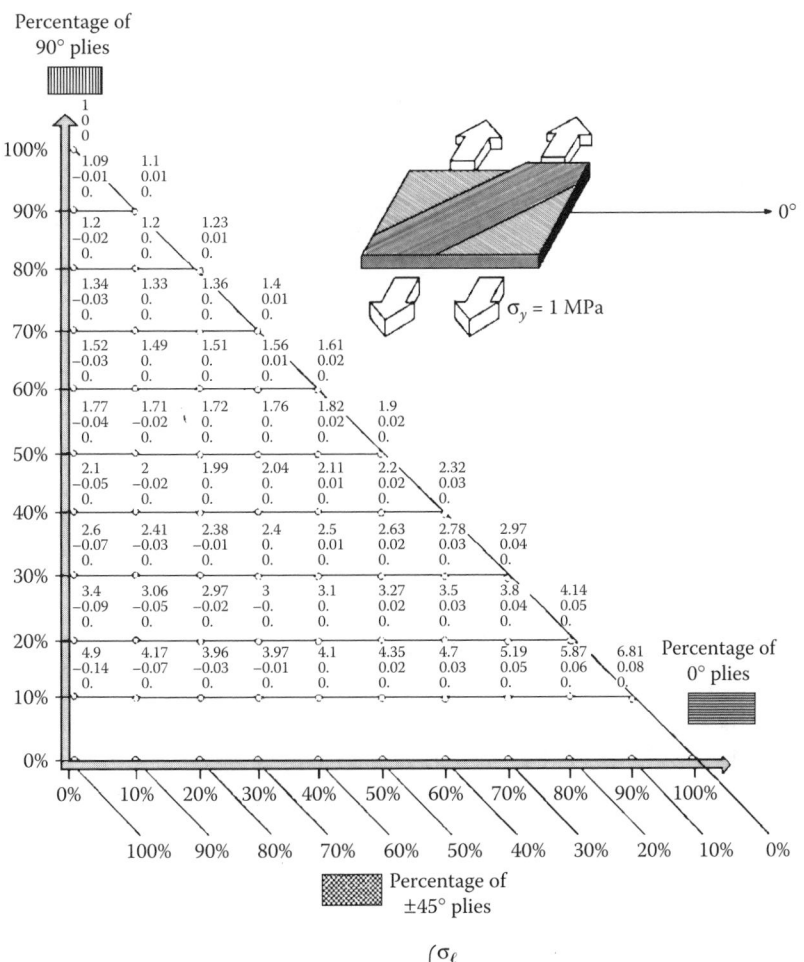

Percentage of
90° plies

Percentage of
0° plies

Percentage of
±45° plies

Stresses in 90° plies, respectively:
as function of the percentage of
plies in directions 0°, 90°, +45°,
−45°, for an applied uniaxial
stress $\sigma_y = 1$ MPa

$$\begin{cases} \sigma_\ell \\ \sigma_t \, (\text{MPa}) \\ \tau_{\ell t} \end{cases}$$

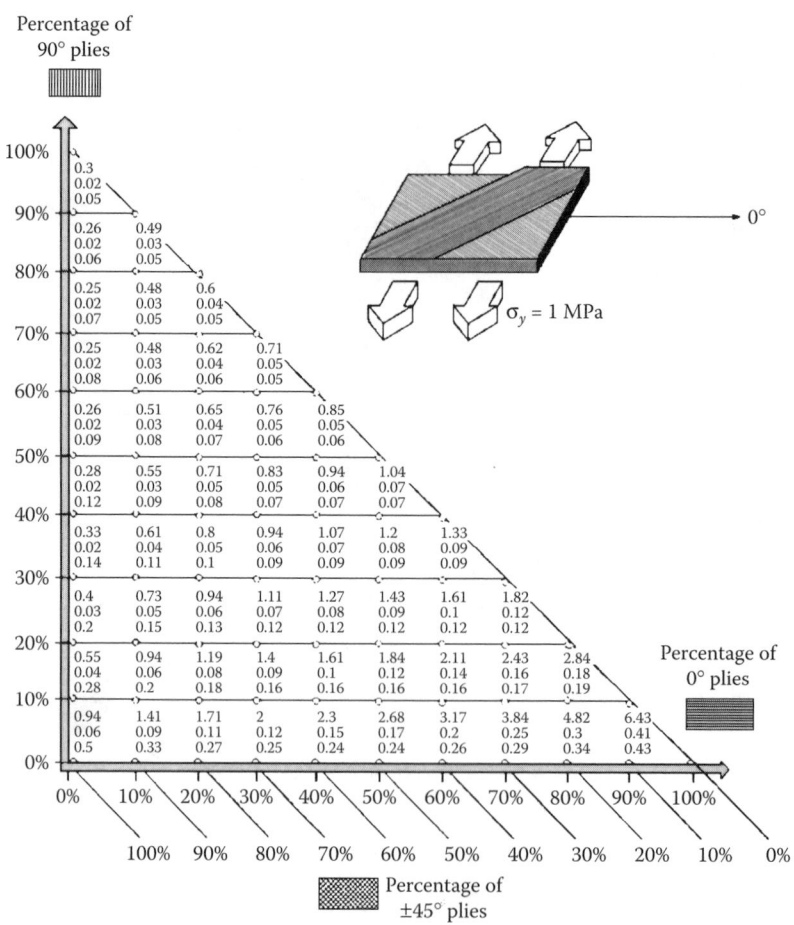

Percentage of
90° plies

Stresses in +45° plies, respectively:
as function of the percentage of
plies in directions 0°, 90°, +45°,
−45°, for an applied uniaxial
stress σ_y = 1 MPa

$\begin{cases} \sigma_\ell \\ \sigma_t \text{ (MPa)} \\ \tau_{\ell t} \end{cases}$

Percentage of
0° plies

Percentage of
±45° plies

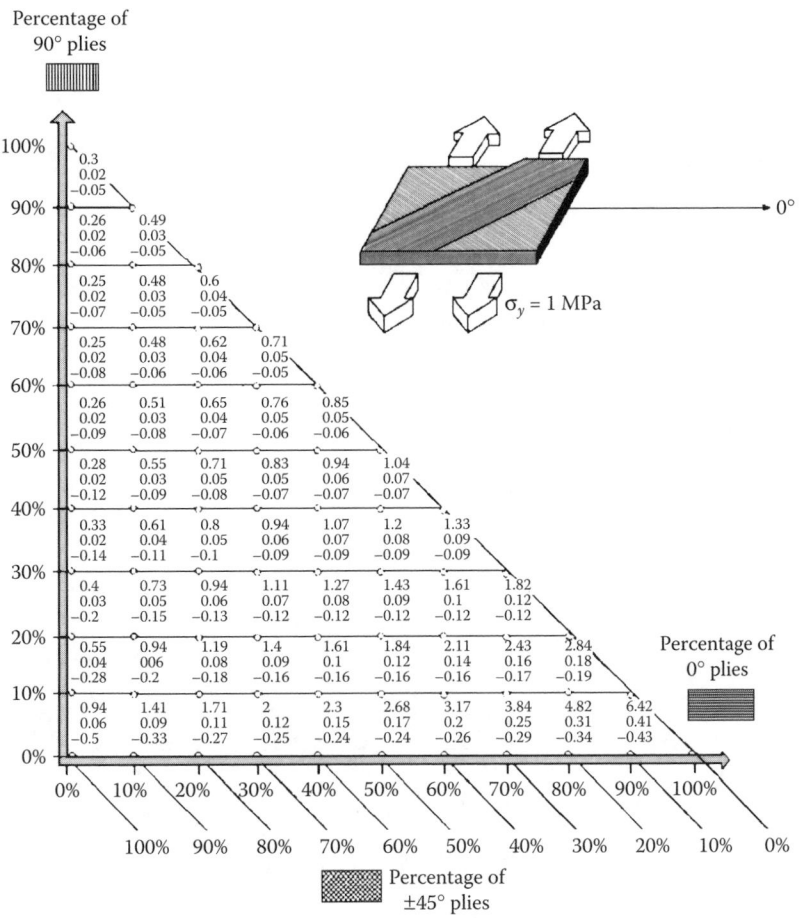

Stresses in −45° plies, respectively: $\begin{cases} \sigma_\ell \\ \sigma_t\,(\text{MPa}) \\ \tau_{\ell t} \end{cases}$ as function of the percentage of plies in directions 0°, 90°, +45°, −45°, for an applied uniaxial stress $\sigma_y = 1$ MPa

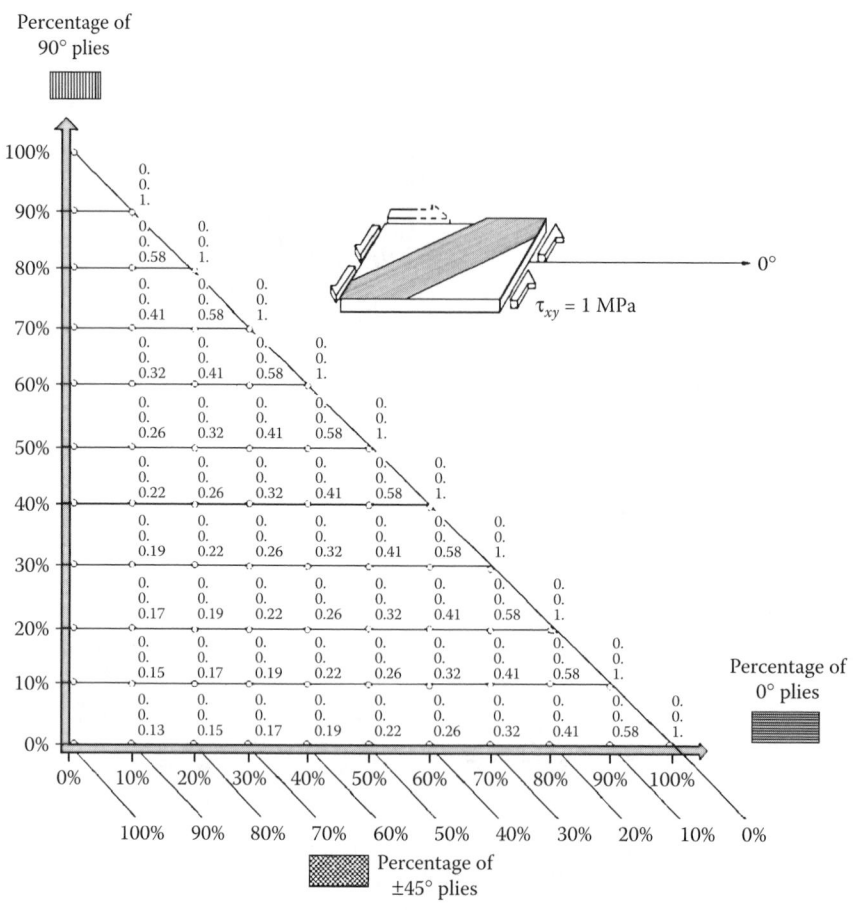

Percentage of
90° plies

100%
90%
80%
70%
60%
50%
40%
30%
20%
10%
0%

0% 10% 20% 30% 40% 50% 60% 70% 80% 90% 100%

100% 90% 80% 70% 60% 50% 40% 30% 20% 10% 0%

Percentage of
±45° plies

Percentage of
0° plies

τ_{xy} = 1 MPa

0°

Stresses in 0° plies, respectively:
as function of the percentage of
plies in directions 0°, 90°, +45°,
−45°, for an applied uniaxial
stress τ_{xy} = 1 MPa

$\begin{cases} \sigma_\ell \\ \sigma_t \ (\text{MPa}) \\ \tau_{\ell t} \end{cases}$

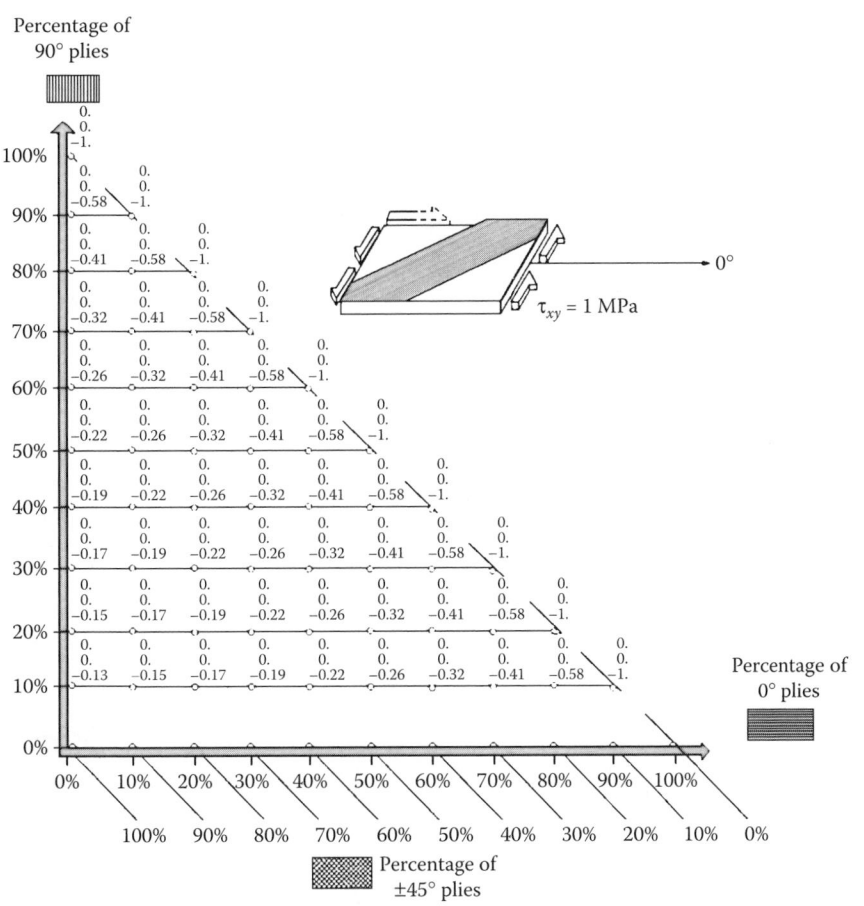

Percentage of
90° plies

Percentage of
0° plies

Percentage of
±45° plies

τ_xy = 1 MPa

0°

Stresses in 90° plies, respectively: as function of the percentage of plies in directions 0°, 90°, +45°, −45°, for an applied uniaxial stress $\tau_{xy} = 1$ MPa

$$\begin{cases} \sigma_\ell \\ \sigma_t \ (MPa) \\ \tau_{\ell t} \end{cases}$$

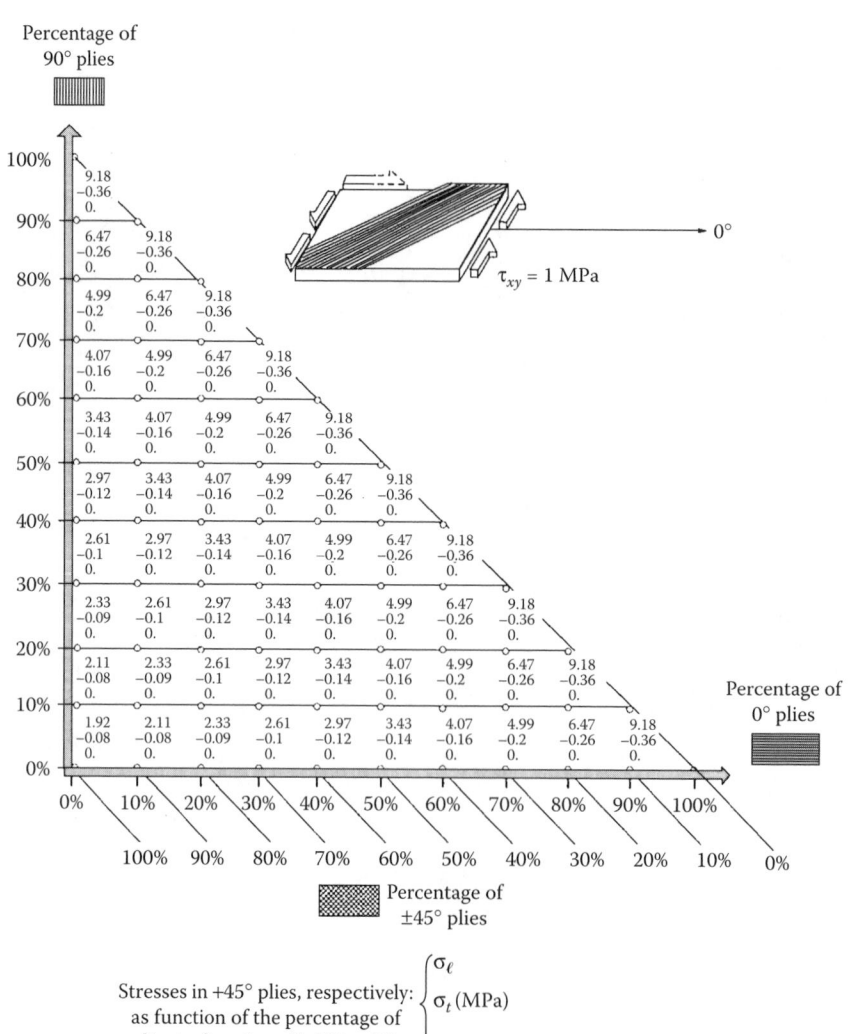

Stresses in +45° plies, respectively: as function of the percentage of plies in directions 0°, 90°, +45°, −45°, for an applied uniaxial stress $\tau_{xy} = 1$ MPa

$$\begin{cases} \sigma_\ell \\ \sigma_t \ (\text{MPa}) \\ \tau_{\ell t} \end{cases}$$

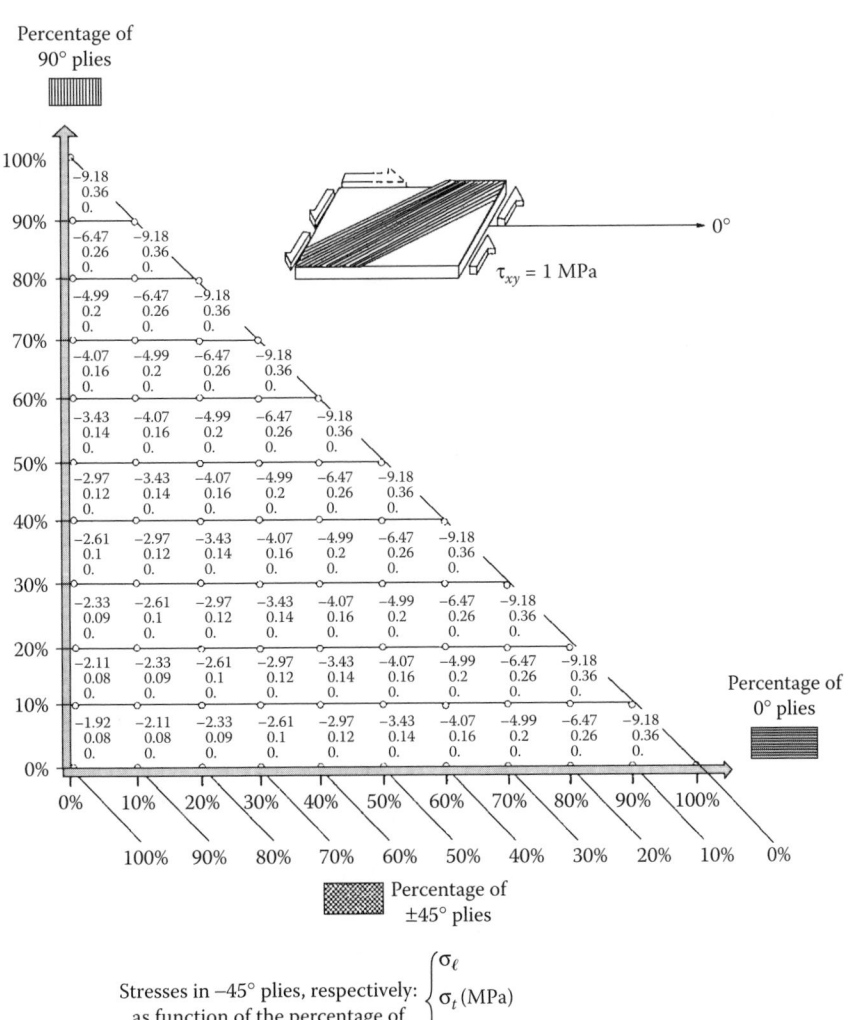

Percentage of
90° plies

Percentage of
0° plies

Percentage of
±45° plies

Stresses in −45° plies, respectively:
as function of the percentage of
plies in directions 0°, 90°, +45°,
−45°, for an applied uniaxial
stress τ_{xy} = 1 MPa

$\begin{cases} \sigma_\ell \\ \sigma_t \text{ (MPa)} \\ \tau_{\ell t} \end{cases}$

APPENDIX 2

BUCKLING OF ORTHOTROPIC STRUCTURES

The problems related to the stability of orthotropic plates and shells are not treated in this book. However, we give in the following a few figures that allow one to estimate the order of magnitude of loads that can lead to buckling due to compression or due to shear in orthotropic panels and tubes.

Rectangular Panels

The following plates allow one to calculate the critical **stress resultants**[1] in compression and in shear for different support conditions.

---

[1] See Sections 5.2.4 or 12.1.1 for the definition of the stress resultants. See relations [12.16] for the definition of the constants C_{11}, C_{22}, C_{12}, C_{33} that appear in the plates.

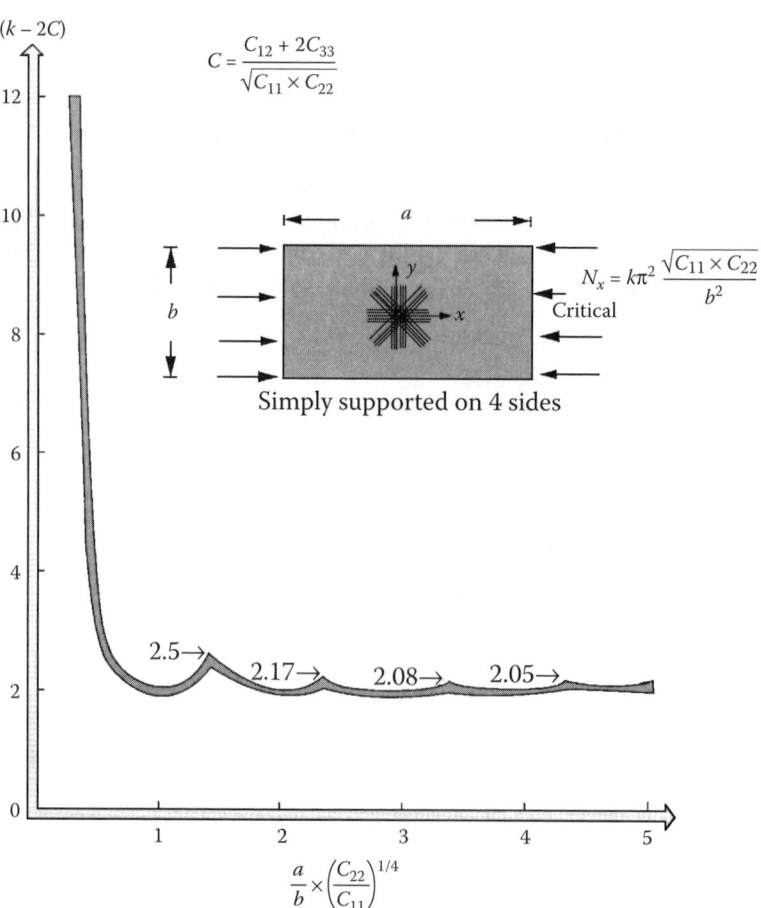

$$(k - 2C)$$

$$C = \frac{C_{12} + 2C_{33}}{\sqrt{C_{11} \times C_{22}}}$$

$$N_x = k\pi^2 \frac{\sqrt{C_{11} \times C_{22}}}{b^2}$$

Critical

Simply supported on 4 sides

$$\frac{a}{b} \times \left(\frac{C_{22}}{C_{11}}\right)^{1/4}$$

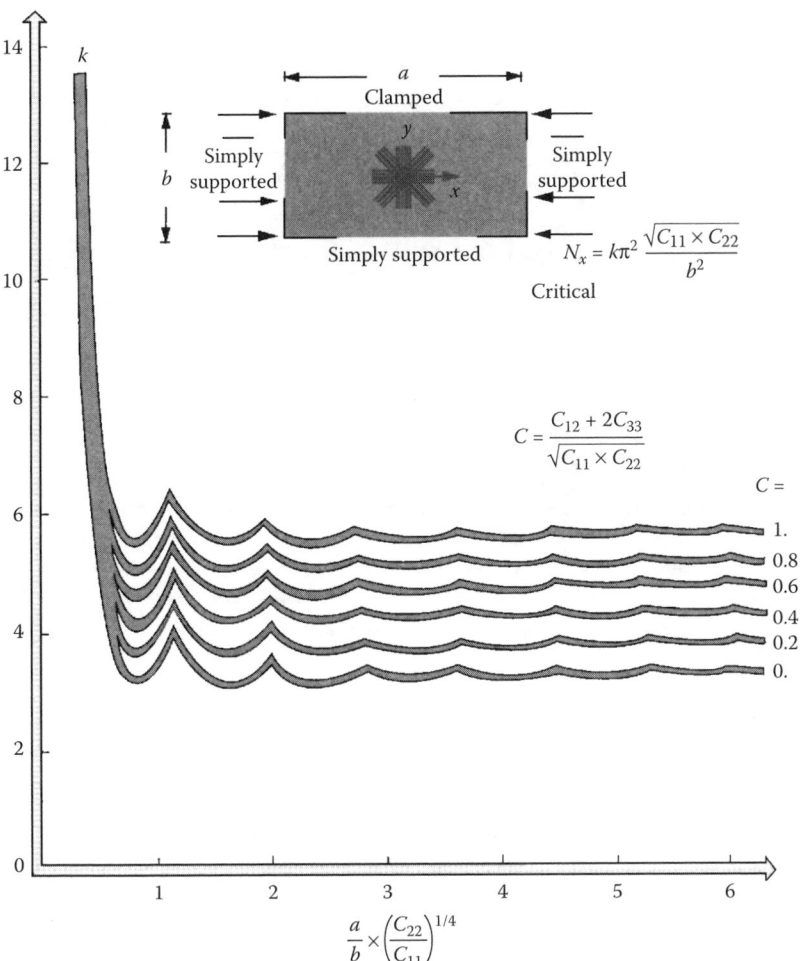

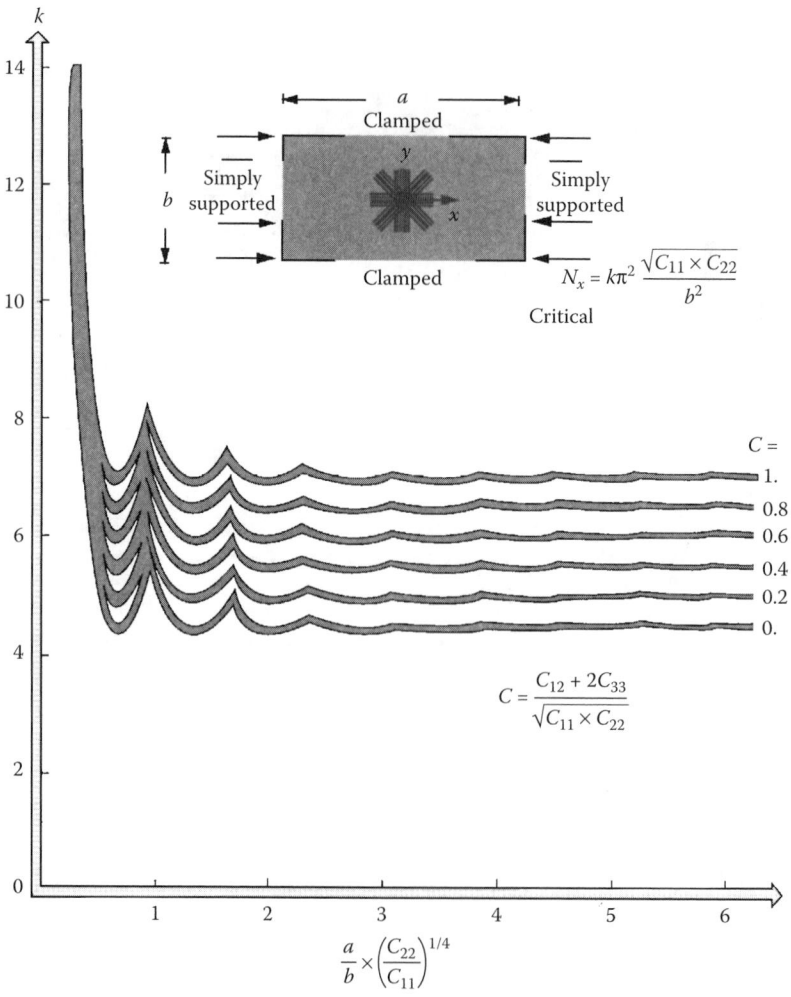

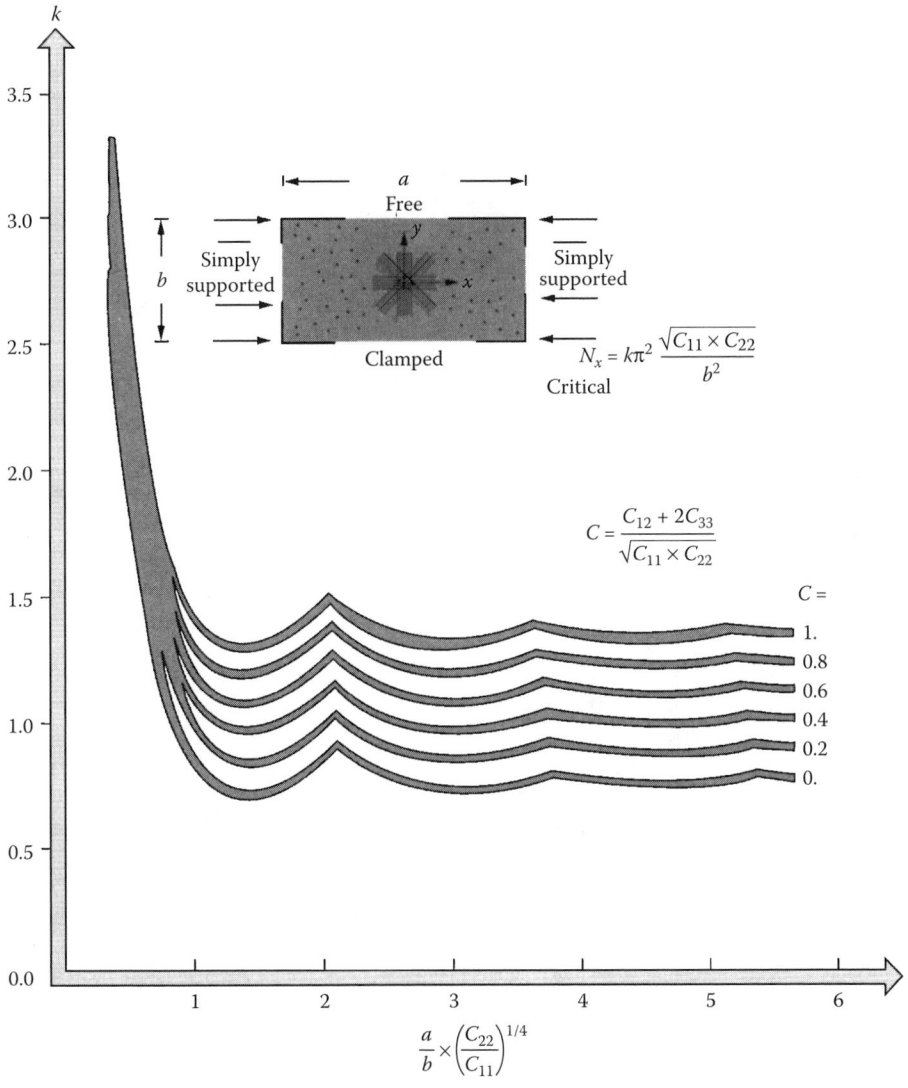

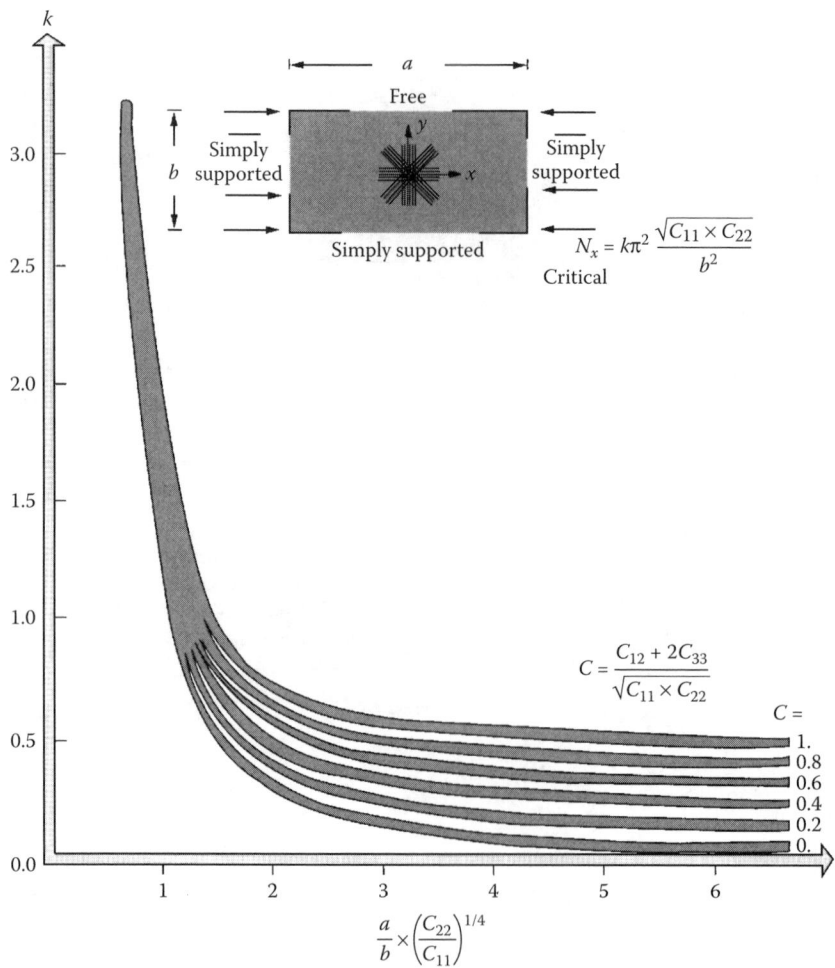

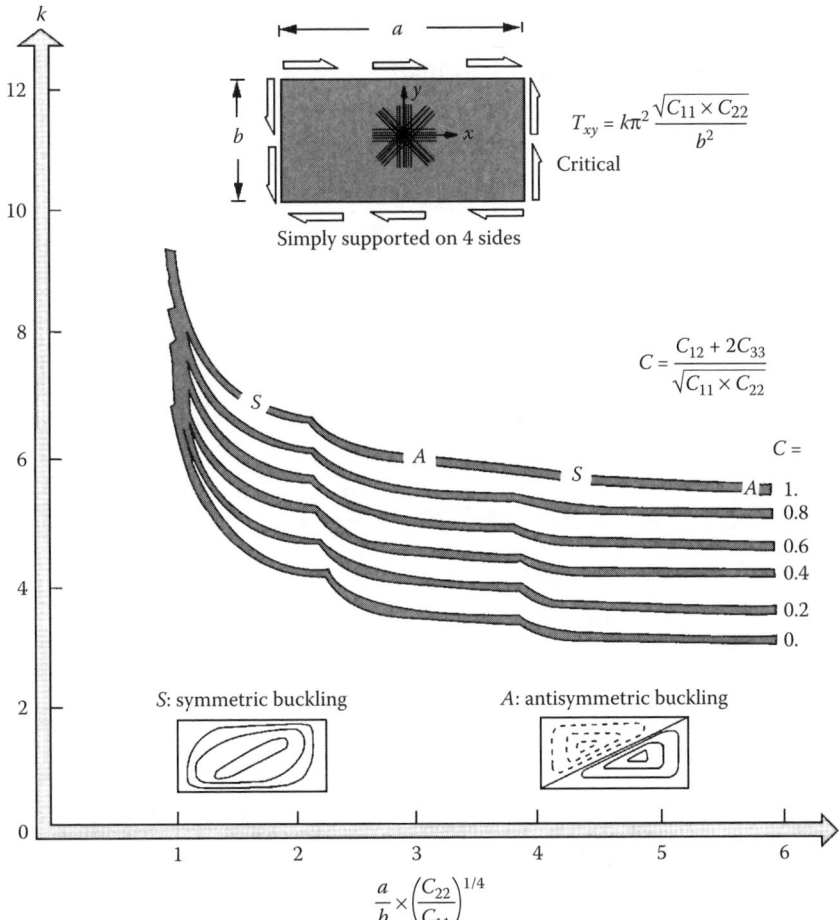

$$T_{xy} = k\pi^2 \frac{\sqrt{C_{11} \times C_{22}}}{b^2}$$

Critical

Simply supported on 4 sides

$$C = \frac{C_{12} + 2C_{33}}{\sqrt{C_{11} \times C_{22}}}$$

$C =$
1.
0.8
0.6
0.4
0.2
0.

S: symmetric buckling   A: antisymmetric buckling

$$\frac{a}{b} \times \left(\frac{C_{22}}{C_{11}}\right)^{1/4}$$

Buckling of Orthotropic Tubes

■ Buckling in bending creating ovalization of the tube:
Bending leads to ovalization of the cross section. The moment of inertia which affects the rigidity in bending decreases (this is the unstable process). The phenomenon is known as the "Brazier effect."

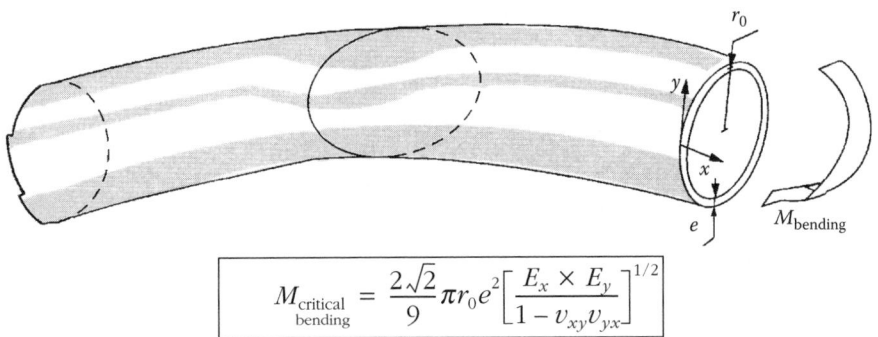

$$M_{\substack{\text{critical} \\ \text{bending}}} = \frac{2\sqrt{2}}{9} \pi r_0 e^2 \left[\frac{E_x \times E_y}{1 - v_{xy} v_{yx}}\right]^{1/2}$$

■ Buckling due to external pressure:

The notations in the above figure are kept. L is the length of the tube or the container that is subject to buckling.

$$p_{\text{critical}} = 0.83 \times \frac{E_y}{1 - 0.1 E_x/E_y} \times \left(\frac{E_x}{E_y}\right)^{1/4} \times \frac{r_0}{L} \times \left(\frac{e}{r_0}\right)^{5/2}$$

■ Buckling due to torsion:

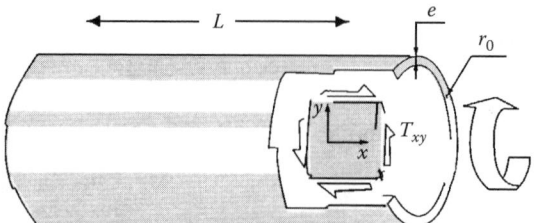

The critical stress resultant in torsion is given by:

$$T_{xy}_{\text{critical}} = \frac{\pi^2}{12}\left[\frac{e^9}{r_0^3 L^2}\right]^{1/4} \times \left[\frac{E_x^3 E_y^5}{(1 - v_{xy} v_{yx})^5}\right]^{1/8}$$

■ Buckling due to axial compression:

This aspect is not considered here, because the occurrence of elastic instability is strongly influenced by the presence of geometric imperfections in the orthotropic cylinder.

BIBLIOGRAPHY

- *Les matériaux composites,* J. Weiss, C. Bord, Vol. 1 and 2, Publications of the Ministry of Industry and Research of Centre d'Études Techniques des Industries Mécanique.
- Les matériaux composites. G. Hellard, P. Lafon, Conference, Dec. 1984.
- Manufacturer's documentation. Vetrotex—Saint Gobain, skis Dynastar, Ciba Geigy.
- *Calcul des structures en matériaux composites.* J.J. Barrau, S. Laroze, Eyrolles-Masson, Paris 1987.
- *Calcul des structures en matériaux composites,* F. Joubert, R. Harry, Mechanical Engineering Laboratory—I.U.T. « A »/C.O.D.E.M.A.C., Bordeaux 1985.
- *Matériaux composites pour structures d'aeronefs,* J. Rouchon, École Nationale Supérieure d'Ingénieurs de construction Aéronautiques, Toulouse, 1980.
- Les applications des composites dans le domaine aéronautique, G. Hellard, G. Hilaire, M. Tores, Aerospatiale publication.
- Fabrication des avions et missiles, M.P. Guibert, Dunod, Paris, 1960.
- *Techniques de l'Ingénieur*—Plastiques—A.3250—K. Gamski.
- Fibres de verre, d'aramide et de carbone, des renforts sur mesure pour composites, R. Kleinholz, G. Molinier, *Vetrotex* journal, No. 59, 1986.
- Résistance des constructions—matériaux composites—G. Hellard, École Nationale Supérieure d'Ingénieurs de Construction Aéronautiques, Toulouse, 1983.
- Initiation aux matériaux composites et à leurs techniques de fabrication—Avions Marcel Dassault—Bréguet Aviation—Edition No. 1, Toulouse.
- *Manuel de techniques de collage,* G. Fauner, W. Hendlich, eds. Soproge S.A., Paris, 1984.
- Recherches sur la théorie des assemblages collés, O. Volkersen—*La Construction Métallique,* No. 4, 1965.
- L'Aéronautique et L'Astronautique, No. 107, 4, 1984.
- Matériaux et structures composites ou l'adaptation aux besoins, P. Lamicq, 7th French Congress in Mechanics, Bordeaux, 1985.
- Publication of specialized press : Air et Cosmos; Aviation magazine; "ça m'intéresse;" Science et vie; Dassault-Bréguet Information; Aérospatiale Information; Vétrotex actualités; Aspects des résines Ciba Geigy; Interavia Revue Aérospatiale.
- *Introduction aux matériaux composites,* R. Naslain, Vol. 2, Matrices métalliques et céramiques, Ed. of C.N.R.S. and of Institute of Composite Materials.

- *Matériaux composites à matrice organique,* G. Chrétien, Technique et documentation Lavoisier, Paris, 1986.
- *Guide pratique des matériaux composites,* M. Geier, D. Duedal, Technique et documentation Lavoisier, Paris 1985.
- Proceedings of the 4th Journées Nationales sur les composites, Ed., Pluralis, Paris, 1984.
- Proceedings of the 5th Journées Nationales sur les composites, C. Bathias, D. Menkes, A.M.A.C., Ed. Pluralis, Paris, 1986.
- Transverse shear deformation in bending of composite beams of any cross section shapes, D. Gay, *Mechanics of structured media,* Part B : pp. 155–171, Elsevier Scientific Publishing Company, 1981.
- Homogèneisation en torsion d'une poutre composite, J.J. Barrau, O. Chambard, M. Nuc, D. Gay-*Tendances actuelles en calcul des structures,* Bastia 1985.
- Les système ressorts-composites, L. De Goncourt, K.H. Sayers, *Composites,* No. 3, May/June 1988, pp. 145–150.
- Composites de grande diffusion, A. Jardon, M. Costes, European conference on Composite Materials, London, July 1987.
- *Conception des véhicules spatiaux,* D. Marty, Masson Ed., Paris, 1986.
- Effets des microfissures de résine sur les propriétés des composites aéronautiques en kevlar/époxyde-F. Turris, H.Y. Loken, R.F. Pinzelli, *Composites* No. 2, March/April 1988, pp. 35–40.
- Conduits PVDF renforcés. Contraintes thermiques et mécanismes de rupture, B. Échalier, P. Bleut, F. Lefebvre, C. Tournut, *Composites,* No. 3, May/June 1987, pp. 183–188.
- Structure coques—Équations générales et stabilité, J. Barbe, E.N.S.A.E., 1983.
- Comment utiliser au meilleur coût les matériaux composites. M. Reyne, Ed. de l'Usine Nouvelle, Paris, 1986.
- Fibres de carbone revêtues de nickel, E.D. Kaufman, *Composites* No. 3, May/June 1988.
- FRANCE—COMPOSITES. French suppliers for the fabrication of organic composites, CEPP, 25 Rue Dagorno, 75012 Paris.
- *Matériaux composites*—Teknea, Marseille, 1989.
- *Manuel de calcul des composites verre/résine,* Manera, Massot, Morel, Verchery, Pluralis, 1988.
- Phénomènes interfaciaux dans les composites à matrice thermoplastique: Application au FIT, M. Guigon, A. Kloster, *Annales des composites* 1989, 1–2, pp. 83–90.
- Développement d'une nouvelle technologie pour fabriquer des poteaux, E. Pfletschinger-*Composites* No. 3, May/June 1987, pp. 67–72.
- Conception et dimensionnement du volet extérieur en fibres de carbone du CN 235, J.L. Duquesne, M. Gonzales, ENSICA, June 1990.
- *Les plastiques,* M. Reyne, Ed. Hermès, Paris, 1990.
- Impact and Residual Fatigue Behavior of ARALL, W.S. Johnson, Advances in Composites Materials and Structures, ASME-AMD, Vol. 82, 1989.

■ Elastic Analysis and Engineering Design formulae for bonded joints—D.A. Bigwood, A.D. Grocombe, *International Journal of Adhesion and Adhesive,* Vol. 9, No. 4, Oct. 1989, pp. 229–242.

■ Conception et essais de tubes composites hautes performances, P. Odru, C. Sparks, J. Schmitt, J.F. Puch, *Matériaux-Mécanique-Électricité,* No. 433, Feb/Mar 1990, pp. 4–6.

■ Metal laminates—A family of Advanced Materials, M.L. Verbruggen, J.W. Gunnik, Plastics/Metals/Ceramics, SAMPE European chapter, Bâle, May 1990, pp. 455–465.

■ General buckling curves for specially orthotropic rectangular plates, E.J. Brunnelle, G.A. Oyibo, *AIAA Journal,* Vol. 21, No. 8, Aug. 1983, pp. 1150–1154.

■ *Technologie des composites,* M. Reyne, Ed. Hermès, Paris, 1990.

■ Interaction effects in fibre composites, V.R. Riley, *Polymer Conference Series,* University of Utah, June 1970.

■ Progiciels de calcul de structures composites; Exemples de réference et de validation, Coordonnateur G. Matheron, ed. Hermès, Paris, 1995.

■ Isotropie de Rigidité et quasi-isotropie de résistance des stratifiés à orientation périodiques, D. Gay, F. Joubert, *Revue des composites et des matériaux avancés,* Vol. 4, No. 2, 1994, pp. 241–261.

■ *Les composites,* M. Reyne, Presses Universitaires de France, Paris, 1995.

■ Analysis for Design of Fiber Reinforced Plastic Vessels and Piping, Hoa, S.V., *Technomic,* 1991.

■ Shear Stresses in Orthotropic Composite Beams, T. Nouri, D. Gay, *International Journal of Engineering Science,* Vol. 32, No. 10, pp. 1647–1667.

■ Influence of Secondary Effects on Dynamical Flexure-Torsion with Non-Uniform Warping in Composite Beams with Any Cross Sectional Shapes Made of Several Orthotropic Phases, F. Tanghe, *Thesis, University of Toulouse III (Paul Sabatier),* December 1999.

■ Dimensionnement des Structures, une Introduction, D. Gay, J. Gambelin, *Hermes Science Publishing Ltd., London,* 1999.

■ Non Uniform Warping Torsion of Orthotropic Composite Beams, F. Tanghe-Carrier, D. Gay, *Archive of Applied Mechanics,* No. 99, 2000.

■ *Structural Modeling and Calculus,* D. Gay, J. Gambelin, I.S.T.E., London, 2007.

INDEX